Advanced Fibrous Composite Materials for Ballistic Protection

Related titles

Lightweight Ballistic Composites
(ISBN: 978-1-85573-941-3)

Military Textiles
(ISBN: 978-1-84569-206-3)

Advances in Military Textiles and Personal Equipment
(ISBN: 978-1-84569-699-3)

Woodhead Publishing Series in Composites Science and Engineering: Number 66

Advanced Fibrous Composite Materials for Ballistic Protection

Edited by

X. Chen

ELSEVIER

AMSTERDAM • BOSTON • CAMBRIDGE • HEIDELBERG
LONDON • NEW YORK • OXFORD • PARIS • SAN DIEGO
SAN FRANCISCO • SINGAPORE • SYDNEY • TOKYO
Woodhead Publishing is an imprint of Elsevier

WP WOODHEAD PUBLISHING

Woodhead Publishing is an imprint of Elsevier
The Officers' Mess Business Centre, Royston Road, Duxford, CB22 4QH, UK
50 Hampshire Street, 5th Floor, Cambridge, MA 02139, USA
Langford Lane, Kidlington, OX5 1GB, UK

ISBN: 978-1-78242-461-1 (print)
ISBN: 978-1-78242-484-0 (online)

British Library Cataloguing-in-Publication Data
A catalogue record for this book is available from the British Library

Library of Congress Cataloging-in-Publication Data
A catalog record for this book is available from the Library of Congress

For information on all Woodhead Publishing publications
visit our website at http://store.elsevier.com/

Contents

List of contributors

M. Al-Haik Embry-Riddle Aeronautical University, Daytona Beach, FL, United states

A.Y. Borujeni Virginia Tech, Blacksburg, VA, United States

F. Boussu ENSAIT – GEMTEX Laboratory, France

D.J. Carr Cranfield University, Wiltshire, United Kingdom

X. Chen University of Manchester, Manchester, United Kingdom

C. Chevalier ISL – French-German Research Institute of Saint-Louis, France

Y. Chu University of Manchester, Manchester, United Kingdom; Zhongyuan University of Technology, Zhengzhou, China

C.R. Cork Nottingham Trent University, Nottingham, United Kingdom

D. Coutellier LAMIH, UMR CNRS 8201, University of Valenciennes, France

C.M. Crawford Cranfield University, Wiltshire, United Kingdom

F. Folgar INTER Materials, LLC, Midlothian, VA, United States

M. Grujicic Clemson University, Clemson, SC, United States

Bohong Gu Donghua University, Shanghai, China

C. Ha-Minh R&D Center, Duy Tan University, Da Nang, Viet Nam; LMT, ENS Cachan, CNRS, Université Paris Saclay, Cachan, France

U. Heisserer DSM Ahead Materials Sciences R&D, Geleen, The Netherlands

M.V. Hosur Tuskegee University, Tuskegee, AL, United States

S. Jeelani Tuskegee University, Tuskegee, AL, United States

C. Kerisit ISL – French-German Research Institute of Saint-Louis, France

A. Klavzar ISL – French-German Research Institute of Saint-Louis, France

B. Lauke Leibniz Institute for Polymer Research e. V., Dresden, Germany

B. Mobasher School of Sustainable Engineering & the Built Environment, Arizona State University, Tempe, AZ, United States

N.K. Naik Indian Institute of Technology Bombay, Mumbai, India

S.D. Rajan School of Sustainable Engineering & the Built Environment, Arizona State University, Tempe, AZ, United States

H.M. Rao Tuskegee University, Tuskegee, AL, United States

S. Rebouillat DuPont International Operations, Geneva, Switzerland

J. Singletary DuPont Protection Technologies, Richmond, VA, United States

D. Sun School of Textiles and Design, Heriot-Watt University, Galashiels, United Kingdom

M. Tehrani University of New Mexico, Albuquerque, NM, United States

H. van der Werff DSM Dyneema, Geleen, The Netherlands

Woodhead Publishing Series in Composites Science and Engineering

1 **Thermoplastic aromatic polymer composites**
F. N. Cogswell
2 **Design and manufacture of composite structures**
G. C. Eckold
3 **Handbook of polymer composites for engineers**
Edited by L. C. Hollaway
4 **Optimisation of composite structures design**
A. Miravete
5 **Short-fibre polymer composites**
Edited by S. K. De and J. R. White
6 **Flow-induced alignment in composite materials**
Edited by T. D. Papthanasiou and D. C. Guell
7 **Thermoset resins for composites**
Compiled by Technolex
8 **Microstructural characterisation of fibre-reinforced composites**
Edited by J. Summerscales
9 **Composite materials**
F. L. Matthews and R. D. Rawlings
10 **3-D textile reinforcements in composite materials**
Edited by A. Miravete
11 **Pultrusion for engineers**
Edited by T. Starr
12 **Impact behaviour of fibre-reinforced composite materials and structures**
Edited by S. R. Reid and G. Zhou
13 **Finite element modelling of composite materials and structures**
F. L. Matthews, G. A. O. Davies, D. Hitchings and C. Soutis
14 **Mechanical testing of advanced fibre composites**
Edited by G. M. Hodgkinson
15 **Integrated design and manufacture using fibre-reinforced polymeric composites**
Edited by M. J. Owen and I. A. Jones
16 **Fatigue in composites**
Edited by B. Harris
17 **Green composites**
Edited by C. Baillie
18 **Multi-scale modelling of composite material systems**
Edited by C. Soutis and P. W. R. Beaumont
19 **Lightweight ballistic composites**
Edited by A. Bhatnagar

20 **Polymer nanocomposites**
Y.-W. Mai and Z.-Z. Yu
21 **Properties and performance of natural-fibre composite**
Edited by K. Pickering
22 **Ageing of composites**
Edited by R. Martin
23 **Tribology of natural fiber polymer composites**
N. Chand and M. Fahim
24 **Wood-polymer composites**
Edited by K. O. Niska and M. Sain
25 **Delamination behaviour of composites**
Edited by S. Sridharan
26 **Science and engineering of short fibre reinforced polymer composites**
S.-Y. Fu, B. Lauke and Y.-M. Mai
27 **Failure analysis and fractography of polymer composites**
E. S. Greenhalgh
28 **Management, recycling and reuse of waste composites**
Edited by V. Goodship
29 **Materials, design and manufacturing for lightweight vehicles**
Edited by P. K. Mallick
30 **Fatigue life prediction of composites and composite structures**
Edited by A. P. Vassilopoulos
31 **Physical properties and applications of polymer nanocomposites**
Edited by S. C. Tjong and Y.-W. Mai
32 **Creep and fatigue in polymer matrix composites**
Edited by R. M. Guedes
33 **Interface engineering of natural fibre composites for maximum performance**
Edited by N. E. Zafeiropoulos
34 **Polymer-carbon nanotube composites**
Edited by T. McNally and P. Pötschke
35 **Non-crimp fabric composites: Manufacturing, properties and applications**
Edited by S. V. Lomov
36 **Composite reinforcements for optimum performance**
Edited by P. Boisse
37 **Polymer matrix composites and technology**
R. Wang, S. Zeng and Y. Zeng
38 **Composite joints and connections**
Edited by P. Camanho and L. Tong
39 **Machining technology for composite materials**
Edited by H. Hocheng
40 **Failure mechanisms in polymer matrix composites**
Edited by P. Robinson, E. S. Greenhalgh and S. Pinho
41 **Advances in polymer nanocomposites: Types and applications**
Edited by F. Gao
42 **Manufacturing techniques for polymer matrix composites (PMCs)**
Edited by S. Advani and K.-T. Hsiao
43 **Non-destructive evaluation (NDE) of polymer matrix composites: Techniques and applications**
Edited by V. M. Karbhari

44 **Environmentally friendly polymer nanocomposites: Types, processing and properties**
S. S. Ray
45 **Advances in ceramic matrix composites**
Edited by I. M. Low
46 **Ceramic nanocomposites**
Edited by R. Banerjee and I. Manna
47 **Natural fibre composites: Materials, processes and properties**
Edited by A. Hodzic and R. Shanks
48 **Residual stresses in composite materials**
Edited by M. Shokrieh
49 **Health and environmental safety of nanomaterials: Polymer nanocomposites and other materials containing nanoparticles**
Edited by J. Njuguna, K. Pielichowski and H. Zhu
50 **Polymer composites in the aerospace industry**
Edited by P. E. Irving and C. Soutis
51 **Biofiber reinforcement in composite materials**
Edited by O. Faruk and M. Sain
52 **Fatigue and fracture of adhesively-bonded composite joints: Behaviour, simulation and modelling**
Edited by A. P. Vassilopoulos
53 **Fatigue of textile composites**
Edited by V. Carvelli and S. V. Lomov
54 **Wood composites**
Edited by M. P. Ansell
55 **Toughening mechanisms in composite materials**
Edited by Q. Qin and J. Ye
56 **Advances in composites manufacturing and process design**
Edited by P. Boisse
57 **Structural integrity and durability of advanced composites: Innovative modelling methods and intelligent design**
Edited by P. W. R. Beaumont, C. Soutis and A. Hodzic
58 **Recent advances in smart self-healing polymers and composites**
Edited by G. Li and H. Meng
59 **Manufacturing of nanocomposites with engineering plastics**
Edited by V. Mittal
60 **Fillers and reinforcements for advanced nanocomposites**
Edited by Y. Dong, R. Umer and A. Kin-Tak Lau
61 **Biocomposites: Design and mechanical performance**
Edited by M. Misra, J. K. Pandey and A. K. Mohanty
62 **Numerical modelling of failure in advanced composite materials**
Edited by P. P. Camanho and S. R. Hallett
63 **Marine applications of advanced fibre-reinforced composites**
Edited by J. Graham-Jones and J. Summerscales
64 **Smart composite coatings and membranes: Transport, structural, environmental and energy applications**
Edited by M. F. Montemor
65 **Modelling damage, fatigue and failure of composite materials**
Edited by R. Talreja and J. Varna

Introduction

1

X. Chen
University of Manchester, Manchester, United Kingdom

1.1 Background

Military confrontations and wars never seem to stop throughout world history. The level of personal and property protection against the threats in the battlefields and in riot situations has been developed in line with the advancement of the attacking weapons. At different times in history, different materials have been used for protection against attacks according to the types of attacking weapons. Among many, leather, fabrics, metal, and wood have played important roles in personal and property protection. In addition, these materials have been manufactured into different shapes with internal structures to optimise the effect of protection. For example, metal has been used as a shell to protect the torso, as well as chain mail for increased armour flexibility (Chen and Chaudhry, 2005; Carother, 1988).

Soldiers were confronted with greater ballistic threats after the invention of the firearm in comparison to other weapons (Tam and Bhatnagar, 2006). Strong and lightweight materials were sought for a new generation of ballistic protection. Along with the invention of synthetic fibre nylon in the 1930s, this much stronger fibre was involved in the creation of the flak vest towards the end of World War II. This technology brought about enhanced protection against firearms, lightweight, and flexibility (Dunn, 2008). Since then, the fibre-based materials have dominated the engineering of personal protective equipment. Along with the progress of textile-reinforced composites, the fibre-based composites also play more important roles as body materials for military vehicles and aircraft. This route for engineering ballistic materials drew much attention when various advanced fibrous materials were invented in modern times, such as aramid fibres (eg, Kevlar and Twaron), ultra-high molecular weight polyethylene (UHMWPE) fibres (Dyneema and Spectra), PBO fibres (eg, Zylon) and PIPD fibres (eg, M5). However, Zylon is reported to be susceptible to hydrolytic and photolytic degradation, and therefore it is not suggested for use in ballistic protection (U.S. Department of Justice, 2005). M5 fibre has similar but less degradation than PBO from exposures to ultraviolet radiation and elevated moisture, while it may exhibit brittleness at ballistic loading rate (Scott, 2006). Such being the case, the fibres for ballistic protection are dominantly the aramid and UHMWPE fibres.

Ballistic protective materials in this context are mainly used for personnel and vehicle protection, for which agile movements of the personnel and vehicles are vital for carrying out their missions. Therefore, high protective performance against ballistic impact and lightweight are probably the two most important requirements. As a matter

Advanced Fibrous Composite Materials for Ballistic Protection. http://dx.doi.org/10.1016/B978-1-78242-461-1.00001-7

of fact, the evolution of modern ballistic materials witnessed the effort and progress in technologies for creating strong and lightweight materials. A modern time soldier in the battlefield could carry up to a 70 kg load, of which 14−17 kg comes from the body armour system (Conference Communication, 2009). There have been campaigns in different countries to work to reduce the weight of the body armours while retaining or even improving the ballistic performance. Ballistic composites are used in military vehicles for territory armies, navies, and air forces, therefore lightweight of the ballistic composites is keenly sought for uncompromised protection at the same time.

1.2 Types of ballistic protective equipment and materials

Ballistic threats refer to those caused by projectiles flying at high velocity, such as bullets, shrapnel, and other similar objects, usually associated with military actions. Ballistic materials are used basically in soft and rigid forms in protection applications.

Soft body armour is made by layering dry ballistic fabrics. This technology is particularly used for body armour used for low levels of ballistic threats, such as NIJ (National Institute of Justice) levels I, IIa, and II, and it offers suitable flexibility for body movement. The most commonly seen soft body armour is for the torso, and there is also soft armour designed to protect neck and groin areas. When facing higher levels of ballistic threats, hard ballistic plates are used in conjunction with the soft armour. The hard ballistic plates could be made of metal, ceramics, and textile composites. Special body protective clothing systems are available for special tasks. The system used for explosive ordnance disposal covers the whole body, and it must be made from soft protective materials to allow free body movement (Scott, 2005).

Military helmets are another piece of important equipment to prevent casualties. Helmets used to be made of steel, which has been replaced by textile reinforced ballistic composites for higher ballistic performance and lightweight. Due to the doubly curved nature of the helmet, protective fabrics are cut into patterns and then patched together to form the helmet shell. The composite helmet represents substantial progress, but the fibre discontinuity in the helmet could be a concern. Chen and his colleagues worked on the formation of helmet shells from a single piece of reinforcing fabric and demonstrated the advantages in improving impact protection (Roedal and Chen, 2007; Zahid and Chen, 2014).

Fibre-based ballistic composites attract much attention for various types of military applications because of the advantages such materials offer. Not only have the ballistic composites demonstrated high mechanical properties including high-velocity impact, the fibre-based composites enable weight savings, high payload and fuel efficiency, high performance, and speed capability. Textile composites, however, are a complex materials system; the mechanical performance and response to ballistic impact are affected by many factors, such as fibre, resin, fibre-resin interface, as well as the construction of the textile reinforcements (Cheeseman and Bogetti, 2003). Investigations into textile composites are carried out based on two approaches: experimental and numerical. The ultimate function of the ballistic composites is to prevent the high-velocity projectile from piercing the material. It is imperative to understand mechanisms for hindering the projectile by taking energy off the projectile. High-performance fibres and compactable resins have been selected for this application. A large volume of the composites are expected to be involved in absorbing energy from the impacting projectile, therefore control over the stress wave propagation and composite delamination will be important engineering issues.

1.3 Projective materials against ballistic impact

1.3.1 Fibre types

The current ballistic gears are made based on the fibre technology where high-performance fibres are manufactured into 2D and 3D assemblies to retard the ballistic threat. Such fibres have significantly higher tensile strength and modulus and lower elongation compared to the ordinary fibres. Due to the nature of impact loading, these fibres are required to be flexible to withstand the transverse loading. Despite the efforts made in generating new and better fibres for ballistic protection, aramid and UHMWPE fibres remain the major players in ballistic body armours and ballistic composites. Table 1.1 shows the high-performance fibres that have been involved in ballistic fabrics and composites.

In some cases, glass and carbon fibres are also used for ballistic protection.

1.3.2 Two-dimensional fibrous assemblies

Ballistic fibres are fabricated in 2D sheet materials for their end applications. The fibres are either made into woven fabrics or unidirectional (UD) sheets, which are then assembled together to form the ballistic panel.

Woven fabrics are usually made from aramid fibres, and they are a very popular type of 2D assembly for ballistic protection because of the good structural integrity and the capability for shock wave propagation between the two sets of yarns arranged perpendicular to one another. The set of yarn running in the length direction of the fabrics is known as the warp yarn and that which goes in the width direction is termed the weft yarn. These two sets of yarns can be interwoven to have different interlacing features. According to the length of float of yarns, plain, twill, and satin fabrics can be produced, each having distinctive structural features. Fabrics made from satin weaves would offer the longest float length of yarns, thus having the lowest level of yarn crimp among the three. Fabrics made from the plain weave, on the other hand, would lead to the most structurally stable fabrics, which have the most crossover points between the warp and weft yarns. The crossover points are found to be effective for shock wave propagation. The structural and mechanical properties of twill fabrics lie in between the plain and satin fabrics. The plain-woven fabrics are mostly used among the three types of woven fabrics for ballistic protection. Yarn linear density and the warp and weft packing densities in the fabric dictate the ballistic performance of the plain-woven fabrics. A plain-woven fabric is shown in Fig. 1.1(a).

Table 1.1 High-performance fibres for ballistic protection

Fibre types	Density (g/cm^3)	Tenacity (N/tex)	Modulus (N/tex)	Elongation (%)
Aramid (Rebouillat, 2001)				
Kevlar® 29	1.44	2.0	49	3.8
Kevlar® 49	1.44	2.1	78	2.4
Kevlar® 149	1.44	1.7	115	1.3
Twaron®	1.44	1.7	60	3.6
Twaron® HM	1.44	2.1	75	2.5
Twaron® HT	1.44	2.4	85	3.3
UHMWPE (van Dingenen, 2001)				
Dyneema® SK60	0.97	2.8	91	3.5
Dyneema® SK65	0.97	3.1	97	3.6
Dyneema® SK71	0.97	3.5	122	3.7
Dyneema® SK75	0.97	3.5	110	3.8
Dyneema® SK76	0.97	3.7	120	3.8
Spectra® 900	0.97	2.6	75	3.6
Spectra® 1000	0.97	3.2	110	3.3
Spectra® 2000	0.97	3.4	120	2.9

UD sheets are made from UHMWPE and aramid fibres in many layers. The fibres in each layer are laid in the same direction, and the fibre direction in the adjacent layers is 90 degree apart. A UD sheet could involve four, six, and even more layers of fibres, and all layers are then laminated into either a soft or a hard sheet. Compared to the woven assemblies, the UD sheets have the advantages in the following two aspects. All fibres are kept straight in the UD sheet, and when being impacted the fibres are able to respond directly to take up the load and to propagate the stress weave. Another obvious advantage of the UD sheets is that they are associated with much reduced back face deformation in the case of soft body armour. Fig. 1.1(b) is a photo of a UD sheet.

1.3.3 Ballistic panels

Ballistic panels for protection are made from numerous layers of the 2D ballistic fabrics or UD sheets. For the soft armour systems, a ballistic panel at a given protective level fails in one of the two situations, the first being that the panel is perforated by the impacting projectile and the other that the panel is not perforated but the back face

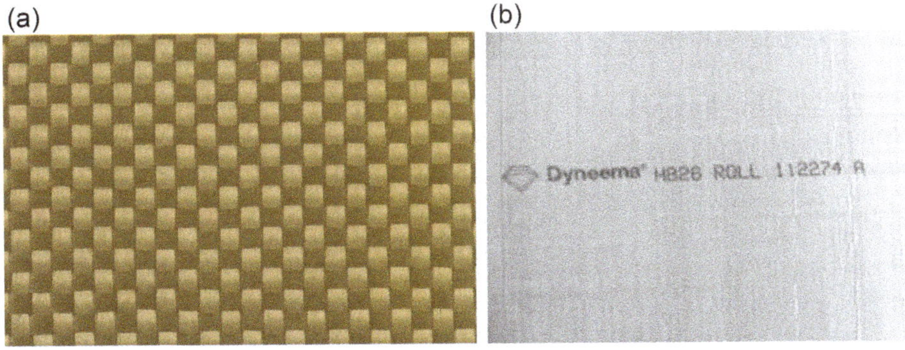

Figure 1.1 Two major types of 2D fibrous assemblies for ballistic protection. (a) Woven fabric from aramid. (b) UD sheet from polyethylene.

signature exceeds a certain depth according to the testing standard, such as the NIJ Standard 0101-06 (U.S. Department of Justice, 2008). The panels are engineered based on the understanding of failure mechanisms of the panel. A simple way to construct the ballistic panel is to layer the same 2D materials up, and this is usually coupled with quilting the layers up for structural integrity. Panels made in such a fashion demonstrated that they are able to satisfy the requirement and provide the needed level of ballistic protection. Some research has confirmed that the quilting lines in the ballistic panel are beneficial in enhancing the ballistic performance (Carr et al., 2012).

Evidences have also been found that failure of a layer in a panel is dependent on the stacking position of the layer. Chen et al. (2013) reported that failure of the front layers of a panel is mainly due to shear strain, and that of the latter layers is mainly because of the tensile strain. Porwal and Phoenix (2008) investigated the ballistic response of a hybrid two-layered flexible armour system based on a computational model, particularly studying the effects of stacking order of the two fibrous layers made from Kevlar® and Spectra®, respectively, which have distinctly different mechanical properties. As observed experimentally by others, their model predicted a large decrease in the V_{50} of the hybrid armour system when Spectra® is the strike layer. Park et al. (2013) examined the effect of the layering sequence of UD/woven fabric hybrid panels on perforation resistance and blunt trauma resistance and found that layer sequence of different layer materials can be used as a parameter to achieve improvement ballistic performance of ballistic panels.

Another approach to achieve improved ballistic performance of a panel has been the consideration of the quasiisotropic panel creation. A recent research at the University of Manchester was focused on the engineering design of quasiisotropic ballistic panels for soft ballistic armour (Wang et al., 2014, 2015a,b) based on experimental and numerical analysis. The principal yarns of a ballistic fabric were found responsible for taking up approximately 70% of the total energy absorbed by the whole ply of fabric, indicating that most of the fabric area was actually idle when the panel was stricken by the impacting projectile. Adopting an approach to lay the fabric layers in distributed angles has led to the approximately 20% increase in the panel's ability to absorb the impact energy. Positive effects of angle-laid fabric layers in composite panels were also demonstrated by Yan and Chen (2014).

Three-dimensional fabrics have also been involved in research and development of ballistic panels. Luan et al. (2013) explored the ballistic performance of 3D layer-to-layer angle interlock woven fabrics composites, and they were able to characterise this type of 3D textile under high-velocity projectile impact. Shi et al. (2011) of the same research group set up finite element (FE) models to simulate the ballistic impact process through 3D orthogonal woven fabrics. Munoz et al. (2015) studied composites made from 3D orthogonal textile reinforcements and found that energy dissipation was not significantly improved by the presence of the z-yarns in the 3D orthogonal composites under high-velocity impact compared to the energy dissipated by 3D woven composites under low-velocity impact, which was around two times higher than that of the 2D counterparts. Sun et al. (2013) investigated 3D networked woven fabrics for their yarn gripping effect and ballistic performance. The research indicated that when properly designed, such 3D networked fabrics are able to distribute the stress effectively through the thickness of the materials.

1.3.4 Ballistic composites

Laminated composite structures fail mainly in two modes: (1) intra-ply failure mode, where the damage happens at the fibres, the polymer matrix, and/or the interface between the fibres and the matrix, and (2) inter-ply failure mode, where the composite fails by delamination of the plies. During the ballistic impact, the composite fails by punching, fibre fracture, matrix cracking, and delamination (Sastry et al., 2014). The ultimate function of fibre-reinforced composites is to stop the projectile from piercing the composites, therefore much effort has been made to absorb the kinetic energy of the projectile to understand the response of the ballistic composites in propagation of the stress waves, fibre and matrix failure, interlaminar delamination, etc. The ballistic composites have been used in military vehicles on land, on sea, and in the air.

1.3.5 Shear thickening fluid and other materials for ballistic protection

The fibre-based technology is the most mature technology for manufacturing ballistic protective materials. However, effort has never stopped in seeking for novel and better solutions for ballistic protection. Shear thickening fluid (STF), also known as dilatant fluid, has drawn attention for research for its suitability for ballistic application.

STF is a non-Newtonian fluid, where the shear viscosity increases with applied shear stresses. The intrinsic properties of the STFs were tunable by varying the dispersing particles from, for example, the spherical silica (SiO_2), to polystyrene-ethylacrylate (PSt-EA) and polymethylmethacrylate (PMMA), and the dispersing medium from, for example, ethylene glycol (EG) to polyethylene glycol 200 (PEG200) and polyethylene glycol 600 (PEG600). The use of SFT for ballistic protection is to wet the fibre (eg, Kevlar) body armour thoroughly with the STF, which could allow the wearer flexibility for a normal range of movement yet provide rigidity to distribute the impact and thus resist the piercing projectiles. There has been some report on the effectiveness of the STF in body armour, but the extra weight added to the armour system poses difficulties for STF to be practically applied into personal protective equipment. More work seems to be necessary to identify solutions to this problem.

There have been other materials and structures that have been used for ballistic protection. Ballistic protective ceramic armour has found application for ballistic protection of personnel, vehicles, and assets, primarily in the form of plates. The ceramics armour serves to enlarge the impact cross-section and to micronise the impact projectile. Hybrid plates made from laminating ceramic and fibre-based layers seemed to be effective for ballistic protection (Liu et al., 2015).

1.4 Engineering design of protective panels

Researchers worked to understand the high-velocity impact mechanisms to engineer and manufacture ballistic materials. The importance of the fibre selection is obvious. For given types of fibres, ballistic materials can be engineered from different perspectives, including the interyarn interface properties, quasiisotropic properties, and hybrid layering.

1.4.1 Fabric creation with controlled interyarn friction

The interactive properties between the fibres and the organisation of fibres into 2D and 3D textile assemblies determine how the ballistic panels would behave in response to ballistic impact. Interyarn friction has been found to play an important role in affecting the fabrics' ability to absorb energy from the projectile (Cunniff, 1992; Duan et al., 2005; Sun et al., 2014; Wang et al., 2015a,b). In particular, Zhou et al. (2014) worked to investigate the influence of yarn gripping on the energy absorption of woven fabrics made from a UHMWPE fibre. FE simulation showed that fabric energy absorption increases with the yarn—yarn coefficient of friction from $\mu = 0$ to $\mu = 0.4$ and decreases thereafter. This is because the impact load is more likely to be concentrated on the primary yarns when yarn—yarn friction is low, which leads the primary yarns to be damaged at an early stage. Certain levels of yarn—yarn friction help to distribute the load to the secondary yarns and enable more materials to be involved in energy dissipation. If the friction exceeds a critical point, which is regarded as $\mu = 0.4$ in this case, the yarn movement is predicted to be overconstrained, decreasing fabric energy absorption and playing a negative role in ballistic performance.

1.4.2 Quasiisotropic ballistic panel

It is well understood that the primary yarns in a woven fabric play the major role in absorbing the impact energy during a ballistic impact event because of the orthotropic nature of the woven fabrics. Effort has been made to achieve quasiisotropic panels. Wang et al. (2015a,b) reported that the ply orientations in a multi-ply ballistic panel significantly affect the energy-absorbing capacity. The angled panels always show better energy-absorbing capacity compared with the aligned panels. For the impact velocity of 500 m/s, the energy absorption in the angled panels increases by 20%, depending on the number of plies. The layering sequence of the angled plies within the panel also plays an important role in absorbing impact energy. The correct positions of the angled plies are required to obtain the maximum energy-absorbing capacity of the panel. For the panels with a large number of plies such as the eight-ply panels, the sequence of the ply group is more important. The panel $[0/22.5/45/67.5]_2$ shows the best impact performance of the panels compared.

1.4.3 Hybrid ballistic panel design

Chen and colleagues investigated the failure mode of ballistic panels analytically (2013) and numerically (2014). The results show that the front layers of fabric in a ballistic panel are more likely to fail due to shear strain caused by the projectile. The rear layers of fabric tend to be stretched to failure and exhibit larger transverse deflection before failure. This indicated that using shear-resistant materials for the front layers and tensile-resistant materials for the rear layers might improve the ballistic performance of the fabric panel.

The properties of woven and UD fabrics made from UHMWPE fibre were characterised for the hybridisation. Both experimental and numerical results show that woven structures exhibit better shear resistance, and UD structures give better tensile resistance and wider transverse deflection upon ballistic impact. Hybrid panels were designed accordingly using these two types of fabric. Practical results demonstrated that placing the woven fabric layers for the impact face and UD fabrics as the rear layers in the panel gave better ballistic protection for the panel than the reverse sequence. It was also found from this particular research that the combination of 25% woven fabric and 75% UD fabric yielded the optimum performance of the hybrid panels (Chen et al., 2014).

1.5 Future materials and technology for ballistic protection

Ballistic protective materials are developed in concert with the development of weapons. There is no doubt that the weapons are made increasingly effective in defeating the target objects, and this demands that the materials used for personnel and facility protective equipment should be more efficient in resisting the ballistic impact. Another trend has been the demand of lightweight ballistic protective materials to enable greater ability for personnel and vehicle mobilisation. New fibres with super mechanical properties will have to be the key for future ballistic materials. Not only should the future ballistic fibres have high strength and high modulus, they will also demonstrate good flexibility, wear comfort, and property stability under various physical and chemical conditions.

Based on the availability of fibres, the construction of 2D and 3D ballistic fabrics will remain one of the focuses for ballistic panel development. Manipulation of inter-fibre and interyarn friction, creation of quasiisotropic structures, and hybridisation of ballistic materials with different features will continue to draw attention and effort with the aim to improve the ballistic protection efficiency.

The search for new technology for improved ballistic protection will continue. Over the past 20 years or so, some new technologies for providing ballistic protection have been witnessed. The technology for creating UD laminates for ballistic application moved away from the traditional weaving technology for making 2D ballistic laminar materials, and it eliminates the crimps in the 2D laminar, which facilitates fast dissipation of the stress shock wave. Due to the construction of the UD laminates, they effectively limit the back face deformation; hence they provide proper protection against blunt trauma accompanied with ballistic impact. The use of shear thickening fluids (dilatant) in ballistic protection opens new possibilities for different materials from fibres to be employed for ballistic protection, and it stimulates new thoughts and ideas in the course of searching for materials and technologies aiming for ballistic protection.

References

Carothers, J.P., 1988. Body Armor, A Historical Perspective. USMC CSC, 1988. http://www.globalsecurity.org/military/library/report/1988/CJ2.htm.

Carr, D.J., Lankester, C., Peare, A., Fabri, N., Gridley, N., 2012. Does quilting improve the fragment protective performance of body armour? Textile Research Journal 82 (9), 883–888.

Cheeseman, B.A., Bogetti, T.A., 2003. Ballistic impact into fabric and compliant composite laminates. Composite Structures 61, 161–173.

Chen, X., Zhou, Y., Wells, G., 2014. Numerical and experimental investigations into ballistic performance of hybrid fabric panels. Composites Part B: Engineering 58 (3), 35–42.

Chen, X., Zhu, F., Wells, G., 2013. An analytical model for ballistic impact on textile based body armour. Composites Part B: Engineering 45 (1), 1508–1514.

Chen, X., Chaudhry, I., 2005. Ballistic protection. In: Scott, R.A. (Ed.), Textiles for Protection. CRC Press, New York.

Conference in Soldier Survivability and Personal Protection (2009), Defence iQ, Le Meridien Piccadilly, London, UK.

Cunniff, P.M., 1992. An analysis of the system effects in woven fabrics under ballistic impact. Textile Research Journal 62 (9), 495–509.

van Dingenen, J.L.J., 2001. Gel-spun high-performance polyethylene fibres. In: Hearle, J.W.S. (Ed.), High-performance fibres. CRC Press, New York.

Duan, Y., Keefe, M., Bogetti, T.A., Cheeseman, B.A., 2005. Modeling the role of friction during ballistic impact of a high-strength plain-weave fabric. Composite Structures 68 (3), 331–337.

Dunn, D.R., 2008. Ballistic testing of textile materials. In: Wilusz, E. (Ed.), Military Textiles. CRC Press, New York.

Liu, W., et al., 2015. Influence of different back laminate layers on ballistic performance of ceramic composite armour. Materials and Design 87, 421–427.

Luan, K., Sun, B.Z., Gu, B.H., 2013. Ballistic impact damages of 3-D angle-interlock woven composites based on high strain rate constitutive equation of fiber tows. International Journal of Impact Engineering 57, 145–158.

Muñoz, R., Martínez-Hergueta, F., Gálvez, F., González, C., LLorca, J., 2015. Ballistic performance of hybrid 3D woven composites: experiments and simulations. Composite Structures 127, 141–151.

Park, J.L., Chi, Y.-S., Kang, T.J., 2013. Ballistic performance of hybrid panels composed of unidirectional/woven fabrics. Textile Research Journal 83 (5), 471–486.

Porwal, P.K., Phoenix, S.L., 2008. Effects of layer stacking order on the V_{50} velocity of a two-layered hybrid armor system. Journal of Mechanics of Material and Structures 3, 627–639.

Rebouillat, S., 2001. Aramids. In: Hearle, J.W.S. (Ed.), High-performance Fibres. CRC Press, New York.

Roedel, C., Chen, X., 2007. Innovation and analysis of police riot helmets with continuous textile reinforcement for improved protection. Journal of Information and Computing Science 2 (2), 127–136.

Sastry, Y.B.S., Budarapu, P.R., Krishna, Y., Devaraj, S., 2014. Studies on ballistic impact of the composite panels. Composite Structures 72, 2–12.

Scott, B.R., 2006. New ballistic products and technologies. In: Bhatnagar, A. (Ed.), Lightweight Ballistic Composites: Military and Law-Enforcement Applications. Woodhead Publishing Limited, Cambridge.

Scott, R.A., 2005. Military protection. In: Scott, R.A. (Ed.), Textiles for Protection. CRC Press, New York.

Shi, W.F., Hu, H., Sun, B.Z., Gu, B.H., 2011. Energy absorption of 3D orthogonal woven fabric under ballistic penetration of hemispherical-cylindrical projectile. Journal of the Textile Institute 102 (10), 875–889.

Sun, D., Chen, X., Mrango, M., 2013. Investigating ballistic impact on fabric targets with gripping yarns. Fibers and Polymers 14 (7), 1184–1189.

Sun, D., Chen, X., Wells, G., 2014. Engineering and analysis of gripping fabrics for improved ballistic performance. Journal of Composite Materials 48 (11), 1355–1364.

Tam, T., Bhatnagar, A., 2006. High-performance ballistic fibers. In: Bhatnagar, A. (Ed.), Lightweight Ballistic Composites: Military and Law-Enforcement Applications. Woodhead Publishing Limited, Cambridge.

U.S. Department of Justice, National Institute of Justice, 2005. NIJ Special Report: Third Status Report to the Attorney General on Body Armor Safety Initiative Testing and Activities. http://ojp.gov/bvpbasi/docs/SupplementII_08_12_05.pdf?popupWindow=Y.

U.S. Department of Justice, National Institute of Justice, 2008. Ballistic Resistance of Body Armor NIJ Standard-0101.06. https://www.ncjrs.gov/pdffiles1/nij/223054.pdf.

Wang, Y., Chen, X., Young, R., Kinloch, I., 2015a. Finite element analysis of effect of inter-yarn friction on ballistic impact response of woven fabrics. Composite Structures. http://dx.doi.org/10.1016/j.compstruct.2015.08.099.

Wang, Y., Chen, X., Young, Y., Kinloch, I., Wells, G., 2014. An experimental study of ply orientation on ballistic impact performance of multi-ply fabric panels. Online publication Textile Research Journal. http://dx.doi.org/10.1177/0040517514566110.

Wang, Y., Chen, X., Young, Y., Kinloch, I., Wells, G., 2015b. A numerical study of ply orientation on ballistic impact performance of multi-ply fabric panels. Composites Part B: Engineering 68, 259–265. http://dx.doi.org/10.1016/j.compositesb.2014.08.049.

Yan, R., Chen, X., 2014. Aramid/epoxy composites with angle-laid reinforcement constructions for ballistic protection. Online publication Journal of Industrial Textiles. http://dx.doi.org/10.1177/1528083714542823.

Zahid, B., Chen, X., 2014. Impact performance of single-piece continuously textile-reinforced helmet shells. Journal of Composite Materials 48 (6), 761–776.

Zhou, Y., Chen, X., Wells, G., 2014. Influence of yarn gripping on the ballistic performance of woven fabrics from ultra-high molecular weight polyethylene fibre. Composites Part B: Engineering 62, 198–204.

ARAMIDS: 'disruptive', open and continuous innovation

2

S. Rebouillat
DuPont International Operations, Geneva, Switzerland

2.1 Introduction

The attractiveness of the aramid-based technologies is reflected by the breadth of their applications and their suitability to address ongoing and new challenges in fields such as safety and security, protection at large, transport, communication, reduced energy and resource consumption. Fig. 2.1(a) provides an actualised synopsis of this dimension.

Aromatic polyamides came as breakthrough materials in commercial applications as early as 1961, with the market launch of the meta-aramid fibre Nomex®[1], which opened new horizons in the field of thermal and electrical insulation. A much higher tenacity and modulus fibre was developed and commercialised, also by DuPont, under the trade name Kevlar®[2] in 1971. Scientists in the fields of liquid crystals, polymers, rheology and fibre processing as well as process and system engineers spent several years prior and during the early stage of its market introduction to establish the basics and fundamental understanding necessary to take full advantage of this new class of high-performance materials. Their outstanding potential, derived mostly from the anisotropy of their superimposed substructures, presenting pleated, crystalline, fibrillar and skin-core characteristics. This achievement is due in part to DuPont science and technology excellence brought in this field by Kwolek [1–4], Blade [5,6], Tanner [7–9], Gabara [10], Yang [11,12] and many others who took an active part in this enterprise starting with the inception of the aramids. Among a rather scattered literature on the matter, four paramount contributions appeared relatively early in four books: 1973, 1983, 1985 and 1989 titled 'High modulus wholly aromatic fibers', Black and Preston [13], 1973, 'The strength and stiffness of polymers', Schaefgen [14], 1983, 'High performance aramid fibers', Jones and Jaffe [15], 1985 and 'Aromatic High-Strength Fibers', Yang [11], 1989.

Another para-aramid, Twaron®[3], similar to Kevlar®, and an aromatic copolyamide appeared on the market towards the end of the 1980s. The aromatic copolyamide was derived from a good fundamental understanding of the earlier work done with rather stiff para-aramid chains and the new space created for more flexible,

[1] Nomex® is a DuPont Registered Trademark.
[2] Kevlar® is a DuPont Registered Trademark.
[3] Twaron® is a registered product of Teijin.

Advanced Fibrous Composite Materials for Ballistic Protection. http://dx.doi.org/10.1016/B978-1-78242-461-1.00002-9

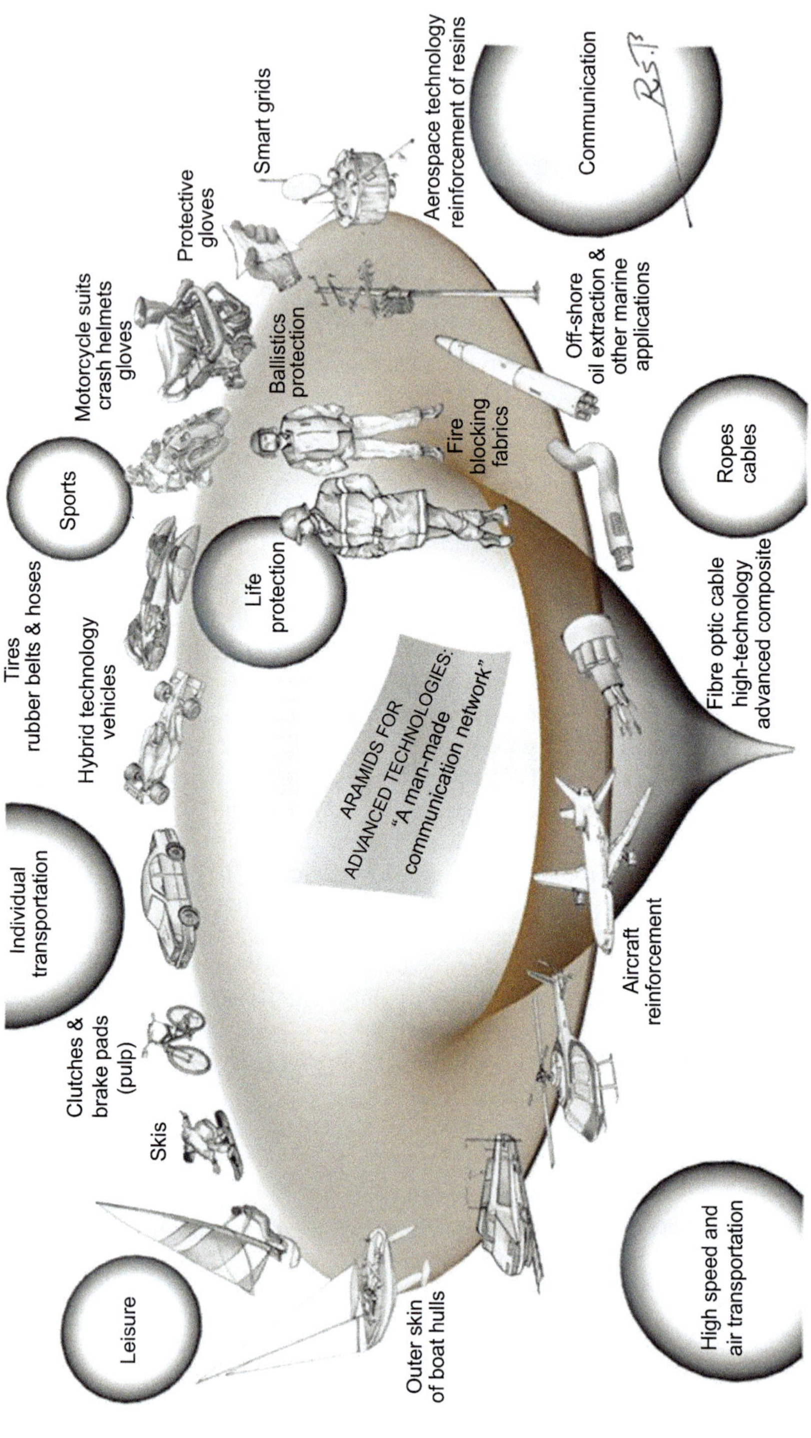

Figure 2.1a Aramid applications: a continuous innovation and the communication network.

high-tenacity fibres. Teijin, after a remarkable scientific interpretation of the prior art by Ozawa [16], who pioneered the development of the aromatic copolyamide fibre, commercialised the Technora®,[4] fibre.

In the last 15 years, tremendous scientific contributions have been made towards establishing the structure–property relationships. Dobb et al. [17,18], Morgan [19], Termonia [20,21], Roche [22], Norholt [23–26] and even more recently Jackson [27] were part of the scientific community bringing analytical, theoretical and modelling tools to approach these rather complex relationships, revealing important characteristics which go beyond the traditional mechanical and chemical interpretations, such as the role which the micro-morphology plays in the biological accessibility of the internal structures.

For the sake of completeness, Monsanto in the 1970s developed, based on an aromatic polyamide-hydrazine composition, an aromatic copolyamide fibre under the code X500 which almost reached the market.

The Russian development history of other related technologies for the manufacture of heat-resistant, high-strength and high-modulus fibres such as SVM®, Armos®, Rusar®, Rusar-S and Rusar-NT aramid fibres is reviewed by Tikhonov et al. [28b]. The onset of this effort is generally associated with the discovery of Tokarev et al. [28c] in 1969. Higher strength and tenacity fibres, from that type of chemistry background, are still expected to emerge as commercial offerings.

In this refreshed review, larger based on Rebouillat's 2001 previous chapter [28d], we will focus on the synthesis, the fibre processing, the structure–property relationships and the applications of the commercial aramids. We will indifferently use the word para-aramid for both the Kevlar® and the Twaron® fibres.

Global estimated demand for aramid fibres (both para and meta) is projected to represent a volume valued at US$4.7 billion [28e].

The EPO [28f] (European Patent office) statistic shows that synthetic fibre industries have enjoyed since 1980 an above average patenting activity as shown, on the excerpt from this EPO's source, on Fig. 2.1(b). This study incorporates 43 selected sectors which data are widely used in macroeconomics. This activity can be compared to pharmaceutical and rubber/plastic industries on that same figure—the deeper the colour, the greater the deviation from average; the first sector is much stronger in terms of patent activity, and the last sector is much weaker than synthetic fibre industry. A noticeable reduction in patent activity is underlined as of 2008 for the synthetic sector.

Fig. 2.1(c) provides the patent activity, in terms of patent publication count per year, for the aramid sector using a sample of 10,000 patents, related to aramids in the context of high-performance fibres, their production and end-uses. Over the last 20 years, that sector has seemed to enjoy a steady growth with peaks and valleys along the way as depicted on that figure.

[4] Technora® is a registered product of Teijin.

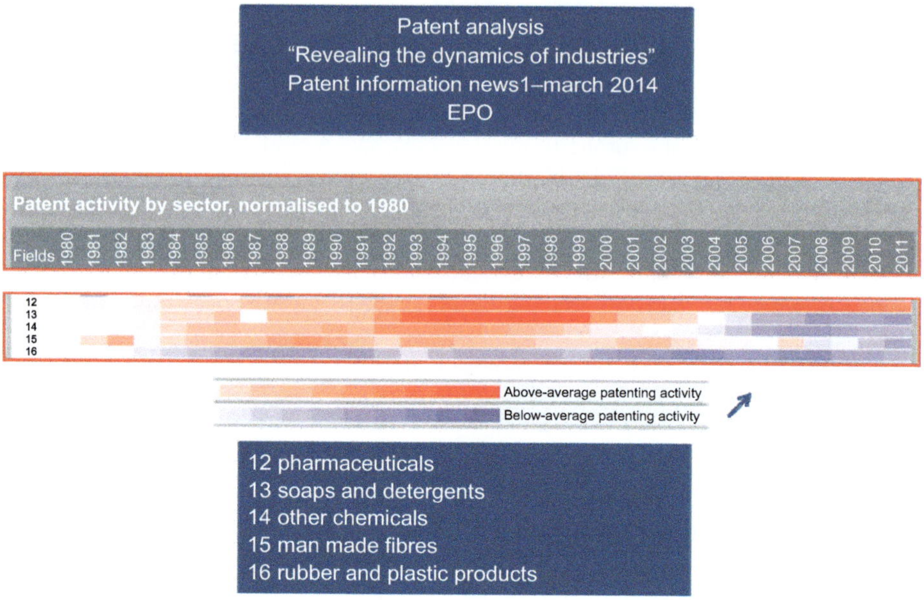

Figure 2.1b Patent analysis revealing some aspects of some industry sectors dynamics.

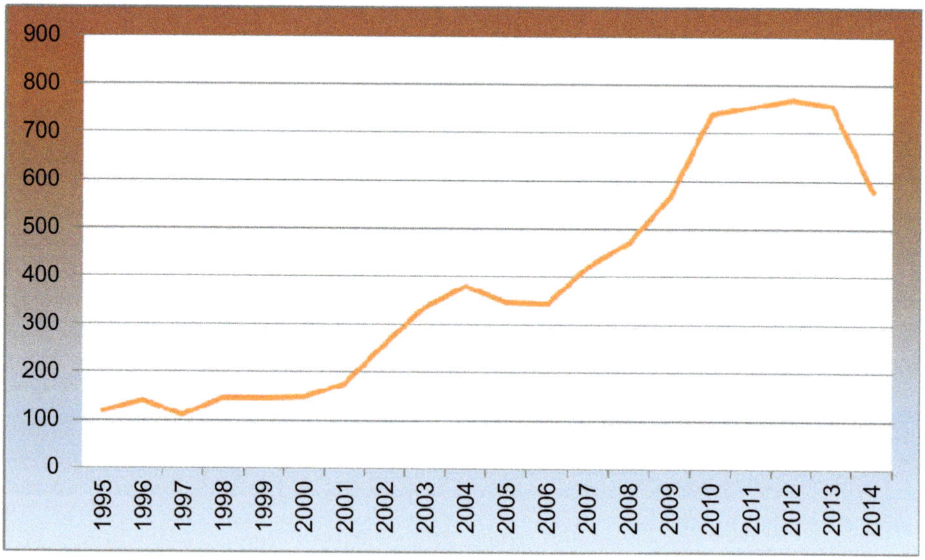

Figure 2.1c The distribution over about 20 years of about 10,000 aramid patent publications (the trend has almost certainly more educational significance than the absolute numbers).

2.2 Polymer preparation

2.2.1 *Basic synthesis*

The following designation was adopted in 1974 by the United States Federal Trade Commission to describe the aromatic polyamide based fibres under the generic term aramid: 'a manufactured fibre in which the fibre forming substance is a long chain synthetic polyamide in which at least 85% of the amide (–CO–NH–) linkages are attached directly to two aromatic rings'.

Aramids are prepared by the generic reaction between an amine group and a carboxylic acid halide group. Simple AB homopolymers may be synthesised according to the scheme below:

$$n\mathrm{NH_2{-}Ar{-}COCL} \rightarrow \mathrm{{-}[\underset{A}{NH}{-}Ar{-}\underset{B}{CO}]{-}}_n + n\mathrm{HCl}$$

A similar reaction scheme also yields AABB homopolymers. For these reactions, several methods [28a,29,30] of preparation can be used such as interfacial polymerisation and low-temperature polycondensation, which remains the most frequently encountered synthesis. Melt or vapour phase polymerisation reactions can also be mentioned.

AABB aromatic polyamides are prepared from various aromatic diamines and diacids or diacid chlorides. This facilitates the generation of aromatic polyamides with extensive variations in molecular structure. The early AABB polymers contained predominantly meta-oriented linkages. The earliest representative of this class is poly-*m*-phenyleneisophthalamide, which was commercialised by DuPont in 1967 as Nomex® aramid fibre. Its chemical formula is as follows:

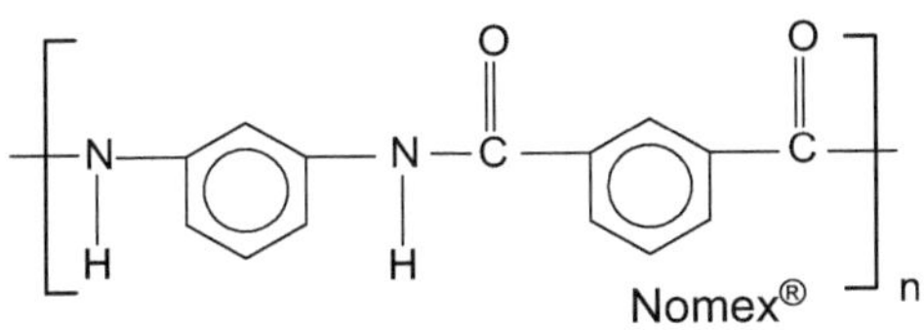

Nomex®

The discovery in 1965 of high-tenacity, high-modulus fibres from liquid crystalline solutions of synthetic para-aromatic polyamides led to the commercial production of Kevlar® aramid fibre by DuPont Co. in 1971 [1]; the corresponding chemical formula is given next:

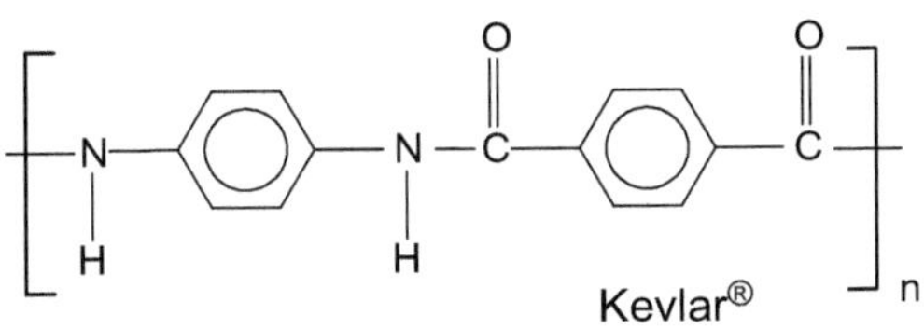

Kevlar®

Kevlar® fibres are poly (*p*-phenylene terephthalamide) (PPTA) [31], the simplest form of AABB para-oriented polyamide.

2.2.2 The aromatic polyamides polymerisation process

Many examples of low-temperature polymerisation for aromatic polyamides and copolyamides can be found in the seminal work by Kwolek [2]. Morgan [32] pointed out several pertinent factors affecting polymer characteristics in low-temperature polycondensation reactions. The most critical ones are listed:

- The solubility–concentration–temperature relationships, which makes the choice of solvent critical.
- The salt concentration at constant polymer concentration, which partly governs the degree of polymerisation and polymer inherent viscosity (η_{inh}).

Rather than detailing generic, mostly empirical, rules governing the polycondensation to achieve the optimum and narrowest molecular weight distribution, we propose next a more specific description of a process based on patent examples and related literature.

Aromatic polyamides of the poly(*p*-phenylene terephthalamide) type are usually synthesised via a low-temperature polycondensation reaction based on *p*-phenylene diamine (PPD) and terephthaloyl chloride (TCl) according to Fig. 2.2. The specifics of the process have been reported by Bair and Morgan [33,34], Kwolek [2] and Blades [5,6]. One methodology involves the dissolution of appropriate quantities of PPD in a mixture of hexamethylphosphoramide (HMPA) and *N*-methylpyrroloidone (NMP), cooling in an ice/acetone bath to 258 K (−15°C) in a nitrogen atmosphere and then adding terephthaloyl chloride accompanied by rapid stirring. The resulting product is a thick paste-like gel. Stirring is discontinued, and the reaction mixture is allowed to stand overnight with gradual warming to room temperature. Work up of the reaction mixture is accomplished by agitating the latter with water in a blender to wash away solvent and HCl. The polymer is collected by filtration. It is important to note that the stoichiometry of the solvent and reactant mixture is important in determining the molar mass of the final product. For instance, Bair and Morgan [33,34] report that a 2:1 ratio by volume of HMPA:NMP produced PPTA product with the largest inherent viscosity (and hence largest molar mass). An optimum reactant (PPD and TCl) concentration was c. 0.25 M. The inherent viscosity was shown to decrease quite rapidly if reactant concentrations less than

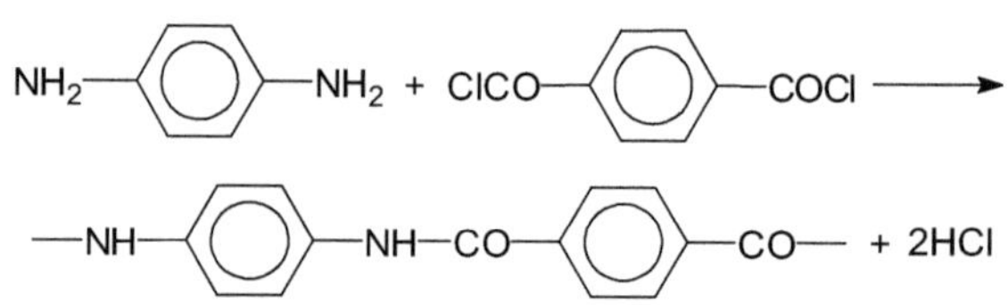

Figure 2.2 PPTA synthesised by low-temperature polycondensation of *p*-phenylene diamine (PPD) and terephthaloyl chloride (TCL).

0.25 M were used. A similar behaviour was observed for reactant concentrations greater than 0.3 M, although the falloff in inherent viscosity was observed to be more gradual in this case. The viscosity falloff at low reactant concentrations can be ascribed to the occurrence of competitive side reactions of the type described by Herlinger and coworkers [35], who examined the solution polymerisation of aromatic diamines and aromatic dicarboxylic acid chlorides in *N,N*-dimethylacetamide (DMAc). The drop-off in inherent viscosity at higher values of the reactant concentration could well be ascribed to a decrease in reactant mobility due to the onset of gelation before a high value of inherent viscosity could be attained. Of course, the polymerisation reaction is exothermic, and a greater quantity of heat will be generated if higher reactant concentrations are used. This could lead to an increased rate in side reactions, which would also depress the value of inherent viscosity attainable.

Alternative synthetic pathways have been reported. In particular, reactions between aromatic dicarboxylic acids and diamines have been investigated since the diacids are reasonably cheap starting materials. The work of Higashi and coworkers is of note in this respect [36–38]. Solvent mixture/salt systems are utilised. In the Higashi approach [37] poly(*p*-phenylene terephthalamide) of high molecular mass was obtained when the polycondensation of terephthalic acid (TPA) and *p*-phenylenediamine (PPD) was carried out in NMP which contained dissolved $CaCl_2$ and LiCl in the presence of pyridine. The synthetic strategy is illustrated in Fig. 2.3.

In earlier work, Higashi and coworkers [36] explored the direct polycondensation reactions of aromatic dicarboxylic acids and diamines by using diphenyl and triaryl phosphates in NMP/pyridine solvent mixtures containing LiCl. The reaction with triaryl phosphates was assumed to proceed via a mechanism outlined in Fig. 2.4. This development arose from earlier work of a fundamental organic chemistry nature reported by Yamazaki and Higashi [39,40]. The work of Preston and Hofferbert [41,42] is also relevant to the synthetic strategy under discussion.

However, the latter strategy was unsuccessful when TPA and diamines were used as polycondensation reagents, since high molar mass aromatic polyamide product was not obtained. This difficulty was subsequently overcome by substituting poly(4-vinlypyridine) in lieu of pyridine.

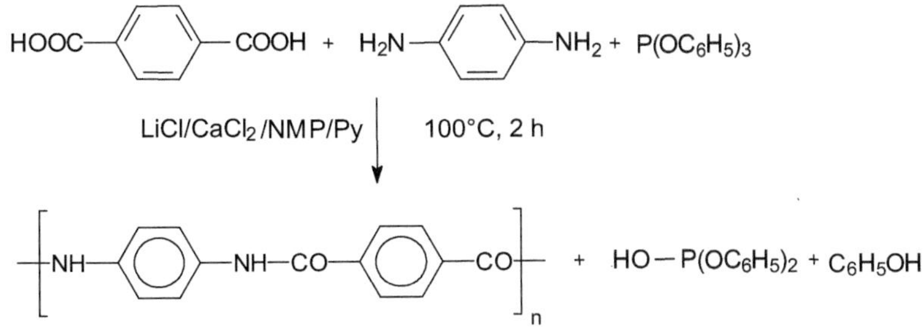

Figure 2.3 The Higashi synthesis: polycondensation of terephthalic acid and *p*-phenylene diamine11.

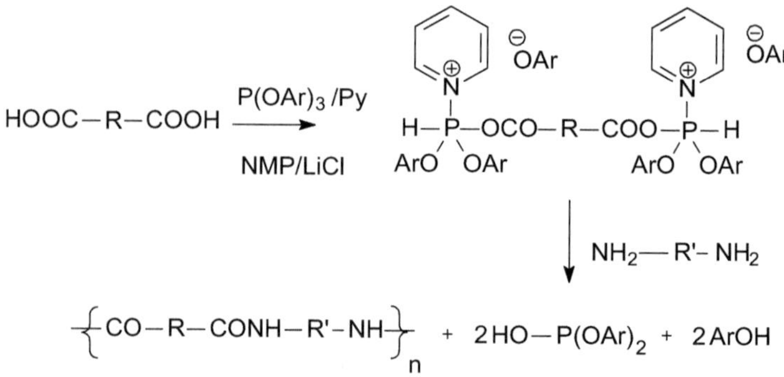

Figure 2.4 The Higashi triaryl phosphite reaction: note the generation of the *N*-phosphonium salt of pyridine, which subsequently undergoes aminolysis.

2.2.3 Copolyamides

The search for aramid copolymers was largely driven by scientific observations made early on by Ozawa [16]. In the pursuit of an ideal synthetic fibre with improved processability, it was recognised that the incorporation of diamines with wider distances between the two amino groups would be favourable to a lower spatial density of hydrogen bonding and rate of crystallisation. Furthermore the incorporation of comonomers in which two phenylene groups are connected with a single atom unit generally improves the fibre drawability.

The resultant species, prepared via the copolymerisation of several aromatic diamines and diacid halides, are referred to as copolymers. The use of copolymers in fibre production is a common technique with aromatic polyamides. These wholly or partially aromatic copolyamides contain alkyl-aryl groups, and others cyclic and heterocyclic groups. Some also contain pivotal substitutions, and some pendent groups.

Teijin has commercialised an aromatic copolyamide Technora®, Fig. 2.5, which exhibits high tensile stress properties as well as high resistance against hydrolysis. The manufacturing process of Technora® [43,44] reacts para-phenylenediamine (PPD) and 3,4′-diaminodiphenylether (3,4′-ODA) (each: 25 mol%) with terephthaloyl chloride (TPC) in an amide solvent such as *N*-methyl-2-pyrrolidone/$CaCl_2$ (10% concentration) to complete the polycondensation. The reaction mixture is neutralised and subjected to spinning into an aqueous coagulation bath. The as-spun fibre is then brought to extraction of solvent, superdrawn at high temperature and passed through finishing to give the final product. The whole process is very simple and

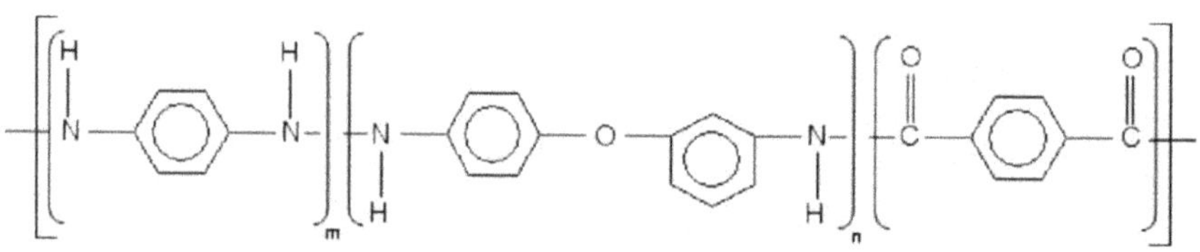

Figure 2.5 Polymer chemical structure of Technora® [43].

straightforward since only a single solvent is used throughout. This greatly simplifies the solvent recovery process and also makes Technora® completely free of residual acid, which avoids later difficulties in the spinning process. Unlike the stiff chain Kevlar® aramid that is spun from a liquid crystalline solution directly to oriented high-tenacity, high-modulus fibres, Technora® is spun from an isotropic solution. Its fibres have tensile stress properties similar to Kevlar® 29. This semi-stiff chain copolymer has a homogeneous morphology, which leads to very good hydrolytic stability while its greater molecular flexibility results in limited modulus.

2.2.4 Other aromatic polyamides

There are many aromatic polyamides not mentioned in this brief overview. They are extensively covered in the monograph by Yang [11].

2.3 Spinning

2.3.1 Solution properties: the 'solubility' challenge

Rigid chain macromolecules such as the aromatic polyaramids exhibit low solubility in many common solvent systems utilised in polymer technology. This is associated with the entropy term in the Gibbs energy of solvation and the enthalpy contribution to the solvation free energy.

The conformation adopted by macromolecules in solution is a well-established area of macromolecular science. However, a somewhat general discussion of the possible states adopted by macromolecules in solution may be presented at this stage with some educational benefit.

A number of common scenarios may be envisaged. A common situation is that of randomly coiled polymer chains. The statistical analysis of such ensembles is well established [45].

However, if the chains are relatively stiff and are linked, then they are ideally described in terms of a random distribution of rods. Now, as the concentration of rod-like macromolecules is increased and the saturation level for a random array of rods is attained, the system will simply become a saturated solution with excess polymer, or more interestingly, if the solvent–polymer relationships are right, additional polymer may be dissolved by forming regions in which the solvated polymer chains approach a parallel arrangement. These ordered regions define a mesomorphic or liquid crystalline state and form a phase incompatible with the isotropic phase.

Continued addition and dissolution of polymer forces more polymers into the ordered state. If the rod-like chains are arranged in an approximately parallel array but not otherwise organised, then the ordered phase is termed nematic. Usually a polymer sample will be polydisperse, and the ordered phase will be organised in microscopic domains of varying size and direction of orientation. Aromatic polyamides form liquid crystal solutions on account of their extended chain structure.

The concentration threshold defining the transition to the liquid crystalline state will depend on the degree of shape asymmetry of the macromolecules, which will be determined as the ratio of their equilibrium length to their diameter, which is termed the axial ratio. It should also be noted that liquid crystallinity can arise from anisotropy of intermolecular forces. But the latter are usually prominent in liquid crystal materials of low molar mass, although they could also be important in polymers, which contain groups such as *p*-phenylene whose polarisabilities are highly anisotropic.

Flory [46a,b–49] has described theoretically the behaviour of hard rod-like polymers in solution and considered in particular the conditions under which the latter macromolecules would produce anisotropic phases. The Flory approach utilises a lattice model. This work followed on from the initial models developed by Onsager [50] and Ishihara [51], which were relatively complex. The simple Flory lattice theory has been extended to consider polydisperse systems [52], mixtures of rod-like particles with random coils [53] and three component systems consisting of a solvent and two solutes, the latter consisting of rod-like molecules with different axial ratios [54]. The problem of biphasic equilibria involving rod-like macromolecules has also been described [55]. Finally, the statistical mechanical analysis has been extended to mixtures of rod-like particles in which flexible joints connect the rods [56].

The behaviour of para-aromatic polyamides in solution has attracted considerable attention, especially with respect to their liquid crystalline behaviour. Both poly (*p*-benzamide) and poly(*p*-phenylene terephthalamide) have received particular attention in this respect. Papkov [57] has provided a useful review of this area. The topic has been discussed in a general way by Morgan [32]. The latter macromolecules are characterised by a stable trans configuration of the amide bond and by a high barrier to rotation around the stable aryl carbon bond.

Bair et al. [58] examined the behaviour of poly(1,4-phenyleneterephthalamide) PPTA in solutions of HMPA/NMP. A strong dependence between the inherent viscosity attained by the polymer solution and the reactant concentration used was found. When dissolved in high-strength sulphuric acid or HMPA/NMP/LiCl solvents, PPTA was found to form an interesting liquid crystalline solution composed of liquid isotropic and anisotropic (mesomorphic) phases at some critical concentration. The effect of polymer intrinsic viscosity on the critical concentration needed to form the anisotropic phase was determined. In the same paper, it was shown that liquid crystalline solutions were formed from poly(chloro-1,4-phenyleneterephthalamide) Cl-PPTA in DMAc/LiCl solvents. In the latter case the liquid crystalline solutions were readily separated into the component isotropic and anisotropic phases. Analysis of the polymer distribution in the respective phases indicated that the distribution depended on polymer concentration, polymer intrinsic viscosity and LiCl content. It was shown that the anisotropic phase always contained a greater amount of polymer than the isotropic phase. Furthermore, the intrinsic viscosity of the polymer present in the anisotropic phase was greater than that found in the isotropic phase. Hence one could conclude that formation of an anisotropic phase occurred by the highest molar mass polymer chains packing into the ordered, more dense fluid. It was also noted that by proper selection of polymer and solvent variables, the extent of anisotropic phase relative to isotropic phase could be varied from zero to essentially 100%.

2.3.2 Spinning of fibres

Production of fibres initially involves heating the spinning solution up to a suitable processing temperature, which is of the order of 80°C for the highly concentrated solutions in 100% (water-free) [87a] sulphuric acid. At this temperature above a polymer concentration of about 10 wt%, the solution state corresponds to a nematic liquid crystalline phase. The concentration limit for the polymer in spinning solution is 20 wt%. If concentrations above this critical limit are used, spinnability is affected due to undissolved material; therefore the resulting fibre has inferior mechanical properties. Because these rod-like polymers are rigid, they oriented themselves with respect to each other forming a nematic phase, illustrated in Fig. 2.6,which shows the orientation angle β with respect to the director $\vec{n}$. This phase is dominated by liquid crystalline domains, which contain aligned polymer chains. The degree of orientation of these polymer chains depends on solution temperature and polymer concentration (Fig. 2.6).

Polymer spinning solutions are extruded through spinning holes and subjected to elongational stretch across a small air gap, illustrated in Fig. 2.7. The spinning holes fulfil an important function. Under shear, the crystal domains become elongated and oriented in the direction of the deformation [3,32,58]. Once in the air gap, elongational stretching takes place. This is effected by making the velocity of the fibre as it leaves the coagulating bath higher than the velocity of the polymer as it emerges from the spinning holes. This ratio is often referred to as the draw ratio, which can be fine-tuned to obtain higher tenacities and moduli with lower elongations and denier. The resulting stretch in the air gap further perfects the respective alignment of the liquid crystal domains. Overall, a higher polymer orientation in the coagulation medium corresponds to higher mechanical properties of the fibre. Because of the slower relaxation time of these liquid crystal systems, the high as-spun fibre orientation can be attained and retained via coagulation with cold water [4]. Essentially, the crystallinity and orientation of the solution are translated to the fibre. These factors allow the production of high-strength,

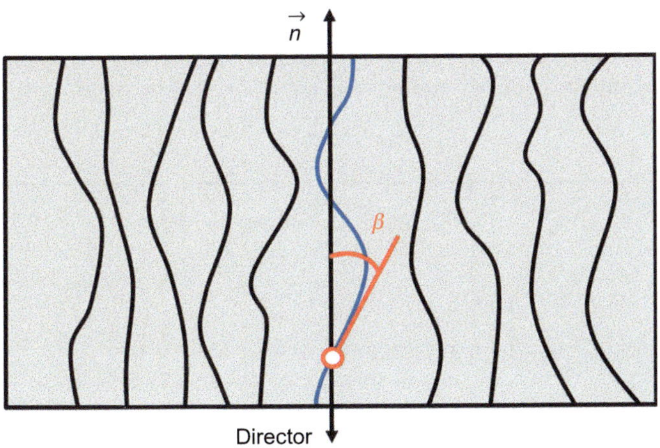

Figure 2.6 Schematic representation of the liquid crystalline solution. The PPTA molecules are indicated by the *thick lines*.

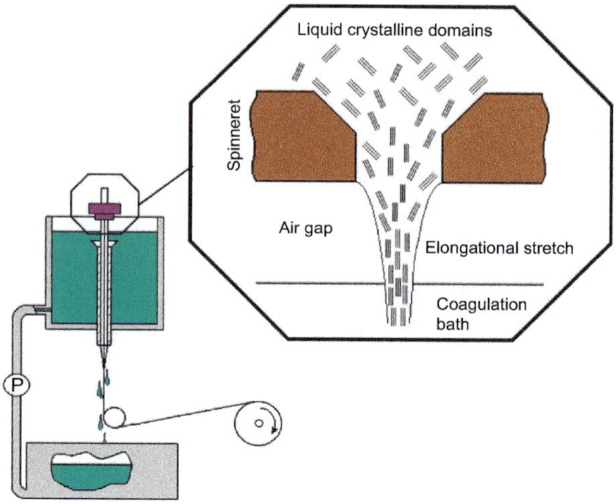

Figure 2.7 Schematic representation of the extrusion of the liquid crystalline solution with the dry-jet wet spinning process.

high-modulus, as-spun fibres. Fibres can exhibit three possible lateral or transverse crystalline arrangements, and these are illustrated in Fig. 2.8. Fig. 2.8(a) represents a fibre with random crystal orientation, Fig. 2.8(b) radial crystal orientation and Fig. 2.8(c) tangential crystal orientation. Interestingly, the radial crystalline orientation can only be brought about using the dry-jet wet spinning process used for para-aramid fibres. This had never been observed before in a synthetic organic fibre.

Present para-aramid products have need of a very high molecular orientation (less than 12 degree) which has an almost direct proportional relationship to fibre modulus. The tenacity of a particular fibre material is also, but not only, governed by this molecular orientation angle. The modulus of the as-spun yarn can be greatly affected by the drying conditions, temperature and tension. Additional orientation inside the solid phase occurs during drying.

Fibres prepared by a dry-jet wet spun process have a noteworthy response to very brief heat treatment (seconds) under tension [59]. These fibres will not undergo

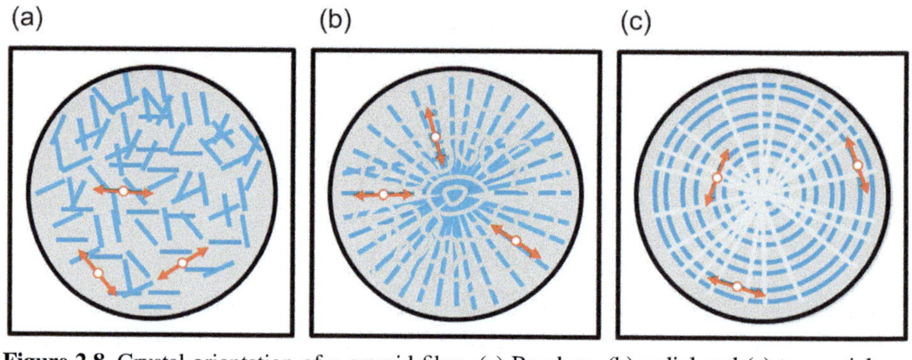

Figure 2.8 Crystal orientation of *p*-aramid fibre. (a) Random, (b) radial and (c) tangential.

drawing in the conventional sense, showing an extension of less than 5% even at temperatures above 500°C, but the crystalline orientation and fibre modulus is increased by this short-term heating under tension. As-spun fibre has an orientation angle of 12–15 degree; these decrease to about 9 degree or less after heat treatment, with the fibre modulus increasing from 64 GPa to over 150 GPa. The applications of these principles led to development of rigid polymer systems forming lyotropic liquids [60]. Molecular orientation, structure and spinning method all affect how aramid fibres respond to this heat treatment. A more recent study [61] of the mechanical change during heat treatment provides a comprehensive set of data regarding the relationships between the annealing time, the final crystallite size, the orientation angle and the tensile modulus. Thermally induced changes in para-aramid fibre surface were evidenced by inverse gas chromatography. Rebouillat [62–64] reported quite extensively on the use of this technique for fibre surface characterisation.

To circumvent the insolubility of rigid molecules in most solvent media, a significant amount of research went into semirigid molecules and their use in fibre production. For example, Black and Preston [13] worked on polymers based on aromatic polyamide-hydrazine structures. The best-known semirigid rod aramid copolymer fibre in use is Technora® developed from the work of Ozawa [65a]. The copolymer is soluble in its polymerisation solvent allowing the reaction mixture to be wet or dry-jet wet spun from an anisotropic solution directly into filaments. The presence of solvent in a reasonably concentrated solution and the inherent rigidity of the molecules avoid molecular entanglement. These spun filaments are washed, neutralised, drawn at a high ratio (10X at 490°C) and dried. As before, the hot drawing process helps to further develop its strength and give it its high properties. The fibre has moderate crystallinity with a random distribution of amorphous regions through its fine structure with its high strength and modulus attributed to its extended chain structure.

Ozawa [16] provides more details regarding the effect of draw ratio and the associated mechanism in the case of aramid copolymers processing. He refers to three modes of drawing. An ordinary draw mechanism would prevail in the case of low draw ratio, below two for Technora®, which is characterised by improved crystal size and uniformity and better orientation of the chains. An intermediate draw mechanism for medium draw ratio (ie, 2 to 14 for Technora® fibre) is dominated by shearing, which yields to chain slips, reduction of the imperfections and therewith higher tenacity, elongation and modulus (TEM) properties. Finally, an excessive draw mechanism, above 15 in the case of Technora®, is associated with the formation of microvoids and microfibrils, and therefore the beginning of decreasing TEM.

A fairly large coverage of spinning variants is proposed in the compilation on aromatic high-strength fibres written by Yang [11].

2.3.3 Aramid types

The properties of aramid fibres depend on the particular spinning and posttreating conditions. Table 2.1(a) and (b) lists some of the fibre categories, which are commercially available, together with their TEM properties [65b].

Table 2.1a Aramid types[a]

Type	Tenacity (mN/tex)	Initial modulus (N/tex)	Elongation at break (%)
Kevlar® 29	2030	49	3.6
Kevlar® 49	2080	78	2.4
Kevlar® 149	1680	115	1.3
Nomex®	485	7.5	35
Twaron®	2100	60	3.6
Twaron® High-Modulus	2100	75	2.5
Technora®	2200	50	4.4

[a]From the manufacturers' leaflets and Refs [3,15,87].

Table 2.1b Strength (S) of Kevlar® yarns at failure

Yarn type	$S^{@failure}$ (GPa)
Kevlar® 29	2.9[a]
Kevlar® 49	2.9[a]
Kevlar® 68	3.1[a]
Kevlar® 119	3.1[a]
Kevlar® 129	3.4[a]
Kevlar® 149	2.3[a]
KM2	3.3[b]

[a]Yang, p. 26 [11].
[b]Yang, p. 219 [12].

2.4 Structure and properties

2.4.1 Characteristics of aramid fibres

It is the unique properties of aramid fibres that set them apart from other fibres. Aramid fibre tensile strength and modulus are significantly higher than earlier organic fibres, and fibre elongation is lower. Aramid fibres can be woven on fabric looms more easily than brittle fibres such as glass, carbon or ceramic. They also exhibit inherent resistance to organic solvents, fuels, lubricants and exposure to flame.

Each type of aromatic polymer gives certain characteristic fibre properties because of its fibre and intrinsic polymer structure. Rather than going through an encyclopaedic description of the properties of the aromatic polyamides, we propose in the following section to review some of the predominant studies, which we believe are important to understand the unique properties of this class of fibre materials. The educational value is tremendous given that this material is such a disruptive innovation model. The superimposed structures, such as the crystallites, the fibrils and the skin-core boundaries, are definitely unique attributes, which can be partially tailored through the fibre process engineering. This is worth more details. It is equally important to provide more insights into the pleated structure and the formation of the hydrogen bonds acting as zippers between the chains. Finally, the mechanical models and the related mathematical analysis are predictive and efficient tools for both the scientist and the engineer to gain fundamental understanding through useful engineering parameters. We propose to focus on these aspects in the following parts.

2.4.2 Structure

It is a truism that the physical properties of macromolecules are determined by their structural characteristics at a molecular level. This is particularly cogent when aromatic polyamides are considered. For instance, as noted by Dobb and McIntyre [68], the tensile modulus of a fibre will be largely determined by the details of the molecular orientation about the fibre axis, and the effective cross-sectional area occupied by single chains, which will, of course, be related to the degree of chain linearity. For instance in poly(*p*-phenylene terephthalamide) the polymer chains are very stiff, which is brought about by bonding of rigid phenylene rings in the para position. In contrast for Nomex® fibres, the phenylene and amide units are linked in the meta position, which result in an irregular chain conformation and a correspondingly lower tensile modulus. Also in poly(*p*-phenylene terephthalamide) the presence of amide groups at regular intervals along the linear macromolecular backbone facilitates extensive hydrogen bonding in a lateral direction between adjacent chains. This in turn leads to efficient chain packing and high crystallinity. A similar situation is encountered for nonaromatic polyamides such as nylon. In the latter material (nylon 6,6 for example) the molecules present in the form of extended planar zigzags joined together in hydrogen-bonded sheets.

Structural studies on poly(*p*-phenylene terephthalamide) fibres have been described by Northolt [23], Haraguchi et al. [66] and Yabuki et al. [67] The supermolecular structure of the high-modulus aramid fibres which are essentially of the poly (*p*-phenylene terephthalamide) type have been reported by Dobb et al. [17,18]. A useful review discussing the relationship between the microstructure and the mechanical properties of aramid fibres has been presented by Northolt [24].

Haraguchi et al. [66] have noted that two crystal modifications of poly(*p*-phenylene terephthalamide) are observed. Northolt [24] has reported on an X-ray diffraction analysis of poly(*p*-phenylene terephthalamide) fibres and proposed a model for the crystal and molecular structure of the PPTA.

When the polymer is spun from a highly concentrated anisotropic solution, the chains form an essentially monoclinic [24] (pseudo-orthorhombic) unit cell with associated parameters $a = 0.787$ nm, $b = 0.518$ nm and c (defining the fibre axis) $= 1.29$ nm.

Approximate values for the orientation angles between the phenylene planes and the amide planes are 38 degree for the PPD segment and −30 degree for the terephthalic segment.

However, fibres formed from anisotropic solutions of lower concentration exhibit a different packing arrangement which is equivalent to a lateral displacement (*b*/2) of chains along alternate 200 planes. It should be noted that both crystallographic forms coexist in fibres spun from solutions of intermediate concentration.

Northolt [24] proposed that hydrogen bonds are formed between adjacent chains lying in the 100 plane. Northolt [24] argued that the chain conformation is primarily governed by competitive intramolecular interactions between the conjugated groups in the chain. Similar interactions are also found for the conformation of benzamide, acetanilide and the terephthalamide molecules. These interactions arise from the operation of a resonance effect, which results from an attempt to stabilise coplanarity of the amide groups and the phenylene groups. A counteracting steric hindrance is also found between the oxygen and the ortho-hydrogen of the PPD segment, and between the amide hydrogen and the ortho-hydrogen of the terephthalic segment.

The regularly positioned amide segments allow reasonably strong hydrogen bonds, which facilitates a proper load transfer between the chains, as found, for example, in a zipper. The hydrogen bonded chains form sheets, which are stacked parallel into crystallites. Between adjacent hydrogen-bonded planes, the interaction largely takes place by van der Waals' forces with some pi-bond overlap of the phenylene segments. This causes the hydrogen-bonded planes to function as slip-planes in a manner analogous to close-packed planes in metals. Note also that bond rotation and hence molecular flexibility is inhibited by the presence both of the aromatic rings and the double-bond nature of the amide group arising from resonance effects.

It should be noted that the microscopic chain structure and chain properties dictate the manner of fibre production. A schematic representation of the microstructure of (a) a semicrystalline polymer such as nylon-6 and (b) poly(*p*-phenylene terephthalamide) is outlined in Fig. 2.9. In each case the fibre axis is vertical.

It should also be noted that poly(*p*-phenylene terephthalamide) fibres are generally characterised by intense but diffuse small-angle scattering features on the equator of the XRD patterns. This effect can be ascribed to the presence of microvoids. Electron microscopy and X-ray data indicate that the voids are rod-shaped with their long axis almost parallel to the fibre axis, having typical widths in the range 5−10 nm and length about 25 nm. The influence of microvoids in para-aramid fibres has been explored by Dobb et al. [68], using a test method originally described by Allen [69].

The surface microstructure of an aramid fibre was also observed directly by Rebouillat et al. by atomic force microscopy [70,71]. The unit cell parameters *b* and *c* of the crystalline structures and the coplanar-like rotation of benzene rings are similar to the reported results, but the registration angle is not constant. This discrepancy is explained as the difference between surface and bulk crystalline structure. At the surface of the polymer, chains tend to be arranged less perfectly, and the surface crystalline structure can be influenced by several parameters such as the fibre manufacturing conditions and posttreatment, adsorbed water and even the possible friction of the Atomic Force Microscopy (AFM) scanning tip.

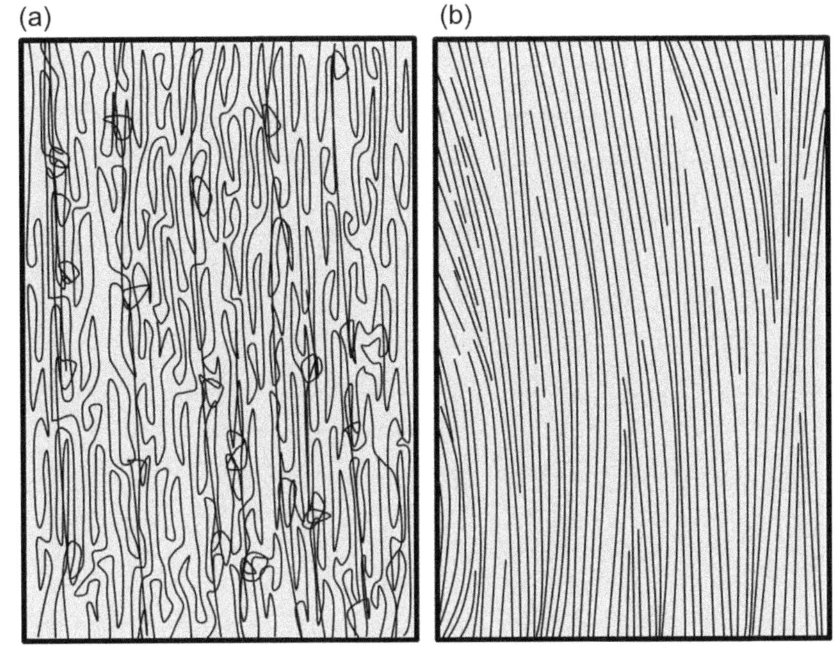

Figure 2.9 Schematic representation of the microstructure of (a) semicrystalline polymers such as nylon-6 and (b) poly-*p*-phenylene terephthalamide. Fibre axis is vertical.

The supermolecular structure of a high-modulus polyaromatic material such as Kevlar® 49 has been reported by Dobb et al. [69]. In this work a combination of electron diffraction and electron microscope dark field image techniques were utilised. Detailed analysis conducted by the authors led to the assignment that there was a uniform distribution of ordered crystalline material throughout the fibre and that the dark field banding observed was a manifestation of changes in crystalline orientation and not of crystalline order. The results reported by Dobb et al. [69] provide strong evidence for a regular pleated structure with the alternating components of each sheet arranged at approximately equal but opposite angles to the plane of the section, as illustrated in Fig. 2.10. Experimental measurements indicate that the angle between adjacent components of the pleat is about 170 degree.

2.4.3 Analysis of mechanical properties

The mechanical properties of aramid materials underlie their significant commercial utilisation in many areas. For instance, the as-spun Kevlar® aramid fibre exhibits over twice the tenacity and nine times the modulus of high-strength nylon [72]. On a weight basis, it is stronger than steel wire and stiffer than glass. Both creep and the linear coefficient of thermal expansion are low, and the thermal stability is high. The latter properties resemble those of inorganic fibres and of course can be attributed to the extended chain morphology, high molar mass and excellent orientation in a thermally

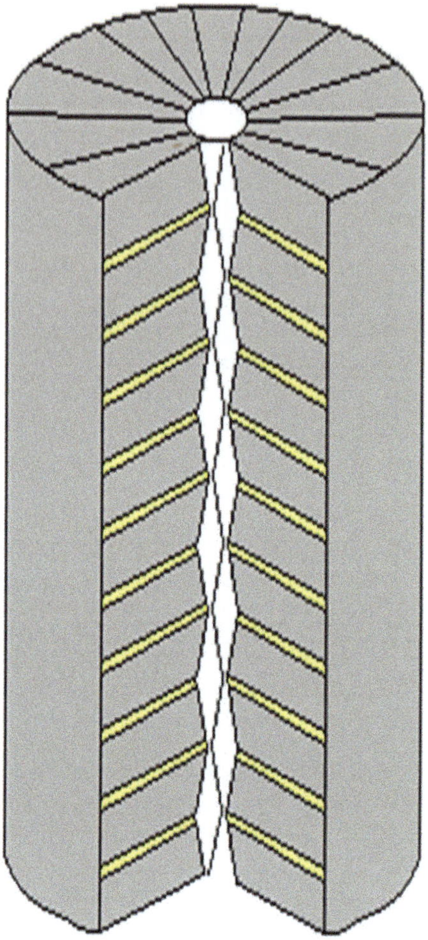

Figure 2.10 Radial pleated structure model of PPTA fibre.

stable structure that does not melt. Para-aramid fibres have utility due to a combination of superior properties allied with features usually associated with organic fibres such as low density, easy processibility and rather good fatigue and abrasion resistance.

The macroscopic mechanical properties of a polymer are determined by the molecular structure of the chain. Northolt [23] has described a polymer chain as a one-dimensional load-bearing conformation. He notes that a material such as polyethylene can be processed to exhibit a large tensile modulus and tensile strength (gel spinning followed by hot drawing of polyethylene afforded a fibre of tensile strength of 3 GPa and modulus of 200 GPa). The downside is that extremely long chains are required to ensure optimum load transfer between adjacent polymer chains due to the fact that intermolecular adhesion in the fibre only arises due to van der Waals forces. Also polyethylene has a very low melting point, which limits practical application. The latter disadvantages are removed if stiff polymer chains linked by hydrogen bonds are used. As previously noted in this report, chain stiffness is achieved when free rotation around interatomic bonds is not allowed, and by necessity, the chain will adopt the linearly extended conformation. Of course the latter idea is manifested in the aromatic aramid materials. As noted by Northolt [23], regularly positioned amide segments in the latter materials allow for medium to strong intermolecular hydrogen bonds, which ensure a proper load transfer between the chains as indeed is found in a zipper. However, the chain properties exhibited by aromatic polyaramids are mainly due to the alignment of chains in one direction, which limits the macroscopic shape of the material to filaments and films. The aromatic polyamides are manufactured in a dry-jet wet spinning process in which the chains are given a narrow orientation distribution around the filament axis. This procedure leads to an initial modulus of c. 70 GPa, which is subsequently increased by short-term heating under tension to c. 140 GPa. Typically the filament strength for a gauge length of 2.5 cm is 4 Gpa with an elongation at break of about 4% and a fracture energy of c. 80 MJ/m^3.

An excellent theoretical approach by Termonia [20,21] led to the calculation, based on the molecular weight of commercial para-aramid fibres, of the ultimate tensile strength of PPTA filaments to be 10 GPa. Another calculation can be estimated from the weakest chemical bond of the PPTA chain divided by the surface area of a cross section of the chain in the crystal lattice. In the latter case a tensile strength of 29 GPa is found. These calculations, for a hypothetical ideal molecular structure and molecular alignment, take into account the weakest bond dissociation energy of the polymer molecule in its most elongated conformation, which as we have underlined in a previous section is a difficult target to attain in view of the solution properties and the spinning conditions. Mechanical modelling further helps to appreciate the attainable improvement from the observable tenacity.

A detailed mathematical model to adequately describe the mechanical properties of an aromatic polyamide fibre such as poly(*p*-phenylene terephthalamide) has been developed by Northolt and van Aartsen [25] and Northolt [26]. The model adopted by Northolt is in good accord with X-ray diffraction studies. On the basis of the mathematical model developed, and if we assume a uniform stress distribution along the fibre, it is possible to show [25,26] that the elastic extension of the aramid fibre is governed by the following expression:

$$S_{33} = \frac{1}{e_3} + A\langle \sin^2 \phi \rangle \qquad [2.1]$$

In this expression, S_{33} denotes the fibre compliance, e_3 is the modulus of elasticity of the chain parallel to its axis of symmetry, and the factor $\langle \sin^2 \phi \rangle$ represents the orientation distribution parameter of the crystallites with respect to the fibre axis. The latter quantity will be zero if the orientation is perfect, and two-thirds if the orientation is random. The factor A represents a measure of the mechanical anisotropy of the crystallite and is given by the following:

$$A = \frac{1}{2g} - \frac{2(1 + \nu_{13})}{e_3} \qquad [2.2]$$

where g denotes the shear modulus in the plane containing the chain axis, and ν_{13} represents the Poisson ratio for a stress along this axis.

Eq. [2.1] has been confirmed experimentally by Northolt [26] for PPTA fibres using dynamic moduli and crystallite orientation measurements. He has shown that the dynamic compliance S_{33} ($S_{33} = 1/E$, where E represents the dynamic tensile modulus) varies linearly with $\langle \sin^2 \phi \rangle$ for three different fibre samples. Very little scatter is obtained over the entire measurement range. Indeed, the expression is valid up to rupture.

From least squares analysis of the data, $A = 0.26 \pm 0.01$ m^2/GN. Typically $e_3 = 240$ GPa and $g = 2$ GPa. It was determined that up to a stress of c. 1.5 GPa, the extension of the fibre (the strain) is brought about mainly by elastic rotation, by some retarded as well as plastic rotation of the crystallites towards the fibre axis, and by elastic extension of the polymer chain itself.

Northolt also derived a relation between the stress σ and the orientation parameter as follows:

$$\langle \sin^2 \phi \rangle = \langle \sin^2 \phi \rangle_0 \exp[-(2A + \lambda)\sigma] \qquad [2.3]$$

where the parameter λ represents both the permanent and the retarded elastic rotation of the crystallites, and $\langle \sin^2 \phi \rangle_0$ denotes the initial value of the orientation parameter. The relation expressed in Eq. [2.3] has been confirmed experimentally. Typical values for λ lie in the range 0.15–0.28 m^2/GN for the aramid fibres examined (four samples). It was noted that λ values tend to be larger for less oriented fibres.

Northolt also determined that for $\sigma > 1.5$ GPa, the fibre extension increases by axial flow, which would in principle be caused by slip between adjacent chains.

The work of Northolt [26] resulted in the generation of an approximate analytical expression for the tensile curve of a fibre, which consists of a linear arrangement of crystallites which have a narrow orientation distribution $\langle \sin^2 \phi \rangle_0$ with respect to the fibril axis. The stress/strain relation is given by Eq. [2.4]:

$$\varepsilon = \frac{\sigma}{e_3} + \frac{\langle \sin^2 \phi \rangle_0}{2}[1 - \exp(-C\sigma)] \qquad [2.4]$$

where $C = 2A + \lambda$. Hence we can conclude that the ductility of the fibres is primarily determined by the initial orientation distribution of the crystallites and by the ability of the crystallites to rotate their symmetry axis towards the stress direction as a result of the relatively low value of the shear modulus g. During fibre extension the latter distribution contracts, and near rupture the chains are oriented almost parallel to the stress direction.

Northolt [26] has also noted that because of the crystallite-orienting mechanism governing the entire tensile deformation process, the presence of structural irregularities may hamper the alignment of the crystallites along the stress direction. This could give rise to premature rupture of the fibre. However, local slip of hydrogen-bonded planes may serve to mitigate the effect of the presence of inhomogeneities.

2.4.4 Some useful comparisons between aromatic polyamides and copolyamides

The levels of tensile properties and thermal stability of Technora® are comparable to highly crystalline para-aramid fibres over standard practical temperature ranges. Technora® has a comparatively high fatigue resistance, which may be the direct result of the flexibility of the copolymer chain and loose crystal structure in the copolymer; this was confirmed using wide-angle X-ray studies [3].

Most of the differences between these two families, the para-aramid and the meta/para copolymers, seem to have been anticipated at the design stage, as described by Ozawa [16] in Table 2.2.

Although mainly derived from hypotheses partly validated through analytical methods, there are several mechanical failure modes, morphology characteristics and chemical accessibility properties which have been proposed by Morgan et al. [19], Panar [73], Jackson et al. [27], Lee et al. [61], Li et al. [74], Fukuda [75], Roche [76], Rao et al. [77], Young [78] and Mooney [79].

Regarding more specifically fibre failure modes and related morphologies, the atlas of fracture, fatigue and durability compiled and commented by Hearle et al. [80] constitutes a unique piece of literature where comparative series of SEM pictures speak by themselves. Supplemented by the necessary information on how breaks occurred, this outstanding collection of data provides quite relevant information which sets the aramids apart from most synthetic fibres due to the multiple and long axial splits in the filaments which are observable when the material is fractured by tensile testing. This mode of failure, which is also observed in the case of high-modulus polyethylene (HMPE) fibres, is easily distinguishable from the brittle tensile fracture of glass, ceramic, carbon and elastomeric fibres, the ductile tensile fracture of nylon, polyester and polypropylene fibres and the granular fracture typical of polybenzimidazole (PBI) or alumina fibres, for example. Finally, the fibrillar failure of cotton may to a certain

Table 2.2 Molecular and polymeric requirements for improved characteristics of HT–HM aromatic fibres [16]

Key improvements	Polymer composition and formulation
Thermal resistance	• Wholly aromatic polyamide • Absence of unstable linkage (urethane, urea, alkylene, etc.)
Solubility	• Copolymer with dissymmetrical units • Inclusion of –O–, –CO–, $-SO_2-$, etc. • Amides rather than esters
Drawing potential	• High molecular weight • Enhanced chain flexibility by incorporating –O–, –CO–, $-SO_2-$, etc., into polymer chain
Dimensional stability	• Rigid molecular chain • Crystallinity

extent show some similarities with the aramids due to the microfibril assembly found in this type of natural fibre, although the very high-strength nature of the para-aramids largely limits this kind of analogy. Hearle et al. [81] also provide more insight into the flex fatigue and abrasion resistance of high-modulus fibres.

Since never commercialised, the experimental X500 series of aromatic copolyamide fibres developed by Monsanto have been subject to detailed research work, which major findings have been disclosed in the extensive technical compilation done by Black and Preston [13] in 1973. The chemical formula of the X500, which is provided below, underlines the larger number of amide functions which shall increase significantly the number of intermolecular chain bonding. This material remains very attractive as a comparative research model to help the understanding of the rather complex behaviours of para-aramids.

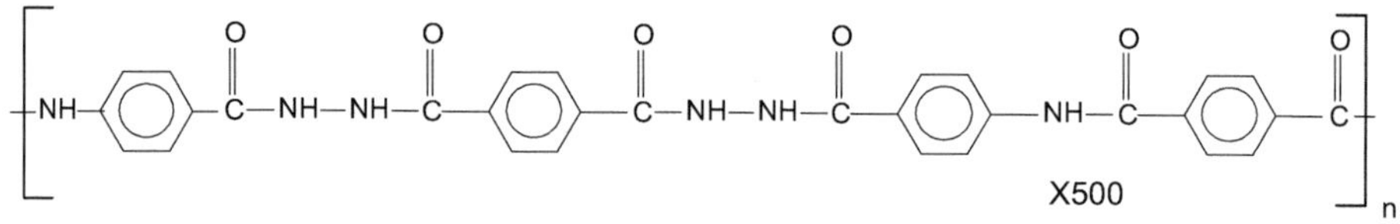

A more recent initiative from a Russian research organisation [82] led to the experimental development of the fibre Armos®[5] based on a copolyamide consisting of para-substituted phenylene and benzimidazole, which shows encouraging mechanical properties. Its chemical formula is shown next:

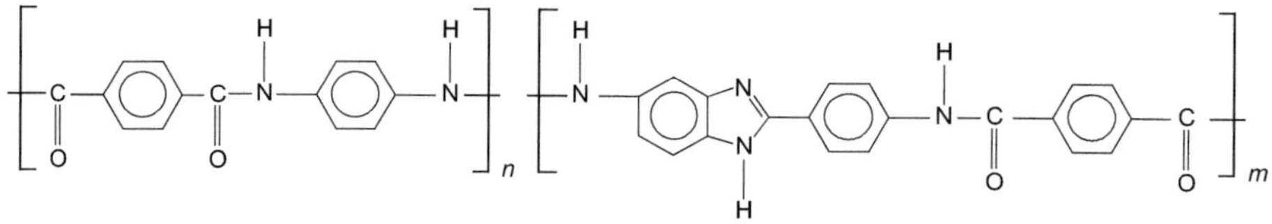

The comparative analysis [84] of the structure and the mechanical behaviour of regular para-aramid and the statistical ARMOS® remains a unique research work outlining the analogies between these two materials, although they differ significantly in terms of crystallinity. Hopefully more studies will confirm the surprising presence of smectic structure as well as nematic mesophase in the noncrystalline phase of the copolyamide.

Recent data deriving from pilot spinning, conducted in 2012 [28b], of the aramid fibres Rusar®, Rusar-S, Rusar-O and Rusar-NTs underlines the following TEM:

- specific rupture load, cN/tex, respectively, not less than 240, 290, 54, 290
- elongation at rupture, %, of respectively, not less than 2.6, 2.5, 3.5, 2.0
- elasticity modulus, GPa (kgf/mm^2), respectively, not less than 140, 160, –, 170–180

Although not always being unanimously accepted, the conclusions presented in these studies offer tremendous opportunities to gain fundamental understanding of

[5] Armos® is a registered product.

structure–property relationships associated with fibre formation and consolidation as well as fibre usage and biological accessibility and degradation.

The skin-core supermolecular structure of the PPTA fibre is probably the most broadly accepted feature. The difference in orientation and alignment of the skin chains versus the core microfibrils, which are substructured by crystallites, has been often used to support a hypothetical fracture model shown in Fig. 2.11. An explanation of this type of propagation can be associated with the fact that the nematic-like structure observable in the spinning solution is rearranged under relaxation effect into a smectic-like structure during the fibre coagulation. The smectic alignment is parallel and stratified as per a two-dimensional order with a very limited number of liaisons across the transverse boundaries, as the ones depicted in the central part of

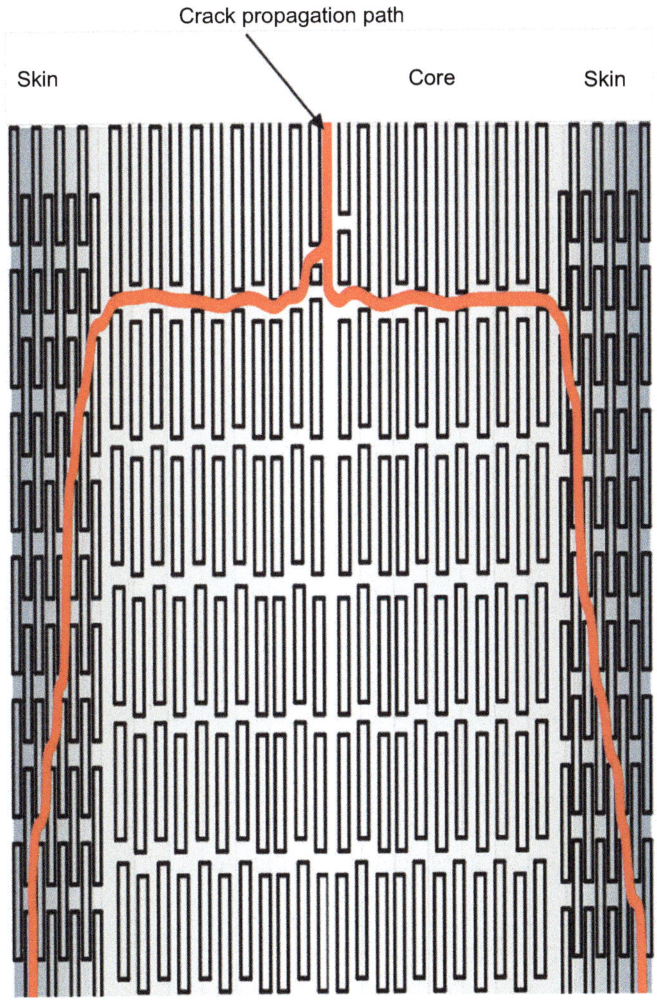

Figure 2.11 Schematic Representation of a fracture model of PPTA fibre.

Fig. 2.11, while the nematic order corresponds to molecules showing a parallel one-dimensional order with overlaps which resembles the skin representation of Fig. 2.11. The type of periodic alignment observed in the core part is promoted by the concentration of chain ends in transverse connecting zones, which are mechanically and chemically vulnerable.

Ozawa [16] suggests that aromatic copolyamides are less susceptible to the formation of smectic structures, therefore offering different mechanical and chemical properties. In the case of Technora® fibres, Ozawa attributes this tendency to the fact that the chains are much more flexible than PPTA ones, and the crystallites are looser; furthermore the process allows more stress dissipation to occur. SEM and X-ray analysis support this position.

There are rather significant differences between PPTA fibres and the aromatic copolyamide Technora® in terms of moisture absorption, permeability of small entities and the related mechanisms. The moisture absorption relative to the unit weight of fibre is about half for the latter (2% vs 4–6% for para-aramids), which Ozawa [16] explains by the density of the hydrophilic amide groups at the surface and the degree of crystallinity. This further supports the observed difference in terms of chemical resistance and higher fatigue performance of the copolyamide fibre of which the level of chain extension and orientation due to the superdrawing tends to retard the migration of small molecules, contributing to a higher apparent chemical resistance. Additional studies regarding the moisture transport phenomena and aqueous accessibility of PPTA crystallites are available [27,83]. These all tend to support and complement, using different approaches, the structure and mechanical properties described before.

The stiffness of the PPTA chain is partly associated with the limited rotation of the carbon–nitrogen bond itself due to the resonance–conjugation existing between the amide and the aromatic groups. Such a situation does not exist in the case of the meta configuration of the Nomex® fibre; therefore the fibre is white versus the golden yellow colour of the para-aramid. The Technora® copolyamide, with a meta and a para molecular configuration, presents a rather golden brown colouration. The conjugated patterns provide a dominantly double-bond character to the carbon–nitrogen liaison in the para configuration of which rotational energy is approximately 10-fold higher than for a single one.

2.4.5 A selection of observed mechanical properties

Typical stress–strain curves of different Kevlar® fibres are provided in Fig. 2.12(a), which clearly outlines the modulus increase from Kevlar® 29 to Kevlar® 149. The respective moduli, brought by various spinning conditions and posttreatments performed on para-aramid precursors are generally considered as intermediate between those of graphite, boron and glass fibres. This linear stress–strain behaviour of para-aramids is typical, compared to most synthetic fibres, of which the tensile behaviours are depicted in Fig. 2.12(b).

Creep is measured either by the length variation under tension or by the stress decrease at constant gauge length. Para-aramids, which exhibit little creep, differ

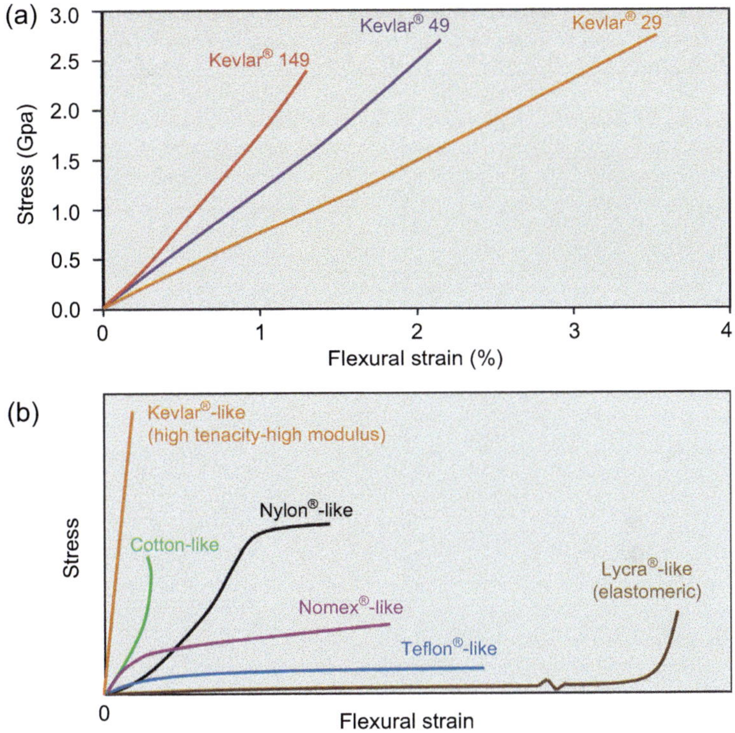

Figure 2.12 (a) Typical stress–strain curves of Kevlar® fibres. Twaron® fibres exhibit similar properties. (b) Typical stress–strain curves of other commercially representative industrial yarns.

significantly from other highly oriented polymeric fibres, such as HMPE fibres, which can break after several days under intermediate load due to their high creep properties, which are associated with a stress slip of molecules compared to a structure tightening in the case of para-aramids. Creep is affected by the temperature, the load relative to the fibre's ultimate strength, the water content and other parameters. There are differences between Kevlar® 49 and Kevlar® 29. The latter tends to show no dependency of the creep amplitude as a function of temperature and load when this load is below 50% of the fibre-breaking strength.

Compressive and shear properties tend to relate very well with the degree of axial orientation and radial intermolecular bonding. It is therefore not too surprising to find that HMPE fibres have been reported [70] to exhibit lower compressive yield stresses than para-aramids. Sikkema [84] gave a compressive strength of 0.6 Gpa for the Twaron® fibres, which is about one-fifth of the tensile strength as mentioned by Yang [11]. Greater than those of conventional synthetic fibres, the shear moduli of para-aramids are lower than their compressive moduli, which to a certain extend is due to the degree of anisotropy of their radial structure.

Table 2.3 **Properties of commercially representative reinforcement fibres[a]**

Material	Density (kg/m^3)	Decomposition melt (°C)	Tenacity (mN/tex)	Initial modulus (N/tex)
Para-aramid standard (Kevlar®, Twaron®, etc.)	1.44	550	2065	55
Para-aramid typical high modulus	1.45	550	2090	77
Nomex®	1.46	415	485	7.5
Technora®	1.39	500	2200	50.3
PA 66	1.14	255	830	5
Steel cord	7.85	1600	330	20
Carbon HT	1.78	3700	1910	134
Carbon HM	1.83	3700	1230	256
E-glass	2.58	825	780	28

[a]From the manufacturers' leaflets and Refs [3,15,16,87].

Like other polymers, aromatic polyamides are sensitive to radiation [11,87a], especially in the range between about 300 and 450 nm. For outdoor applications, proper protection from radiation is necessary to maintain good mechanical properties.

The preceding discussions of the structural and mechanical relationships provide a more precise and complete picture of the properties of the related polymers. Out of the very large numbers of reports on the subject matter, general tendencies are rather broadly, although not unanimously, accepted among scientists. This situation is largely favourable to fruitful scientific debates, which support the uniqueness of the aromatic polyamides. One may see an opportunity here, still as a future direction, for the translation of this immense knowledge into end-use science, which could help the fundamental understanding of the design parameters and the properties of parts manufactured from these polymers in various forms. For the sake of offering comparative basic data, some of the most frequently encountered materials used in advanced fibre technologies are compared in Table 2.3.

2.5 Applications

2.5.1 Preliminaries and systems engineering

From an educational standpoint, the scientific activity is generally known to precede or react to technological innovation; it is sometimes valuable, in retrospective, to glance at the ratio of publications over a significant time span.

Thomson Reuters Web of Science™ has been used for the keyword searches of publications, including patents, related to the classification proposed in Table 2.4. The 1500 most pertinent publications (including patents), over a period starting from 1980, were selected. The patents ratio, in the publication rate, is dominant, which underlines the application technology strength of the related industry. Fig. 2.12(c) provides means to position the end-use applications (vs publication rate), which will be

Table 2.4 Aramid market segments and key attributes [85]

End use	End-use systems	Key attributes
Composites	Fabrics for aircraft and containers Pressure vessels Ship building Sporting goods Plastics additives Civil engineering	Light weight High strength High modulus Good impact strength Wear resistance Hybrid tailoring ability
Protective apparels	Heat resistance workwear Fire blankets Flame-retardant textiles Cut-protective gloves Cut-protective seat cover layers	Heat resistance Flame retardation Cut resistance Chemical resistance Blending ability
Tyres	Truck and aircraft tyres High-speed tyres Motorcycle tyres Bicycle tyres	Low density Weight saving High tenacity Dimensional stability Low shrinkage Puncture resistance Fatigue and creep balance Hybrid cord tailoring ability
Mechanical rubber goods (MRG)	Conveyor belts Transmission belts Hoses for automotive, including turbo Hydraulic hoses Hoses in offshore Umbilicals	High strength High modulus Dimensional stability Thermal resistance Chemical resistance
Friction products and gaskets	Brake linings Clutch facings Gaskets Thixotropic additive Industrial paper	Fibre fibrillation Heat resistance Chemical resistance Low inflammability Mechanical performance Compounding ability

Continued

Table 2.4 Continued

End use	End-use systems	Key attributes
Ropes and cables	Aerial optical fibre cable Traditional optical fibre cable Electro-cable Mechanical construction cable Mooring ropes	High strength High modulus Dimensional stability Low density Corrosion resistance Good dielectric properties Heat resistance
Life protection	Bulletproof vests Helmets Property protection panels Vehicle protection Strategic equipment shielding	High tenacity High energy dissipation Low density and weight reduction Comfort Potential for higher tenacity Blending ability

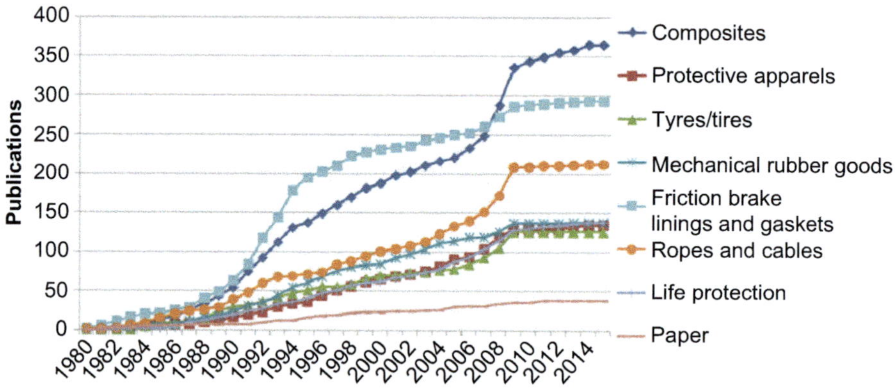

Figure 2.12c Science and technology activity in various aramid end-use applications over time. (Relative positioning has probably more teaching value than absolute numbers.)

discussed in the current section. Relative positioning has probably more teaching value than absolute numbers.

The previous section of this chapter illustrated the role of molecular chain extension, orientation and structure in the fibre characteristics. The transition from the chemical structure to a physical structure, such as a polymer, can be considered as being part of the molecular engineering as described by Tanner et al. [7]. A further step is the transition from the given physical structure to a product form, such as a fibre, and can be defined as part of the process engineering. A later step encompasses the integration of the product, such as the aramid fibre, into an advanced structure or an advanced technology, which according to the definition of Tanner et al. [7] is part of the system engineering.

The present section takes into consideration a selected, mostly commercially dominant, range of applications of the aramids with an attempt to underline some of the system engineering aspects. This approach may provide not only a better understanding of the major reasons for these fibres to contribute to the best balanced performance of the system, but also may stimulate new ideas in the way these fibres are used in the considered system or how they could be used in a new one. As an example, the understanding of how a para-aramid fibre can be shaped into an optimised fabric pattern to resist a fragment impact can provoke new ideas regarding the engineering of these fibres to offer cut and heat protection in other sectors where the kinetics may be much closer to static conditions.

This is a system approach, which forces clarity of rationale to utilise the advantageous properties of the fibre. These can be used to minimise the effect of some features of the fibre as well as finding possible synergistic combinations of properties with the other system components. These approaches led to the development of some of the aramid fibre applications outlined in this section, illustrated in Table 2.4 [85].

The system engineering approaches have likely powered, from the very inception of the high-performance fibres in general, the continuous disruptive innovation [87b] associated with this technological enterprise, the market development and the societal adoption, the overall benefits and wellness.

2.5.2 Ballistic and life protection

The dynamic mechanical properties of the aromatic polyamide fibres, as well as their visco-elastic behaviour, which can be tailored via suitable resin reinforcement, are perfectly adapted to their use in impact-resistant systems being at low or high velocities. As early as back in the 1970s, now celebrating the 50th year anniversary of Kevlar®, this opportunity was rapidly recognised and translated into concrete applications such as for the life protection of military personnel and then civilians, as well as the protection of strategic vital equipment.

This type of application implies the use of a fairly large variety of materials, reinforced or not, in various forms, exposed to a fairly wide range of threats of different geometrical shapes being directed at the target with variable dynamic impact profiles in terms of energy and velocity.

The diversity of components and parameters renders the task of the scientists, the designers and the engineers extremely complex, furthermore multidisciplinary. Rather than falling into descriptive details regarding how aramids are used in this area, we want to pursue a more scientific approach favouring the comprehension of some selected physical phenomena related to the fibre structure–property relationships and the dynamics involved. Therefore our approach does not pretend to be exhaustive but should stimulate a reflection educational mode rather than an empirical guessing methodology. One needs nonetheless, to remain humble in front of the enormous challenge, which is yet to be faced to further enhance the ever evolving life protection requirements and the associated fundamental understanding.

Armour systems are generally classified in two groups: soft and rather rigid composite structures. For the sake of conciseness, associated with the risk of possibly over simplifying, we propose a single approach to summarise the rather multifaceted dynamics of ballistic impact and relevant fibre property relationships. The absorption of the kinetic energy of the projectile is related to the wave propagation, transversely and longitudinally, and the penetration stress—strain profile which largely depends on the frictional energy dissipation and matter deformation and motion therewith.

Although there exist separate basic concepts and approaches for fibre and textile fabrics, and their interpretation in the case of embedded textile and/or fibres, it is generally admitted that the crossover tightness for the textile, the delaminating effect obtainable by adhesion tailoring in the composite armour, and the looseness of the yarn ends at the boundaries are key elements in terms of the wave propagation both longitudinally and transversely. A rather simplified physical explanation relates to the reflection of the wave at the boundaries, which if they are too tight or compact tend to promote a series of multiple reverberations back to the impact region, amplifying the local stresses and further helping the projectile penetration.

For the textile designer and engineer there are some practical consequences which can be deduced from the aforementioned. As long as the reverberation effect can be optimised by means of uniform tension, fabric density and crossover compactness, one can consider that the number of overlapping entangled zones and the number of yarns are the key variables affecting the ballistic performance of the textile structure. As a point of reference, it is generally accepted that up to 50% of the impact energy can be absorbed through wave propagation in the secondary yarn networks, which are the ones neighbouring the fibre bundles directly impacted by the projectile. Multidirectional assemblies, made of judiciously oriented lightly cross-linked unidirectional layers, are now available for ballistic soft protection. The wave reverberation and dissipation aspects remain equally essential compared to traditional woven fabric assemblies mostly discussed in this chapter section.

Regarding the fibre and the aforementioned ballistic event, a dominating selection parameter lies with the fact that the velocity of the wave propagation is directly proportional to the square root of the modulus and inversely proportional to the square root of the fibre density. This, of course, is quite favourable to the aramid fibres in general since, for example, the velocity of propagation in the para-aramids is about 8000 m/s (ie, fourfold higher than for nylon) [86]. This relates directly to the volume and the intermolecular cohesion, which can participate with the wave propagation and energy dissipation. In this example, it is up to four times the volume of para-aramid which can be involved in this process compared to nylon. One can easily conceive that the trivial approach of energy potential, based on the stress—strain curve integral, is by far under-predictive of such phenomena, especially in the case of nylon, which would be superior to para-aramid if such a limited approach was used. This is today not even a matter of comparison since the aramids are dominantly used in ballistic protection.

The propagations of the transverse and longitudinal waves have common physics, although the first one more directly affects the local deformation and penetration of the matter being impacted at the ballistic strain rate. A debate still prevails within the community of ballistic experts, associated physicists, mathematicians and hopefully more and more rheologists and tribologists. This debate originated with the necessity to associate a rate-dependent or independent effect in the analysis of the impact effectiveness. By rate-independent effect, one has to understand that the fibrous structure would deform and respond to the impact variable rate of stress and strain as if the governing parameters were the ones corresponding to their static properties. One can easily agree with the fact that all materials, including fibres, exhibit rate-dependent properties. Then should the dynamic visco-elastic properties of the fibre, and of its direct neighbouring environment, be taken into account? Undoubtedly, yes. To what extent? Attempts [87a] have been reported and reveal extreme complexities in their interpretations. In this discussion, which may generate opportunities for end-use science fundamental understanding programmes, it appears clearly that high-modulus fibres with oriented molecular structures, in which internal process stresses have been dissipated by relaxation and/or drawing with the minimum boundary stress gradients, are by far the ones being the least affected by variable strain rate. Nylons are known to have tensile properties highly dependent on strain rate. On the other side a superior dynamic modulus, which may contribute to a higher wave propagation speed, is not by itself always the answer. Carbon fibre brittleness is detrimental to its ballistic performance.

Finally, once the maximum energy, allowable by a given structure, has been extracted from the incoming projectile, there are still numerous aspects to be considered to assess the probability of total penetration. Among these the deformation mode and frictional characteristics of both the matter and the projectile itself have attracted the most attention. Beyond and during the initial contact, the protective layers deform depending in part on their respective shapes and geometry. Here again, scenarios are multiple and one shall not over simplify. For example, the well-known deformation cones are rarely conical but pyramidal in the case of woven fabrics. This anisotropy has to be integrated right away at the design level and on assembly blueprints of the armour structure being developed. Similarly, idealised projectiles, mostly seen as blunt curve shapes, are by definition too ideal and may be misleading in view of the more complex multifaceted shapes encountered [88]. Self-explanatory fracture descriptions, such as brittle, radial, fragmenting or spalling, plugging and petaling or the blend of, correspond to given modes of penetration, which descriptions are beyond the scope of this study. In these latter cases, more semiempirical analysis based on fractals coupled with finite element and continuum equation computing are as probably the most predictive and of reasonable help to detect the impact characteristics and their relationships with the armour system strengths and weaknesses.

Prior to the eventual penetration of the impacting threat and after the cone has been extended to its maximum strain, which is dictated by the fibre elongation and/or the fabric distortion and/or eventually the fibre reinforced matrix delamination, the projectile further loses energy to surpass the boundary frictional forces at the fabric–projectile interface. Many components are to be considered in the frictional force balance, such as the impacting velocity and the morphology of the penetrating fragment or bullet, the fabric pattern and its volume density, the fibre surface characteristics and the tribological properties of all these elements. Once the maximum energy dissipation has occurred, either the projectile energy has been overcome and the threat is stopped, or the projectile reaches the target if the residual kinetic energy is still sufficient. Of course, during this ballistic event, considerable heat has been generated at the impact zone. In this case, para-aramids are more suitable due to their intrinsic thermal resistance compared to fusible polymer such as nylon and polyethylene. Furthermore the formation of a glassy layer associated with the polymer melting can facilitate the penetration at lower speed. Such aspects may put in light the advantages of certain categories of fibres which offer tremendous frictional resistance, such as the para-aramids, compared to others which may present better potential for wave propagation than for frictional dissipation, such as nylons.

For obvious reasons the theoretical physics of a ballistic event can only be simplistic regardless of the degree of mathematical sophistication brought into the modelling steps. Reality checks against well-designed experiments are more than recommendable in view of the ever-growing complexity and changing nature of the ballistic threats and related dynamics. The rather well-known V_{50} ballistic velocity limit is accepted as the performance parameter. V_{50} is defined as the statistical minimum velocity at which 50% of the projectiles will pass through the tested structure while the other 50% will be stopped by this ballistic barrier. Three books [89–91] are to be mentioned that cover in more detail impact dynamics and penetration mechanics. A six sigma/DMAIC (Define, Measure, Analyse, Improve and Control) type of analysis can still be beneficial given the reproducibility and repeatability aspects; unavoidably to be coupled with the predictability of an ever changing ballistic event environment.

As a conclusion of this section, one can outline that para-aramids are quite suitable in this domain of application where for a number of years now multilayered jackets are produced and successfully used, saving numerous lives. Helmets, tank panels and other military devices are reinforced with para-aramids, which is also the fibre of choice in these applications for now several decades.

Past, present and future in this field are driven by man ingenuity, artificial intelligence and biomimicking [93b,94a].

The multigenerational development of protection, for men, equipment and civil and military environments, proved to be part of a continuous, to a certain extend disruptive, innovation starting with the first body armour launched around 1970, followed by the Personnel Armour System for Ground Troops, including vest and helmet, around 1980, and the Interceptor Body Armour, late 1990s.

Those key steps constituted the starting points of a series of new and improved technologies, almost every 5 years, with essential evolutionary attributes to meet the search for lower weight, more comfortable and flexible equipment addressing more specific threats and multiple threats (pics, knife, bullets, fragments, etc.) taking into account particular injuries such as trauma reduction.

From an advanced material standpoint innovation occurred thanks to the use of the following: smaller linear density yarn, such as 300 denier; higher strength and tenacity yarns reaching 30 g/denier; finer square woven fabric grades with tighter constructions such as correctional purpose grades; more formable and flexible impregnated advanced textiles; and unidirectional (UD) and multiaxial fibre arrangements converted into impregnated layers. Combinatorial approaches of systems comprising well-engineered assemblies of the aforementioned technologies is part of that continuous innovation; examples include the following:

1. the association of various yarns with various TEM profiles,
2. the association of various fabric designs based on UD and plain square fabrics,
3. the association of 1,2 with ceramic insert plates, polymer interlayered glass and
4. the association of 1,2,3 with various resins with dissimilar crosslinking profiles, with creep/thixotropic and other rheological attributes, with nano-components (mineral and organic)... only a part of this evergreen innovation chain.

Fig. 2.12(d) below may provoke some reflections without further rhetoric; for example nature and technology may further converge with the advent of nano-observations. The principles of biomimicry are well summarised by Vullings

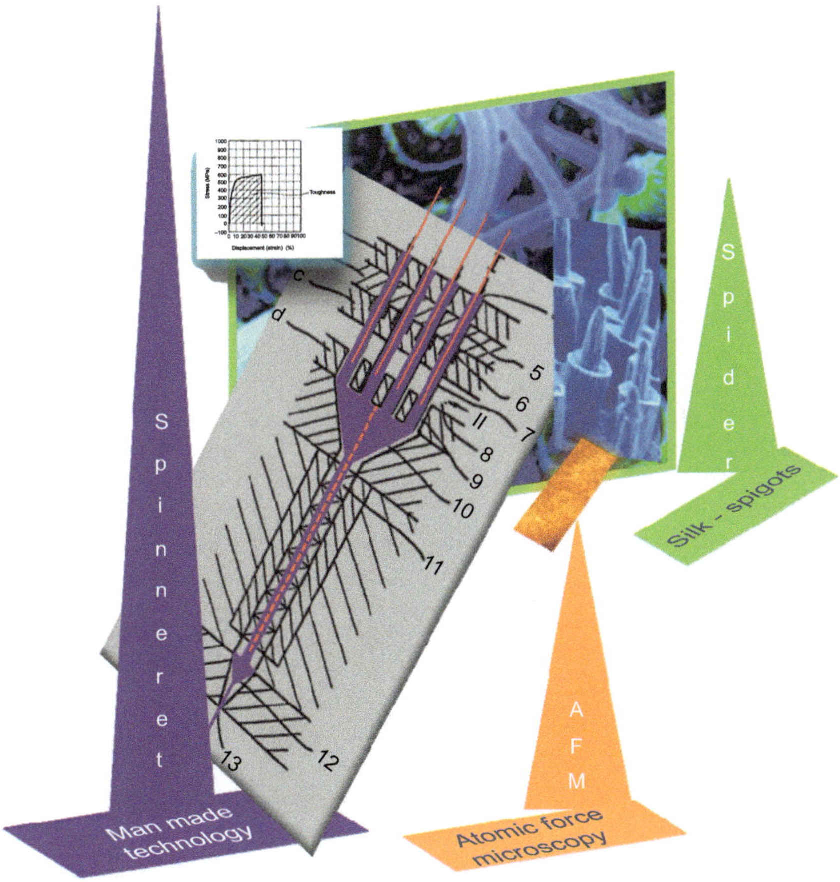

Figure 2.12d Biomimicry [93b], combinatorial bigger data [94a], crowd sourcing, nano-observations [71,72], etc., can raise confidence and take aramid innovation beyond NIH frontiers.

and Heleven in their 'Not Invented Here', NIH, visualisation booklet: 'Nature runs on sunlight'. Their first selected principle can be easily replaced by 'Nature runs on Sunshine and Confidence' which can pave the way to the cross-industry innovation beyond the frontier of the NIH turf battles.

2.5.3 Protective clothing with a focus on fire protection

It is well established that aramid polymers, particularly meta-aramids, exhibit high-temperature resistance and stability. This gave a strong incentive to develop these materials for specific industrial heat-resistant applications. Poly-*m*-phenyleneisophthalamide (Nomex®) was the first aramid fibre developed with good thermal stability at exposure temperatures as high as 500°C and long-term stability in environmental temperatures as high as 220°C.

One of the attributes among these, which have further promoted this type of application, is associated with a peculiar physical behaviour initially described by Tanner [7] as follows. When meta-aramid fabrics are rapidly heated in a flash fire, the vaporised moisture and degradation gases expand the softened polymer. This expanded material forms carbonaceous insulating foam up to 10 times the thickness of the original layer. When the flame is removed from the fibre, it self-extinguishes. Initially, these fibres seemed ideal for use in clothing, which can benefit from this intumescence. However, extensive studies showed that the fibre would shrink when heated to temperatures above its glass-transition threshold, causing tear. This would expose the wearer to direct flames causing serious injuries. Further work led to the development of a material using intimate staple blends of the meta-aramid fibre Nomex® with the para-aramid Kevlar®. The *p*-aramid fibre maintains a strong structural integrity while the meta-aramid provides the insulating foam-like barrier, which also contributes to the stress translation of the reinforcing network.

A specific equipment, the Thermoman®, initially developed for this very purpose and now more widely used for protective clothing flame resistance testing, has been and remains very instrumental to the evaluation of newly engineered blends and designs. The corresponding instrumented Thermoman® manikin is equipped with a large number of thermo-sensors (>100) and connected to a computer which delivers a comprehensive mapping of the burn level profiles outlining the sensitive zones which can then be further engineered to meet more and more stringent specifications. This type of technology has found uses for Nomex®, Kevlar® and *spunnized* yarns of Technora® in protective garments such as flight and industrial coveralls, firefighter's turnout coats and welder's and racecar driver's apparel.

Hand injuries count heavily in the safety statistics. Metal sheet forming, glass manufacturing handling and food processing are only but a few areas where hand protection is a standard. Formerly made of cotton, synthetic commodity fibres, metal and other basic materials, gloves are now more and more manufactured out of aramids, mostly para-aramid, either alone or in blends. The advantages are several, including the cut and puncture resistance through adequate coating, the abrasion resistance, the inherent thermal and insulating properties, against heat and cold and the tremendous weight saving, which is part of a bigger comfort offering. In this application area the synergism gain from the association of various materials

[95a,b] is a typical system engineering approach where the benefits are highly visible. A mitt designed for high-temperature protection is made of multiple layers such as a para-aramid in a knitted or woven form as a framework, an aluminium reflective foil and a meta-aramid or fibreglass woven or felt. The best balance of performance offered by the intrinsic properties of the aramids, as described in the more fundamental sections, can be reached without major compromise and most of the time with an added comfort, associated with the lightweight and breathability of fibre structures, and the dexterity. Aramids are usefully compared with glass and steel, in various forms and blends, in terms of their fundamental and practical contributions to mechanical and cut performance, and tribology by Rebouillat [95a,b].

2.5.4 Advanced composites

Advanced composites reinforced with aramid fibres have found significant applications in areas where strength-to-weight and stiffness-to-weight ratio of these composites makes them more attractive for use over conventional materials like aluminium and steel. This has been particularly evident where systems engineering has been used to incorporate fibres with high properties into structural components of aircraft. Not only must these components sustain the full spectrum of stresses and strains incurred in flight and remain structurally sound, they must also survive a catastrophic impact. Carbon fibres are not perfectly suitable since they are unyielding and undergo brittle fracture despite their stiffness and compressive strength. This brittle behaviour is attributed to their rigid coplanar ring structure.

Worth mentioning is that the carbon-based structures, including hybrids, are currently evolving considerably with a lower cost spectrum, integrated sustainability criteria, a holistic value chain and a rethought manufacturing, such as the cell and modular concepts. Aramids maintain their recognised role and specificity as part of the industry of interest.

The usually expected significant reduction in composite compressive strength with the increase of the modulus has not been observed in para-aramid fibres [92]. Para-aramid fibres in a resin mix have structural features that provide excellent damage tolerance. Under compressive strain (0.5%), *p*-aramid molecules tend to buckle rather than fracture. This is in part due to the molecular rotation of the amide carbon to nitrogen bonds, which allow configurational changes without resulting in bond cleavage [8,9]. The result is that although damaged the structure will continue to sustain a load, and the material collapses accordion-like upon stress. This material ductility compared to carbon fibre materials can be illustrated by the flexural stress–strain behaviour of the unidirectional epoxy material composites and of aluminium; this is illustrated in Fig. 2.13.

These considerations brought about the development of hybrid composites of para-aramid and carbon fibre, which are used in commercial airliners and helicopters today. The more favourable weight to strength ratio of Kevlar® 49/epoxy composite is the key reason for their use for floorings, doors fairings and radomes in airliners. Filament-wound pressure vessels developed using Kevlar® 49 have been used to successfully contain high-pressure gases in space ships and aircraft. The flammability resistance is also of great obvious advantage for these applications.

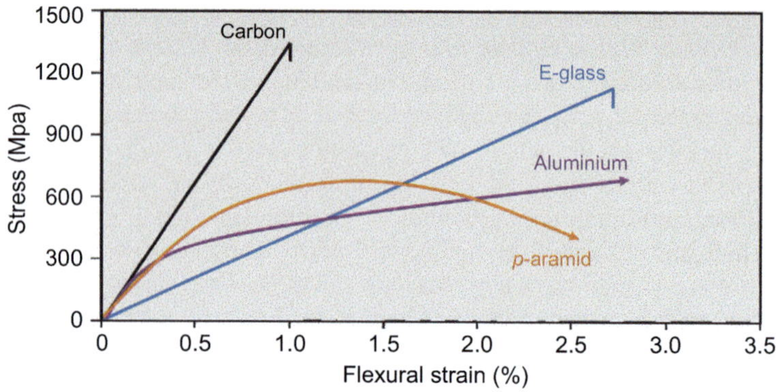

Figure 2.13 Stress/strain behaviour for unidirectional reinforced epoxy matrix composites.

2.5.4.1 Nanocomposites, grafting and nanotechnologies—a likely bridge to reaching smart high-performance fibres and systems?

Back in 1990–92, Burch and Sweeny [94b,c] further explored the related scientific aspects of previous work regarding the deprotonation of aramid polymers. This has resulted in Kevlar® thin films prepared by casting a Kevlar® polyanion organic solution. Later in the mid-1990s, Rebouillat [94c], US6045907, brought the fibre-grafted dimension of that work to industrially applicable technologies. The grafting process of this invention utilises a strong base material, and it has been determined that potassium tert-butoxide is preferred because it will function in the water-containing polymer material on which the grafts are to be made. While the grafting of this invention could be conducted on unshaped PPD-T polymer, it has been concluded that the grafting is more efficiently accomplished on the fibre form. Grafting on the fibre form provides grafted moieties on and near the fibre surface rather than deep within the fibre structure. Subsequent grafted and alkylated aramid fibres served to prove the benefit in term of adhesion performance of rubber goods.

Kevlar® fibre surface structure was primarily and directly observed in the filament configuration by the tapping mode AFM, Fig. 2.12(d), by Rebouillat et al. [71,72]. The microfibrils feature was observed with an average width of 500 nm, composed of various types of periodical units of an average size 50 nm in a pleating appearance. At the less crystalline spot on the Kevlar® fibre surface, the periodical organisations exhibit the skin-core-like differentiation. In contrast, at the highly crystalline spot, the periodicity is more uniformly arranged in a network manner.

From 2012–13 [94d], functionalised nanometre-scale aramid structures, nanofibres and nanosheets were obtained by deprotonating Kevlar® yarns by potassium hydroxide in dimethyl sulfoxide as per previously mentioned methods; yielding in that manner, dispersions of nanometre-scale aramid fibres. Nanostructured aramid materials with tailorable properties can be synthesised similarly to Rebouillat's work on grafted filaments. Nanostructures can be, for example, crosslinked by glutaraldehyde [94d]. Those reactive nanocomposites prove to be a successful concept within an open continuous innovation process that took about 10 years since the industrial demonstration on filament and the early feasible AFM nanoscale observations.

Similarly and recently (2012), Kevlar®-like aerogels, defined as low-density, three-dimensional assemblies of nanoparticles, based on the reaction of multifunctional carboxylic acids and isocyanates have been described [94e]. The reported resulting materials display a high compressive strength, a high specific energy absorption, a performant thermal conductivity and can be subject to a pyrolysis process at 800°C yielding electrically conducting carbon.

As earlier established by Burch and Rebouillat, bulk aramid macroscale fibres can be effectively split into what later revealed to be aramid nanofibres, by dissolution in dimethyl sulfoxide under alkaline conditions. Fan et al. [94f], 2012, showed 'that aramid nanofibre-functionalised graphene sheets can be prepared by adding a graphene oxide dispersion to a nanofibre solution in dimethyl sulfoxide and reducing with hydrazine hydrate'. The nanofibres then absorbed and adsorbed onto the surface of the graphene nanosized sheets are dispersible in *N*-methyl-2-pyrrolidone and are blended and used as nano fillers for the reinforcement of polymers and fibres, possibly.

Microscopy shows that the carbon nanotubes are well disseminated in the composite films and are enfolded with *m*-aramid chains [94g]. The films exhibit a surprising electric heating performance and a prompt temperature response at high electric power voltages [94h].

In summary, aramids (meta and para) can be nanosized, grafted and harmonised with graphene structures, as per carbon fibre technology (Rebouillat [94i]), possibly leading the way to smart materials able to store, vehicle and interpret signals involving local or remote data from environmental and situational knowledge databanks. This may seem to be still a far-reaching goal, definitely part of an open innovation scheme, a likely disruptive step. Nanocellulose, and technology therewith (Rebouillat [94j]) can also play a role in the paper forms.

2.5.5 Other important applications and future trends

There are several other end-use technologies where aramids provided significant benefits. These are listed in Table 2.4 [87a], which categorises most applications in seven classes, outlining by the same token the key benefits. For the sake of uniformity, we have adopted after minor modifications the rather well-accepted classification described in the recent literature [87a]. One will notice that each specific advantageous property can almost seamlessly be related to the polymer or the fibre attributes which we have described before in other sections of this chapter. The system engineering approach, which we have attempted to illustrate with the selected end-use segments described earlier in this section, has been adopted in most cases in Table 2.4 [87a] and can be revisited in the literature [7,10,93]. Technora® copolyamide fibre is also used in several applications listed in this table. The associated Fig. 2.1(a) provides a concrete illustration of the synthetic dynamic network established since the inception of the aramids. This network outlines the dynamic integration of aramids in advanced and modern technologies with a constant adaptation to the new challenges associated with various generations of product. There are end-use market segments, which have been quite rethought because of the aramids, such as the asbestos replacement by para-aramid pulps. There are other areas, which will continue to evolve because of the ever-growing stringent requirements for energy saving in transportation, for example. The aramid contribution in this sector is outstanding. Clearly, communications, including transportation as well as transmission, leisure and sports, life protection and health and safety in general have been tremendously improved and adapted to modern technologies in part because of the aramids. The present chapter has brought, hopefully, a refreshing scientific description of the state-of-the-art knowledge with an effort to stimulate the desire to go beyond the knowledge and applications of today. There are properties of the aramids, which can still be optimised via the system approach. These materials have the potential to participate in the solutions of tomorrow, continuing to evolve with smart protection of people, environment and strategic equipment as well as transportation and energy rationalisation (distribution, smart grids, etc.). Crowd sourcing, bigger data analysis approaches can undoubtedly complement the massive engineering, material science and biotechnology expertise available worldwide.

One way to appreciate the great accomplishments done, recognising the work left, seizing the opportunity and embracing the challenges facing the high-performance fibres adventure is to considerer what technology domains can be reasonably reached and what sort of interrelationships are already in place. In Fig. 2.14, data are symbolised from about 10,000 of the most relevant patents regarding the applications of aramids (applications listed in Table 2.4 or depicted on Fig. 2.1(a)). The proposed Fig. 2.14 representation provides a visual relationship between the 10 main actors as patent applicants (DuPont, Teijin, etc., left part: applicants) in the fields of aramids and close companion fibres and clusters (right part: LSA clusters, based on Latent Vectorial Semantic Analysis, https://en.wikipedia.org/wiki/Latent_semantic_analysis); those clusters comprise the IPC (International Patent Classification: codes which represent the fields of interest for each patent). The particular IPC range encompasses between 1500 and 2500 IPC/Clusters for the batch of 10,000 patents considered. The black zone is associated with all actors other than the 10 main actors. One can easily conclude that there are many fields of interest that are not directly linked to the main actors' prime 'focus'. This is definitely an indication of open innovation chances. Those open fields can also be seen as remaining opportunities to further develop aramids and partner yarns. This conclusion is limited to the analysis conducted, which has educational orientation value and may be interpreted differently depending on the particular viewpoint and the observation angle.

Similarly, the representation of liaisons, inter-patent words and concepts cooccurrence in particular, between the most appearing nodes (ie, keywords/concepts of selected patent portion, such as claims, specifications, abstract or title) helps one to understand the existing and future span of applications. Fig. 2.15 provides, for educational value again, an example of such a representation. On that figure, nodes labelled 1, 2, 3, 4 and 5 (representing textile technology and fibres, processing and methods, composite and materials) are heavily connected. This tends to reflect that the textile expertise, composite and material know-hows mostly remain the focus and interconnections in this 10,000 patent batch; this sample represents the most

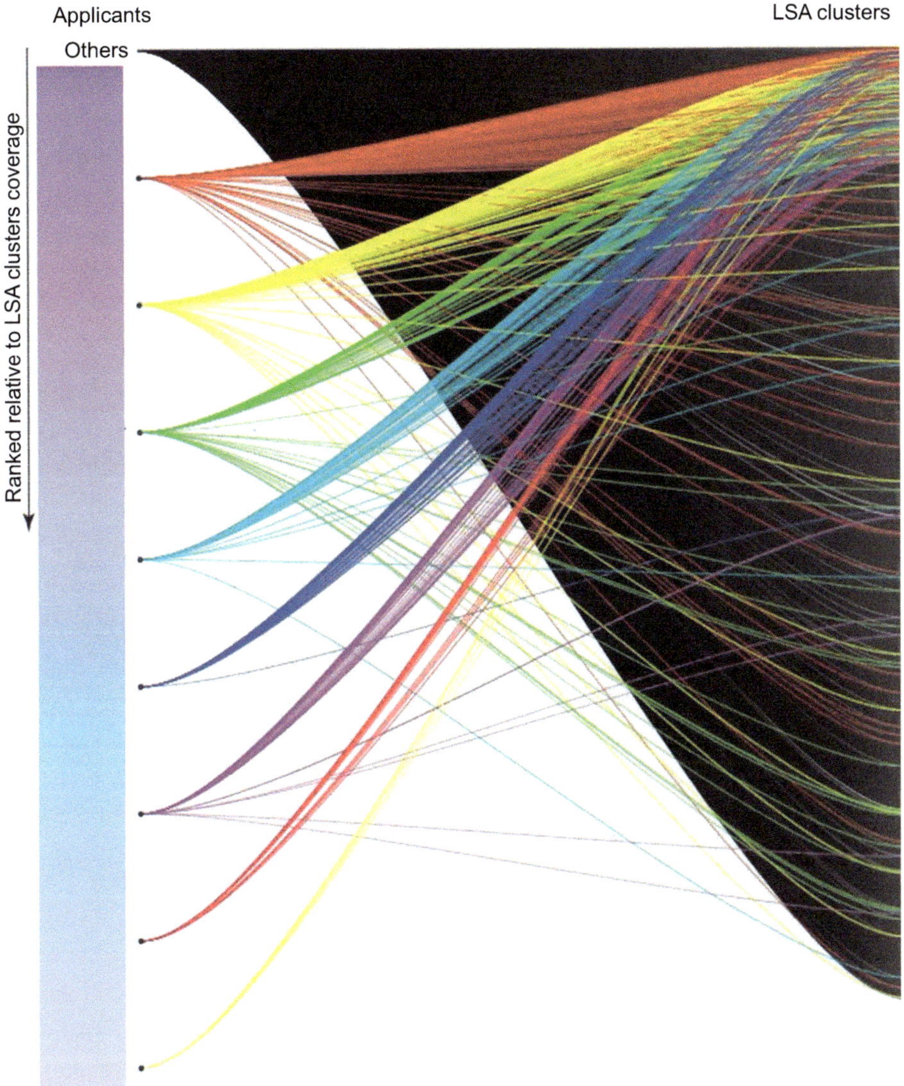

Figure 2.14 IP focus of 10 main players.

relevant patents regarding the application of aramids as listed in Table 2.4 or depicted on Fig. 2.1(a).

There is likely more to be done out of aramids and related products, possibly and for example, in the fields of smart materials and systems able to accommodate knowledge processing on mobile units for protection and resource saving. The bigger data and crowd sourcing happening is a vector towards that innovation prospective. One can further exploit and translate the hereby proposed visualisations [94a] to fine-tune the identification of innovation areas within the current range of applications and beyond.

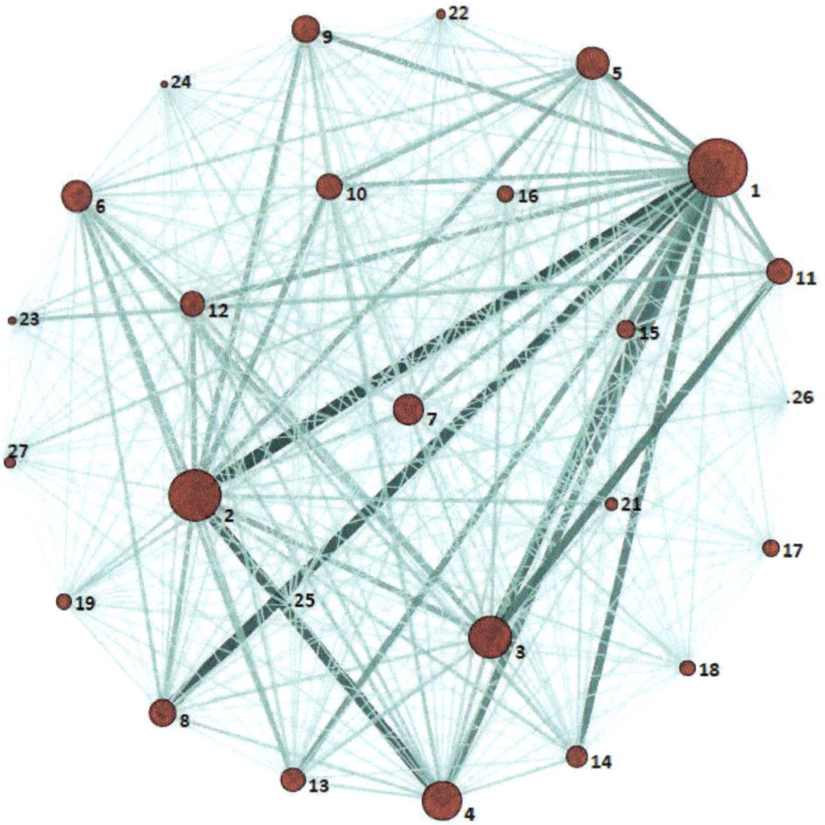

Figure 2.15 Cooccurrence chart based on a set of 10,000 patents publications related to aramids and companion materials.

Disclaimer

This article, refreshment of the existing chapter [28d] on the matter, is primarily for educational purposes. Selected cases are strictly illustrative. Neither the author nor the illustrator assumes any liability for any errors or oversights, for how this article or its contents are utilised or interpreted or for any consequences resulting directly or indirectly from the usage of it.

This information is for general knowledge. For critical applications, requirements should be checked with the suppliers. For additional guidance, legal or any other, seek advice from the appropriate qualified professionals; this study can by no means substitute for legal, technical and managerial expert advice.

The opinions expressed by the writer in this article do not necessarily represent the viewpoints of the company the author is employed by.

Acknowledgements

Special thanks are due to the following people:

- Vlodek Gabara for his useful, continuous, direct and indirect suggestions on the matter.
- In memory of Stephanie Kwolek (1923–2014), eminent laureate of the National Medal of Technology and Innovation in 1996; 'For her contributions to the discovery, development and liquid crystal processing of high-performance aramid fibres (Kevlar®)'. As a pioneering polymer scientist she offered a 'solubility' processing challenge to the engineer community. This has produced the best in the field, protecting life and resources via disruptive innovation.
- Benoît Steffenino for part of the chapter illustrations and the artistic codesign of the new Fig. 2.1(a), of which the original format has become a reference in the illustration of aramids' 'disruptive' influence in many fields.

Sources of further information and advice

A list of further references, mostly less than 10 years old, is provided below. The selection is not based on citation success criteria but rather on educational and ideation-stimulating potential (about 100 references, papers and patents are classified as per sectors of Table 2.4).

Composites & aramids

Aircraft

Closson AW. Laminar composite for armor, shielding, helmet, aircraft and vehicles, has aramid or olefin fiber layer bonded with eutectic impact absorbing adhesive resin or adhesive composition layer and ionomer layer. 2008. WO2008105929-A2; WO2008105929-A3.

Hufenbach W, Gude M, Ebert C. Hybrid 3D-textile reinforced composites with tailored property profiles for crash and impact applications. Compos Sci Technol 2009;69(9):1422–6.

Kain TR. Composite honeycomb sandwich panel for fixed leading edges on aircraft, has facing ply of woven fabric impregnated resin which extends over erosion strip and adhered to mandrel outside the trim margin of panel. 2001. US6180206-B2; US6180206-B1.

Kilchert S, Johnson AF, Voggenreiter H. Modelling the impact behaviour of sandwich structures with folded composite cores. Compos Part A-Applied Sci Manuf 2014;57:16–26.

Limas M. Thermosetting or thermoplastic device for the fabrication of pipes e.g. aircraft air conditioner pipe formed of a stack of fold with composite materials, comprises a structured fold, a rigid fold, a reinforcement unit, and a sealing unit. 2008. EP1878562-A1; FR2903342-B1; US8047237-B2.

Motojima S. Fiber reinforced composite of thermosetting resin used as components in aircraft, motor vehicle and structure material is formed by vulcanizing the blended mixture comprising thermosetting resin and fiber of carbon, aramid or glass. 2000. JP11309783-A; EP1063255-A1; US6258422-B1.

Perciballi WJ. Ceramic armor for providing bullet protection to vehicles, ships and aircraft, has aramid fiber composite layer along with ceramic facing and adhesive layers each of specific thickness. 2002. US6408733-B1; US6408733-C1.

Qaiser MH, Umar S, Nauman S. Development and characterization of fatigue resistant aramid reinforced aluminium laminates (ARALL) for fatigue critical aircraft components. In: 13th International symposium on advanced materials (Isam 2013); 2014. p. 60.

Simpson J, Boller C. Performance of SMA-reinforced composites in an aerodynamic profile. In: McGowan AMR, editor. Smart structures and materials 2002: industrial and commercial applications of smart structures technologies; 2002. p. 416–26.

Sinmazcelik T, Avcu E, Bora MO, Coban O. A review: fibre metal laminates, background, bonding types and applied test methods. Mater Des 2011;32(7):3671–85.

Thill C, Etches JA, Bond IP, Potter KD, Weaver PM, Wisnom MR. Investigation of trapezoidal corrugated aramid/epoxy laminates under large tensile displacements transverse to the corrugation direction. Compos Part A-Appl Sci Manuf 2010;41(1):168–76.

Xie M, Lachapelle DG, Mitchell SC. Article i.e. fan casing, for aircraft engine, has body comprising composite material with toughened region and untoughened region, where toughened region comprises toughening agent, and toughened flange coupled with toughened region. 2008. WO2008130739-A1; US7914256-B2; EP2146897-B1; JP5431302-B2.

Xing L, Liu L, Huang Y, Jiang D, Jiang B, He J. Enhanced interfacial properties of domestic aramid fiber-12 via high energy gamma ray irradiation. Compos Part B Eng 2015;69(0): 50–7.

Automotive/automobile/vehicle

Aktay L, Johnson AF, Holzapfel M. Prediction of impact damage on sandwich composite panels. Comput Mater Sci 2005;32(3–4):252–60.

Bahner H, Bizjak W, Weisshaupt HP. Mine protection, for armored vehicles, comprises flexible crash elements with a honeycomb structure combined with metal or aramid plates. 2003. WO2003058151-A1; EP1463916-A1.

Beraud JM, Bruyere A. Pre-impregnated composite material fabricating method for e.g. automobile industry, involves applying torsion on yarn to compensate length differences presenting trajectories of yarn on both sides of width measured in parallel to surface. 2009. FR2917330-B1.

Cheng L, Li B, Zheng Y. Polypropylene-aramid fiber composite material used in fields of avigation, aerospace and military, contains polypropylene, aramid fiber, polypropylene-maleic anhydride-styrene graft copolymer and antioxidant. 2009. CN101358011-B.

Closson AW. Laminar composite for armor, shielding, helmet, aircraft and vehicles, has aramid or olefin fiber layer bonded with eutectic impact absorbing adhesive resin or adhesive composition layer and ionomer layer. 2008. WO2008105929-A2.

Du Y, Yan N, Kortschot MT. Light-weight honeycomb core sandwich panels containing biofiber-reinforced thermoset polymer composite skins: fabrication and evaluation. Compos Part B-Eng 2012;43(7):2875–82.

Hoppel CPR, Burns BP, Newill JF. Energy absorption apparatus for use in a vehicle floor to reduce the amount of force transmitted to a seated occupant from a mine blast. 2002. US1833-H.

Hufenbach W, Gude M, Ebert C. Hybrid 3D-textile reinforced composites with tailored property profiles for crash and impact applications. Compos Sci Technol 2009;69(9):1422–6.

Kim SJ, Cho MH, Lim DS, Jang H. Synergistic effects of aramid pulp and potassium titanate whiskers in the automotive friction material. Wear 2001;251:1484–91.
Kumar M, Satapathy BK, Patnaik A, Kolluri DK, Tomar BS. Hybrid composite friction materials reinforced with combination of potassium titanate whiskers and aramid fibre: assessment of fade and recovery performance. Tribol Int 2011;44(4):359–67.
Kumfu S, Nhuapeng W, Thamjaree W, Tunkasiri T. Mechanical properties and microstructure of aramid/Al(2)O(3)/epoxy resin laminated composites. In: Tunkasiri T, editor. Smart materials; 2008. p. 389–92.
Kwitek BJ. Conformable composition useful in soft and strong consumer, industrial or military product comprises soft elastomer compound; and a filler composed of aramid fibers, where the fibers are fibrillated and dispersed in a base polymer. 2009. US2009005510-A1.
Motojima S. Fiber reinforced composite of thermosetting resin used as components in aircraft, motor vehicle and structure material is formed by vulcanizing the blended mixture comprising thermosetting resin and fiber of carbon, aramid or glass. 2000. EP1063255-A1; US6258422-B1.
Muzzy JD, Holty DW, Eckman DC, Stoll JR. Thermoplastic composite useful in manufacturing automotive products, e.g. battery trays, comprises matrix of recycled thermoplastic material and high modulus fibers. 2001. US6271270-B1.
Ozturk B, Ozturk S. Effects of resin type and fiber length on the mechanical and tribological properties of brake friction materials. Tribol Lett 2011;42(3):339–50.
Patnaik A, Kumar M, Satapathy BK, Tomar BS. Performance sensitivity of hybrid phenolic composites in friction braking effect of ceramic and aramid fibre combination. Wear 2010; 269(11–12):891–9.
Pincerato NF, Santos RD. Tie bar made of composite for utilization in mechanical systems, specifically for application in motor vehicles, comprises rod constructed of fiber reinforced polymeric materials, coupled to metallic terminals. 2003. US2003160414-A1; BR200200820-B1.
Savage G. Formula 1 composites engineering. Eng Fail Anal 2010;17(1):92–115.
Shi S, Sun Z, Hu X, Chen H. Flexural strength and energy absorption of carbon-fiber–aluminum-honeycomb composite sandwich reinforced by aluminum grid. Thin-Walled Struct 2014;84(0):416–22.
Soahib A, Miao Z, Irfan Y. Analysis of cure reaction kinetics of aramid/toughened-epoxy prepreg system aiming at processing of composite material for structural applications. J Polym Eng 2007;27(8):565–82.
Sun Z, Hu X, Sun S, Chen H. Energy-absorption enhancement in carbon-fiber aluminum-foam sandwich structures from short aramid-fiber interfacial reinforcement. Compos Sci Technol 2013;77(0):14–21.
Swolfs Y, Gorbatikh L, Verpoest I. Fibre hybridisation in polymer composites: a review. Compos Part A-Appl Sci Manuf 2014;67(0):181–200.

Container

Defenstech. Blast resistant refuse container has water in-between the molded outer container and funnel-shaped inner container formed of a polymer resin layered with blast resistant fiber. 2007. WO2007106229-A2.
Lee HK, Kim JC, Kang SW, Lee CK. Composite material formed of micro-pore for reinforcing ultra-low temperature liquefied gas container and manufacturing method thereof. 2007. KR2006012348-A.

Mosallam AS. Stress profile changing method of composite material used in construction industry, involves providing several pre-stressing elements that are used to pre-stress composite material. 2007. US2007175583-A1.

Rebouillat S, Liksonov D. Fluid–structure interaction in partially filled liquid containers: A comparative review of numerical approaches. Comput Fluids 2010;39(5):739–46. http://dx.doi.org/10.1016/j.compfluid.2009.12.010.

Pressure vessel

Aceves SM, Martinez-Frias J, Garcia-Villazana O. Analytical and experimental evaluation of insulated pressure vessels for cryogenic hydrogen storage. Int J Hydrogen Energy 2000; 25(11):1075–85.

Grau DL, Qiu XS, Sankar BV. Relation between interfacial fracture toughness and mode-mixity in honeycomb core sandwich composites. J Sandw Struct Mater 2006;8(3):187–203.

Hu H, Li S, Wang J, Zu L. Structural design and experimental investigation on filament wound toroidal pressure vessels. Compos Struct 2015;121(0):114–20.

Krikanov AA. Composite pressure vessels with higher stiffness. Compos Struct 2000;48(1–3): 119–27.

Wang X, Liang G, Zhang W, Du N, Yang J. Study on properties of pressure vessels based on CF/AF hybrid fiber reinforced epoxy composites. Sci Eng Compos Mater 2006;13(2):131–7.

Zu L, Koussios S, Beukers A. Optimal cross sections of filament-wound toroidal hydrogen storage vessels based on continuum lamination theory. Int J Hydrogen Energy 2010;35(19): 10419–29.

Sporting goods/Sport equipment

Adams MT, Aspenns GD. Composite article of manufacture e.g. fiberglass, marking method for national defense application, involves binding printed porous material between layers of reinforcing material with resin material to form composite material. 2008. US2008193639-A1.

Bhatnagar A, Wagner LL, Arvidson BD, Grunden BL, Bhatnargar A. Molded helmet for military and non military e.g. law enforcement and sporting, purposes, has fibrous layers with network of high tenacity fibers in resin matrix that comprises polyolefin fibers and agamid fibers. 2008. WO2008101138-A1; EP2111128-B1; CN101662963-B; MX300720-B; US2013212763-A1; KR1453293-B1.

Dimario CN, Falone T, Vito RA. Vibration reducing material for use as protective pad to form grips for e.g. golf club, has aramid material layer distributing vibration to facilitate dampening, where aramid material layer is located between foam and elastomeric layers. 2009. US2009075026-A1; US8545966-B2.

Frati MG. Composite material for sole insert for protective shoes, comprises layer of textile fibers which is chosen from aramid fibers, polyethylene-ultra high molecular weight polyethylene fibers, carbon fibers and/or glass fibers. 2006. EP1700687-A1.

Hida M, Tateda S, Shimoyama N. Molded part consisting of macromolecular composite material for aircraft/automobile/electrical appliance/sporting goods/leisure goods and its production. 2001. JP2001106799-A.

Vito RA, Falone T, Dimario C, Carmen D, Robert AV, Thomas F. Material for regulating vibration in sporting equipment, has elastomer layers, and cloth layer formed from several woven aramid fibers disposed between elastomer layers. 2004. US2003228816-A1; TW306810-B1.

Xin C, Dong J, Xin Z, Wei G, Gu B. Laid fabric composite sheet for use in field of e.g. national defense, aviation and sporting goods has at least two orthogonally overlapped laid fabrics that consist of reinforcing fiber, bonding adhesive and base fabric. 2009. CN201169100-Y.

Plastic additive/polymer additive

Fakhar A, Razzaghi-Kashani M, Mehranpour M. Improvements in tribological properties of polyoxymethylene by aramid short fiber and polytetrafluoroethylene. Iran Polym J 2013; 22(1):53–9.

Markarian J. Strengthening compounds through fibre reinforcement. Reinf Plast 2007;51(2): 36–9.

Suzuki H, Sekiguchi S, Krizan TD. Resin composition for preparing articles such as seal ring useful for sealing lubricant oil fluid in automatic transmission assembly comprises polymer, graphite additive material of round shape having specific surface area and/or fiber. 2006. WO2006128127-A2; US2007021547-A1; EP1883673-A2.

Civil engineering

Böer P, Holliday L, Kang THK. Independent environmental effects on durability of fiber-reinforced polymer wraps in civil applications: a review. Constr Build Mater 2013; 48(0):360–70.

Çavdar A. Investigation of freeze–thaw effects on mechanical properties of fiber reinforced cement mortars. Compos Part B Eng 2014;58(0):463–72.

Fangueiro R, Rana S. Novel fibrous materials for civil engineering applications. In: ICEM15: 15th international conference on experimental mechanics; 2012.

Garden HN. Use of advanced composites in civil engineering infrastructure. Proc Inst Civ Eng-Struct Build 2004;157(6):357–68.

Hollaway LC. The evolution of and the way forward for advanced polymer composites in the civil infrastructure. Constr Build Mater 2003;17(6–7):365–78.

Ozbakkaloglu T, Akin E. Behavior of FRP-confined normal- and high-strength concrete under cyclic axial compression. J Compos Constr 2012;16(4):451–63.

Wang Y-f, Wu H-l. Size effect of concrete short columns confined with aramid FRP jackets. J Compos Constr 2011;15(4):535–44.

Wang Y-f, Ma Y-S, Wu H-l. Reinforced high-strength concrete square columns confined by aramid FRP jackets. Part I: experimental study. Steel Compos Struct 2011;11(6): 455–67.

Wu H-L, Wang Y-F. Impact resistance of reinforced concrete floor slabs strengthened with FRP. In: Isiss 2009: innovation & sustainability of structures, vols. 1 and 2; 2009.

Protective apparels/protective clothing and aramids

Heat-resistant clothing

Bae H-S, Kim M-Y. Performance evaluation of protective clothing materials for welding in a hazardous shipbuilding industry work environment. Fash Text Res J 2013;15(3):452–60.

Duran K, Bahtiyari I, Atav R. Protective nonwoven technical textiles. Tekstil Ve Konfeksiyon 2007;17(3):174–7.

El Aidani R, Nguyen-Tri P, Malajati Y, Lara J, Vu-Khanh T. Photochemical aging of an e-PTFE/NOMEX® membrane used in firefighter protective clothing. Polym Degrad Stab 2013;98(7):1300–10.

Hatano T, Matsumura M. Woven knitted goods for fabric of heat-resistant protective clothing, has heat-resistant para-aramid fiber with greater weight ratio and lesser surface area ratio, compared to that of meta-aramid fiber. 2006. JP4447476-B2.

Ishimaru H, Akamatsu T. Heat-resistant textile for heat-resistant protective clothing, contains preset amount of para-type aramid fiber having specific single fiber size and limiting oxygen index. 2007. JP2007077537-A.

Ishimaru H, Akamatsu T. Heat resistant textile for protective clothing, is made of para type aramid fibers containing carbon black. 2008. JP4690806-B2.

Murashova VE, Voloshchik TE. Non-woven fire resistant material comprises aramid fibers and polyester fibers. 2009. RU2361973–C1.

Okuie T, Wada N. Heat-resistant protective clothing for fire-resistant garments, has fabric with surface layer, heat barrier, and intermediate layer which contains high and low heat contraction fiber threads including meta and para-aramid fiber. 2009. JP2009280942-A.

Osaki M, Murakami S, Takigaura Y, Fukuda H. Flame-retardant undergarments or clothes, consists of modified aramid fiber textile obtained by adhering bifunctional monomer and/or its polymer, and specific monomers, polymerizing monomer components and modifying aramid fiber textile. 2004. JP4055848-B2.

Padleckiene I, Abraitiene A, Valaseviciute L, Krauledas S, Puodziuniene R, Jovaisiene L. Mechanical and thermal properties of meta-aramid, viscose fr and polyester fr knitted materials. In: Baksys B, Bargelis A, Bockus S, Fedaravicius A, Leonavicius M, Ziliukas A, et al., editors. Mechanika 2013: proceedings of the 18th international conference; 2013. p. 174–8.

Plavanescu S, Nedelcu D, Lohan MN. Considerations on the micro-indentation and differential scanning calorimetry of arboform reinforced with aramid fibres. In: Carausu C, Cohal V, Doroftei I, Wrobel A, Nedelcu D, editors. Modern technologies in industrial engineering; 2014. p. 718–23.

Se HJ, Park SH. Core spun yarn used for manufacturing safety clothing, includes core made of heat-resistant fiber chosen from aramid fiber, polybenzimidazole fiber, and poly-acrylonitrile fiber, and core wrapper which encompasses and coats core. 2009. KR2008062319-A.

Watanabe S, Kamishima K. Protection textile for superheated-steam protective clothing e.g. fire-fighting clothing, has surface layer with heat-resistant waterproof film mounted on meta-aramid fiber textile fabric. 2007. JP2007009380-A.

Fire blankets

Accioly Lins L. Protective blanket for projecting fire-arm, is composed of kevlar, which has maximum thickness between eight millimeter and twelve millimeter with finishing of nylon in exterior. 2008. BR200603217-A.

Bass D, Sparks B, Hope D, Dawson W, Edwards W, Brian S, et al. Flame-resistant fiber blend for manufacture of barrier fabric, comprises amorphous silica fibers and fiber chosen from flame-resistant fibers and/or binder fibers. 2007. WO2007061423-A2; EP1861524-B1; JP5312794-B2.

Takahashi F, Abbott A, Murray TM, T'Ien JS, Olson SL. Thermal response characteristics of fire blanket materials. Fire Mater 2014;38(6):609–38.

Flame retardant textiles/clothing

Chang C. Flameproof container for protecting delicate goods e.g. cash bill, during fire disaster, has box body and lid including inner cushion, and outer cladding covering box body and lid, where outer cladding is formed of flameproof cloth. 2009. US2009205542-A1.

Ertekin M, Kirtay E. Burning behaviour and mechanical properties of fabrics woven with ring spun aramid and flame retardant polyester yarns. Tekstil Ve Konfeksiyon 2014;24(3):259–65.

Flambard X, Bourbigot S, Kozlowski R, Muzyczek M, Mieleniak B, Ferreira M, et al. Progress in safety, flame retardant textiles and flexible fire barriers for seats in transportation. Polym Degrad Stab 2005;88(1):98–105.

Kruse JP. Thermally insulative garment i.e. fire-fighter turnout coat, for protection from physical harm and protection from heat during fire, has pair of heat sensors that monitor variance in temperature and signal temperature change. 2009. US2009188017-A1.

Mettananda CVR, Crown EM. Effects of oily contamination and decontamination on the flame resistance of thermal protective textiles. Fire Mater 2011;35(5):329–42.

Muto T, Iwaki T, Sasa K. Flame-retardant aramid fiber structure such as textile, is formed by bonding phenoxy phosphazene(s) and aromatic diphosphate to fiber which comprises aramid fiber structure. 2007. JP4619187-B2.

Osaki M, Murakami S, Takigaura Y, Fukuda H. Flame-retardant undergarments or clothes, consists of modified aramid fiber textile obtained by adhering bifunctional monomer and/or its polymer, and specific monomers, polymerizing monomer components and modifying aramid fiber textile. 2004. JP4055848-B2.

Padleckiene I, Abraitiene A, Valaseviciute L, Krauledas S, Puodziuniene R, Jovaisiene L. Mechanical and thermal properties of meta-aramid, viscose fr and polyester fr knitted materials. In: Baksys B, Bargelis A, Bockus S, Fedaravicius A, Leonavicius M, Ziliukas A, et al., editors. Mechanika 2013: proceedings of the 18th international conference; 2013. p. 174–8.

Wang Y, Li J, Zhang X, Han J, Huang Z. Total protective performance of outer fabrics in firefighter clothing. In: Textile bioengineering and informatics symposium proceedings, vols. 1 and 2; 2008.

Zhang H, Shen L. The Development of the multi-functional wool/aramid 1313 blended fabric. In: Jiang Z, Han J, editors. Advances in materials and materials processing IV; 2014. p. 585–91. Pts 1 and 2.

Zhu R, Guckert D, Lovasic SL, Parry R. Intimate blend of staple fibers for yarn used in flame retardant fabric for flame retardant article of clothing, comprise specific amount of aramid staple fiber(s), flame retardant cellulosic staple fiber(s), and polyamide staple fiber(s). 2005. US2005025962-A1; WO2005019512-A2; EP1649088-A2.

Cut-protective gloves

Blake K. Yarn for protective garment, comprises resilient core retained by retention wrap, and cut resistant yarn having tenacity wrap to define cut resistant combination and cut resistant combination is provided in side of tenacity wrap. 2008. GB2446866-A; WO2008102130-A1.

Hardee F, Radhakrishnan G, Carrillo M, Narasimhan D, Nelson CN. Cut, oil and fire resistant composite yarn for use in flexible, cut, oil and fire resistant glove, has bottom and top wraps preventing unraveling of core sheath during passage of composite yarn through knitting machine. 2009. US2009183296-A1; WO2009094412-A1; EP2234513-A1.

Hanyon WJ, Chapman MR. Heat and cut resistant yarn for article of manufacture e.g. clothing, jump suit, glove, or sock, comprises fire retardant and heat resistant strand, strengthening fibers, and strengthening filament. 2005. US2005025950-A1; US7087300-B2.

Prickett LJ. Cut resistant fabric for cut resistant articles such as glove, comprises yarn comprising staple fiber blend which contains preset amount of fiber, and aramid fiber mixture comprising primary aramid fiber, and secondary aramid fiber. 2008. US2008085411-A1; WO2008045441-A2; JP2010506060-W; EP2097566-B1; KR1394876-B1.

Prickett LJ. Stain-masking cut resistant fabric useful for making article e.g. glove, knit, aprons comprises yarn containing intimate blend of lubricating fiber, and three aramid fibers each having specific filament linear density. 2008. WO2008045440-A2; US7358203-B1; EP2126171-B1; JP5091244-B2; KR1394749-B1.

Rebouillat S, Moore SD, Steffenino B. Cut-resistant yarns and method of manufacture. 2009. CN101522970B; EP2102397B1; US7638193; WO2008045492A2.

Zhu R, Prickett LJ, Baron MR. Fiber blend with cut/abrasion resistance, for protective garments such as gloves and aprons, comprises uniform mixture of cotton, nylon and p-aramid fibers, both nylon and p-aramid fibers having specific fiber length and linear density. 2001. US6254988-B1; WO200198568-A2; EP1297205-B1; KR655831-B1; JP4786857-B2.

Tyres/tyres and aramids

Truck tyres

Chang Hwan K, Yeong Hyeon G. Rubber composition used for tire tread of bus and truck, contains carbon black, silica, aramid fiber, stearic acid-group organic metal as dispersing agent of crude rubber containing natural rubber and synthetic rubber. 2009. KR894453-B1.

Corn M, Moon M. Non-pneumatic tire for vehicle e.g. truck, has curved springs constructed of composite material and embedded within crown portion and sidewalls of tire body, where each of curved springs comprises ends terminating within beads of tire body. 2009. US2009301625-A1.

Datta RN. Mechanistic study on the role of sulfurized para-aramid short fibers in rubber to brass adhesion. Rubber Chem Technol 2007;80(2):296–310.

Kim TE. Tire tread rubber composition for truck or bus employing rubber material comprising natural rubber and butadiene rubber, and aramid monofilament. 2007. KR2007000860-U; KR2007000860-A; KR693449-B1.

Nakagawa T, Noda T. Bladder for tire vulcanization comprises reinforcement suppressing element formed by sandwiching small fishing net round braid of aramid fiber between rubber sheets, extended in tire width direction at bladder cap portion. 2006. JP2006103123-A; JP4507808-B2.

Rebouillat S, Rozant O, Fleury F. Composite cord and method of making and support structute for a tire containing the same. 2012. US2011/044123.

Villanueva RD. Pneumatic tire i.e. aircraft tire, has carcass reinforcement structure comprising low modulus chippers that are sandwiched between high modulus up ply portions, where low modulus is less than specified value. 2009. BR200804608-A2; US7926533-B2; EP2055508-B1.

Yoon GY. Radial tire improving durability for truck or bus comprises belt is formed in a radial tire for a truck or a bus and upper and lower parts of the belt edge are surrounded by an aramid ply. 2004. KR2004037501-A.

Aircraft tyres

Baldwin J, Gartland RJ, Westgate WK, Baldwin DW, Baldwin JDW. Cable for use as reinforcement for pneumatic tire e.g. aircraft tire, has aramid and nylon yarns, having respective predetermined linear densities, which are cabled together. 2009. US2009090447-A1; EP2045379-B1.

Herbelleau Y, Guerinon B, Roux P. Pneumatic tire for aircraft, etc. has two sets circumferentially aligned reinforcing elements of high elastic modulus separated by very high modulus layer. 2002. WO200200456-A; WO200200456-A2; US6951234-B2; EP1299256-B1.

Sheng B. Meridian aircraft tire, has airtight layer, tire body cord layer and belt ply, where tire body cord layer is made by winding nylon cord and belt ply is made by winding nylon aramid composite cord. 2009. CN101323372-A.

Van Den Nieuwenhuyzen K, Esnault P, Cogne M. Tire for heavy vehicle, has additional reinforcement with two sets of parts having breaking strengths such that strength of one set of parts is lesser than that of other set of parts, where parts are alternately arranged in direction. 2008. WO2008098639-A1; EP2121353-B1; US8418738-B2; JP5518487-B2.

High-speed tyres

Baldwin JDW, Gartland RJ, Westgate WK. Cable for use as reinforcement for pneumatic tire e.g. aircraft tire, has aramid and nylon yarns, having respective predetermined linear densities, which are cabled together. 2009. US2009090447-A1; EP2045379-B1.

Oh YS, Kwon SY, Jeong JC, Kim WC, Jeon OH, Chung I, et al. Cellulose-based fibers for use in tire cord for high-speed driving tire comprise cellulose and polymer including polysiloxane, polyacrylic acid, polyacrylamide, m-aramid, and/or polyvinyl alcohol/polystyrene copolymer. 2009. WO2009031869-A2; EP2185753-B1; KR1316019-B1; US8584440-B2.

Yokokura H, Hiroyuki Y. Pneumatic tire has carcass ply consisting of rubberized cloth, parallely arranged with para-aramid fiber cord. 2008. WO2008062862-A1; JP4771923-B2; EP2085250-B1; US8584724-B2.

Motorcycle tyres

Bordoz F. Tire for two-wheeled vehicle e.g. motorcycle, has complementary layer constituted by aramid reinforcements, where orientation of reinforcements in side wall is symmetric to orientation of reinforcements in another side wall. 2009. WO2009063012-A1; EP2219886-B1; US8505603-B2.

Hirano T. Pneumatic tire for two-wheeled vehicle such as motorcycle, has reinforced belt having specified width at its center area, embedded with aramide cord set at specified angle with respect to tire peripheral direction. 2005. JP2005231529-A; JP4382519-B2.

Matsumura S, Sadahiko M. Motorcycle tire for e.g. forest, has tread constituted by crosslinked rubber, where tread has base and set of blocks erected almost outward in radial direction from base, and cord formed by twisting yarn. 2008. EP1992502-B1; US7975742-B2; JP5059646-B2.

Shibamoto S. Pneumatic tire for motorcycle, has belt ply including center portion, where side portions are positioned in axial direction and ratio of perimeter of center portion to that of side portion is of specific percentage. 2008. EP1859961-B1; JP4956049-B2.

Suzuki S. Pneumatic tire for motorcycle, has band including band ply with band cord wound spirally in circumferential direction, where absolute value of angle formed by band cord with respect to equator plane is equal to or smaller than five degrees. 2008. EP1987963-A2; JP2008296898-A; US2008271830-A1.

Bicycle tyres

Shen J. High-strength puncture-proof tire for e.g. airplane, has inner layer provided with buffer layer, where buffer layer is formed by high-strength natural silk canvas, chemical fiber canvas, steel wire canvas and/or canvas weaved by fibers. 2009. CN201136412-Y.

Mechanical rubber goods and aramids

Conveyor belts

Dimascio RJ, Schisler RC. Method of treating aramid fiber cord used as tensile reinforcement, involves contacting aramid fiber cord with composition containing liquid epoxy resin, optionally reactive diluent, and curing agent. 2006. EP1676873-B1; US8211502-B2.

Harikae S. Adhesion of aramid fiber cord and rubber composition for composite material of fiber cord and rubber composition, involves adhering process liquids on fiber cord, performing norm rising, vulcanizing and integrating composite material. 2006. JP2006045703-A; JP4506339-B2.

Miyao H, Yamashina T. Conveyor belt has aramid cores whose junction portions are stitched using aramid thread. 2002. JP2002372106-A.

Yoshida K, Tachibana H, Yoshida Y, Nakamoto Y. Belt used as transmission belt and conveying belt, comprises textile material containing polyester fiber, cotton fiber, polyamide fiber, vinylon fiber, polyallylate fiber, glass fiber and/or aramid fiber adhered to rubber material. 2008. WO2008023556-A1; JP2009270583-A.

Transmission belts

Fujimoto H, Takahashi H. Transmission belt, such as V-ribbed belt for motor vehicle, has core wire having cord of primary twisted aramid fiber yarn as center yarn, and secondary twisted polyethylene terephthalate fiber yarn. 2005. JP2005256961-A.

Hintze C, Shirazi M, Wiessner S, Talma AG, Heinrich G, Noordermeer JWM. Influence of fiber type and coating on the composite properties of epdm compounds reinforced with short aramid fibers. Rubber Chem Technol 2013;86(4):579–90.

Sato Y, Sato M, Tokunaga T, Furukawa S. Transmission belt, e.g. V-belt for scooter, comprises belt body which is molded from stock rubber, chopped aramid fibers, and chopped polyester fibers, where chopped polyester fibers are longer than the chopped aramid fibers. 2004. WO2004015300-A1; TW271479-B1; JP4133595-B2; US7927243-B2; EP1537343-B1.

Sawada M, Noguchi T. Adhesion for aramid fiber and rubber composition for transmission belt/tire. 2001. JP2000355882-A.

Takehara T, Hasaka H, Kimura T, Kunihiro Y, Takada T. A rubber based drive belt containing poly-p-phenylene-benzobisoxazole and aramid fibers and a compression region useful for belt drive transmissions. 2004. US2004005447-A1; US6866922-B2.

Hoses/pipes for automotive/automobile

Cleveland RL, Davis AR. Flexible tube for transporting fluid in automotive coolant system, has tubular structures of respective polymeric material, in which the second structure has reinforcement elements dispersed in the second material. 2005. US7281547-B2.

Li L, Xiao Y. Production of aramid fiber-silicone rubber cloth for preparing air hose for automobiles, by adding additives into silicone rubber, adding solvent and coupling agent, stirring, scrapping obtained paste on aramid fiber cloth, and sulfidizing. 2009. CN101328650-A; CN101328650-B.

Nam SU. Water hose for automobiles, comprises ethylene-propylene-dienemonomer and aramid fiber in specific weight percentage as main materials. 2002. KR2001054615-A.

Hoses in offshore umbilicals and turbo charger hoses

Branch M, Gray H, Schelhaas D. Fluid transfer hose reinforced with hybrid aramid yarn. 2008. US2008/0072984-A1.

Jahnsen OF, Baalerud P. Composite hybrid riser for fluid transport between seabed and offshore installation. 2000. WO9957413-A1; GB2353309-B; US6612370-B1.

Reinhart K, Osen P, Semb G. Reinforced rubber hose. 2000. WO200070256-A1; NO312483-B1.

Friction brake linings and gaskets and aramids

Clutch facings

Gardner TH, Lamport RA. Paper composition for friction facings of clutch plate(s) and brake shoes comprises fiber(s), mixture of abrasive, inorganic filler and dry lubricant, and preset amount of binding resin. 2003. US2003041524-A1; US6534565-B1.

Gurunath PV, Bijwe J. Potential exploration of novel green resins as binders for NAO friction composites in severe operating conditions. Wear 2009;267(5–8):789–96.

Kusaka S, Sasaki Y. Frictional component e.g. brake pad for disc brakes of passenger car, has reinforcement fiber selected from the group consisting of dry and wet aramid pulps, wood and acryl group pulps. 2003. JP2002285143-A; US2003022961–A1.

Levit MR, Amma A. Preparing para-aramid pulp useful as reinforcement material involves mixing never-dried meta-aramid fibrid, para-aramid fiber and water to form uniform slurry; co-refining or refining the slurry; and removing water from the refined slurry. 2007. WO2007075575-A2; US7740741-B2; EP1974093-B1; KR1359265-B1.

Marumoto Y. Manufacture of aramid dope for manufacturing aramid raw material used for aramid fiber for garments, involves contacting aramid waste textile with amide group solvent, mixing and heating slurry at preset temperature. 2008. JP5193439-B2.

Phipps M, Vyletel G. Friction material useful, e.g., in brake pads, brake linings, clutch facings, comprises a binder, a reinforcement fiber, e.g. aramid pulp, and a friction modifier comprising flat layered titanate particles and baryte particles. 2007. WO2007106880–A2; US2007219289-A1; WO2007106880-A3.

Subasi O, Mollaoglu H, Topuz A. The production and testing of composite friction materials. Mater Test 2010;52(4):222–6.

Zhang N, Chen L, Yin Y, Han Y, Wang Z. Effects of load and sliding distance on the wear properties of hybrid friction materials. In: Han YF, Pan F, Tang JM, Zhou CG, editors. Advanced structural materials; 2011. p. 401–5.

Gaskets

Amma A, Mulcahy KA, Periyasamy M, Schwartz JT, Abdou M, Tannenbaum HP. Fibrous pulp useful in articles e.g. filters, papers, sheet materials comprises polypyridobisimidazole fibers of specific dimension entangled within fibrous structures derived from another polymer e.g. aramid, polybenzazole. 2007. WO2008085149-A2; EP1963572-A2; US7727358-B2; EP1963572-B1.

Hashimoto S, Horiuchi T. Joint sheet used for gasket used for industrial pipe, is formed using material containing expanded graphite, aramid fiber and rubber binding material. 2008. JP2008201988-A; JP4734271-B2.

Hendriks AJJ, Tiecken JDC, Grotendorst H, Journee R, Oldenzeel ME, Rena J. Aramid fibrils for making papers, gaskets and breaking lines have low Canadian Standard Freeness and low specific surface area. 2005. WO2005059211-A1; EP1694890-B1; US7629047-B2; KR1116598-B1.

Scholer FR, Wrobleski J. Sealant composition used by plumbers to form tough, pliable film on threaded pipe, comprises engineered fiber, comprising e.g. polyamide, melamine, para- and meta-aramid, polysulfone, polyester, and/or polypropylene, and inorganic minerals. 2007. WO2006110679-A2; EP1874961-A2; US7624990-B2.

Yamada S. Gasket contains sealant made of aramid fiber and highly efficient structural materials, adhered to front and back surfaces of steel sheet. 2005. JP2005155667-A.

Thixotropic additive/agent

Sato H, Onozato A, Kotani H. Reinforcement structure construction method for concrete structure, involves mixing thixotropic applicator agent with impregnation adhesive resin. 2000. JP2000034841-A.

Industrial paper

Khan S, Levit MR. Honeycomb structure for sandwich panel useful in e.g. aerospace industry, comprises honeycomb cells having cell walls that are formed from paper comprising specific amounts of para-aramid pulp, para-aramid floc and aramid fibrids. 2009. US2009214818-A1; WO2009108672-A2; EP2247787-B1.

Ropes and cables and aramids

Optical fibre cable

Bai L. Armored cable for use in optical fiber connector, has layer of light, strong para-aramid synthetic fiber added between PVC tube and mesh grid, and light, strong para-aramid synthetic fiber added between mesh grid and stainless steel tube. 2008. CN201044004-Y.

Chae I, Choi S, Jun Y, Lee J, Seo K, Jeon Y, et al. Tight buffer optical cable for indoors and outdoors, has aramid yarn applied externally to multiple optical fiber units and optical fiber, and coating layer is coated on wrapped surface of central bar. 2009. KR969483-B1.

Chen B, Zhang H. Micro single-core cable, has light strong para-aramid synthetic fiber layer longitudinally installed at periphery of optical fiber unit, and jacket layer coated on periphery of light strong para-aramid synthetic fiber layer. 2009. CN201212920-Y.

Chen X, Qiu W, Yang H. Tightly sleeved optical fiber cable for third generation network baseband remote technique to transmit optical signal remote radio frequency unit, has sleeved layers whose peripheries are covered with aramid fiber yarn reinforcing elements. 2009. CN101458379-A.

Chen X, Qiu W, Yang H, Zhu Y. Distributed temperature sensing optical cable for distributed temperature monitoring system, has metal soft tube lined by aramid strength layer that is coated by metal braid ply, where metal braid ply is lined by flame retardant jacket. 2009. CN101458375-A.

Chitrangad B. Para-aramid i.e. poly-phenylene terephthalamide, yarn for use as reinforcement of fiber optic cable, has set of fibers with specific orientation angle, where fibers are made of para-aramid with specific tenacity. 2009. WO2009048770-A2; EP2198078-A2; US7976943-B2; JP5216094-B2.

Graveston MG, Pedder J, Weimann PA. Optical fiber cable for indoor applications in which a cable jacket surrounds a high-strength polyaramid yarn that surrounds a fiber buffer encasement containing optical fibers encased in ultraviolet light cured acrylate matrix and layer. 2008. US7720338-B2.

Helton J, Joder R, Keller DA, Moore J, Raynor C, Rhodes W, et al. Multi-tight buffer fiber optic cable i.e. 24 fiber cable, has tight buffer fiber optics layers helically wound in unidirectional direction at same lay length, and aramid yarns layer provided between layers of buffer optical fibers. 2009. EP2056146-A2.

Lu J. Aramid optical cable reinforced core comprises aramid fiber and outer coating coated to aramid fiber, where outer coating is obtained by curing photocurable binder comprising activated monomer and photosensitizer. 2009. CN101546018-A.

Lv Z. Non-metal indoor-outdoor armored soft cable, has optical fiber line placed at center of protective shell that includes non-metal soft tube layer, and aramid fiber layer formed outside soft tube layer. 2008. CN201096933-Y.

Mower TM. Sheave-bending and tensile fatigue of aramid-fiber strength members for communications cables. Int J Fatigue 2000;22(2):121−35.

Okura M. Manufacturing method of optical cable, involves performing extrusion-coating of plastic casings on outer periphery of aramid fiber bundle. 2007. JP2006317490-A.

Park IC. Cable integrating optical fiber and electric wire used for communication network, comprises optical cable and optical fiber jacketed by inserting aramid-yarn to increase tractive force. 2003. KR2003033334-A.

Rebouillat S, Pfister F. Fiber coated with water blocking material. 2001. WO0031752; EP1133774B9; US6891003.

Su Young L. Optical cable for optic patch cord, has two tension protective layers made of aramid fibrous layer for covering outer side of buffer and subunit jacket, respectively, and outer jacket for covering outer side of one of protective layers. 2009. KR2008105763-A.

Tokuda M, Uraka S, Katsuya T, Chino S, Yanagisawa K. Metal composite cable for underground pipeline, has metal cable core integrated with optical-fiber cord by common casing and reinforcement layer formed in surface of optical fiber core provided with aramid fibers. 2005. JP2004342579-A.

Yoshitaka M, Imaizumi M, Hirano T. Optical fiber cable for television, has two aramid fiber bundles which are wound in different directions and separated by aramid fiber layer. 2005. JP2005037641-A.

Electronic cable/wire

Yoshida M, Yoshino Y. Shielded cable for portable audio-video equipment, has aramid fiber-made wire which is knitted into multiple strand of braided shield, such that tensile-strength characteristics of wire is greater than strand. 2006. JP2006164830-A; JP5124902-B2.

Mechanical construction cable

De Angelis C. Determination of end of safe service life of aramid lift cable. 2000. US6392551-B1; EP1022376-B1.

Rebouillat S, Liksonov D, Courgey A. New experimental device to test the dynamic behavior of fiber assemblies and fibrous composite structures with a focus on larger industrial-scale-Like samples. J Appl Polym Sci 2012;123(3):1708−17. http://dx.doi.org/10.1002/app.34626.

Tomko M, Kmet S, Soltys R. Vibrations of an aramid anchor cable subjected to turbulent wind. Adv Eng Softw 2014;72(0):39–56.

Mooring ropes

Davies P, Baron P, Salomon K, Bideaud C, Labbe JP, Toumit S, et al. Influence of fibre stiffness on deepwater mooring line response. In: OMAE 2008: proceedings of the 27th international conference on offshore mechanics and arctic engineering, vol. 1; 2008.

Davies P, Reaud Y, Dussud L, Woerther P. Mechanical behaviour of HMPE and aramid fibre ropes for deep sea handling operations. Ocean Eng 2011;38(17–18):2208–14.

Gilmore J, Stenvers D, Chou R, Mts I. Some recent developments of rope technologies - further enhancements of high performance ropes. In: Oceans 2008, vols. 1–4; 2008. p. 82–8.

Huang W, Liu H, Lian Y, Li L. Modeling nonlinear creep and recovery behaviors of synthetic fiber ropes for deepwater moorings. Appl Ocean Res 2013;39:113–20.

Liu H, Huang W, Lian Y, Li L. An experimental investigation on nonlinear behaviors of synthetic fiber ropes for deepwater moorings under cyclic loading. Appl Ocean Res 2014;45: 22–32.

Van Zonneveld W, Van Leeuwen M. Endless winding technology for lightweight synthetic mooring lines. In: Oceans 2009, vols. 1–3; 2009. p. 2201–9.

Life protection and aramids

Bulletproof vests

Castano VM, Rodriguez R. Nanotechnology for ballistic materials: from concepts to products. Mater Technol 2013;47(3):267–71.

Grineviciute D, Abraitiene A, Sankauskaite A, Tumeniene DM, Lenkauskaite L, Barauskas R. Influence of chemical surface modification of woven fabrics on ballistic and stab protection of multilayer packets. Mater Sci-Medziagotyra 2014;20(2):193–7.

Hamada K, Wakatake M, Kosuga K. Dyeing of aramid fiber involves dyeing aramid fiber by acid dye in presence of specific electrolyte compound. 2006. JP4544573-B2.

Lefebvre M, Boussu F, Coutellier D. Influence of warp interlock parameters on the ballistic behavior of a structure. In: Binetruy C, Boussu F, editors. Recent advances in textile composites; 2010. p. 322–8.

Se Won S. Bulletproof helmet for defending ball bomb, has high intensity part provided in intermediate layer, and adhesive and material powder combination forming aramid and high density polyethylene fiber material. 2009. KR2008107850-A.

Stempien Z. Effect of velocity of the structure-dependent tension wave propagation on ballistic performance of aramid woven fabrics. Fibres Text East Eur 2011;19(4):74–80.

Helmets

Closson AW. Laminar composite for armor, shielding, helmet, aircraft and vehicles, has aramid or olefin fiber layer bonded with eutectic impact absorbing adhesive resin or adhesive composition layer and ionomer layer. 2008. WO2008105929-A2.

Kim HJ. Hybrid bulletproof helmet, comprises hybrid bulletproof helmet, intermediate layer having plural bulletproof polyethylene layers and outer surface layer stacked on outer surface of intermediate layer. 2003. KR2003013888-A.

Lee CB, Park TH, Yeo SY. Para-aramid paper used in stab resistant protection suit, bulletproof jacket, bulletproofing helmet, electronic circuit board and speaker board, comprises para-aramid pulp and short para-aramid fiber of specified weight per unit area value. 2009. KR2009110031-A.

Sabesan S. Rendering article, e.g. ballistic fabric, antimicrobial, comprises contacting article with solution comprising chitosan agent consisting of chitosan, chitosan salts or chitosan derivatives. 2005. WO2005012630-A2; US2005181024-A1; EP1649101-A2; KR2006054344-A; CN1823194-A; JP2007500798-W; WO2005012630-A3.

Sarron JC, Dannawi M, Faure A, Caillou JP, Da Cunha J, Robert R. Dynamic effects of a 9 mm missile on cadaveric skull protected by aramid, polyethylene or aluminum plate: an experimental study. J Trauma-Injury Infect Crit Care 2004;57(2):236−42.

Se Won S. Bulletproof helmet for defending ball bomb, has high intensity part provided in intermediate layer, and adhesive and material powder combination forming aramid and high density polyethylene fiber material. 2009. KR2008107850-A.

Shockey DA, Erlich DC, Simons JW. Ballistic barrier for aircraft, has fabric layer capable of absorbing kinetic energy of foreign projectile, which is set at inner side of outer shell of aircraft, and composed of ballistic materials. 2005. US6951162-B1.

Skiba J. Helmet system player engaged in contact sports, e.g. football, comprises helmet shell having inner and outer surfaces reinforced with bonded net or mesh of long length fibers, inner helmet, and strap within helmet shell. 2005. US2005015855-A1; WO2005009159-A2; US6931671-B2; WO2005009159-A3.

Tham CY, Tan VBC, Lee HP. Ballistic impact of a KEVLAR® helmet: experiment and simulations. Int J Impact Eng 2008;35(5):304−18.

Property protection panels

Farquhar T, Marlowe R. A novel lightweight sandwich panel with substantial resistance to ballistic penetration. In: Jones N, Brebbia CA, editors. Structures under shock and impact IX; 2006. p. 377−87.

Gower HL, Cronin DS, Plumtree A. Ballistic impact response of laminated composite panels. Int J Impact Eng 2008;35(9):1000−8.

Grineviciute D, Abraitiene A, Sankauskaite A, Tumeniene DM, Lenkauskaite L, Barauskas R. Influence of chemical surface modification of woven fabrics on ballistic and stab protection of multilayer packets. Mater Sci-Medziagotyra 2014;20(2):193−7.

Jeong W, Yoon YN, Lim DY, Yoo E, Kim J. investigation of a relationship between fabric structure and the bullet-protection efficiency using lab-scale dropping projectile. Sen-I Gakkaishi 2012;68(12):314−8.

Karahan M, Kus A, Eren R. An investigation into ballistic performance and energy absorption capabilities of woven aramid fabrics. Int J Impact Eng 2008;35(6):499−510.

Ramadhan AA, Abu Talib AR, Mohd Rafie AS, Zahari R. High velocity impact response of Kevlar-29/epoxy and 6061-T6 aluminum laminated panels. Mater Des 2013;43(0): 307−21.

Vehicle protection

Bahner H, Bizjak W, Weisshaupt HP. Mine protection, for armored vehicles, comprises flexible crash elements with a honeycomb structure combined with metal or aramid plates. 2003. WO2003058151-A1; EP1463916-A1.

Boogh L, Pont N, Rebouillat S, Rolland L. Polymer blend useful for ballistic applications. 2012. EP2580280A1; US20120264342.

Browne AL, Johnson NL. Motor vehicle body, has lightweight side impact protection baffler with high elongational stiffness and high strain-at-failure. 2000. US6220652-B1; EP1043181-B1.

Closson AW. Laminar composite for armor, shielding, helmet, aircraft and vehicles, has aramid or olefin fiber layer bonded with eutectic impact absorbing adhesive resin or adhesive composition layer and ionomer layer. 2008. WO2008105929-A2.

Follo BJ, Isley FP. Composite assembly for providing antiballistic protection comprises confined particulate structural layer; confined material in compartments of structural layer; and fiber-reinforced structural structural layer in layers of structural layer. 2008. US2007293107-A1; WO2007146259-A2.

Jin L, Hu H, Sun B, Gu B. A simplified microstructure model of bi-axial warp-knitted composite for ballistic impact simulation. Compos Part B-Eng 2010;41(5):337–53.

Johnson NL, Browne AL, Jones GL. Side-impact protection system for motor vehicle body, comprises first and second L-shaped rigid structures, synthetic cable attached to second portions of structures, and deployment device connected with first portions of structures. 2005. US6893079-B1.

Johnson RD, Ross MK, Luther R. Armored cab for vehicle e.g. tactical trucks, has walls formed of high hard steel and different thicknesses. 2006. WO2006085926-A2; EP1766321-A2; US7770506-B2.

Ravid M, Hirschberg Y. Armor for ballistic protection against armor piercing projectiles comprises main plate including ballistic material layer and auxiliary plate of ballistic material, in front of main plate and spaced from it by area free of additional layers. 2005. EP1574810-A1; US7513186-B2.

Scheifele J, Schwerdt A. Armored door for a passenger car, with armor being integrated in door housing, and both armor and door housing being made of plastics, with at least one side of door housing having an integral composition. 2001. DE10002447-A1.

Shockey DA, Erlich DC, Simons JW. Ballistic barrier for aircraft, has fabric layer capable of absorbing kinetic energy of foreign projectile, which is set at inner side of outer shell of aircraft, and composed of ballistic materials. 2005. US6951162-B1.

Singletary J, Chang K, Scott B, Squillacioti R. Properties of cross-plied unidirectional aramid fiber laminates for a new detailed military specification: MIL-DTL-32378. In: Ballistics 2011: 26th international symposium on ballistics, vols. 1 and 2; 2011.

Strategic equipment shielding

Makris A, Lavallee BC, Rozon DP, Nerenberg JG, Myles JE, Smith M, et al. Protective suit for covering the limbs and body of an operative to shield them from the effect of explosive device such as antipersonnel mines. 2001. WO200164063-A; EP1267651-A; US6240558-B1; WO200164063-A2.

Paper and aramids

Filtration

Molina S, Carretero P, Teli SB, de la Campa JG, Lozano ÁE, de Abajo J. Hydrophilic porous asymmetric ultrafiltration membranes of aramid-g-PEO copolymers. J Membr Sci 2014; 454(0):233–42.

Electrical insulation

Jeon KS, Nirmala R, Navamathavan R, Kim KJ, Chae SH, Kim TW, et al. The study of efficiency of Al2O3 drop coated electrospun meta-aramid nanofibers as separating membrane in lithium-ion secondary batteries. Mater Lett 2014;132(0):384–8.

Kato M, Katsumata H, Yogura M. Aramid paper used for aramid resin film laminate, and sheet for electric insulation material, transformer, motor and generator, contains aramid fibrid and short fiber as main component, and is formed in shape of paper. 2009. JP2009138312-A; JP4617513-B2.

Kawka DW, Anderson DW. Laminate useful in electrical insulation of transformer, comprises aramid nonwoven sheet and polyester resin having specific overall thickness, and elongation at break and average tear load in both cross and machine directions. 2004. WO2004031466-A1; US2004071952-A1; EP1546446-A1.

Levit MR. Aramid paper for use as e.g. basic material for honeycomb cores or reinforcement for printed circuit boards, comprises p-aramid pulp and floc. 2005. US2005230072-A1; WO2005103376-A1; EP1756360-A1; EP1756360-B1.

Kang BS, Fenyvesi G, Rebouillat S. Multilayer structure useful for electrical insulation. 2013. WO 2013/033044.

Rebouillat S, Knowlton S. Dielectric heat-transfer fluid. 2009. CA2754291A1; CN102365343A; EP2411481A2; US9185826; US20140036447; WO2010111698A2.

Rebouillat S, Thoonen GF, Rochette F, Steffenino B. Liquid compositions used as insulating and heat transfer means, electrical devices containing said compositions and preparation methods for such compositions. 2011. CN103988266A; EP2764520A2; US20130099182; WO2013052956A3.

Rolink B, Visser R, Diedering F. Electrical insulating paper. 2014. WO2014079761-A1.

Ul-Haq S. Influence of moisture on dielectric strength in polyamide (aramid) paper. In: IEEE, editor. Annual report conference on electrical insulation and dielectric phenomena; 2003. p. 325−8.

Yokura M, Inoue T. Aramid paper used as electrical insulating sheet for transformers and generators, has thermal bonding face which thermally bonds with another sheet of aramid paper containing polyester film, provided on surface of aramid paper. 2008. EP1873307-A2; JP4607826-B2; US7927461-B2; EP1873307-B1; TW352766-B1.

References

[1] Kwolek SL. Am Inst Chem, Chem 1980;57(11):9.

[2] Kwolek SL. U.S. Patent Office, Pat No 3 819 587. 1974.

[3] Kwolek SL, Morgan PW, Schaefgen JR, Gulrich LW. Macromolecules 1977;10:1390.

[4] Kwolek SL. U.S. Patent Office, Pat No 3 671 542, 1972, Pat No 3 819 587. 1974.

[5] Blades H. U.S. Patent Office, Pat No 3 767 756. 1973.

[6] Blades H. U.S. Patent Office, Pat No 3 869 429. 1975.

[7] Tanner D, Fitzgerald JA, Riewald PG. In: Lewin M, Preston J, editors. High technology fibers Part B. New York: Publisher Marcel Dekker, Inc.; 1989.

[8] Tanner D, Fitzgerald JA, Knoff WF, Pigliacampi JJ, Riewald PG. In: Proceedings of the international symposium on fiber science technology, Hakone, Japan; 1985.

[9] Tanner D, Fitzgerald JA, Knoff WF, Pigliacampi JJ. J Mater 1986;38(3):21−5.

[10] Tanner D, Gabara V, Schaefgen JR. In: Lewin M, editor. Polymers for advanced technologies. Israel: V.C.R. Publishers; 1987.

[11] Yang HH. Aromatic high-strength fibers. USA: Wiley & Sons; 1989.

[12] Yang HH. Kevlar aramid fiber. New York: Wiley & Sons; 1993.

[13] Black WB, Preston J. High modulus wholly aromatic fibers. New York: Publisher Marcel Dekker; 1973.

[14] Schaefgen JR. In: Zachariades AE, Porter RS, editors. Strength and stiffness of polymers. New York: Marcel Dekker; 1983.

[15] Jones RS, Jaffe M. In: Lewin M, Preston J, editors. High performance aramid fibers. New York: Publisher Marcel Dekker; 1985.

[16] Ozawa S, Matsuda K. In: Lewin M, Preston J, editors. High technology fibers Part B. New York: Publisher Marcel Dekker, Inc.; 1989.

[17] Dobb MG, Johnson DJ, Saville BP. J Polym Sci, Polym Phys Ed 1977;15:2201.
[18] Dobb MG, Hindeleh AM, Johnson DJ, Saville BP. Nature 1975;253:189.
[19] Morgan RJ, Pruneda CO, Steele WJ. J Polym Sci, Polym Phys Ed 1983;21:1757–83.
[20] Termonia Y. J Mater Sci 1987;22:504.
[21] Termonia Y. J Mater Sci 1987;22:1733.
[22] Roche EJ, Wolfe MS, Suna A, Avakian P. J Macromol Sci Phys 1985:141–57.
[23] Northolt MG. Eur Polym J 1974;10:799.
[24] Northolt MG, Chapoy LL. London: Elsevier Applied Science Publishers; 1985. p. 299–310.
[25] Northolt MG, Van Aartsen JJ. J Polym Sci Polym Symp 1977;58:283.
[26] Northolt MG. Polymer 1980;21:1199.
[27] Jackson CL, Schadt RJ, Gardner KH, Chase DB, Allen SR, Gabara V, et al. Polymer 1994; 35:1123–31.
[28] [a] Magat EE, Strachan DR. U.S. Patent Office, Pat No 2 708 617, 1955.
[b] Tikhonov, et al. Russian aramid fibres: past–present–future*. Fibre Chem May 2013; 45(1).
[c] Tokarev AV, Kudryavtsev GI, et al. USSR Inventor's Certificate No. 53132, priority dated 24.04.69.
[d] Rebouillat S. Aramids. In: Hearle JWS, editor. High performance fibres. Sawston: Woodhead Publishing Limited; 2001. p. 23.
[e] http://www.reportlinker.com/p02232396/Aramid-Fibers-Para-and-Meta-A-Global-Market-Overview.html.
[f] http://www.epo.org/service-support/publications.html?id=97.
[29] Morgan PW. Condensation polymer by interfacial and solution methods. New York: Wiley Interscience; 1965.
[30] Shin H. U.S. Patent Office, Pat No 4 009 153. 1977.
[31] Penn L, Newey HA, Chiao TT. J Mat Sci 1976;11:190.
[32] Morgan PW. Macromolecules 1977;10(6):1381.
[33] Bair TI, Morgan PW. U.S. Patent Office, Pat No 3 673 143. 1972.
[34] Bair TI, Morgan PW. U.S. Patent Office, Pat No 3 817 941. 1974.
[35] Herlinger H, Hoerner HP, Druschke F, Denneler W, Haiber F. J Appl Polym Sci 1973; 21:201.
[36] Higashi F, Goto M, Nakano Y, Kakinoki H. J Polym Sci, Polym Chem Ed 1980;18(851): 1099.
[37] Higashi F, Ogata S, Aoki Y. J Polym Sci, Polym Chem Ed 1982;20:2081.
[38] Higashi F, Akiyama N, Ogata S. J Polym Sci, Polym Chem Ed 1983;21:913.
[39] Yamazaki N, Higashi F. Tetrahedron Lett 1972;49:5047.
[40] Yamazaki N, Higashi F. Tetrahedron Lett 1974;30:1323.
[41] Preston J, Hofferbert WL. J Appl Polym Sci 1979;24:1109.
[42] Preston J, Hofferbert WL. J Polym Sci, Polym Symp 1978;65:13.
[43] Ozawa S. Polym J, Jpn 1987;19:199.
[44] Kazama K. Preprints for plastics engineering forum. Soc Polym Sci Jpn 1981:3.
[45] Flory PJ. Statistical mechanics of chain molecules. New York: Wiley Interscience; 1969.
[46] [a] Flory PJ. Proc Roy Soc Lond Ser A 1956;234:73.
[b] Flory PJ. J Polym Sci 1961;49:105.
[47] Flory PJ. Adv Polym Sci 1984;59:1–36 [and references therin].
[48] Flory PJ, Chapoy LL. London: Elsevier Applied Science Publishers; 1985.
[49] Flory PJ, Ronca G. Mol Cryst Liq Cryst 1979;54:289–311.
[50] Onsager L. Ann NY Acad Sci 1949;51:627.

[51] Ishihara A. J Chem Phys 1951;19:1142.
[52] Flory PJ, Abe A. Macromolecules 1978;11:1119.
[53] Flory PJ. Macromolecules 1978;11:1138.
[54] Abe A, Flory PJ. Macromolecules 1978;11:1122.
[55] Flory PJ, Frost RS. Macromolecules 1978;11:1126–34.
[56] Flory PJ. Macromolecules 1978;11:1141.
[57] Papkov SP. Adv Polym Sci 1984;59:76–102.
[58] Bair TI, Morgan PW, Killian FI. Macromolecules 1977;10:1396.
[59] Blades H. U.S. Patent Office, Pat No 3 869 430, 1975.
[60] Allen SR, Filippov AG, Farris RJ, Thomas EL, Zachariades AE, Porter RS. The strength and stiffness of polymers. New York: Publisher Marcel Dekker; 1983. p. 357–71.
[61] Lee K-G, Barton Jr R, Schultz JM. J Polym Sci, Part B 1995;33:1–14.
[62] Rebouillat S, Escoubes M, Gauthier R, Vigier A. Polymer 1995;36:4521–3.
[63] Rebouillat S, Escoubes M, Gauthier R, Vigier A. J Appl Polym Sci 1995;58:1305–15.
[64] Rebouillat S, Donnet JB, Guo H, Wang TK. J Appl Polym Sci 1998;67:487–500.
[65] [a] Ozawa S. Polym J, Tokyo 1987;19:119.
[b] Mulkern TJ, Raftenberg MN. Kevlar KM2 yarn and fabric, strength under quasi-static tension. 2002. ARL-TR-2865.
[66] Haraguchi K, Kajiyama T, Takayanagi M. J Appl Polym Sci 1979;23:915.
[67] Yabuki K, Ito H, Ota T. Sen'I Gakkaishi 1975;31:T524.
[68] Dobb MG, Park CR, Robson RM. J Mater Sci 1992;27:3876–8.
[69] Allen SR. J Mater Sci 1987;22:853–9.
[70] Rebouillat S, Donnet JB, Wang TK. Polymer 1997;38:2245–9.
[71] Rebouillat S, Peng JCM, Donnet JB. Polymer 1999;40:7341–50.
[72] Zachariades AE, Porter RS. The strength and stiffness of polymers. New York: Publisher Marcel Dekker; 1983. p. 327.
[73] Panar M, Avakian P, Blume RC, Gardner KH, Gierke TD, Yang HH. J Polym Sci, Polym Phy Eds 1983;21:1955–69.
[74] Li L-S, Allard LF, Bigelow WC. J Macromol Sci-Phy 1983;22:269–90.
[75] Fukuda M, Ochi M, Miyagawa M, Kawai H. Text Res J 1991;61:668–80.
[76] Roche EJ, Wolfe MS, Suna A, Avakian P. J Macromol Sci-Phys 1986;24:141–57.
[77] Rao Y, Farris RJ. J Appl Polym Sci 2000;77:1938–49.
[78] Young RJ. J Text Inst 1995;86:360–81.
[79] Mooney DA, MacElroy JMD. Water vapour sorption in polyamide materials. Department of Chemical Engineering, University College Dublin; 1994.
[80] Hearle JWS. Fibre failure and wear of materials. New York: Wiley & Sons; 1989.
[81] Hearle JWS, Wong BS. J Mater Sci 1977;12:2447–55.
[82] Levchenko AA, Antipov EM, Plate NA. Macromol Symp 1999;146:145–51.
[83] Rebouillat S, Escoubes M, Gauthier R. J Adhes Sci Technol 1996;10:635–50.
[84] Sikkema DJ. In: New developments HM/HS fibres, polymer fibres 2000 conference, UK; 2000.
[85] Caesar HM. Twaron products BV. Chem Fiber Int 2000;50:161–4.
[86] Roylance D. Text Res J 1977;47:679.
[87] [a] Smith JC, Blandford JM, Schiefer HF. Text Res J 1960;30:752.
[b] Rebouillat S. A science & business equation for collaborative corporate innovation. Business strategy, IP strategy, R&D strategy: an all-in-one business model. A review with a bio-technology & Green chemistry focus. Int J Innov Appl Stud 2013;4(1):1–19.
[88] Backman ME, Goldman W. Int J Eng Sci 1978;16:1.
[89] Lyons WJ. Impact phenomena in textiles. MIT Press; 1963.

[90] Laible RC, editor. Ballistic materials and penetration mechanics. Elsevier; 1980.
[91] Zukas JA, Nicholas T, Swift H, Greszczuk LB, Curran DR. Impact dynamics. New York: Wiley-Interscience; 1982.
[92] Kumar S. Structure and properties of high performance polymeric and carbon fibers—an overview. Publisher Sampe Qquarterly; 1989. p. 3—8.
[93] [a] Chiao CC, Chiao TT. Handbook of composites. New York: George Lubin; 1982.
[b] Rebouillat S, Lapray M. Bio-inspired and bio-inspiration: a disruptive innovation opportunity or a matter of 'semantic'? A review of a 'stronger than logic' creative path based on curiosity and confidence. Int J Innov Appl Stud 2014;6(3):299—325.
[94] [a] Lapray D, Rebouillat S. "Bigger Data" visualization to visual analytics: a path to innovation. "Happening, definitely! Misleading, possibly?" A review of some examples applicable to IP discovery. Int J Innov Appl Stud August 2014;7(4):1251—73.
[b] Burch RR, Sweeny W, Schmidt HW, Kim YH. Preparation of aromatic polyamide polyanions: a novel processing strategy for aromatic polyamides. Macromol/Am Chem Soc 1990;23(4):1065—72.
[c] Rebouillat S. "Grafted poly(*p*-phenylene terephthalamide) fibers". 2000. US6045907.
[d] Cao K, Siepermann CP, Yang M, Waas AM, Kotov NA, Thouless MD, et al. Reactive aramid nanostructures as high-performance polymeric building blocks for advanced composites. Adv Fund Mater 2013;23(16):2072—80.
[e] Chidambareswarapattar C, Mohite DP, Lari-more ZJ, Lu II, Sotiriou-Leventis C, Leventis N. One pot synthesis of multifunctional aramid aerogels. MRS Proc 2012;1403: 1403.
[f] Fan J, Shi Z, Zhang L, Wang J, Yin J. Aramid nanofiber-functionalized graphene nano-sheets for polymer reinforcement. Nanoscale 2012;4(22):7046.
[g] Jeong YG, Jeon GW. Microstructure and per-formance of multiwalled carbon nanotube/m-aramid composite films as electric heating elements. ACS Appl Mater Interface 2013;5(14):6527—34.
[h] Jeon GW, Jeong YG. Electric heating films based on m-aramid nanocomposites containing hybrid fillers of graphene and carbon nanotube. J Mater Sci 2013;48(11):4041—9.
[i] Rebouillat S, et al. In: Deckert M, editor. Carbon fibers. CRC Press; 1998. ISBN-10: 0824701720, ISBN-13: 978—0824701727.
[j] Rebouillat S, Pla F. State of the art manufacturing and engineering of nanocellulose: a review of available data and industrial applications. J Biomater Nanobiotechnol 2013;4: 165. http://dx.doi.org/10.4236/jbnb.2013.42022.
[95] [a] Rebouillat S, Steffenino B, Miret-Casas A. Aramid, steel, and glass: characterization via cut performance testing, of composite knitted fabrics and their constituent yarns, with a review of the art. J Mater Sci 2010;45(19):5378—92. http://dx.doi.org/10.1007/s10853-010-4590-5.
[b] Rebouillat S. Tribological properties of woven para-aramid fabrics and their constituent yarns. J Mater Sci 1998;33(13):3293—301.

High-performance ballistic fibers: ultra-high molecular weight polyethylene (UHMWPE)

3

H. van der Werff
DSM Dyneema, Geleen, The Netherlands

U. Heisserer
DSM Ahead Materials Sciences R&D, Geleen, The Netherlands

3.1 Introduction

3.1.1 Development of high-strength UHMWPE fibers

Ultra-high molecular weight polyethylene (UHMWPE) fibers are the strongest and lightest fibers available, due to their combination of high strength and low density. The basis for the development of high-strength UHMWPE fibers was laid down in the 1960s by the discovery in the DSM research laboratory of a new fibrillar crystal form for UHMWPE called shish-kebabs (Pennings and Kiel, 1965). In the 1970s, UHMWPE fibers were obtained directly from stirred solutions (Zwijnenburg, 1978) with a tensile strength of 2.9 GPa and a Young's modulus of 101 GPa. Discovery of the so-called gel-spinning process by DSM and Pennings (Smith and Lemstra, 1979; Kalb and Pennings, 1979) enabled commercial-scale production of high-strength and high-modulus UHMWPE fibers. DSM's fibers and its composites are branded Dyneema®, and a technology license was granted to, at that time, Allied Signal (now Honeywell), who uses the brand name Spectra®. The essence of the gel-spinning process for the production of high-strength and high-modulus UHMWPE fibers is schematically given in Fig. 3.1.

In the gel-spinning process, a suspension of UHMWPE powder in a solvent is fed to an extruder that dissolves the UHMWPE powder by heating. The spinning solution is fed to the spinneret and extruded through spinning holes. Fiber formation can be established by enforcing UHMWPE crystallization by cooling. The thus obtained fiber is subjected to solvent removal (by evaporation or extraction) and extensive hot drawing. Due to the reduction of the number of entanglements per polymer chain as a result of the dissolution and the low interaction between polyethylene chains, very high draw ratios can be achieved, leading to extremely high molecular orientation and, thus, high fiber strength and modulus.

Despite the thermal limitation inherent to polyethylene, UHMWPE fibers have found a very wide range of applications: ballistic protection, marine and offshore,

Advanced Fibrous Composite Materials for Ballistic Protection. http://dx.doi.org/10.1016/B978-1-78242-461-1.00003-0

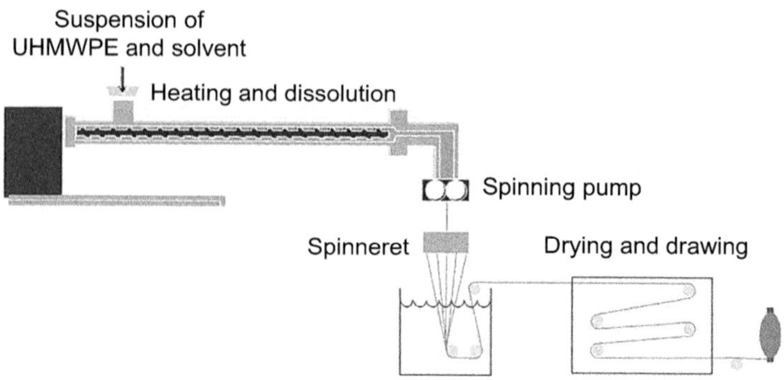

Figure 3.1 Schematic representation of the gel-spinning process.

cut-resistant textiles, sports, fishing, and medical. An extensive review on this is given by Vlasblom and van Dingenen (2009).

3.1.2 Specific tensile strength

The basis for the extreme performance of the UHMWPE fiber lies in the combination of its high strength and its low density of typically 970–980 kg/m^3, and it is expressed by the specific tensile strength. Normally, the tensile strength of fiber is expressed in N/m^2 and equals the force at break (in N), F_b, divided by the cross-sectional area A (in m^2):

$$\text{Tensile strength} = \frac{F_b}{A} \quad [3.1]$$

The specific tensile strength is the tensile strength of the fiber divided by the density ρ of the fiber:

$$\text{Specific tensile strength} = \frac{\text{tensile strength}}{\text{density}} = \frac{F_b}{A \cdot \rho} \quad [3.2]$$

A fiber filament can be envisioned as a cylinder, with a length L (in m) and cross-sectional area A (in m^2). By definition, the density ρ (in kg/m^3) of a cylinder is shown in Eq. [3.3]:

$$\rho = \frac{\text{mass}}{\text{volume}} = \frac{m}{V} = \frac{m}{A \cdot L} \quad [3.3]$$

Simple rearrangement of Eq. [3.3] results in the following:

$$A \cdot \rho = \frac{m}{L} \quad [3.4]$$

This can be inserted in Eq. [3.2] of specific tensile strength:

$$\text{Specific tensile strength} = \frac{F_b}{A \cdot \rho} = \frac{F_b}{m/L} \quad [3.5]$$

The result of Eq. [3.5] is that the experimental determination of the specific tensile strength of a fiber can be extremely easy. The force at break is readily determined with a tensile tester. Instead of very elaborate and difficult determinations of the cross-sectional area and the density of a fiber, one can simply determine the mass per length of a fiber (with a balance and a ruler), and use Eq. [3.5]. The international unit for mass per length is *tex*, which is 1 g/km, or *dtex*, which is 1 g per 10 kilometers. The typical value for the tensile strength of commercial UHMWPE fibers is 3.6 GPa. Using a density of 980 kg/m^3, the corresponding specific tensile strength is 35 cN/dtex. In a similar manner, the specific modulus can be expressed in cN/dtex too.

3.1.3 Ballistic applications of UHMWPE fibers

High-strength fibers can be used in ballistic applications in principle in the form of woven fabrics or in composites. UHMWPE fibers are generally used, for ballistic applications, in composites only. Woven fabrics of UHMWPE fibers have lower ballistic performance than UHMWPE fiber composites. A possible explanation for this is the very low friction coefficient of UHMWPE fibers (Vlasblom and van Dingenen, 2009, p. 450), which could facilitate the movement of a UHMWPE filament, upon projectile impact, aside from the projectile, and thus to less engagement of the fiber in the ballistic impact process (Prevorsek et al., 1988).

In typical UHMWPE ballistic composite sheets manufactured by DSM Dyneema, unidirectional (UD) layers of fibers are impregnated with a thermoplastic matrix (typically 20 w% or less), and stacked in 0 degree/90 degree arrangement (see Fig. 3.2), or

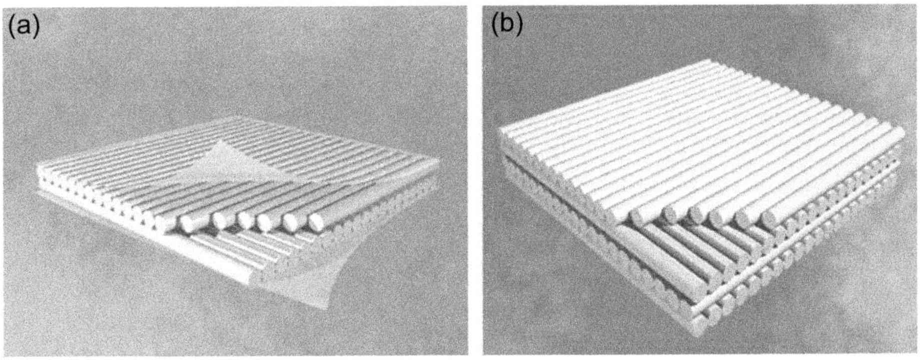

Figure 3.2 Schematic representation of UD sheets intended for (a) flexible and (b) stiff applications.

in general, in [0 degree/90 degree]$_n$ arrangements with n varying between one and maximally three for the different product types. For flexible applications like ballistic vests, the UD layers are covered on both sides with a thin polymer film (Fig. 3.2(a)). For stiff applications like helmets and panels, stacks of fiber composite sheets without the polymer film (Fig. 3.2(b)) can be pressed at higher temperatures and pressures into stiff, flat panels or curved shapes. In the following, we will refer to the composite sheets, consisting of [0 degree/90 degree]$_n$ fiber layers with n between one and three and possibly covered with polymer films, as *UD sheets*, as this is the common designation in the market (although the composites are de facto bidirectional).

In general, UHMWPE fiber composites can be used for personal protection (bullet-resistant vests, helmets, and inserts) and vehicle protection (land, air, and water). An impression of several applications made with Dyneema® composites is given in Fig. 3.3.

3.2 Mechanical properties

To understand the ballistic performance of a fiber-based composite, knowledge about the mechanical properties of the fiber-based composite is required. The question then is "which composite mechanical properties are relevant for the ballistic impact process?" The answer to this question is not simple. It is difficult to determine the geometrical scale relevant for the ballistic performance. Tentatively, one could postulate that for the local penetration process the fiber filament properties are relevant and that for macroscopic events, like the amount of deformation of the target back face, larger scale mechanical properties of the whole composite are relevant.

As, according to our experience, the tensile properties of the fiber and its composite are especially important, we will discuss in this chapter the development and change of tensile properties going from the fiber filaments to ultimately thick, hot-pressed fiber composite panels.

3.2.1 Filaments

UHMWPE yarns consist of a large number of filaments. For example, a 1760 dtex Dyneema® yarn will consist of 780 filaments, each filament typically with a diameter of 17 μm. The determination of reliable UHMWPE filament tensile properties is not straightforward. The UHMPWE filaments have the challenging combination of small diameter, high strength, high modulus, and a very low coefficient of friction. These properties quickly result in clamping problems (slippage and/or failure in the clamps) during filament tensile testing. Additionally, there will be variations in filament thickness along and between filaments. To avoid these problems, DSM Dyneema has developed, together with Textechno GmbH in Germany, special clamps to be used in the Textechno Favimat filament tensile tester. The Favimat determines the single filament linear density by the vibroscopic method (ASTM D1577) and subsequently carries out the tensile test (ASTM D3882) on exactly the same piece of single filament. In this

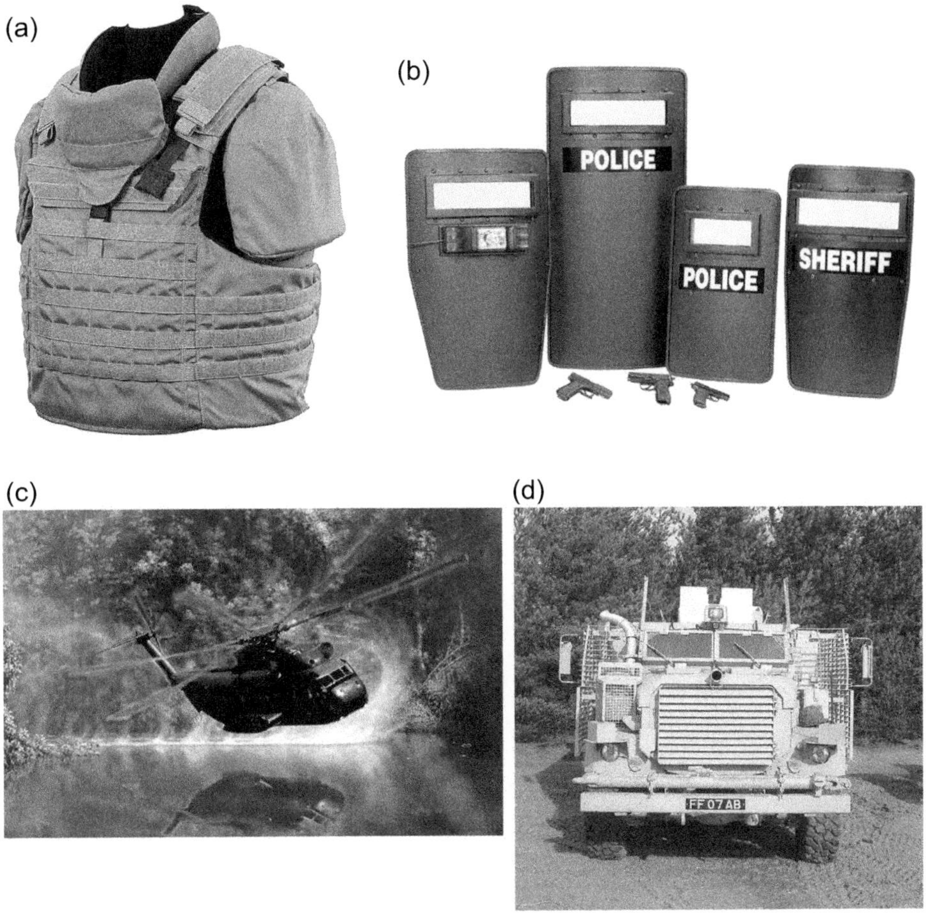

Figure 3.3 Applications made with Dyneema® composites for ballistic protection: (a) military protective vest, protection level NIJ01.01.04 level IIIA, (b) riot shields for law enforcement for protection against hand gun and rifle bullets, (c) inside integrated ballistic armor in the CH-53 helicopter of the German army, and (d) protection against roadside bombing of the Mastiff vehicle of the UK army.

way, specific tensile strengths can be tested very reliably and straightforwardly on single filaments. The final clamping situation in the Favimat is given schematically in Fig. 3.4.

Both on the upper and lower clamp, the single UHMWPE filament is clamped between Plexiglas jaw faces, and subsequently wrapped around a ceramic pin. The upper clamp is connected to a load cell. Prior to tensile testing, the vibroscope determines the filament linear density by measuring the resonating frequency of the transversal vibration of the filament induced by sound. Tensile testing is subsequently carried out by displacement of the lower clamp. For determination of modulus and strain,

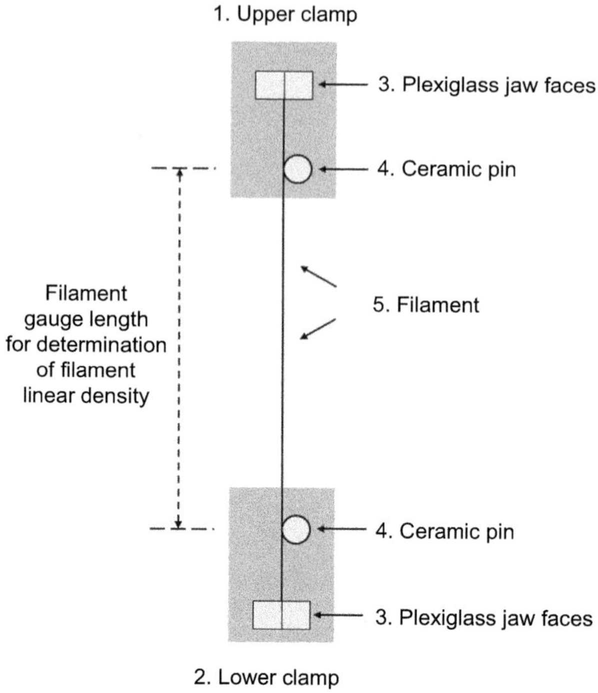

Figure 3.4 Schematic clamping setup used for the tensile testing of UHMWPE single filaments.

the whole sample length between the upper and lower Plexiglas jaw faces needs to be taken into account. A representative filament stress—strain curve is given in Fig. 3.5. The UHMWPE fiber stress—strain curve is a curve with a continuously declining slope (van der Werff and Pennings, 1991). Fig. 3.5 clearly shows an initial inclining slope,

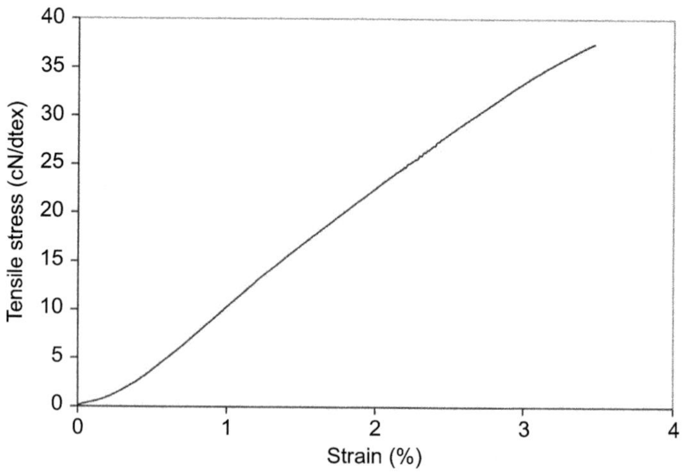

Figure 3.5 Stress—strain curve of a single filament from a 1760 dtex Dyneema® SK75 yarn.

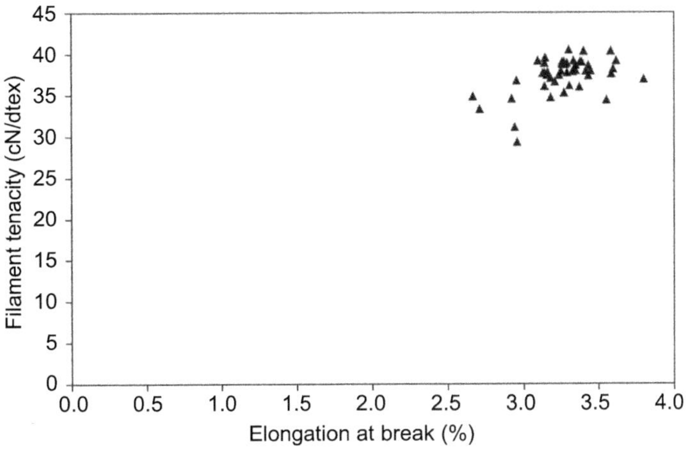

Figure 3.6 Filament tenacity versus elongation at break for 48 filaments isolated from a single yarn.

but this is totally due to the slackness introduced by the wrapping of the filaments around the pins.

The results of 48 single filament tensile tests (gauge length = 50 mm, tensile test speed = 25 mm/min) of filaments isolated from a Dyneema® 1760 dtex f780 SK75 yarn are given in Fig. 3.6. The average filament tenacity was 37.5 cN/dtex (with a standard deviation of 2.2 cN/dtex). The variation of filament fracture properties is often described in the literature with a so-called Weibull distribution (Kelly and McMillan, 1986). The cumulative failure distribution function $F(\sigma)$ for the Weibull distribution is given by Eq. [3.6]:

$$F(\sigma) = 1 - \exp\left[-L\left(\frac{\sigma}{\sigma_0}\right)^m\right] \quad [3.6]$$

where σ is filament stress at break, σ_0 is a normalizing constant, L is the filament length, and m is the dimensionless parameter called the Weibull modulus. The higher the value is of the Weibull modulus m, the less the variation is in the filament stress at break.

Given a set of N measurements, the cumulative failure probability F for certain stress σ_i is approximated by Eq. [3.7]:

$$F(\sigma_i) = \frac{N_i}{N+1} \quad [3.7]$$

where N_i is the number of filaments that have failed at or below the stress σ_i. Based on Eq. [3.6], the plot of $\ln(-\ln(1-F(\sigma_i)))$ versus $\ln(\sigma_i)$ should give a straight line with slope m. For the data from Fig. 3.6, this plot is given in Fig. 3.7.

The value of the Weibull modulus m, derived by linear regression from the data in Fig. 3.7, is 18.0. This value is almost 2× higher than those reported on Kevlar® single

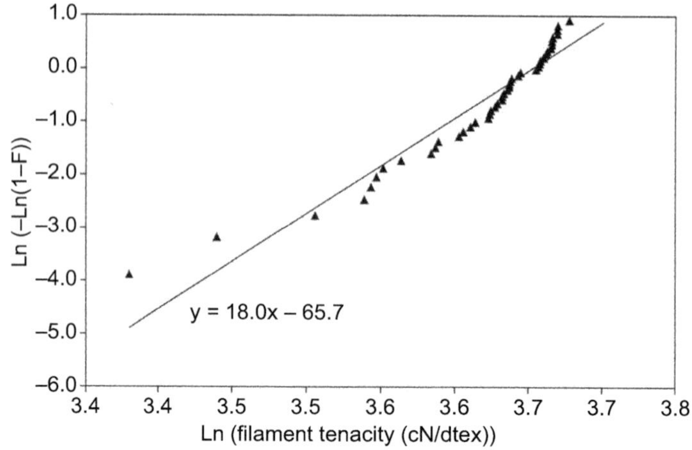

Figure 3.7 Weibull plot for the filament tenacity.

filaments (Wagner et al., 1984), but it is similar to the value reported by Northolt (Northolt et al., 2005).

High-performance fibers are by design highly anisotropic and mainly applied for their exceptional longitudinal properties. In ballistic applications, the loading will occur in a complex multiaxial stress state at very high strain rates. Fig. 3.8(a) shows that Dyneema® SK76 fibers are insensitive to higher strain rates as investigated by Hudspeth et al. (2012). This is also found by Russell et al. (2013) on SK76 yarns. It is known that UHMWPE fibers creep, and for this deformation mode there exist good predictive models (see Govaert et al., 1993). Recently, new Dyneema® grades with creep rates reduced by orders of magnitude were developed that open up new

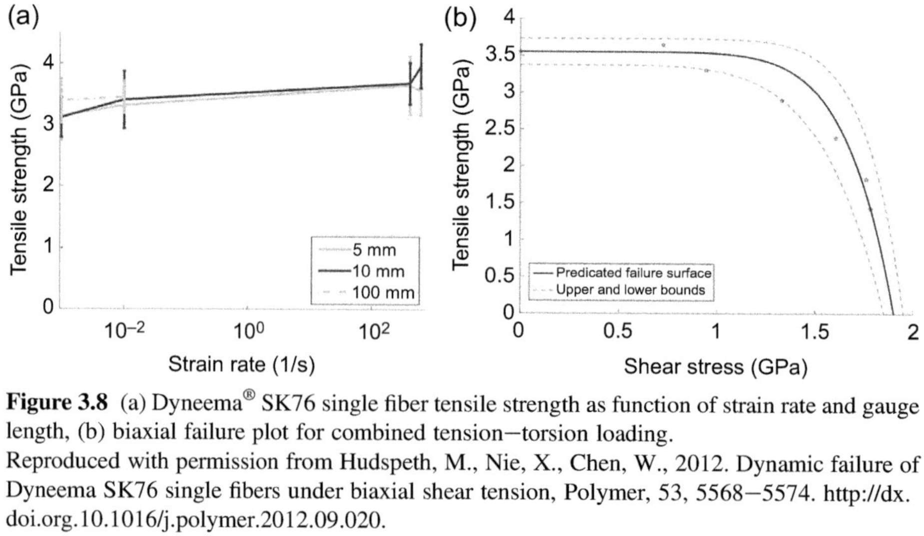

Figure 3.8 (a) Dyneema® SK76 single fiber tensile strength as function of strain rate and gauge length, (b) biaxial failure plot for combined tension−torsion loading.
Reproduced with permission from Hudspeth, M., Nie, X., Chen, W., 2012. Dynamic failure of Dyneema SK76 single fibers under biaxial shear tension, Polymer, 53, 5568−5574. http://dx.doi.org.10.1016/j.polymer.2012.09.020.

applications in permanent mooring (Vlasblom et al., 2012). However, in ballistic applications, creep behavior is irrelevant. The biaxial failure envelope for combined torsional shear and longitudinal tensile stress in Fig. 3.8(b) shows that the SK76 fibers keep their longitudinal strength up to very high shear stresses. This behavior is not achieved by aramid fibers (Deteresa et al., 1984).

Longitudinal compressive strength of single fibers is investigated by the loop/elastica method (eg, Fukuda et al., 1999) or by the tensile recoil method (Allen, 1987; and refined by Newell and Gustafson, 2001). Attwood et al. (2015) determined the compressive strength of single SK76 fibers by both methods and report a compressive strength of about 880 MPa for the knot test and 340 MPa for the recoil test before fiber kinking occurs. They remark that the factor 2.5 higher value for the knot test is not uncommon for high-performance fibers as a stress gradient is acting from inside to outside of the loop. This is much higher than the compressive strength of 90 MPa found by Knijnenberg (2009) (the Dyneema® grade is not specified) and shows how delicate this measurement method is. Nevertheless, in composite applications, the limiting factor is not compressive fiber strength but macroscopic buckling. A new collective microbuckling mode under macroscopic bending in Dyneema® composites is reported in Liu et al. (2014).

In transverse compression, Dyneema® fibers can fibrillate due to their low transversal Van der Waals bonding. This deformation mode, however, protects the longitudinal load carrying capacity as is visually demonstrated in Fig. 3.9 (Marissen, 2011). In ballistic protection applications, this property is advantageous because it ensures retention of longitudinal tensile strength of the fibers under very high transverse pressures aroused by the projectile impact.

3.2.2 Yarns

The mechanical properties of UHMWPE yarns used for the production of composites for ballistic protection are in industrial practice determined by tensile testing of twisted yarns. Like with the UHMPWE filaments, the combination of high strength and low friction coefficient gives rise to practical issues such as slippage in the clamps and thus incorrect tensile strength determination. Dyneema® yarns can be tested according to industrial standards (ISO 2062, ASTM D885, ASTM D2256) with special

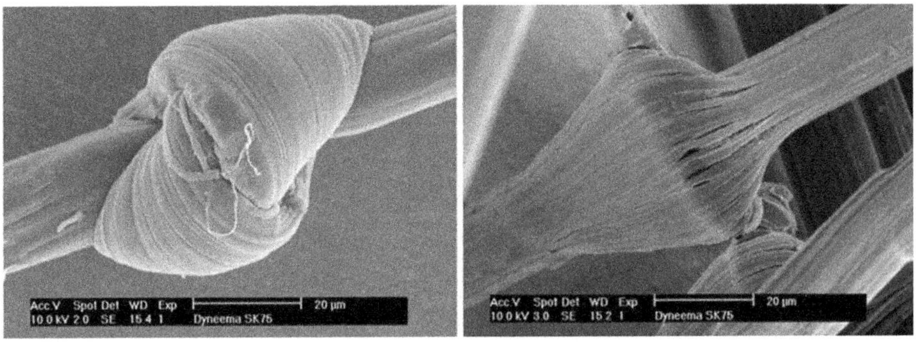

Figure 3.9 Knot in a Dyneema® SK75 fiber and pulling the fiber over a sharp blade.

pneumatic cord and yarn grips (eg, by Instron®) equipped with a curved half capstan to overcome specimen failure in the gripping area. Yarns with a linear density of 1760 dtex are tested after applying a twist of 40 turns per meter. A typical stress–strain curve of a twisted yarn is given in Fig. 3.10.

The results of 30 yarn tensile tests on one 1760 dtex Dyneema® SK75 yarn (gauge length 500 mm, tensile test speed 250 mm/min) are shown in Fig. 3.11. The average yarn tenacity, modulus, and elongation at break were 35.9 cN/dtex, 1199 cN/dtex, and 3.59% with standard deviations of respectively 1.3 cN/dtex, 14 cN/dtex, and 0.17%. The average yarn tenacity is only 4.3% lower than the average filament tenacity. Although the test conditions are different, it is therefore very reasonable to

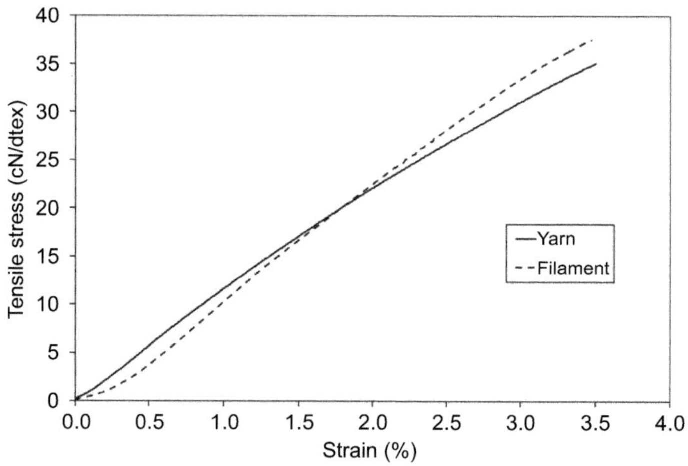

Figure 3.10 Stress–strain curves of a twisted 1760 dtex Dyneema® SK75 yarn and a representative single filament.

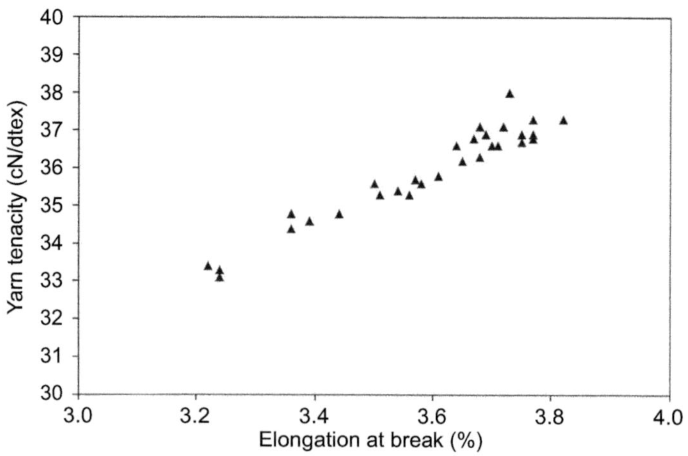

Figure 3.11 Yarn tenacity versus elongation at break.

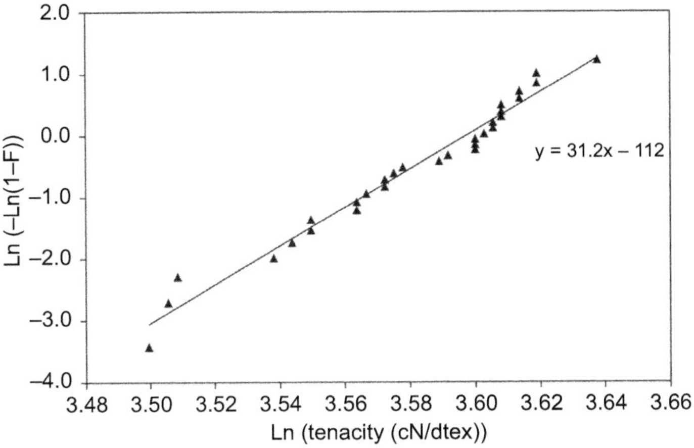

Figure 3.12 Weibull plot for the yarn tenacity.

use Dyneema® yarn tenacity data as an approximation for filament tenacity. The latter will be most relevant for establishing relations between fiber properties and ballistic performance of Dyneema® composites. Clearly, in the yarn tensile test, the variation in measured yarn tenacity is directly linked to the variation in elongation at break.

The Weibull plot is shown in Fig. 3.12. The Weibull modulus m obtained by linear regression is 31.2, so it is significantly higher than the value for filament tensile strength.

3.2.3 Single yarn shooting

Given the difficulty of valid high strain rate tensile tests an attractive method is to shoot at single yarns and extract the behavior under ballistic conditions. There exists a significant body of literature from the 1950s onward mainly on nylon, then the high-performance manmade fiber. An extensive overview is given by Phoenix and Porwal (2003). A complete theory describes the two kinds of waves in the yarn and their geometrical relations. We investigated shooting on 1760 dtex SK76 yarn with 40 twists/m with 1.1 g cylindrical projectiles (5.56 mm diameter, fragment-simulating projectile (FSP) impacting with flat side) and saddle-shaped projectiles. Fig. 3.13 shows schematically the experimental setup (Fig. 3.13(a)) and the longitudinal and transversal wave phenomena (Fig. 3.13(b)).

When a yarn (possibly with pre-tension T_0 induced by a weight w) is transversely impacted by the projectile, a longitudinal and a transversal wave begin to run out simultaneously from the point of impact. The longitudinal wave travels at the speed of sound, C_L, which is the square root of longitudinal modulus over density, sqrt(E/ρ), into the undisturbed region and induces an extra strain (on top of the pre-strain) in the fiber as depicted schematically in Fig. 3.13(b). Behind this wave front, material moves at a particle speed U toward the point of impact causing the additional strain $\varepsilon_d = U/C_L$. This longitudinal straining generates extra length that is taken along

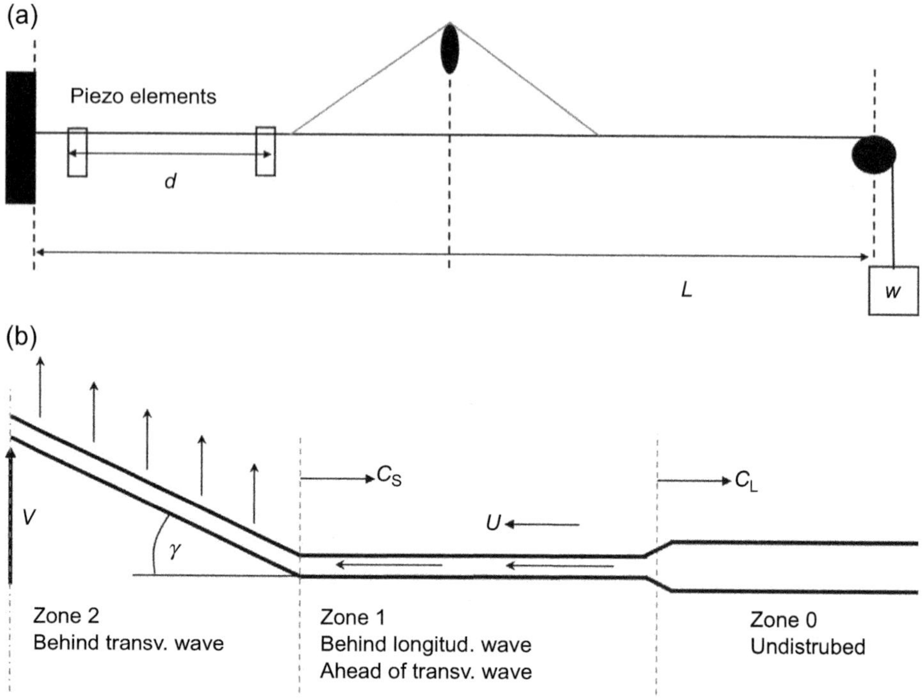

Figure 3.13 Transversely impacted yarn: (a) Setup with a length L of 3 m and attached weight w. (b) The transverse (cone) and longitudinal wave define three zones.

with the moving projectile causing the slower transversal (tent or cone) wave to move at a velocity of C_S. The angle γ stays constant over time as is clearly visible in the multiexposure high-speed photographs shown in Fig. 3.14. At the front of the transversal wave, the particles suddenly change their direction of motion from perpendicular to parallel to the direction of motion of the projectile. The transversal wave speed depends on the yarn density, the total tension and the strain in zone 1 behind the

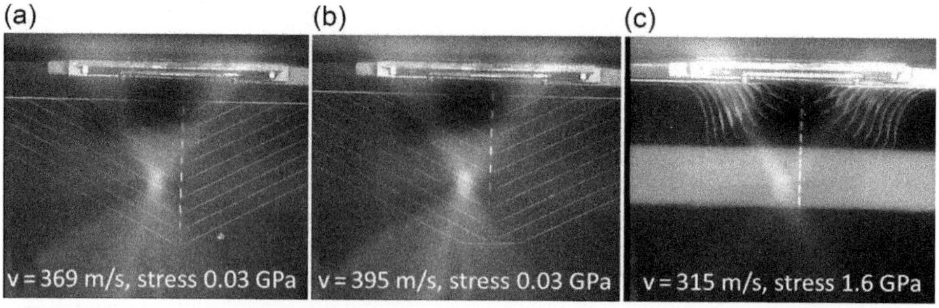

Figure 3.14 Typical multiexposure pictures. Indicated is the impact velocity and the pre-stress in the yarn introduced by the attached weight. (a) No break, (b) projectile slips off the yarn, (c) break (the white region on the picture is an exposure artifact). The exposure time is 500 ns; the interframe time is 200 ns.

longitudinal wave. Interestingly, the strain is constant over the kink as derived by Wang (2007). Using the experimental setup, we can vary the attached weight and the velocity of the projectile. The two piezo elements along the yarn register the advent of the longitudinal sound wave. With the known distance, *d,* and the time between the signals, Δt, the speed of sound, C_L, can be determined.

Independently, the theory as is recapitulated in consistent notation by Wang (2007) connects 11 variables with 6 equations. Three variables are known from the setup (pre-tension and pre-strain and yarn density). If two more are measured, there are six unknowns for six equations, that is, the system is determined. Easily accessible pairs of variables to measure are, for example, $\{V, C_L\}$, the projectile velocity V and the longitudinal sound speed C_L as derived from the piezo transducers, or $\{V, C_S\}$ with the transverse wave speed C_S in lab coordinates or $\{V, \gamma\}$ using the cone angle γ. The modulus derived from the three approaches is very similar for our data set.

Using the speed of sound C_L as determined with the piezo elements, Fig. 3.15 plots the sonic yarn modulus $\left(E = C_L^2\rho\right)$ as a function of applied stress and projectile velocity. Remarkably, the modulus is independent from the projectile velocity which is in line with the observation that Dyneema® yarns show hardly any strain rate sensitivity for elevated rates, as discussed previously. The key parameter is the applied stress. Plotting the applied stress versus the average sonic modulus, we obtain Fig. 3.16. Extrapolating to no applied stress, we obtain a modulus of 120 GPa that is also found in quasistatic tensile tests. On the other hand, the sonic modulus rises quickly with stress to level off at around 210 GPa.

It is known from the theory of rigid rod molecule fibers that the modulus is strongly dependent on the orientation of the molecules to the macroscopic load direction (Northolt et al., 2005). Although Dyneema® fibers contain already highly oriented molecules (misalignment smaller than five degrees), applying already small stresses may cause further crystal rotation and thus an increase in modulus. The weak van der Waals interchain bonds are beneficial for this. The exact mechanism for this increase in sonic modulus with applied stress for UHMPWE fibers remains to be determined however.

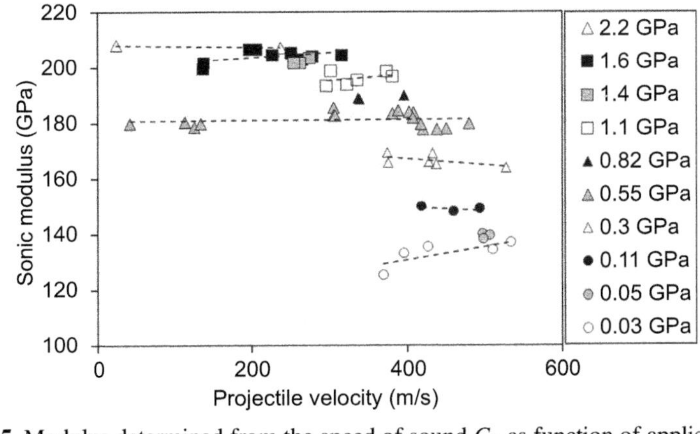

Figure 3.15 Modulus determined from the speed of sound C_L as function of applied stress and projectile velocity.

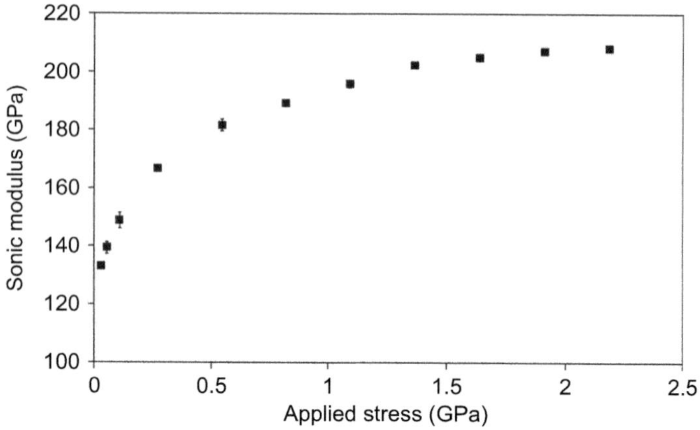

Figure 3.16 Sonic modulus as a function of applied stress.

The classical theory that works very well to describe the geometric relations of surviving yarns predicts strain as a function of projectile velocity (Smith et al., 1958). Conforming to the theory, at and above a certain so-called critical velocity, the strain induced by the projectile will be equal to or higher than the actual strain break of the yarn, and thus projectile impact on the fiber at and above the critical projectile velocity results in immediate yarn failure (Smith et al., 1962; Roylance, 1977). If Eq. [3.13] from Roylance (1977) is converted from textile to metric parameters, we obtain Eq. [3.8] for the critical velocity V (in m/s):

$$V = \sqrt{\varepsilon \cdot \frac{E}{\rho} \cdot \left\{ 2\sqrt{\varepsilon \cdot (1+\varepsilon)} - \varepsilon \right\}} \qquad [3.8]$$

For a Dyneema® SK76 yarn, using a modulus E of 120 GPa, a density ρ of 980 kg/m^3 and failure strain ε of 2.8%, the corresponding critical projectile velocity V according to this theory is 1033 m/s. However, the observation is that yarns fail immediately at a lower velocity of 518 m/s (see Fig. 3.17). This observation spun off some research and is partly explained by strain concentration under the flat projectile. The strain waves do not only run along the yarn outside the projectile but also under the projectile where they meet and double the strain (Phoenix et al., 2011). In the experiment, saddle projectiles managed to reduce the stress concentration and did increase the critical velocity by 18%, however not the doubling that would be expected. Based on our findings, an additional effect was put forward by Walker and Chocron (2011), who argue that next to the stress concentration, yarn bouncing off the projectile can explain a further reduction. Together these mechanisms are estimated to explain half of the gap in critical velocity between experimental and theoretical values. Recent publications by Hudspeth et al. (2015a,b) argue that there is a multi-axial stress state in the bent fiber that is not taken into account by the classical theory that only assumes axial stresses.

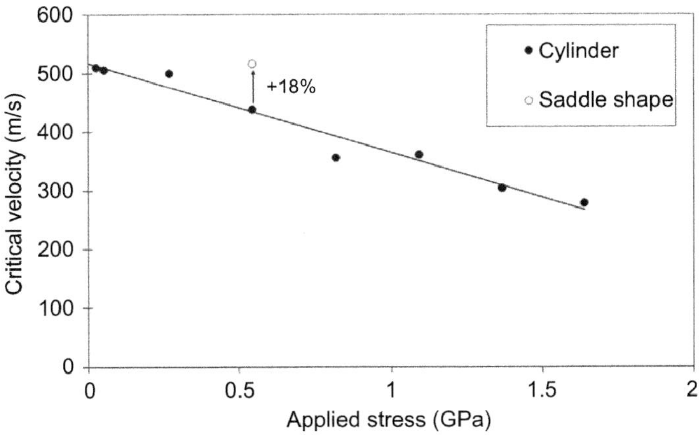

Figure 3.17 Critical velocity for a Dyneema® SK76 yarn as a function of applied stress. It is interesting to note that the line extrapolates to zero critical velocity at a stress of 3.4 GPa, which is almost identical to the quasistatic tensile strength.

3.2.4 Fiber composites

UHMWPE fibers are processed with a low amount of thermoplastic matrix (less than 20 wt%) into UD sheets, as shown in Fig. 3.2. The mechanical tensile testing of these composites has the same challenges as the filament and yarn testing due to the high strength, stiffness, and low coefficient of friction. Additionally, for valid tensile tests of a composite, the whole cross section of the sample must be loaded cooperatively to determine the intrinsic strength of the composite. The chemical inertness and absence of polarity of the UHMWPE fibers, together with the soft thermoplastic matrix, result in a low interlaminar shear strength of the composite that renders loading the whole cross section cooperatively a challenging task. This means that the standard method of using dog bone–shaped specimens (ISO 527) and pressurized grips acting on the sample surfaces gives misleading results, that is, an apparent strength that is biased by the test. In Fig. 3.18, the tensile strength of Dyneema® HB26 as determined by different methods and sample thicknesses illustrates this point. The lower line shows apparent strength values of ISO dog bones of different thickness that were water jet cut from HB26 composites pressed at 165 bar. The pressurized grips with corrugated surface that were used are only able to transfer the tensile load in part of the sample's cross section due to the low interlaminar shear strength. The region adjacent to the surfaces is loaded most and fails first. This mechanism is reflected in the observation that the apparent strength increases with decreasing sample thickness.

The individual stress–strain curves in Fig. 3.19 show a plateau of progressive failure. Only the very thin ply tests exhibit abrupt termination indicating cooperative loading and failure of the whole cross section. In an attempt to achieve more uniform loading throughout the whole thickness, several authors developed nonstandard geometries with enlarged clamping sections and bolts through the thickness. For example, a 6-mm-thick HB26 specimen reaches an apparent strength of about 800 MPa

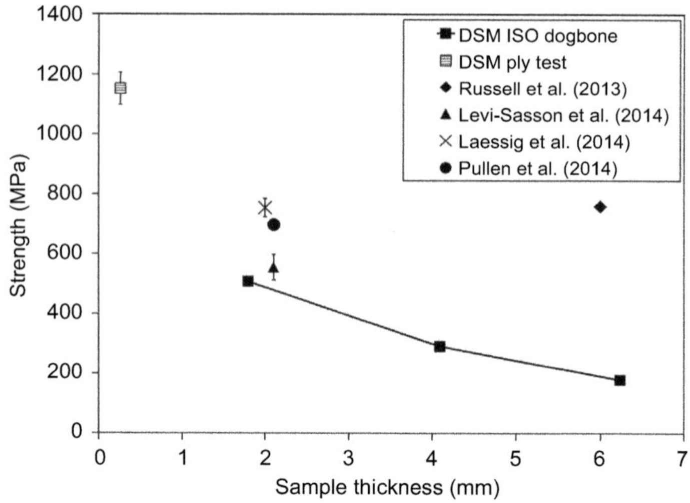

Figure 3.18 Tensile strength of HB26 composites as a function of sample thickness and specimen/clamp configuration.

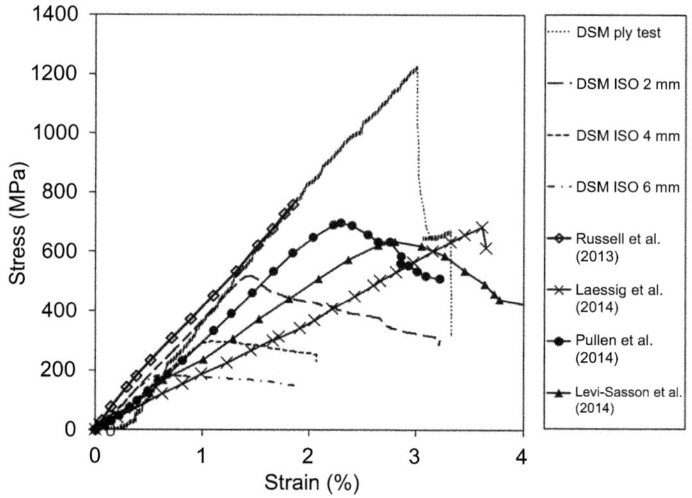

Figure 3.19 Representative stress–strain curves for Dyneema® HB26 obtained by different methods.

(Russell et al., 2013; Laessig et al., 2015), while the same thickness with pressure clamps achieves only about 200 MPa. An account of the quest for valid tensile and shear tests is given by Levi-Sasson et al. (2014), who opt for a bow-tie specimen shape with bolts reporting strength of about 550 MPa on average on 2.1-mm-thick HB26 samples. An interesting observation of this work is that correctly stacked specimens, that is, having continuous [0 degree/90 degree] alternation between all consecutive fiber layers, resulted in a higher apparent strength (550 MPa) than that observed for an incorrectly stacked specimen (450 MPa). This indicates that the [90 degree/90 degree] interface in the symmetry plane with reduced shear strength influences the macroscopic composite tensile strength. It is known from ballistic observations that the limit velocity (V50) is reduced if stacking faults are introduced. This is probably linked to the out-of-plane compressive strength of cross-plied UHMWPE composites that was shown to be dependent on the in-plane fiber strength but also on the layer thickness with thin layers resulting in higher out-of-plane compressive strength (Attwood et al., 2014).

An order of magnitude estimation for the tensile strength of the composite starts with the SK76 fiber strength of 3.6 GPa; in a cross-ply orientation, only half of the fibers are oriented in load direction (factor 0.5) and a knockdown for approximately 80% fiber, 20% matrix (factor 0.8) resulting in 1.44 GPa. Tensile tests on HB26 sheets ($[0 \text{ degree}/90 \text{ degree}]_2$), with a thickness of about 260 μm (areal density 261 g/m^2), on dedicated textile clamps (eg, Zwick roller grips type 8564), exploiting the capstan principle, achieve an average strength of 1.15 GPa (Figs. 3.18 and 3.19). This is the apparent intrinsic material strength that probably should be used in models. If the measured force is normalized by the linear density of only the fibers running in load direction, we find 28 cN/dtex that compares to the 36 cN/dtex of the fiber, as determined in normal yarn tensity testing. It is well known that in hierarchical fibrous structures there are strength losses due to bundle effects.

3.3 Mechanism of ballistic penetration

Ballistic performance of UHMWPE fiber composites, in practical applications, will basically concern the resistance to penetration upon ballistic impact of a projectile onto a target *pack* (a flexible stack consisting of loose UD sheets, generally stitched at corners) or a target *panel* (obtained by hot pressing of a stack of UD sheets). An additional property of interest, especially for body armor applications, is, in the case of a stopped projectile, the extent of bulging at the back of the target, as this may lead to trauma by blunt force. Typically, sets of such tests are standardized (eg, National Institute of Justice, Standard 0101.06 for body armor).

The resistance to penetration upon ballistic impact is generally assessed by a V50 determination (the projectile impact velocity at which there is a probability of 50% for total target penetration) via prescribed protocols, invariably with multiple shots on one target pack or panel. It is known that the ballistic performance of a pack of UD sheets will vary with the shot sequence number (van Es, 2010). Better mechanistic information can thus be obtained in our view by execution of only one shot per sample and, in the case of stopped projectiles, by examination of the depth of penetration (DOP) into the target. We have carried out such experiments on panels and packs of Dyneema® UD materials, and we will report the results here.

For the experiments, we used a metal frame with a wide aperture (70 or 140 mm wide). Typically, 200 × 200 mm target samples were placed with the target center onto the aperture center and kept in position by clamps at the four corners. Packs were always stitched at all corners prior to testing to ensure pack integrity during handling before ballistic testing. As a projectile, we used stainless steel ball bearings with a diameter of 5.54 mm and a weight of 0.70 g. Invariably, even at high projectile impact velocities, these projectiles did not show any measurable deformation due to the impact on the Dyneema® UD targets. The projectiles were fired perpendicularly to the target plane in the center of the target sample, with only one shot per sample. Projectile velocity before impact and velocity after target penetration (residual velocity) were determined by infrared light screens.

In case of packs and stopped projectiles, the number of penetrated UD sheets in a pack was determined by manual counting. Invariably, the projectile was stopped before an intact UD sheet, irrespective of the degree of penetration into the pack. In case of panels and stopped projectiles, the remaining target thickness directly under the projectile was measured using a caliper. The DOP was then straightforwardly calculated as the original sample thickness minus the remaining thickness after partial project penetration.

For packs, Dyneema® SB71 sheets were used. SB71 consists of six orthogonally stacked fiber layers ([0 degree/90 degree]$_3$), covered at top and bottom with a thin polyethylene film. The areal density of the SB71 is 190 g/m^2. Target samples consisted of corner-stitched packs of 25 SB71 UD sheets (total areal density 4.75 kg/m^2). The results in Fig. 3.20 suggest a remarkably simple penetration mechanism.

The first observation is that, for the stops, there is a linear relation between the number of penetrated UD sheets, N_p, and the projectile impact velocity, V_p. This points to a mechanism where a certain amount of impulse, and not energy, is lost by the projectile upon penetration of one UD sheet, and where the loss of projectile impulse is independent of the projectile impact velocity on a UD sheet inside the pack. The penetration mechanism is therefore a very gradual process. The loss of projectile velocity per penetrated layer, ΔV_p, observed in Fig. 3.20 is 16.7 m/s. At projectile velocities immediately below the ballistic limit, the projectile is stopped before the very last UD sheet.

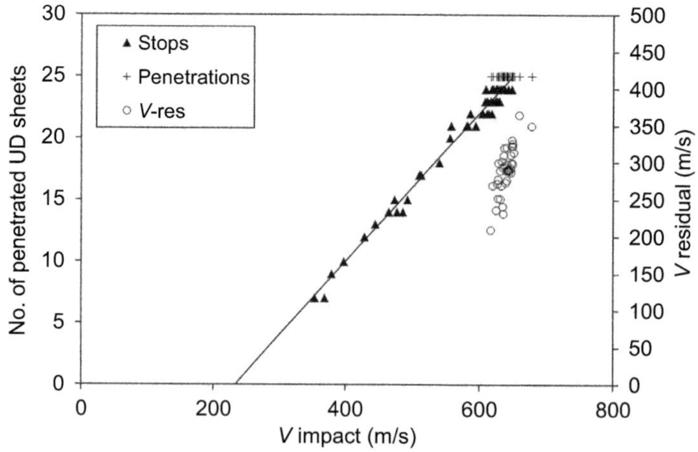

Figure 3.20 Number of penetrated Dyneema® SB71 UD sheets (for stops and penetrations) and residual velocity (for penetrations) versus projectile impact velocity.

The second observation is that there seems to be a minimum projectile velocity, V_0, to start penetration of the SB71 pack. At projectile impact velocities below V_0 the projectile probably will simply bounce back from the target. This phenomenon could, however, not be investigated experimentally as the firing equipment could not generate such low projectile velocities. V_0 can be determined by linear extrapolation of the linear regression line, and it was 233 m/s.

The third observation is that, for penetrations, there is a minimal residual projectile velocity, and the residual velocity increases with increasing projectile velocity. The lowest observation residual velocity was 209 m/s. This value is just somewhat lower than that of V_0 (233 m/s). The necessity of the existence of a minimal residual velocity can be straightforwardly explained. When the projectile hits the last UD sheet at the back of the pack, the projectile will be decelerated in time and the strain in the fibers in the last UD sheet will increase in time. Numerical simulations and analytical treatments of a projectile impacting a single membrane (van der Werff et al., 2010; Phoenix and Porwal, 2003; see Section 3.4.2.1) have shown that, after the beginning of the impact process, the membrane strains pass very early in the projectile deceleration process through a maximum. So the critical moment, where the membrane will fail, is very early in the projectile deceleration curve. This means that upon membrane failure the projectile velocity will be definitely nonzero but lower than the original impact velocity.

This phenomenon could be corroborated by stills from high-speed video recordings (Fig. 3.21). Here, the back of the target pack is viewed perpendicularly to the projectile direction. Fig. 3.21 shows two different experiments with exactly the same projectile impact velocity of 648 m/s. The experiment in Fig. 3.21(a) led to a perforation, in Fig. 3.21(b) to a stop. Fig. 3.21(a) shows the moment were the projectile penetrates the last ply. The bulging at the back side is then only about 5 mm; the projectile residual velocity was 299 m/s. Fig. 3.21(b) shows the moment where the bulge and

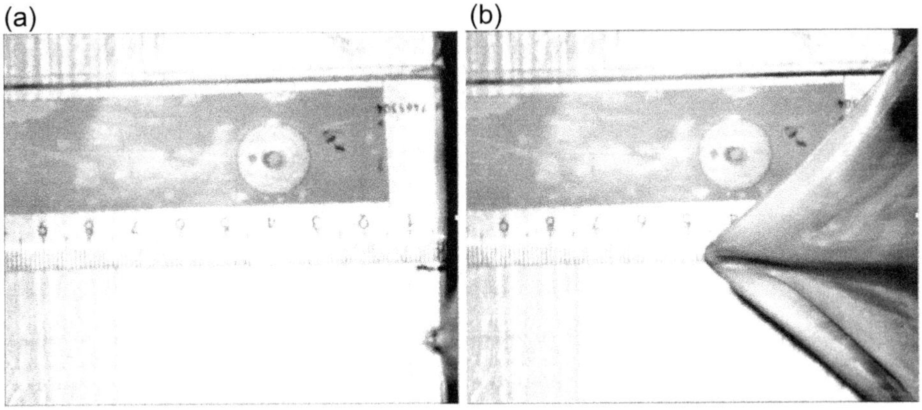

Figure 3.21 Stills from the high-speed video recordings of (a) the bulge at the moment of penetration and (b) the bulge at full stop of the projectile (b). Projectile impact velocities for both (a) and (b) are 648 m/s. The unit of the reference ruler is centimeters.

projectile velocity is zero. The amount of bulging at the back side is 47 mm. Inspection of the high-speed video recording of Fig. 3.21(b) further revealed that the bulge and projectile velocity at the position of 5 mm bulging at the back side was about 260 m/s. So clearly, the critical moment for penetration of a ply is very early in the projectile deceleration process.

Qualitatively, similar behavior (existence of V_0, linear increase of the number of penetrated sheets with increasing projectile impact velocity, high lowest projectile residual velocity) was observed in a semianalytical theoretical study by Porwal and Phoenix (2005) on stacks of membranes with *p*-aramid-like mechanical properties.

Under conditions similar to the experiments discussed in Figs. 3.20 and 3.21, the DOP has been studied for impact on hot-pressed panels of Dyneema® HB26. HB26 consists of four orthogonally stacked fiber layers ([0 degree/90 degree]$_2$) only and has a total areal density of 261 g/cm^2. Experiments were carried out on hot-pressed panels of an areal density of 12.5 kg/m^2 (48 UD sheets, 12.9 mm thickness). The results are given in Fig. 3.22, and the observed behavior is identical to that of Fig. 3.20. Also for the hot-pressed panels, there is a minimum velocity to start panel penetration V_0 (211 m/s) and a linear relation between the DOP and the projectile impact velocity. The loss of projectile velocity per amount of penetrated material is 83 m/s per mm penetration, or 22 m/s per penetrated UD sheet. The ratio of ΔV_p and the UD sheet areal density is correspondingly 84 m/s per kg/m^2 penetrated areal density for HB26 and 88 m/s per kg/m^2 for SB71, which is fairly comparable.

We have observed that the obtained values for V_0 and ΔV_p will depend slightly on sample geometry (thickness, size), but in all examined cases, the behavior, as observed in Figs. 3.20 and 3.22, has been demonstrated. Given these seemingly simple and straightforward characteristics, development of an engineering-like analytical model

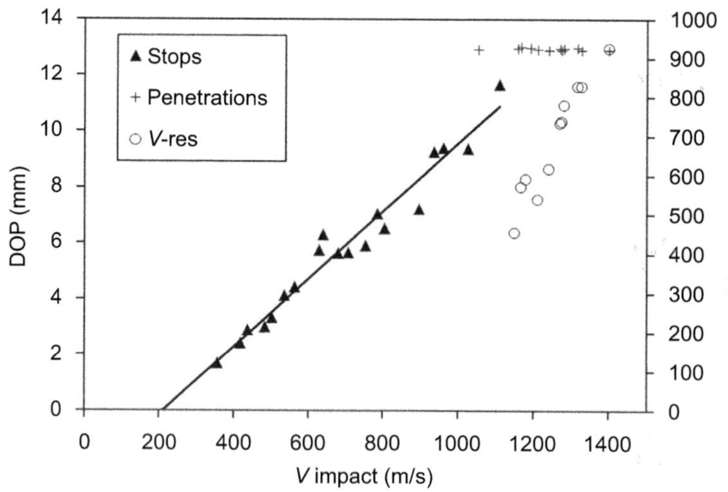

Figure 3.22 DOP (for stops and penetrations) and residual projectile velocity (for penetrations) versus projectile impact velocity for Dyneema® HB26 panels with an areal density of 12.5 kg/m^2 (12.9-mm thickness).

of the penetration process of UHMWPE fiber composites should be possible and can probably be extended to composites based upon other fibers.

Pictures of cross sections of samples after shooting for different projectile impact velocities are given in Fig. 3.23 (for a panel with an areal density of 25 kg/m^2). The projectile certainly does not create plugs that are shoved downward in front of the projectile. During penetration, the projectile seems to shove aside broken UD material, which results in bulges at the impact side of the panels around the projectile channel of penetration. At larger DOP, delamination occurs only at the end of the penetration channel, which can be explained by the fact that most of the deceleration process to a full projectile stop needs to occur via bulging (analogous to Fig. 3.21(b)), which then, depending on the degree of panel penetration, can result in delamination.

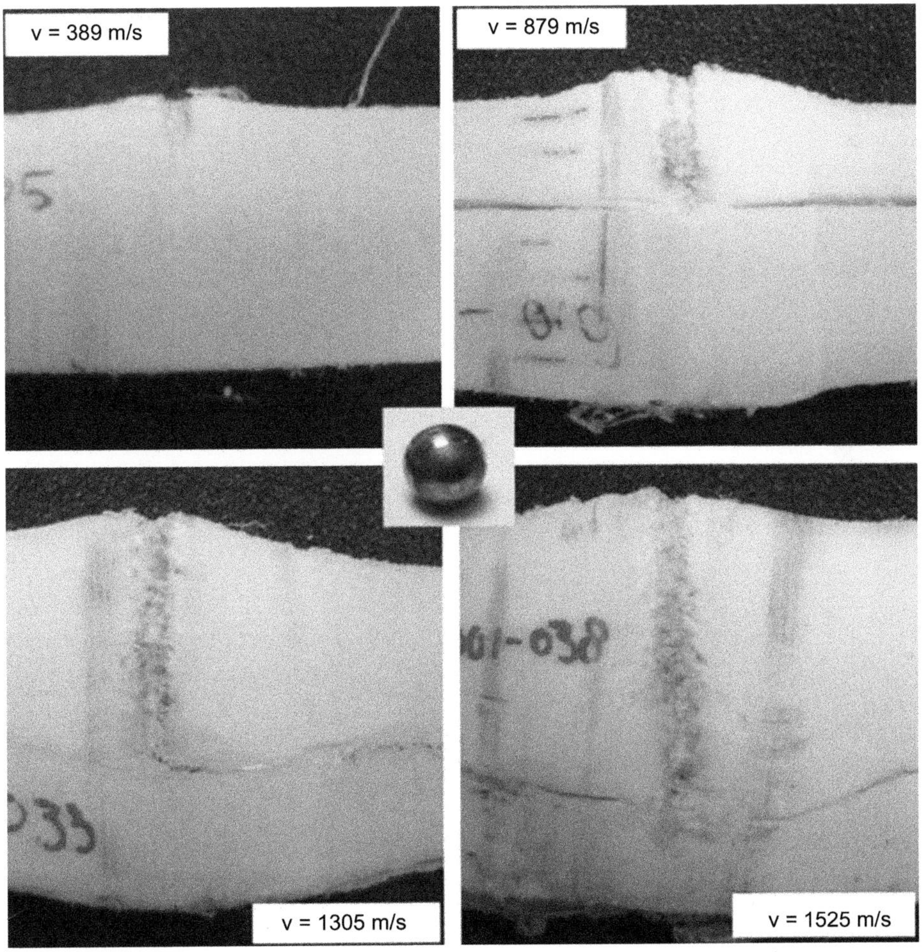

Figure 3.23 Cross sections of Dyneema® HB26 panels (areal density 25 kg/m^2) shot with different projectile velocities. The center picture shows the projectile.

Based upon the results given here, the following general relation can be postulated between the ballistic limit V_{BL} of a nondeforming projectile onto a UD target with areal density AD:

$$V_{BL} = V_0 + AD \cdot \Delta V_P \qquad [3.9]$$

The kinetic projectile energy absorbed per areal density of the target, E_{abs}, is, in industrial practice, an often used parameter. Using Eq. [3.9], the E_{abs} can also be expressed as Eq. [3.10]:

$$E_{abs} = \frac{1}{2} m_P \left(\frac{V_0^2}{AD} + 2 \cdot V_0 \cdot \Delta V_P + \Delta V_P^2 \cdot AD \right) \qquad [3.10]$$

where m_P is the mass of the projectile.

These relations have been applied to practical V50 data for 1.1 g FSP for two different Dyneema® UD types (SB21 and HB80) as shown in Fig. 3.24. SB21 is tested in corner-stitched packs, HB80 as hot-pressed panels. All evaluations used clay as a backing material and multiple shots per sample to determine the V50. The experimental data agree very well with the postulated relations, and it shows that the observed mechanism of the penetration process is basically valid and can thus be used to further examine and rationalize ballistic performance of different fiber-based composites.

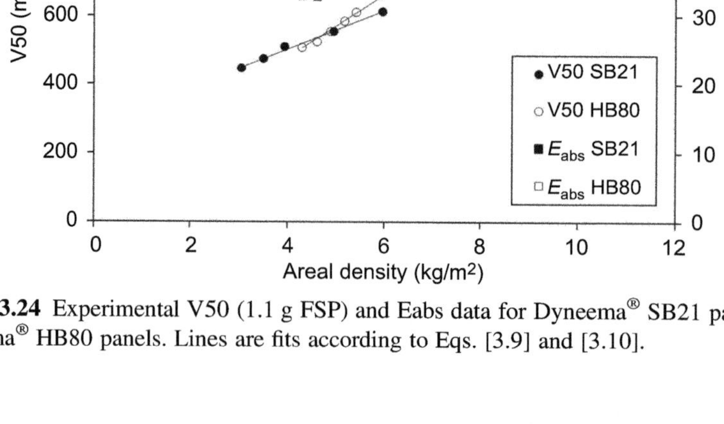

Figure 3.24 Experimental V50 (1.1 g FSP) and Eabs data for Dyneema® SB21 packs and Dyneema® HB80 panels. Lines are fits according to Eqs. [3.9] and [3.10].

3.4 Ballistics models

Basic reviews on the modeling of ballistic impact on fiber-based composites are in our view (Cheeseman and Bogetti, 2003; Abrate, 2011; and for a general mathematical framework Ben-Dor et al., 2005). Mathematical models to describe the penetration phenomena may be classified as empirical, engineering type, numerical and analytical, and examples of all types are discussed in these review papers. Engineering models make assumptions about the relevant energy absorption mechanisms and quantify the contributions, see e.g. Karthikeyan and Russell (2014). If they are cast in analytic form the relations and governing parameters become apparent. Numerical models, on the other hand, try to avoid limiting assumptions as much as possible and start with the basic conservation laws and observe the evolution of the system. For interpretation of numerical models, sensitivity studies are important. Often, simplified analytical models or design maps are derived from numerical models.

3.4.1 *Engineering and analytical models*

Cunniff (1999) reported an elegant dimensionless analysis for ballistic impact on fiber-based targets and showed that a single curve could relate normalized experimental V50 values of different fiber-based targets, impacted by a projectile, to the ratio of the areal density of the projectile, $AD_{projectile}$, and the areal density of the target, AD_{target}. The dimensionless fiber property indicating the ballistic performance of a fiber, denoted here as Ω, relevant for this relation was found to be the product of fiber specific toughness (ie, elastic energy to break in a tension test per kg of material) and the speed of sound in the fiber. A similar ballistic performance indicator was proposed earlier by Prevorsek et al. (1988). This ballistic performance indicator combines the energy absorption capacity of the fiber and the speed to spread out the absorbed energy away from the point of impact. Assuming a linear stress—strain curve for the fiber, with strength σ (in N/m^2), modulus E (in N/m^2), elongation at break ε and density ρ (in kg/m^3), the expression for the ballistic performance indicator Ω is shown in Eq. [3.11]:

$$\Omega = \frac{\sigma \cdot \varepsilon}{2 \cdot \rho}\sqrt{\frac{E}{\rho}} \qquad [3.11]$$

As Ω has the unit of m^3/s^3, Cunniff (1999) stated the following relation:

$$\frac{V_{50}}{\sqrt[3]{\Omega}} = f\left(\frac{AD_{target}}{AD_{projectile}}\right) \qquad [3.12]$$

He found as a single, unique function f to be valid for all investigated fiber systems. The same relation was derived theoretically by Phoenix and Porwal (2003) for an isotropic target membrane. Rearrangement of Eq. [3.12] using tex-based units for strength (cN/dtex) and modulus (N/tex) leads to Eq. [3.13]:

$$\sqrt[3]{\Omega}\ (\mathrm{m/s}) = 171\frac{\sigma(\mathrm{cN/dtex})^{2/3}}{E(\mathrm{N/tex})^{1/6}} \qquad [3.13]$$

Next to the intrinsic mechanical properties of the fiber, details of the design and properties of the fiber composite (matrix content, mechanical properties and thickness of the individual fiber layers) will have drastic influences on ballistic performance of the target. Numerous publications have investigated aspects of possible relations between ballistic performance and composite properties. For example, Karthikeyan et al. (2013) showed that the interlaminar shear strength of the composite strongly influences ballistic performance. Interestingly complete analytical and numerical models, incorporating fiber and matrix properties and fiber layer thickness, for out-of-plane compression properties of Dyneema® composites were developed by Attwood et al. (2014). However, complete, detailed and unambiguous understanding of the relation between fiber composite properties and ballistic performance has certainly not been achieved yet.

3.4.2 Numerical models

Contrary to engineering models, numerical models aim to use as few assumptions as possible to describe the phenomena of interest. The idea is to put the relevant physics in the model, at least Newton's law, and observe the system response. Depending on the complexity of the question, sophisticated constitutive laws, contact algorithms and multi-physics approaches might be appropriate, and there are commercial and academic hydrocodes for this. In the subsequent chapter, we will treat a simple bead-spring numerical model that gives important information on relevant fiber properties for improved ballistic performance. Furthermore it spun off some (semi) analytic modeling to capture and rationalize the phenomena that evolved (Li, 2011; Zhou, 2014). Subsequently, a state-of-the-art 3D hydrocode model is presented.

3.4.2.1 Bead-spring model

One of the most basic numerical models that allows interesting studies discretizes a cross-ply armor plate as a single membrane. Conceptual mass points (beads) having three translational degrees of freedom are connected by springs in a square chessboard-like manner (Fig. 3.25). This setup reflects the orthogonal arrangement of the fibers in the armor plate, and the springs are given appropriate properties. Exploiting symmetry, a 40 × 40 cm plate is modeled as a quarter 20 × 20 cm domain with 2001 × 2001 beads and the right boundary conditions. The mass of the beads is related to the areal density of the armor considered (areal density times the square grid area, here $0.1 \times 0.1\ \mathrm{mm}^2$). The impact of a flat cylindrical projectile is modeled by imposing an initial velocity (after momentum exchange) on the beads covered by the projectile, whose mass is increased according to the mass portion of the projectile

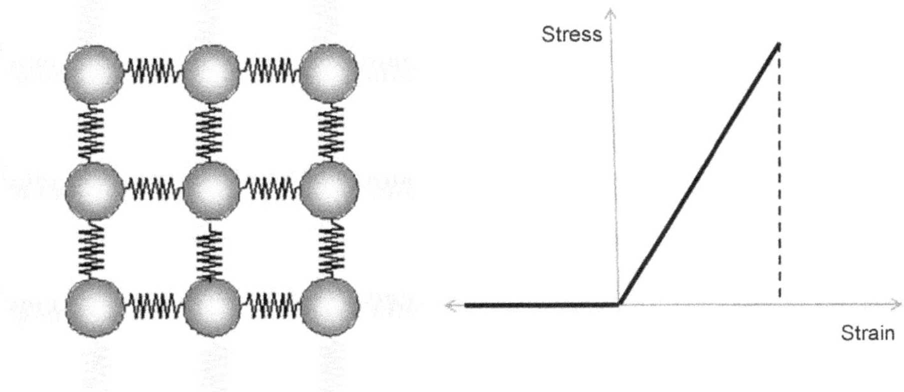

Figure 3.25 Schematic representation of the target (left) and the stress–strain curve of the springs connecting the beads (right).

acting on them. As constitutive law, linear elastic behavior of the springs is specified in tension with a modulus derived from the fibers and a failure strain ε_{break}. No compressional stiffness is assumed. After far field boundary conditions are prescribed, that is, the outmost beads are fixed in all three space directions, a simple leap-frog time-stepping scheme can advance the positions and velocities of all beads via Newton's law ($F = ma$) in time (see van der Werff et al., 2010).

Similar implementations can be traced back to Roylance et al. (1973) and Roylance and Wang (1979, 1981). A summary and extensions to multilayers and fiber slippage are given by Roylance et al. (1995). A recent multilayer version incorporating viscoelastic material behavior, yarn slippage, decrimping and bending resistance of the fabric is reported by Joo and Kang (2007, 2008).

The following material properties were used to simulate the ballistic performance of composites based on Dyneema® fibers. The strength of these Dyneema® fibers was set to 3.43 GPa with a modulus of 122 GPa (defining a strain at break of 0.0281). Density was set to 980 kg/m^3. Dyneema® fiber composites consist of [0 degree/90 degree] stacked fiber layers containing a matrix. For the simulations, the matrix content was set to 17 w% and the matrix density also to 980 kg/m^3. Using these boundary conditions, it is straightforward to calculate that the modulus of the springs should be equal to $(1 - 0.17) \times 122$ GPa / 2 = 50.6 GPa. The strain at break of the springs is still 0.0281. Simulations were carried out around a reference situation of an FSP (1.1 g mass, 5.74 mm diameter) impacting a target with an areal density of 4.89 kg/m^3. In the simulation, the projectile was modeled with a flat tip, while experimentally the FSP has the well-known beveled face. The simulations showed that a projectile with a velocity of 568 m/s will be just stopped. At 569 m/s, the projectile penetrates. Experimentally, Dyneema® composites with fiber properties, matrix content and areal density comparable to the numerical model have a V50 of 573 m/s for 1.1 g FSP, so the agreement is extremely good for this condition. The development of projectile velocity versus time after impact is shown in Fig. 3.26.

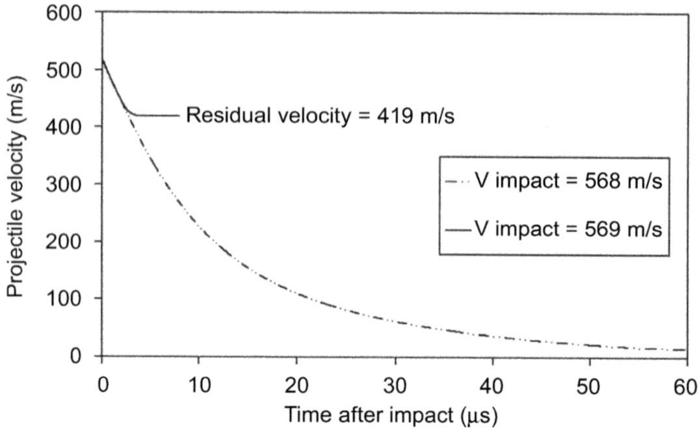

Figure 3.26 Projectile velocity versus time after impact for the reference situation at projectile impact velocities of 568 m/s (stop) and 569 m/s (penetration).

The observation then is that the numerical model implies a step function for the residual velocity as a function of projectile impact velocity. The reason for this becomes clear if one looks at the maximum strains in the springs directly adjacent to the beads covered by the projectile. Fig. 3.27 shows that the strain in the target, at the perimeter of the projectile, goes through a maximum very quickly and then decreases gradually again with time. The critical moment is the time after impact where these strains reach their maximum. For the reference situation, this is at 1.94 μs after impact. The projectile velocity has then only dropped to about 419 m/s, which will be approximately the residual velocity when penetration occurs, since no further work is done on the projectile to decelerate it. Fig. 3.27 shows some spikes at the critical moment, both under and

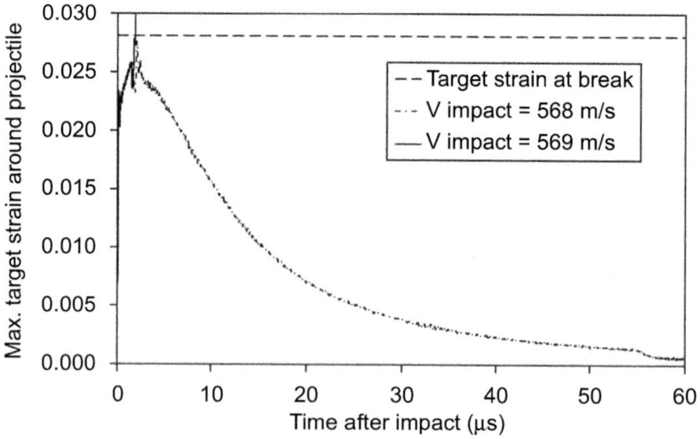

Figure 3.27 Maximum strain in the target directly around the projectile versus time after impact for the reference situation at impact velocities of 568 m/s (stop) and 569 m/s (penetration). The critical moment for perforation is at 1.94 μs after impact.

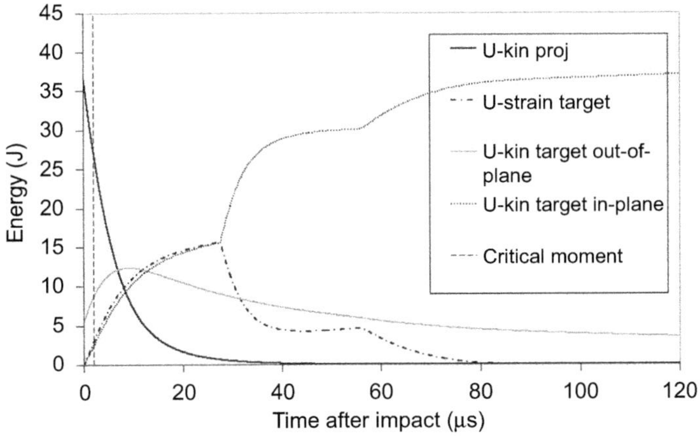

Figure 3.28 The development of the various energies (see text) versus time after impact for the reference situation. The dotted vertical line shows the critical moment for penetration.

above the ballistic limit. This is because at the ballistic limit a few springs can break without leading to a total penetration. In simulations with a few m/s less impact velocity, these spikes disappear.

Using the numerical model, details of the actual stopping mechanism of the projectile by the target can be investigated. Fig. 3.28 shows the development, with time after impact, of several energies of the projectile and target:

- U-kin proj = kinetic energy of the projectile
- U-strain target = the total strain energy in the target
- U-kin target out of plane = the total kinetic energy in the target (excluding the projectile covered point masses) from z-velocity components only (so velocities parallel to projectile velocity)
- U-kin target in plane = the total kinetic energy in the target (excluding the projectile covered point masses) from x- and y-velocity components only (so velocities perpendicular to projectile velocity)

The conservation of the total energy of the whole system during the simulation was within 4% of the original projectile kinetic energy. The decrease of the projectile kinetic energy is very rapid at the beginning of the stopping process, and it levels off slowly to almost zero at longer times after impact. Fig. 3.28 shows that the critical moment, that is, the time at which the maximum strains have developed in the target directly around the projectile, occurs very early in the stopping process (below 2 μs). The projectile kinetic energy has decreased by then only by about 20%. During the beginning of the stopping process, the target takes up the projectile energy mainly by the kinetic energy of the out-of-plane movement. So the formation and behavior of the transversal wave is most important until the critical moment. Reflection of the longitudinal sound wave occurs at the outer target edge at 28 μs and subsequently at the point of impact at 56 μs. The thus observed speed of sound (7246 m/s) along the fiber directions agrees well with the theoretical value (square root of modulus over density) of 7188 m/s. Fig. 3.28 shows that, due to these reflections, energy is increasingly absorbed by in-plane kinetic energy of the target.

The numerical model can be very straightforwardly used to investigate the influence of fiber strength and modulus on the ballistic limit. Simulations were carried out where, with respect to the reference situation, only the value of either the fiber strength or the fiber modulus were doubled or halved, while all other parameters were kept constant. The effect of these variations on the ballistic limit is given in Fig. 3.29. Next to the

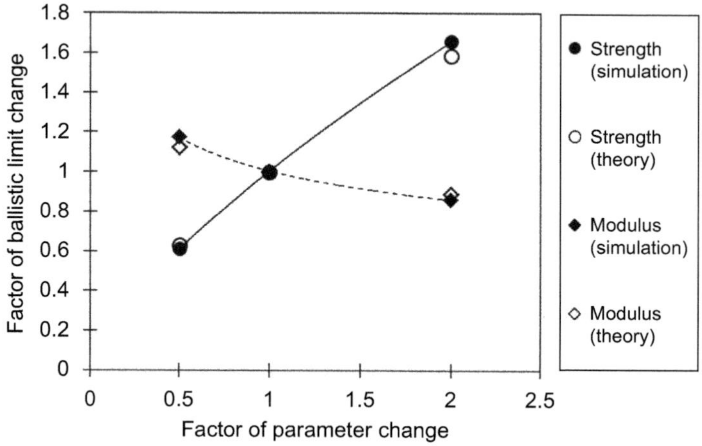

Figure 3.29 Sensitivity of the ballistic limit to independent changes in fiber strength and modulus.

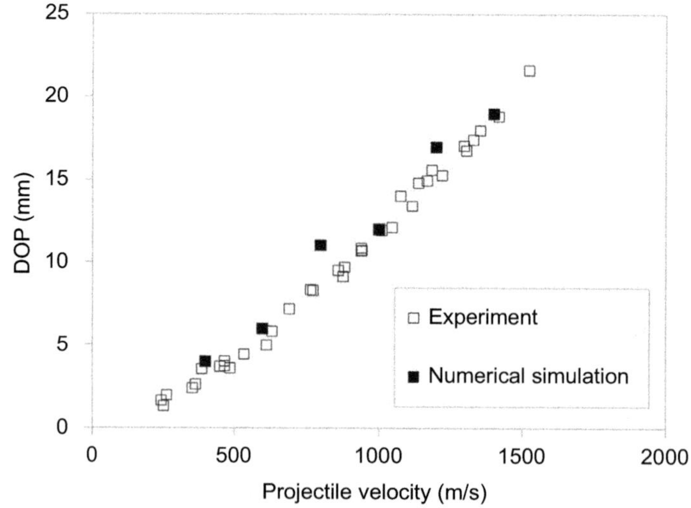

Figure 3.30 Depth of penetration: experimental and numerical results for Dyneema® HB26 with areal density of 25 kg/m^2 and 5.56-mm-diameter steel projectile.

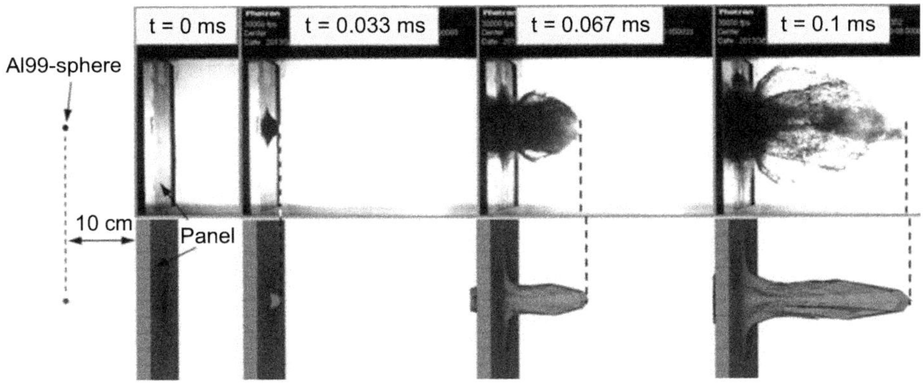

Figure 3.31 Comparison of the back side bulge of experiment and simulation after impact of an 6-mm-diameter aluminum sphere with 6591 m/s on a Dyneema® HB26 panel (areal density = 15 kg/m^2).

simulation results, theoretical expectations according to Eqs. [3.12] and [3.13] were calculated. Interestingly, the results of the simulations completely agree with the theoretical estimations. The ballistic limit increases strongly with increasing fiber strength and decreases slightly with increasing fiber modulus. Based on these results, higher strength Dyneema® fibers were indeed produced. The ballistic performance of the composites of these fibers will be reported in Section 3.5.

3.4.2.2 Hydrocode models

More sophisticated software than the bead-spring model discussed previously allows a more realistic investigation of the physical phenomena. For historical reasons the explicit finite element software used in impact and penetration modeling is known under the name *hydrocodes*. Those codes can be traced back to the 1950s and were first used to investigate impacts where the yield strength of the materials is exceeded by orders of magnitudes. Consequently, neglecting the material strength and considering only hydrodynamic effects is a good approximation. Later, more complex constitutive relations incorporated shear strength, plasticity and strain rate effects. For a historical survey and current developments consult Zukas (2004).

To model the ballistic performance of a composite armor like Dyneema® a number of approaches are possible. For most applications, it is essential to capture the two scales of wave speeds involved: the sound speed along the fibers is in the order of 10 km/s (see Section 3.2.3), while the speed of sound through the thickness direction of the composite is in the order of 2 km/s (Hazell et al., 2011). One could resolve the physical constituents, fiber and matrix, and discretize on that level. An example is the so-called digital element analysis where individual yarns are resolved (Cheeseman et al., 2006). At the other end of the spectrum are models where the composite is modeled as a virtual continuum with direction-dependent material properties and failure modes. An example of this approach is introduced in this section based on Lässig et al. (2015). To resolve fiber bundles as proposed by Chocron et al. (2014) is an in-between step.

The model reported here is a general nonlinear plasticity and shock compression model in the sense that the main energy absorption mechanisms during an impact event are addressed. The energy of the projectile is absorbed by breaking the strong fibers while perforating layers (governed by maximum in-plane strength). During that process, the localized kinetic energy of the impactor is spread out to an increasing volume of the target (strain waves governed by the orthotropic stiffness matrix). Besides this, permanent plastic deformation (orthotropic yield function) and delamination (controlled by fracture toughness) absorb energy. The model uses a continuum 3D finite element framework developed for high-velocity impact into orthotropic materials by Riedel et al. (2003). Choosing the concept of orthotropic material behavior is justified by the cross-ply construction (Fig. 3.2). As a consequence of the macroscopic approach, we model the effective behavior of a continuum with direction-dependent properties. The four basic phenomenological blocks of the constitutive model are the following:

- linear elastic orthotropic stiffness
- orthotropic plasticity with hardening
- polynomial equation of space that is coupled to the orthotropic stiffness (Anderson et al., 1994)
- direction-dependent damage and failure

As far as possible the parameters of the constitutive model are determined by basic static and dynamic flyer plate tests on the composite. Next to the tensile tests on the composites along the fibers (see Section 3.2.4), the tensile test in ±45 degree direction determines the nonlinear plastic response that is captured by the orthotropic yield function with an effective plastic stress–strain curve. The important plastic through-thickness response was probed by a nonstandard fork tool compressing the composite in thickness direction. Mode-I fracture toughness tests are a challenge due to fiber bridging and the soft matrix material, but with thick enough samples, this issue was solved and a fracture toughness of 790 J/m^2 derived. The calibration of the polynomial equation of state (important for high-velocity impacts) uses experimental flyer plate data where an HB26 sample is shot against a well-characterized steel plate. The back surface velocity–time traces as recorded by a laser VISAR system are matched in a numerical substitute model by fitting the parameters of the equation of state. The complete characterization and resulting parameter set of Dyneema® HB26 for the use in the commercial hydrocode Ansys Autodyn is available in Laessig et al. (2015). The contributions of the constitutive building blocks were conclusively demonstrated by switching off the plasticity and nonlinear equation of state, respectively. These changes lead to a significantly different residual velocity for the benchmark cases shown later in Fig. 3.32.

As a first validation case, the DOP tests of HB26 with areal density of 25 kg/m^2 using the same fixture (70-mm-diameter hole) and 5.5-mm-diameter steel projectiles as in Section 3.3 are compared to the numerical results in Fig. 3.30. A quarter-symmetric model with Lagrangian hexahedral elements refined to the impact point is applied. One element layer is 1 mm thick, and the projectile is resolved by six elements over the radius. An equivalent geometric erosion strain of 150% allows

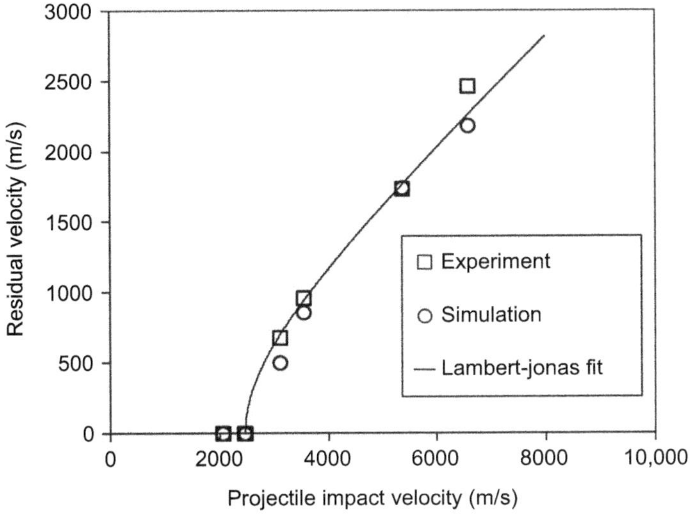

Figure 3.32 Validation at high velocities: residual velocities of the projectile after impacting a Dyneema® HB26 panel (areal density = 15 kg/m^2).

numerical element deletion. It is evident that the DOP trend is closely followed. The early observed linear relation between DOP and projectile velocity (see Section 3.3) is less valid for the extremely thick samples used in Fig. 3.30.

A comparison of experimental and numerical snapshots, Fig. 3.31, and the residual velocity for a 6-mm-diameter aluminum projectile into HB26 of areal density 15 kg/m^2 for high-velocity impact is shown in Fig. 3.32. Not only the overmatch cases are predicted with a good accuracy but also the stop cases below the ballistic limit.

The sensitivity of the model with regard to material parameters was investigated also in Laessig et al. (2015) with higher in-plane strength resulting in a higher ballistic limit as predicted by Cunniff (Fig. 3.29) and experimentally found with the next-generation fibers (Section 3.5). The advantage of the complex model is that also the sensitivity to other parameters like fracture toughness and through-thickness properties can be assessed.

3.5 Next-generation Dyneema® fibers and their composites

The UHMWPE fiber system is unique, compared to, for example, *p*-aramid fibers, in its remaining potential to improve tensile strength and modulus very substantially. Tensile strength of commercially available UHMWPE fibers are traditionally around 3.5 GPa or lower, whereas on lab scale, tensile strengths up to about 7 GPa have been reported by various research groups (Savitskii et al., 1984; Pennings et al., 1990; Wang and Smith, 1999). In the preceding chapters, the relevance of tensile strength for ballistic performance has been demonstrated clearly by empirical, numerical and analytical techniques. DSM Dyneema, therefore, decided to make the next generation of Dyneema® fibers with increased tensile strength and its fiber-based composites for ballistic protection commercially available, positioned as the Dyneema® Force Multiplier Technology platform.

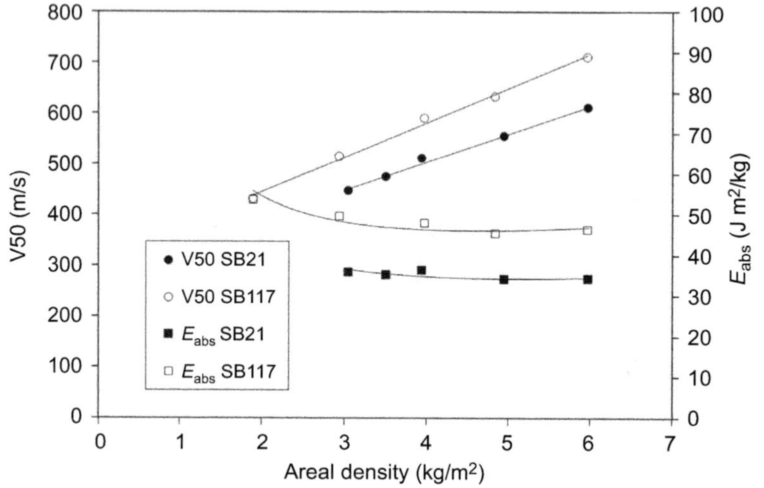

Figure 3.33 V50 for 1.1 g FSP and E_{abs} versus areal density of corner-stitched packs of state-of-the-art Dyneema® SB21 and newly developed Dyneema® SB117 from the Dyneema® Force Multiplier Technology platform.

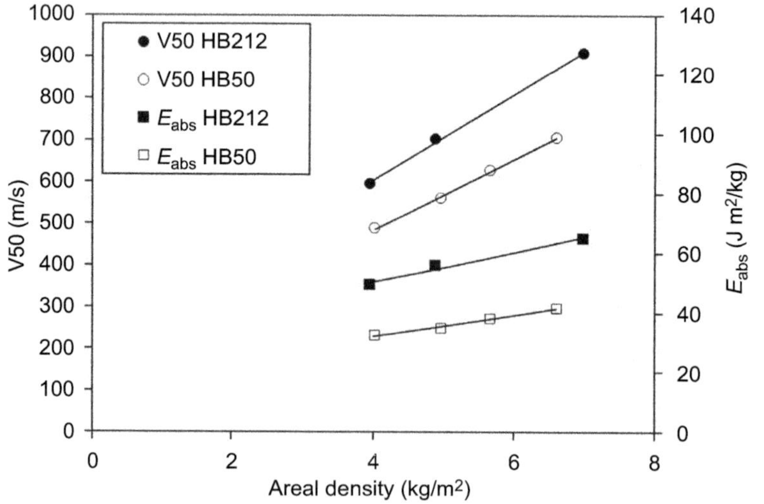

Figure 3.34 V50 for 1.1 g FSP and E_{abs} versus areal density of hot-pressed panels of state-of-the-art Dyneema® HB50 and newly developed Dyneema® HB212 from the Dyneema® Force Multiplier Technology platform.

The improvement in tensile strength of the Dyneema® fibers is about 20%. The obtained improvements of the ballistic performance of the next-generation Dyneema® composites are shown in Fig. 3.33 (packs) and Fig. 3.34 (panels). The lines in Figs. 3.33 and 3.34 are fitted and drawn according to Eqs. [3.9] and [3.10]. The values obtained for V_0 and ΔV_p are given in Table 3.1. The improvements demonstrated by Figs. 3.33 and 3.34 are very substantial. At equal V50, the next-generation Dyneema® fibers offer the possibilities to reduce weights, compared to state-of-the-art products, by around 25% or more. This step in performance for weight-dominated technical material applications is drastic.

The remaining potential of the UHMWPE fiber system to even improve substantially upon this is depicted in Fig. 3.35, where fiber tensile strength is plotted versus fiber modulus for different fiber systems. The dotted lines are lines with constant cubic root of Ω (in m/s) according to Eq. [3.13]. Points connected by such lines are basically points with equal theoretical ballistic performance. The higher the value of Ω, the

Table 3.1 Experimental values of V_0 and ΔV_p, according to Eq. [3.9], for Dyneema® UD types for 1.1 g FSP

Dyneema® UD type	V_0 (m/s)	ΔV_p (m/s) per penetrated areal density (kg/m^2)
SB117	310	78
SB21	282	55
HB212	196	102
HB50	143	85

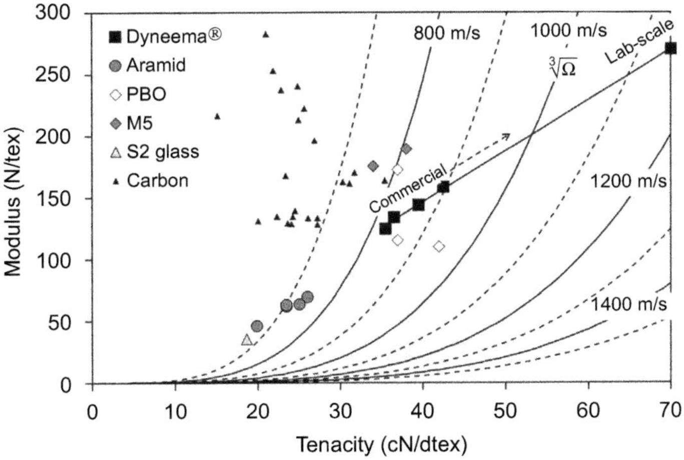

Figure 3.35 Modulus versus tensile strength for various high-performance fibers. Lines are drawn with the indicated constant value of the cubic root of Ω (according to Eq. [3.13]).

higher the ballistic performance is. From Fig. 3.35, it is evident that the ballistic potential of the UHMWPE fiber system is totally beyond that of other high-performance fibers, and it is to be expected that UHMWPE fibers and its composites will play an even larger role in ballistic armor systems in the future.

References

Abrate, S., 2011. CISM courses and lectures. Impact Engineering of Composite Structures, vol. 526. Springer. http://dx.doi.org/10.1007/978-3-7091-0523-8.

Allen, S.R., 1987. Tensile recoil measurement of compressive strength for polymeric high performance fibres. Journal of Materials Science 22, 853–859. http://dx.doi.org/10.1007/BF01103520.

Anderson, C., Cox, P., Johnson, G., Maudlin, P., 1994. A constitutive formulation for anisotropic materials suitable for wave propagation computer programs—II. Computational Mechanics 15, 201–223. http://dx.doi.org/10.1007/BF00375030.

Attwood, J.P., Khaderi, S.N., Karthikeyan, K., Fleck, N.A., O'Masta, M.R., Wadley, H.N.G., Deshpande, V.S., 2014. The out-of-plane compressive response of Dyneema® composites. Journal of the Mechanics and Physics of Solids 70, 200–226. http://dx.doi.org/10.1016/j.jmps.2014.05.017.

Attwood, J.P., Fleck, N., Wadley, H., Deshpande, V., 2015. The Compressive Response of Ultra-high Molecular Weight Polyethylene Fibres and Composites. International Journal of Solids and Structures 71, 141–155. http://dx.doi.org/10.1016/j.ijsolstr.2015.06.015.

Ben-Dor, G., Dubinsky, A., Elperin, T., 2005. Ballistic impact: recent advances in analytical modeling of plate penetration dynamics – a review. Applied Mechanics Reviews 58, 355–371. http://dx.doi.org/10.1115/1.2048626.

Cheeseman, B.A., Bogetti, T.A., 2003. Ballistic impact into fabric and compliant composite laminates. Composite Structures 61, 161–173. http://dx.doi.org/10.1016/S0263-8223(03)00029-1.

Cheeseman, B.A., Yen, C.F., Scott, B.R., Powers, B., Bogetti, T.A., 2006. From filament to fabric packs – Simulating the performance of textile protection systems. In: 25th Army Science Conference, Orlando (FL), 27th–30th November 2006.

Chocron, S., Nicholls, A.E., Brill, A., Malka, A., Namir, T., Havazelet, D., van der Werff, H., Heisserer, U., Walker, J.D., 2014. Modeling unidirectional composites by bundling fibers into strips with experimental determination of shear and compression properties at high pressures. Composite Science and Technology 101, 32–40. http://dx.doi.org/10.1016/j.compscitech.2014.06.016.

Cunniff, P.M., 1999. Dimensionless parameters for optimization of textile-based body armor systems. In: Proceedings of the 18th International Symposium on Ballistics, San Antonio (TX), 15th–19th November 1999, pp. 1303–1310.

Deteresa, S.J., Allen, S.R., Farris, R.J., Porter, R.S., 1984. Compressive and torsional behaviour of Kevlar 49 fibre. Journal of Material Science 19, 57–72. http://dx.doi.org/10.1007/BF02403111.

Fukuda, H., Yakushiji, M., Wada, A., 1999. A loop test to measure the strength of monofilaments used for advanced composites. Advanced Composite Materials 8, 281–299. http://dx.doi.org/10.1163/156855199X00272.

Govaert, L.E., Bastiaansen, C.W.M., Leblans, P.J.R., 1993. Stress-strain analysis of oriented polyethylene. Polymer 34, 534–540. http://dx.doi.org/10.1016/0032-3861(93)90546-M.

Hazell, P.J., Appleby-Thomas, G.J., Trinquant, X., Chapman, D.J., 2011. In-fiber shock propagation in Dyneema®. Journal of Applied Physics 110, 043504. http://dx.doi.org/10.1063/1.3622294.

Hudspeth, M., Nie, X., Chen, W., 2012. Dynamic failure of Dyneema SK76 single fibers under biaxial shear tension. Polymer 53, 5568–5574. http://dx.doi.org/10.1016/j.polymer.2012.09.020.

Hudspeth, M., Li, D., Spatola, J., Chen, W., Zheng, J., 2015a. The effects of off-axis transverse deflection loading on the failure strain of various high-performance fibers. Textile Research Journal, 0040517515588262, first published on September 8, 2015. http://dx.doi.org/10.1177/0040517515588262.

Hudspeth, M., Chen, W., Zheng, J., 2015b. Why the Smith theory over-predicts instant rupture velocities during fiber transverse impact. Textile Research Journal, 0040517515586158, first published on July 15, 2015. http://dx.doi.org/10.1177/0040517515586158.

Joo, K., Kang, T.J., 2007. Numerical analysis of multi-ply fabric impacts. Textile Research Journal 77 (6), 359–368. http://dx.doi.org/10.1177/0040517507076746.

Joo, K., Kang, T.J., 2008. Numerical analysis of energy absorption mechanism in multi-ply fabric impacts. Textile Research Journal 78 (7), 561–576. http://dx.doi.org/10.1177/0040517507085194.

Kalb, B., Pennings, A.J., 1979. Spinning of high molecular weight polyethylene solution and subsequent drawing in a temperature gradient. Polymer Bulletin 1, 871–876. http://dx.doi.org/10.1007/BF00283221.

Karthikeyan, K., Russell, B.P., Fleck, N.A., Wadley, H.N.G., Deshpande, V.S., 2013. The effect of shear strength on the ballistic response of laminated composite plates. European Journal of Mechanics A/Solids 42, 35–53. http://dx.doi.org/10.1016/j.euromechsol.2013.04.002.

Karthikeyan, K., Russell, B.P., 2014. Polyethylene ballistic laminates: failure mechanics and interface effects. Materials and Design 63, 115–125. http://dx.doi.org/10.1016/j.matdes.2014.05.069.

Kelly, A., MacMillan, N.H., 1986. Strong Solids, third ed. Clarendon Press, Oxford.

Knijnenberg, A., 2009. Compressive Failure Behaviour of Novel Aramid Fibres (PhD thesis). Delft University of Technology, The Netherlands. uuid:f0228918–d5c2-46ed-adf0-976dd825050a.

Laessig, T., Nguyen, L., May, M., Riedel, W., Hiermaier, S., Heisserer, U., van der Werff, H., 2015. A non-linear orthotropic hydrocode model for ultra-high molecular weight polyethylene in impact simulations. International Journal of Impact Engineering 75, 110–122. http://dx.doi.org/10.1016/j.ijimpeng.2014.07.004.

Levi-Sasson, A., Meshi, I., Mustacchi, S., Amarilio, I., Benes, D., Favorsky, V., Eliasy, R., Aboudi, J., Haj-Ali, R., 2014. Experimental determination of linear and nonlinear mechanical properties of laminated soft composite material system. Composites: Part B 57, 96–104. http://dx.doi.org/10.1016/j.compositesb.2013.09.043.

Li, W., 2011. Models for Projectile Impact into Hybrid Multi-layer Armor Systems with Axisymmetric or Biaxial Layers with Gaps (PhD thesis). Cornell University.

Liu, G., Thouless, M.D., Deshpande, V.S., Fleck, N.A., 2014. Collapse of a composite beam made from ultra high molecular-weight polyethylene fibres. Journal of the Mechanics and Physics of Solids 63 (1), 320–335. http://dx.doi.org/10.1016/j.jmps.2013.08.021.

Marissen, R., 2011. Design with ultra strong polyethylene fibers. Materials Sciences and Applications 2, 319–330. http://dx.doi.org/10.4236/msa.2011.25042.

Newell, J.A., Gustafson, J.M., 2001. An improved interpretation of recoil compressive failure data for high-performance polymers. High Performance Polymers 13, 251–257. http://dx.doi.org/10.1088/0954-0083/13/4/303.

Northolt, M.G., den Decker, P., Picken, S.J., Baltussen, J.J.M., Schlatmann, R., 2005. The tensile strength of polymer fibres. Advances in Polymer Science 187, 1–108. http://dx.doi.org/10.1007/b104207.

Pennings, A.J., Kiel, A.M., 1965. Fractionation of polymers by crystallization from solution, III. On the morphologogy of fibrillar polyethylene crystals grown in solution. Kolloid Zeitschrift und Zeitschrift für Polymere 205 (2), 160–162. http://dx.doi.org/10.1007/BF01507982.

Pennings, A.J., Roukema, M., van der Veen, A., 1990. Further studies on the high-speed gel-spinning of ultra-high molecular weight polyethylene. Polymer Bulletin 23, 353–359. http://dx.doi.org/10.1007/BF01032453.

Phoenix, S.L., Porwal, P.K., 2003. A new membrane model for the ballistic impact response and V50 performance of multi-ply fibrous systems. International Journal of Solids and Structures 40, 6723–6765. http://dx.doi.org/10.1016/S0020-7683(03)00329-9.

Porwal, P.K., Phoenix, S.L., 2005. Modeling system effects in ballistic impact into multi-layered fibrous materials for soft armor. International Journal of Fracture 135, 217–249. http://dx.doi.org/10.1007/s10704-005-3993.

Phoenix, S.L., Heisserer, U., van der Werff, H., van der Jagt-Deutekom, M.J., 2011. Modelling and Experiments on Ballistic Impact into UHMWPE Yarns Using Flat and Saddle-nosed Projectiles. Internal report by DSM Dyneema and TNO, The Netherlands.

Prevorsek, D.C., Harpell, G.A., Kwon, Y.D., Li, H.L., Young, S., 1988. Ballistic armor from extended chain polyethylene fibers. In: Proceedings of the 33rd International SAMPE Symposium and Exhibition, 7th–10th March 1988, pp. 1685–1696.

Riedel, W., Harwick, W., White, D.M., Clegg, R.A., 2003. Advanced Material Damage Models for Numerical Simulation Codes, Report No. I 75/03, Final report to ESA/ESTEC, Contract no. 12400/97/PA(SC), CCN No. 2.

Roylance, D., Wilde, A., Tocci, G., 1973. Ballistic impact of textile structures. Textile Research Journal 43 (1), 34–41. http://dx.doi.org/10.1177/004051757304300105.

Roylance, D., 1977. Ballistics of transversely impacted fibers. Textile Research Journal 47 (10), 679–684. http://dx.doi.org/10.1177/004051757704701007.

Roylance, D., Wang, S.S., 1979. Penetration Mechanics of Textile Structures. Technical report Natick/TR-80/021. U.S. Army Natick R&D Command.

Roylance, D., Wang, S.S., 1981. Influence of fibre properties on ballistic penetration of textile panels. Fibre Science and Technology 14, 183–190. http://dx.doi.org/10.1016/0015-0568(81)90010-5.

Roylance, D., Chammas, P., Ting, J., Chi, H., Scott, B., 1995. Numerical modeling of fabric impact. In: Proceedings of the National Meeting of the American Society of Mechanical Engineers (ASME), San Francisco, October 1995.

Russell, B.P., Karthikeyan, K., Deshpande, V.S., Fleck, N.A., 2013. The high strain rate response of ultra high molecular-weight polyethylene: from fibre to laminate. International Journal of Impact Engineering 60, 1–9. http://dx.doi.org/10.1016/j.ijimpeng.2013.03.010.

Savitskii, A.V., Gorshova, I.A., Demicheva, V.P., Frolova, I.L., Shmikk, G.N., 1984. Model of the orientation reinforcement of polymers and the preparation of high-strength polyethylene fibres. Polymer Science USSR 26 (9), 2007–2016. http://dx.doi.org/10.1016/0032-3950(84)90442-8.

Smith, J.C., McCrackin, F.L., Schiefer, H.F., 1958. Stress-strain relationships in yarns subjected to rapid impact loading. Part V: wave propagation in long textile yarns impacted transversely. Textile Research Journal 28 (4), 288–302. http://dx.doi.org/10.1177/004051755802800402.

Smith, J.C., Blandford, J.M., Towne, K.M., 1962. Stress-strain relationships in yarns subjected to rapid impact loading. Part VIII: shock waves, limiting breaking velocities, and critical velocities. Textile Research Journal 32, 67–76. http://dx.doi.org/10.1177/004051756203200109.

Smith, P., Lemstra, P.J., 1979. Preparing Polyethylene Filaments. UK patent application GB2051667.

van der Werff, H., Pennings, A.J., 1991. Tensile deformation of high strength and high modulus polyethylene fibers. Colloid & Polymer Science 269, 747–763. http://dx.doi.org/10.1007/BF00657441.

van der Werff, H., Heisserer, U., Phoenix, S.L., Modelling of ballistic impact on fiber composites. In: Proceedings of the Personal Armour Systems Symposium (PASS) (2010), September 13th–17th, Quebec, Canada.

van Es M.., 'Change of ballistic resistance of composite armor during V50 testing', Personal Armour Systems Symposium (PASS) (2010), September 13th–17th, Quebec, Canada.

Vlasblom, M.P., van Dingenen, J.L.J., 2009. The manufacture, properties and applications of high strength, high modulus polyethylene fibers. In: Bunsell, A.R. (Ed.), Handbook of Tensile Properties of Textile and Technical Fibres. Woodhead, Cambridge, pp. 437–485. http://dx.doi.org/10.1533/9781845696801.2.437.

Vlasblom, M.P., Boesten, J., Leite, S., Davies, P., 2012. Development of HMPE fiber for permanent deepwater offshore mooring. In: OTC 23333, Paper at the Offshore Technology Conference, Houston (TX), 30th April–3rd May, 2012. http://dx.doi.org/10.4043/23333-MS.

Wagner, H.D., Phoenix, S.L., Schwartz, P., 1984. A study of statistical variability in the strength of single aramid filaments. Journal of Computational Materials 18, 312–338. http://dx.doi.org/10.1177/002199838401800402.

Walker, J.D., Chocron, S., 2011. Why impacted yarns break at lower speed than classical theory predicts. Journal of Applied Mechanics 78 (5), 051021. http://dx.doi.org/10.1115/1.4004328.

Wang, J., Smith Jr., K.J., 1999. The breaking strength of ultra-high molecular weight polyethylene fibers. Polymer 40, 7261–7274. http://dx.doi.org/10.1016/S0032-3861(99)00034-8.

Wang, L.L., 2007. Foundations of Stress Waves. Elsevier.

Zhou, R., 2014. Effects of Crimp and Slip on Laminar and Woven Fabrics Subjected to Ballistic Impact (PhD thesis). Cornell University.

Zukas, J.A., 2004. Introduction to hydrocodes. Studies in Applied Mechanics 49. http://dx.doi.org/10.1016/S0922-5382(04)80001-1. Elsevier.

Zwijnenburg, A., 1978. Longitudinal Growth, Morphology and Physical Properties of Fibrillar Polyethylene Crystals (Ph D thesis). University of Groningen.

Fabrics and composites for ballistic protection

4

D.J. Carr, C.M. Crawford
Cranfield University, Wiltshire, United Kingdom

4.1 Introduction

Diverse personnel wear personal protective equipment that prevents injury, for example, military personnel, police officers, and security guards. This equipment includes body armour and helmets; exactly what these items protect the user from is dependent upon the threat analysis that has been conducted (Lewis, 2006; Croft and Longhurst, 2007a,b). Body armour provides protection to the critical organs of the torso; these are usually defined as the heart, lungs, liver, spleen and kidneys, although exactly what is protected may differ among users and is related to the tasks they perform and the risk analysis conducted (Tobin and Iremonger, 2006; Dixon and Croft, 2007). Military body armour generally protects the user from fragments, which are the major cause of injury in general warfare, and from high-velocity (rifle) bullets (Ryan et al., 1991; Breeze et al., 2011; Horsfall, 2012; Carr and Lewis, 2014). Fragments may originate from traditional munitions, for example, artillery shells, mortars and grenades, or from improvised explosive devices (IEDs). In comparison, police body armour provides protection from sharp weapons and low-velocity (handgun) bullets; although some officers additionally wear protection from high-velocity (rifle) bullets (eg, Authorised Firearms Officers, Specialist Firearms Officers) (Tobin and Iremonger, 2006; Croft and Longhurst 2007a,b; Horsfall, 2012; Carr and Lewis, 2014). Military personnel also wear helmets; these provide protection to the brain from fragments and nonballistic impacts (Breeze et al., 2013; Carr et al., 2014). Some specialised military and police personnel may wear helmets that provide a level of protection from bullets. This chapter summarises the fabrics and composites used in typical personal ballistic protection, discusses failure mechanisms and provides references for further sources of information.

A mature and extensive literature exists in peer-reviewed academic journals and conference proceedings discussing ballistic protective fabrics and composites. However, distinction must be made between testing conducted using projectiles and velocities of interest to designers of military vehicles, body armour and helmets (for example), and research conducted in laboratories using relatively slow gas guns and similar apparatus (typically 150–200 m/s) due to the strain-rate effect. This latter group of experiments are not usually considered to be of great use to designers of body armour and helmets.

Advanced Fibrous Composite Materials for Ballistic Protection. http://dx.doi.org/10.1016/B978-1-78242-461-1.00004-2

Military body armour and helmets will typically provide protection from an approximate 1 g metallic fragment travelling at between 400 and 600 m/s (different countries provide different levels of protection for different users).

In body armour, this fragmentation protection is achieved by the use of multiple layers of fabrics manufactured from synthetic-polymer fibres such as para-aramids and ultra-high molecular weight polyethylenes (UHMWPEs); these fibres were introduced in chapters ARAMIDS: "disruptive", open and continuous innovation and High-performance ballistic fibers: ultra-high molecular weight polyethylene (UHMWPE). High-velocity (rifle; typically 5.56- and 7.62-mm-calibre ammunition with muzzle velocities of 920 ± 15 and 830 ± 15 m/s, respectively) bullet protection is provided by the use of plates which are usually ceramic faced (alumina, silicon carbide, boron carbide) and composite backed (para-aramid or UHMWPE). In some cases a 100% UHMWPE plate might be used. In comparison, police body armour is usually designed to provide protection from sharp weapons and low-velocity (handgun) bullets (in the UK 9 mm Luger and R357M3; 365 ± 10 and 390 ± 10 m/s, respectively), and it typically uses a fabric solution combined with mail (commonly referred to as chain mail) or fabrics impregnated with a polymeric material (Horsfall, 2012). High-velocity (rifle) bullet protection is provided using plates in a similar manner to military armour.

The brain is protected by a helmet which has a composite shell, a nonballistic impact protective liner, suspension and size adjustment systems, comfort pads and a retention system (eg, Breeze et al., 2013; Carr et al., 2014). The retention system usually incorporates a chin-cup and closure system (chinstrap), a size adjustment system, and a mounting system onto the helmet shell. Helmet security, and thus protection to the brain, relies on the strength of the retention system (including chinstrap), which is usually made from webbing. Such helmets are primarily worn by the military, but some police officers wear them as well. The majority of helmet shells are manufactured from woven para-aramid fabric reinforced composites, but UHMWPE composites (0°/90° lay-up) are also used. Older style military helmet shells (eg, the UK's Helmet Combat General Service Mk6) contain plain-woven nylon 6,6 fabric composites (Marsden, 1994).

Vehicle armour that contains composites aims to provide protection from medium-calibre weapons ($\geq$12.7-mm calibre); such protection will require the use of ceramic-faced and composite-backed or ceramic-faced and metallic-backed combinations, or a 100% metallic solution. However, 100% composites may be used in relatively lightweight vehicles to provide protection from 7.62-mm-calibre bullets and/or spall.

This chapter summarises the fabrics and composites used in typical personal ballistic protection, discusses failure mechanisms and provides references to further sources of information.

4.2 Fibres and fabrics

4.2.1 Fibres

The main fibre types used in modern body armour and helmets are para-aramids and UHMWPEs; these fibres are discussed in chapters ARAMIDS: "disruptive", open and continuous innovation and High-performance ballistic fibers: ultra-high molecular weight polyethylene (UHMWPE), respectively. However, there are two other fibre types that should be mentioned: (1) glass fibre and (2) nylon 6,6.

Glass fibre (GF) composites are used in the front plates that are incorporated into many explosive ordinance disposal (EOD) suits, in vehicle armour, some Force Protection applications and solutions in areas such as bank teller stations. The fibre reinforcement is usually in the form of a plain-woven fabric. GF is an inorganic amorphous synthetic fibre produced in several grades, with silica being the major component. Filament GF was introduced in 1938 by the Owens-Corning Fiberglass Corporation. Glass is heated, drawn and cooled; annealing may be used to further increase strength. The primary GF used in composites that provide ballistic protection is S2-GF which is manufactured by AGY (http://www.agy.com/products/s2/index.htm). GF composites are chosen for their combined structural and ballistic protective properties.

Nylon was first synthesised by Wallace Carothurs at the Du Pont Experimental Station in 1935 (Richards, 2004). Nylon is a semicrystalline thermoplastic polymer with a glass transition temperature of 46–57°C, a melting temperature of 185–265°C, and a crystallinity of approximately 50% (depending on grade) (Mills, 1986; Morton and Hearle, 1997). In terms of commercial production, nylon 6 and nylon 6,6 are the most important polyamide fibres. The fibres are typically melt-spun and drawn; both have an α-structure, the length of the repeat unit in nylon 6 being half of that in nylon 6,6 due to the differences in chemical structure (Lewin and Pearce, 1985; Morton and Hearle, 1997). Nylon 6,6 is manufactured from hexanedioic acid $[(CH_2)_4(COOH)_2]$ and hexane-1,6-diamine $[H_2N(CH_2)_6NH_2]$. The chemical reaction that is used to manufacture nylon 6,6 is both a step-growth and a condensation reaction. In a step-growth reaction the growing chains react with each other to form longer chains, whilst in a condensation reaction a by-product is formed (water). Hexanedioic acid and hexane-1,6-diamine are combined in solution forming a *nylon salt*, which is then purified; polymerisation occurs in an autoclave with a nitrogen atmosphere. The molten polymer is usually extruded as a flat ribbon, quenched in cold water and cut into small pieces for further processing, for example, fibre production.

Plain-woven nylon 6,6 fabrics (ballistic nylon) were used by several governments to provide ballistic protection in body armour (eg, the UK MoD in Combat Body Armour, in combination with para-aramid fabric) and to manufacture the first composite helmets issued to British Armed Forces from the 1970s (Helmet, Parachutist, Lightweight; Helmet, Combat, General Service Mk6) (Marsden, 1994; Carr and Lewis, 2014). However, in the 21st century, nylon 6,6 is rarely used in body armours and helmets.

4.2.2 Fabrics

In modern military body armour, multiple layers of woven para-aramid fabrics are typically used to provide protection from fragments, although some nations use cross-ply UHMWPE. The armour is a compromise of protection offered, effect on the burden of the soldier (particularly dismounted; mass, stiffness, increased bulk, thermophysiological loading) and effect on human performance (physical and mental) (Carr and

Lewis, 2014). High-sett, plain-woven, balanced fabrics reportedly provide the best protection from fragmentation threats (Cunniff, 1992; Shim et al., 1995; Cheeseman and Bogetti, 2003). However, this choice of fabric architecture was also influenced by early difficulties weaving high-tenacity para-aramid filament yarns, and by the relatively low cost of plain-woven fabrics. Typically, the para-aramid fabrics are ~220 g/m^2; earlier nylon 6,6 fabrics were 290 g/m^2. The design of military body armour varies among countries and manufacturers, and details are usually classified. However, the literature suggests that the fragment protective element is typically constructed from between 15 and 40 layers of fabric and weighs up to 9 kg (Scott, 2005; Carr and Lewis, 2014).

In police body armour, para-aramid fabrics and UHMWPE are also typically used; sometimes a layer of para-aramid felt or piece of thick foam (typically ~3 mm) is included to reduce the back-face signature, which is the back-face deformation that occurs due to the impact event (see failure mechanisms) (Carr et al., 2013). Protection from sharp weapons is achieved by the use of very low linear density fibres/yarns and high-sett fabrics, mail, or by the use of laminated fabrics (Horsfall, 2012).

4.3 Composites

4.3.1 Introduction

A composite material can be defined as 'two or more physically distinct and mechanically separable materials…made by mixing the separate materials in such a way that the dispersion of one material in the other can be done in a controlled way to achieve optimum properties…the properties are superior, and possibly unique in some specific respects, to the properties of the individual components' (Hull, 1988). Composites are used in ballistic protective helmets, plates, EOD suits, ballistic shields, vehicles and some Force Protection applications.

4.3.2 Composite fibre architecture

Fibre architecture in composites used for ballistic protection has been influenced by fabric structure used in soft body armour design; thus many of the composites used for ballistic protection contain similar woven fabrics as those that are used in body armour. Thus, para-aramid and glass-based composites for ballistic protection generally contain plain-woven fabrics, usually combined with a phenolic poly-vinyl butyral or epoxy resin. In comparison, UHMWPE fibres are used in a cross-ply (0°/90°) lay-up with a thermoplastic resin to reduce fibre damage during production (Hockenberger, 1998), although early research did consider woven UHMWPE fibres (Carr, 1996; Morye et al., 1999). What is important is that a ballistic protective composite should have a high volume fraction of fibre with relatively weak inter-ply adhesion, that is, not necessarily a structurally good composite; matrix content may be as low as 10% w/w (Shephard, 1987; Brown and Egglestone, 1989; Hsieh et al., 1990; Lin et al., 1990; Segal, 1991). This is because delamination is the major mechanism by which energy is dissipated away from the impact point in ballistic protective composites (Shephard 1987; Morye et al., 1999; Sharma et al., 1999; Taylor and Carr, 1999; Cheeseman and Bogetti, 2003); failure mechanisms are discussed in Section 4.4. Processing parameters are important, but they are rarely explored in the open literature. The effect of hot-pressing time on the ballistic protective properties of Kevlar®-phenolic/PVB composites suggested that longer pressing times reduced the performance, presumably due to increased interlaminar bonding (Akdemir et al., 2008).

Limited information is provided in the peer-reviewed literature regarding optimum variables for composites used for ballistic protection; selected topics are discussed below.

4.3.2.1 Reinforcement type

The effects of fabric type (unidirectional, cross-ply, plain-woven, 3/1 twill, Fraglight® a needle punched felt), manufacturing route (hot compaction, pre-preg pressing) and fibre type (Dyneema®, Spectra®) on fragmentation protection of UHMWPE composites have been reported (Morye et al., 1999). Not all combinations were available for testing, particularly for Spectra® composites. Mass per unit area was normalised (4.7 kg/m^2) and ballistic protective performance measured. The fragment protective performance of 3/1 twill weave (press), unidirectional (hot compaction) and Fraglight® (hot compaction) composites was lower than other composites; failure was localised in the case of the pressed 3/1 woven fabric with minimal delamination. In the case of the unidirectional and Fraglight® hot compacted composites, the fibre easily separated during perforation. For the remaining composites the primary failure mechanisms appeared to be via fibre failure with some delamination. Composites containing nonwoven fibres demonstrated improved fragmentation protective performance over those containing woven fabrics, probably due to reduced interference of wave propagation.

4.3.2.2 Ply orientation

The ballistic impact performance of two S2-epoxy GF unidirectional ply lay-ups (cross-ply $[0°_3/90°_3]_{2S}$; quasi-isotropic $[0°_3/45°_3/90°_3]_S$) has been reported (Sevkat et al., 2009). Unfortunately, different velocities were used to impact the two specimen types. Cross-ply laminates were not perforated at velocities of 120 and 298 m/s; at 465 m/s perforation occurred. In comparison, quasi-isotropic laminates were perforated at 390, 423 and 442 m/s.

4.3.2.3 Hybrids

Hybrid solutions for improved impact performance have been discussed in the literature, often as a solution to improve the impact performance of carbon fibre composites, which is poor (Nettles and Lance, 1993; Larsson and Svensson, 2002; Muhi et al., 2009; Pandyr et al., 2012). The use of carbon fibre composites is of interest in a number of military applications when high strength or stiffness is required, but the performance of carbon fibre hybrids cannot compete with 100% UHMWPE or 100%

para-aramid composites. The use of hybrids as a possible solution for military helmets has been discussed (para-aramid/UHMWPE; nylon 6,6/UHMWPE; nylon 6,6/para-aramid) (Morye, 1998). Stacking sequence, represented by the incorporation of Kevlar® fabric in GFRP, affected the performance when tested with different shaped projectiles (Muhi et al., 2009). Impact resistance increased when the Kevlar® fabric layer was moved from the strike face to layer number four in the laminate; the authors related this to an increase in toughness. However, the strike velocity used during testing, 176 m/s, was slow compared to typical ballistic protective composite uses.

4.3.2.4 Through-thickness reinforcement

The performance of impact-tolerant composites can be improved through various through-thickness reinforcement mechanisms such as pinning, stitching, rods, 3D weaving, knitting, and braiding (Mouritz et al., 1997; Kuo and Lee, 1998; Jancar, 2000; Rugg et al., 2001; Mouritz, 2003; Cartié et al., 2004). These various mechanisms work by restricting delamination growth. However, in ballistic impacts, delamination formation is a major source of energy dissipation. Specifically, the mode I interlaminar fracture toughness is increased by through-thickness reinforcement. Mouritz demonstrated this effect for plain-woven GF composites stitched with Kevlar® 49 thread for perforating impacts 5.56 mm ammunition; delamination was reduced for stitched composites (Mouritz, 2001). As with fabrics, stitching can also be used to reduce back-face signature during a ballistic impact onto a composite (Ahmead et al., 2008).

4.3.2.5 Manipulation through thickness

The incorporation of an impendence layer caused by stacking two composites together rather than producing a single composite of equivalent thickness improved performance by reducing back-face delamination, suggesting that other failure mechanisms became dominant (Nunes et al., 2004).

4.3.2.6 Particulate reinforcement

Incorporation of a particulate reinforcement (eg, sand or conch shell) can reportedly increase ballistic performance (Sabet et al., 2008; Liu et al., 2010). Improved performance was attributed to increased fracture surfaces due to the reinforcement and crack stopping.

4.4 Failure mechanisms

Discussion of fabric and composite failure mechanisms resulting from a ballistic impact event is compounded by the different projectiles, materials, specimen dimensions and impact velocities used in work described in the literature.

The response of yarns, fabrics and composites to ballistic impact has been discussed in the peer-reviewed literature for over 60 years from both a theoretical and experimental perspective. A comprehensive summary of the key literature has recently been published; the pertinent points follow (Carr et al., 2012). A longitudinal wave is formed in the yarn due to the impact, whilst in the strained region a transverse wave forms (Smith et al., 1956). The waves interact and reflect at clamping points (or yarn crossovers in fabrics). When a woven fabric is impacted, longitudinal waves pass along the impacted warp and weft yarns and can result in yarn pull-out; a pyramidal or cone-shaped through-thickness deformation occurs (Cunniff, 1992). Fibre and yarn rupture occurs when the strain at the impact point exceeds the failure strain. At higher impact energies, fibres and yarns fail in a shear or plug mode (Carr, 1999).

A woven fabric pack has a higher ballistic protective level than a composite containing the same type and number of woven fabric layers due to the pinning of yarn crossover point by the matrix materials in the composite and enhanced reflection of waves (Shephard, 1987; Brown and Egglestone, 1989; Cheeseman and Bogetti, 2003). Fabrics and/or composites with no crossovers minimise wave reflection; thus noncrimp fabrics and cross-ply laminates are widely used to manufacture both soft body armour and composites that provide ballistic protection, and these structures provide improved levels of protection compared to woven fabric reinforcement (Morye et al., 1999). Improving the level of protection offers the possibility of the same protection at an overall reduced mass, or an improved level of protection at the same mass.

The addition of the matrix material to form the composite minimises the transient deformation (and hence reduces the contribution of that mechanism to failure) that occurs during the impact event; clearly this is of interest when protecting vulnerable structures such as the cranium. High-speed video demonstrates that this transient deformation is mostly elastic; a smaller plastic deformation is observed post-testing (Taylor and Carr, 1999; Morye et al., 2000). Modelling suggests that the nonperforating acceleration of this deformation is a major energy dissipation mechanism (Sharma et al., 1999; Morye et al., 2000). Transient deformation can be reduced by increasing the number of reinforcing plies used to manufacture the composite (Tarim et al., 2002; Carrillo et al., 2012).

Fibre/yarn/tow pull-out occurs on the strike and rear faces of the composite along with associated matrix cracking (Woodward et al., 1994; Taylor and Carr, 1999; Mouritz, 2001; Carrillo et al., 2012). Various fibre failure mechanisms have been described resulting from ballistic impacts on both fabrics and composites, for example, fibrillation of para-aramid and poly(*p*-phenylene-2,6-benzobisoxazole) fibres, and shear and melting of nylon 6,6 and UHMWPE fibres (Susich et al., 1958; Carr, 1999; Mitchell and Carr, 1999).

A major failure mechanism observed in composites providing ballistic protection is delamination; this is directly related to the poor mode I interlaminar fracture toughness, which is a desirable feature of such materials (Shephard, 1987; Woodward et al., 1994; Morye et al., 1999; Sharma et al., 1999; Morye et al., 2000). Material is compacted near the strike face, with the formation of a plug of composite material travelling ahead of the projectile (Sharma et al., 1999). The projectile slows during penetration of the composite and delaminations form more readily due to enhanced bending, particularly in thicker sections (Woodward et al., 1994; Gellert et al., 2000). Therefore, the delaminated area increases through the thickness of the composite towards the rear face (Woodward et al., 1994; Taylor and Carr, 1999; Gellert et al., 2000; Mouritz, 2001; Sevkat et al., 2009).

More recently, improved results have been reported in the literature by the use of a thermoplastic rather than a thermosetting resin; an associated change in failure mechanism was noted from fibre shear to tensile failure of fibres (Carrillo et al., 2012; Nayak et al., 2012). There is also some evidence that tougher resins lead to improved performance during nonperforating bullet impacts (Nunes et al., 2004). This is the reason that the first composite helmets contained a phenolic resin that was toughened with PVB.

Sources of further information and advice

Journals

Textile Research Journal
Journal of the Textile Institute
International Journal of Impact Engineering
Composites Science and Technology
Composites Part A

Conferences

Personal Armour Systems Symposia
International Symposia on Ballistics
Textile Institute World Congress
International Conference on Composite Materials
European Conference on Composite Materials

References

Ahmead, M.R., Ahmad, W.Y.W., Salleh, J., Samsuri, A., 2008. Effect of fabric stitching on ballistic impact resistance of natural rubber coated fabric systems. Materials and Design 29, 1353–1358.

Akdemir, A., Candan, C., Sahin, O.S., 2008. Effects of production parameters and conditioning upon ballistic characteristics of para aramid light armors. Journal of Composite Materials 42, 2051–2061.

Breeze, J., Baxter, D., Carr, D.J., Midwinter, M.J., 2013. Defining combat helmet coverage for protection against explosively propelled fragments. Journal of the Royal Army Medical Corps 161, 9–13.

Breeze, J., Gibbons, A.J., Shieff, C., Banfield, G., Bryant, D.G., Midwinter, M.J., 2011. Combat-related craniofacial and cervical injuries: a 5-year review from the British military. The Journal of Trauma Injury, Infection, and Critical Care 71. http://dx.doi.org/10.1097/TA.0b013e318203304a.

Brown, J.R., Egglestone, G.T., 1989. Ballistic properties of composite materials for personnel protection. Australia, Materials Research Laboratory, DSTO, Melbourne.

Carr, D.J., 1996. Assessment of UHMWPE composites as candidate materials for UK military helmets. In: European Composite Materials Conference-7, London, UK.

Carr, D.J., 1999. Failure mechanisms of yarns subjected to ballistic impact. Journal of Materials Science Letters 18 (7), 585–588.
Carr, D.J., Horsfall, I., Malbon, C., 2013. Is behind armour blunt trauma a real threat to users of body armour? A systematic review. Journal of the Royal Army Medical Corps. http://dx.doi.org/10.1136/jramc-2013–000161.
Carr, D.J., Lankester, C., Peare, A., Fabri, N., Gridley, N., 2012. Does quilting improve the fragment protective performance of body armour? Textile Research Journal 82, 1157–1163.
Carr, D.J., Lewis, E.A., 2014. Chapter 6 Ballistic protective clothing and body armour. In: Wang, F. (Ed.), Protective Clothing: Managing Thermal Stress. Woodhead Publishing and The Textile Institute, Manchester, UK.
Carr, D.J., Starling, G., de Wilton, T., Horsfall, I., 2014. Tensile properties of military chin-strap webbing. Textile Research Journal 84, 655–661.
Carrillo, J.G., DGamboa, R.A., Flores-Johnson, E.A., Gonzalez-Chi, P.I., 2012. Ballistic performance of thermoplastic composite laminates made from aramid woven fabric and polypropylene matrix. Polymer Testing 31, 512–519.
Cartié, D.D.R., Conx, B.N., Fleck, N.A., 2004. Mechanisms of crack bridging by composite and metallic rods. Composites Part A 35 (article in press 10 September 2004).
Cheeseman, B.A., Bogetti, T.A., 2003. Ballastic impact into fabric and compliant composite laminates. Composite Structures 61 (1–2), 161–173.
Croft, J., Longhurst, D., 2007a. HOSDB Body Armour Standards for UK Police (2007) Part 2: Ballistic Resistance Publication No. 39/07/B. Home Office Scientific Development Branch, Sandridge, St Albans.
Croft, J., Longhurst, D., 2007b. HOSDB Body Armour Standards for UK Police. Part 3: Knife and Spike Resistance. Home Office Scientific Development Branch, St Albans, United Kingdom.
Cunniff, P.M., 1992. An analysis of the system effects in woven fabrics under ballistic impact. Textile Research Journal 56, 45–60.
Dixon, C., Croft, J., 2007. Body Armour Good Practice and Quality Framework Publication No. 44/07. St Albans, Home Office Science.
Gellert, E.P., Cimpoeru, S.J., Woodward, R.L., 2000. A study of the effect of target thickness on the ballistic resistance of glass reinforced composites. International Journal of Impact Engineering 24, 445–456.
Hockenberger, A.S., 1998. The effects of weaving on fibre performance in ballistic fabrics made from high-performance polyethylene fibres. Journal of the Textile Institute 89 (3), 595–604.
Horsfall, I., 2012. Key issues in body armour: threats, materials and design. In: Sparks, E. (Ed.), Advances in Military Textiles and Equipment. Woodhead Publishing and The Textile Institute, Manchester, UK.
Hsieh, C.Y., Mount, A., Jang, B.Z., Zee, R.H., 1990. Response of polymer composites to high and low velocity impact. In: 22nd International SAMPE Technical Conference. SAMPE, Boston, MA.
Hull, D., 1988. An Introduction to Composite Materials. Press Syndicate of the University of Cambridge, Cambridge, UK.
Jancar, J., 2000. Impact behaviour of short glass fiber reinforced thermoplastic polyurethane. Polymer Composites 21 (3), 369–376.
Kuo, W.-S., Lee, L.-C., 1998. Impact response of 3-D woven composites reinforced by consolidated rods. Polymer Composites 19 (2), 156–165.

Larsson, F., Svensson, L., 2002. Carbon, polyethylene and PBO hybrid fibre composites for structural lightweight armour. Composites Part A 33, 221–231.

Lewin, M., Pearce, E.M. (Eds.), 1985. Handbook of Fiber Science and Technology: Volume IV Fiber Chemistry. Marcel Dekker, Inc, New York, USA.

Lewis, E.A., 2006. Between Iraq and a hard plate: recent developments in UK military personal armour. In: Personal Armour Systems Symposium 2006. The Royal Armouries, Leeds, UK.

Lin, L.C., Bhatnager, A., Chang, H.W., 1990. Ballistic energy absorption of composites. In: 22nd International SAMPE Technical Conference. SAMPE, Boston, MA.

Liu, S., Wang, J., Wang, Y., Wang, Y., 2010. Improving the ballistic performance of ultra high molecular weight polyethylene fiber reinforced composites using conch particles. Materials and Design 31, 1711–1715.

Marsden, A., 1994. UK body armour and helmets. In: Personal Armour Systems Symposium (PASS94), Colchester, UK.

Mills, N.J., 1986. Plastics Microstructure, Properties and Applications. Edward Arnold, London, UK.

Mitchell, C.A., Carr, D.J., 1999. Post failure examination of a new body armour textile by the use of an environmental scanning electron microscope. In: Institute of Physics Conference Series, vol. 161, pp. 103–106.

Morton, W.E., Hearle, J.W.S., 1997. Physical Properties of Textile Fibres. The Textile Institute, Manchester, UK.

Morye, S.S., 1998. High Performance Polymer Composites for Ballistic Protection (Ph.D. thesis). University of Leeds.

Morye, S.S., Hine, P.J., Duckett, R.A., Carr, D.J., Ward, I.M., 1999. A comparison of the properties of hot compacted gel-spun polyethylene fibre composites with conventional gel-spun polyethylene fibre composites. Composites Part A 30, 649–660.

Morye, S.S., Hine, P.J., Duckett, R.A., Carr, D.J., Ward, I.M., 2000. Modelling of the energy absorption by polymer composites upon ballistic impact. Composites Science and Technology 60, 2631–2642.

Mouritz, A.P., 2001. Ballistic impact and explosive blast resistance of stitched composites. Composites Part B 32, 431–439.

Mouritz, A.P., 2003. Comment on the impact damage tolerance of stitched composites. Journal of Materials Science Letters 22, 519–521.

Mouritz, A.P., Leong, K.H., Herszberg, I., 1997. A review of the effect of stitching on the in-plane mechncial properties of fibre-reinforced polymer composites. Composites Part A 28A, 979–991.

Muhi, R.J., Najim, F., de Moura, M.F.S.F., 2009. The effect of hybridization on the GFRP behavior under high velocity impact. Composites Part B 40, 798–803.

Nayak, N., Sivaraman, P., Banerjee, A., Madhu, V., Dutta, A.L., Mishra, V.S., Chakraborty, B.C., 2012. Effect of matrix on the ballistic impact of aramid fabric composite laminates by armor piercing projectiles. Polymer Composites 33, 443–450.

Nettles, A.T., Lance, D.G., 1993. On the enhancement of impact damage tolerance of composite laminates. Composites Engineering 3 (5), 383–394.

Nunes, L.M., Paciornik, S., d'Almeida, J.R.M., 2004. Evaluation of the damaged area of glass-fiber-reinforced epoxy-matrix composite materials submitted to ballistic impacts. Composites Science and Technology 64, 945–954.

Pandyr, K.S., Dharmane, L., Pothnis, J.R., Ravikumar, G., Naik, N.K., 2012. Stress wave attenuation in composites during ballistic impact. Polymer Testing 31, 261–266.

Richards, A.F., 2004. In: McIntyre, J.E. (Ed.), Nylon Fibres. Synthetic Fibres: Nylon, Polyester, Acrylic, Polyolefin. Woodhead Publishing Ltd./The Textile Institute, Manchester.
Rugg, K.L., Cox, B.N., Massabo, R., 2001. Mixed mode delamination of polymer composite laminates reinforced through the thickness by z-fibers. Composites Part A: Applied Science and Manufacturing 33A (2), 177–190.
Ryan, J.M., Cooper, G.J., Haywood, I.R., Milner, S.M., 1991. Field surgery on a future conventional battlefield: strategy and wound management. Annals of the Royal College of Surgeons of England 73 (1), 13–20.
Sabet, A.R., Beheshty, M.H., Rahimi, H., 2008. High velocity impact behavior of GRP panels containing coarse-sized sand filler. Polymer Composites 29, 932–938.
Scott, R.A. (Ed.), 2005. Textiles for Protection. Woodhead Publishing Limited, Cambridge, UK.
Segal, C.L., 1991. High performance organic fibres, fabrics and composites for soft and hard armour applications. In: 23rd International SAMPE Technical Conference. Kiamesha Lake, NY.
Sevkat, E., Liaw, B., Delale, F., Raju, B.B., 2009. A combined experimental and numerical approach to study ballistic impact response of S2-glass fiber/toughened epoxy composite beams. Composites Science and Technology 69, 965–982.
Sharma, N., Carr, D.J., Kelly, P.M., Viney, C., 1999. Modelling and experimental investigation into the ballistic behaviour of an ultra high molecular weight polyethylene/thermoplastic rubber matrix composite. In: 12th International Conference on Composite Materials (ICCM-12) Paris, France.
Shephard, R.G., 1987. The Use of Polymers in Personal Ballistic Protection. Polymers in Defence. The Plastics and Rubber Institute, London.
Shim, V.P.W., Tan, V.B.C., Tay, T.E., 1995. Modelling deformation and damage characteristics of woven fabric under small projectile impact. International Journal of Impact Engineering 16 (4), 585–605.
Smith, J.C., McCrackin, F.L., Schiefer, H.F., Stone, W.K., Towne, K.M., 1956. Stress-strain relationships in yarns subjected to rapid impact loading: part IV: transverse impact tests. Textile Research Journal 26, 821–828.
Susich, G., Dogliotti, L.M., Wrigley, A.S., 1958. Microscopical study of a multilayer nylon body armour panel after impact. Textile Research Journal 28, 361–377.
Tarim, N., Findik, F., Uzun, H., 2002. Ballistic impact performance of composite structures. Composite Structures 56, 13–20.
Taylor, S.A., Carr, D.J., 1999. Post failure analysis of 0°/90° ultra high molecular weight polyethylene composite after ballistic testing. Journal of Microscopy 196 (2), 249–256.
Tobin, L., Iremonger, M., 2006. Modern Body Armour and Helmets: An Introduction. Argros Press, Canberra, Australia.
Woodward, R.L., Egglestone, G.T., Baxter, B.J., Challis, K., 1994. Resistance to the penetration and compression of fibre-reinforced composite materials. Composites Engineering 4, 329–341.

Ballistic damage of hybrid composite materials

5

M. Al-Haik
Embry-Riddle Aeronautical University, Daytona Beach, FL, United states

A.Y. Borujeni
Virginia Tech, Blacksburg, VA, United States

M. Tehrani
University of New Mexico, Albuquerque, NM, United States

5.1 Introduction

Layered structural composites with polymer matrix are frequently reinforced with glass, carbon, and Kevlar fibers to give the additional strength and stiffness. These composites have been used in many structures (eg, vehicles and airplanes) for their significantly high specific stiffness and strength and their manufacturing flexibility, enabling achievement of properties difficult to attain with single-phase materials. However, typical structural composites have also been known for their limited deformability and ductility, which is related to the poor composite energy absorption to failure (Langdona et al., 2007). Limited energy absorption by composites was related to the absences of plasticity mechanism and the dominant debonding at weak fiber/matrix interfaces (Tsai and Prakash, 2005). These limitations necessitate developing efficient composite design methodologies to enhance the composite behavior under high strain rate loading environments such as blast events. The combination of elastic fibers in a viscoelastic matrix restricts the total energy absorption of the composite under high-energy environments during blast/impact events. The amount of energy absorption in the composite is strongly dependent on the fracture mechanism of the composite material (Zhandarov et al., 2002; Shyr and Pan, 2003). Taha and Shrive (2001) showed experimentally that controlling the fibers' aspect ratio can enhance the energy absorption characteristics of structural composites. This is based on the fact that the critical fiber length (l_c) is the fiber length beyond which fiber will be ruptured rather than debonded. The critical fiber length (l_c) can be calculated as

$$l_c = \frac{\sigma_f \ d}{2 \ \tau_u} \qquad [5.1]$$

where σ_f is the fiber rupture strength, d is the fiber diameter and τ_u is the ultimate shear strength of fibers embedded in the matrix. If fibers shorter than l_c are used in the matrix, fiber debonding will be the dominant mechanism. Kau (1990) showed that the fiber length in composite could alter the failure mode. For shorter fibers, the

Advanced Fibrous Composite Materials for Ballistic Protection. http://dx.doi.org/10.1016/B978-1-78242-461-1.00005-4

damage consisted of an even mixture of fiber breakage and fiber pull-out (under dead weight impact long at 2 m/s), while for a composite comprising an equal number of longer and shorter fibers the damage shifted predominantly to breakage of the longer fibers.

Moreover, researchers showed that the total energy absorbed by the composite up to fracture (E_f) for fibers shorter than or equal to l_c can be computed as (Kelly, 1973)

$$E_f = \frac{1}{24}\pi d \tau_u l^2 \qquad [5.2]$$

where l is the fiber length used in the composite. Therefore, the energy absorbed to failure by a fiber-reinforced polymeric (FRP) can be maximized by increasing the shear strength between the fiber and the matrix. The inelastic properties of fiber composites both in the fiber direction and transverse to the fiber are significantly different and depend on the level of fiber-matrix interfacial shear strength (Daniel and Ishai, 2005; Madhukar and Drzal, 1992a). While the fiber strength–dominated properties of the composite (such as the longitudinal tensile, compressive and flexural strength) showed only moderate sensitivity to changes in interfacial shear strength, the off-axis strength properties of the composite (such as the transverse tensile and flexural strength properties, in-plane and interlaminar shear strength) were shown to be highly sensitive to the fiber-matrix adhesion level. Furthermore, modes I and II fracture toughness also changed significantly with varying the fiber-matrix adhesion (Madhukar and Drzal, 1992b).

The mechanisms of fiber-matrix adhesion include chemical bonding, van der Waals bonding, and mechanical interlocking. Chemical bonding offers the largest bonding force, provided that the density of chemical bonds across the fiber-matrix interface is sufficiently high. This density can be increased by chemical treatments of the fibers or by enhancing the chemical interaction at the interface through increasing the fiber surface area. On the other hand, mechanical interlocking between the fibers and the matrix can significantly contribute to the fiber-matrix adhesion if the fibers can form a 3D network (Chung, 1994). In addition to the role of the fiber bridging effect in toughening the composite matrix, the ability of the fibers to enhance the FRP stiffness helps to increase the fracture toughness of the composite for the direct relation between fracture toughness K_{IC} and the material stiffness E as follows:

$$K_{IC} = \alpha\left(\frac{E}{H}\right)^{1/2}\left(\frac{P}{c^{3/2}}\right) \qquad [5.3]$$

where H is its hardness and P and c are the applied load and radial crack width in Vickers hardness test, respectively (Abrate, 1994). The previous expression does not account for the fiber-bridging effect explained by Eqs. [5.1] and [5.2]. Nevertheless, Eq. [5.3] states the direct relation between fracture toughness and stiffness in materials. The effect of fiber stiffness on the composite stiffness is limited in most exiting composites because of the small volume fraction of fibers in composites and the relatively small difference between the stiffness of the fibers and the matrix (eg, carbon fibers are only 10 times stiffer than cement matrix and usually constitute 2–3% in volume).

The dominant failure mechanisms in composite laminates subjected to impact loading comprise complex combinations of energy absorption mechanisms such as delamination predominantly caused by mode-II shear, matrix cracking caused by transverse shear, and translaminar fracture in terms of fiber fracture and kinking (Abrate, 1998). The mechanical properties of the fibers and matrix (particularly the failure strains, interface properties, fiber configuration, and stacking sequence in angle-ply laminates) play important roles in determining the impact damage resistance of composites (Lopresto et al., 2006; Bibo and Hogg, 1996). Response of conventional laminated structures to blast loading has been studied (Fleck and Deshpande, 2004), but experimental observations on the real-time response of composite and layered structures under blast are limited (Tekalur et al., 2008). Significant improvement in energy absorption of multilaminate FRPs was reported utilizing plain stitching (Kang, 2004). Several investigations were devoted to developing computational tools to simulate composite behavior under impact loads (Morozov et al., 2003; Batra and Hassan, 2008; Donadona et al., 2008).

A great body of research has been devoted to improving the through-thickness properties of FRPs (Trabelsi et al., 2010; Khan and Kim, 2011). Improvements in the interlaminar properties of FRPs was achieved via different approaches such as stitching (Trabelsi et al., 2010), braiding (Gu and Li, 2005), fiber surface treatment (Chung, 2003), and interleaving with toughened polymers (Aksoy and Carlsson, 1992). However, these enhancements are achieved at the expense of inadequate in-plane mechanical properties. On the other hand, utilizing nanomaterials as reinforcements (Vlasveld et al., 2005; Hudnut and Chung, 1995) offers the opportunity to enrich the fiber-matrix interface and matrix-dependent properties of FRPs with minimal weight penalty.

5.2 Three-phase hybrid composites

Utilizing nanoscale reinforcements in FRPs (Ajayan et al., 2003) offers the opportunity to enrich the fiber-matrix interface and enhance the matrix-dependent properties with minimal weight penalty. The extraordinary mechanical, electrical, and thermal properties of carbon nanotubes (CNTs) (Bernholc et al., 2002; Dresselhaus et al., 2001; Endo et al., 1997; Gojny et al., 2004; Treacy et al., 1996; Miyagawa and Drzal, 2004) have motivated researchers to utilize them as a filler phase in composite materials to improve the properties of the host matrix (Gorga and Cohen, 2004; Papakonstantopoulos et al., 2005; Stephan et al., 2000).

5.2.1 Hybrid composites based on matrix reinforcement with nanofillers

The extraordinary mechanical properties of carbon nanotubes (CNTs) have motivated researchers to utilize them as a filler phase in composite materials to improve the mechanical properties of the host matrix. The strength, elastic modulus, and the fracture properties of CNTs are orders of magnitudes higher than most common composites used in civilian and military applications. For example, a cast composite film of polystyrene and carbon nanotubes (5.0% vol.) exhibits a 100% increase in modulus and 25% increase in the strength of polystyrene (Thostenson and Chou, 2002). Moreover, CNTs reinforcements increase the toughness of polymer matrices enabling them to absorb higher impact energies (Dalton et al., 2003; Gojny et al., 2004; Gorga and Cohen, 2004).

To benefit from the remarkable properties of both carbon fiber–reinforced plastic (CFRPs) and CNTs, multiscale, three-phase, carbon fiber/carbon nanotube-reinforced polymer composites were developed (Khan and Kim, 2011; Qian et al., 2010). Several methodologies for integrating CNTs into FRPs have been investigated: grafting, attachment, or growth of CNTs on the fibers (Luhrs et al., 2009; Safdari and Al-Haik, 2012; Tehrani et al., 2013a,c; Boroujeni et al., 2014; Tehrani et al., 2014), spraying of CNTs on prepregs (Joshi and Dikshit, 2012), electrophoretic deposition of CNTs on fabrics (An et al., 2012), addition of CNTs to the polymer sizing formulation of fibers (Warrier et al., 2010), and dispersing the CNTs in the polymer matrix. Among all these approaches, mixing CNTs with the matrix in a composite is, to date, the most facile and economically compatible route.

Much interest has been directed to examining the impact strength of hybrid CNT–FRP composites. Soliman et al. (2012a,b) incorporated pristine and functionalized CNTs by dispersing them into the matrix polymer of a CFRP and showed improvement in the matrix-dominated properties such as interlaminar shear strength (ILSS) and low-velocity transverse impact resistance. Functionalized CNTs have been shown to produce even higher ILSS and impact resistance than pristine CNTs due to stronger interfacial bonding between the CNTs and the matrix. In an experimental study by Inam et al. (2010), the addition of 0.1 wt% CNTs to the matrix of a glass fiber–reinforced composite improved the flexural modulus by 35% and the flexural strength by 5%. Improvement on the absorbed impact energy was reported to be 6%, and the mode-I interlaminar toughness was reduced by 23%. The effect of a higher loading of CNTs (ie, 0.5 wt%) on the impact properties of a CFRP was studied by Kostopoulos et al. (2010). He showed for low-impact energies (2–8 J) that no significant effects on energy absorption and the delamination area were observed. However, at higher impact energies (12–20 J), moderate improvement on the absorbed energy was reported. Moreover, the delamination area for the CNT-reinforced FRPs was less than that for composites with no CNTs. Safdari and Al-Haik (2012) showed—computationally—that the addition of CNTs to the polymer matrix of a laminated FRP can significantly increase the impact resistance of the composite. Grujicic et al. (2008a,b) reported improvement on the impact resistance of CNT-epoxy matrix composite reinforced with E-glass fiber mats as a result of adding a multiwalled carbon nanotube (MWCNT) layer sandwiched in between the lamina. Improvements in the out-of-plane mechanical properties of the CNTs comprising FRPs could be attributed to the additional energy required for pull-out or rupture of the CNTs, CNT toughened matrix, and stronger fiber-matrix adhesion as a result of the CNTs addition.

5.2.2 Hybrid composites based on fibers sheathed with nanofibers

The dispersion of CNTs and eliminating the inherent tendency of CNTs to form agglomerates are two major problems associated with mixing CNTs with polymers (Advani and Fan, 2007). Alternatively, growing CNTs on desired surfaces and infusing them with a polymer has overcome these issues. CNTs can be grown over different substrates such as silicon, silica, and alumina (Yamamoto and Wardle, 2008). To make use of surface-grown CNTs in carbon FRPs (CFRPs), CNTs should be directly grown on carbon fiber surfaces. Nevertheless, in situ growth of CNTs on fibers gives rise to several complications. The first challenge is that the transition metals, used as catalyst for CNT synthesis, can easily diffuse into the fiber substrate. The second challenge stems from similar environments and conditions in which different carbon structures, rather than CNTs, can be grown (Zhu et al., 2003). Catalytic chemical vapor deposition (CCVD) has been used to grow carbon nanofilaments and CNTs on carbon fiber utilizing nickel, iron, cobalt, or palladium as the catalyst (Gibson, 2010; Sinnott and Andrews, 2001; Luhrs et al., 2009; Al-Haik et al., 2010; Boskovic et al., 2005; Chen et al., 1999; Garcia et al., 2008; Phillips et al., 2007). The operation temperature range for CCVD is between 700°C and 1200°C, making CCVD destructive to the carbon fiber substrate itself (Zhang et al., 2009; Westwood et al., 1996), thus resulting in unexpectedly low values for mechanical properties of the CFRP. For example, Thostenson and Chou (2002) utilized the CCVD method to grow CNTs on carbon fibers at 700°C and fabricated composites based on them. They reported significant degradation in matrix/fiber interface due to exposure to the elevated temperature growth conditions. In a similar work, Zhang et al. (2009) used the CCVD method to grow CNTs on a polyacrylonitrile (PAN)-based carbon fiber fabric (T650) substrate at 700–800°C and reported a 40% reduction in the on-axis strength of the manufactured composites. However, while the fiber-dominated mechanical properties decrease due to elevated temperatures of the CCVD process, matrix-dominated properties, for example, ILSS, showed improvement (Askari and Ghasemi-Nejhad, 2011; Zhu et al., 2003). In spite of being destructive to the carbon fibers, growing CNTs via CCVD on temperature-resistant fibers has been reported to achieve substantial improvements in mechanical properties. Veedu et al. (2006) grew CNTs on SiC woven yarns, which are highly resistant to high temperatures, using the CCVD method. They reported improvements of 348% in mode-I fracture toughness, 140% in flexural strength, 424% in flexural toughness, and 514% in damping capability of the hybrid composites compared to the ones made from raw SiC fabrics.

To minimize the destructive effect of elevated temperatures on carbon fibers during the CCVD process, a recently developed method—graphitic structure by design (GSD)—has been proposed for growing MWCNTs over carbon fibers at 550°C (Phillips et al., 2007; Tehrani et al., 2013c).

5.3 Energy absorption of hybrid composites

5.3.1 Composites based on reinforced epoxy matrix

Commercial high-strength PAN-based plain-woven carbon fabric in 3000-filament count tows (HexTow AS2C supplied by Hexcel Inc.) was utilized as the prime reinforcement. The chosen matrix material was Aeropoxy™ manufactured by PTM&W Industries, Inc. Aeropoxy is a medium viscosity, unfilled, light amber laminating resin that is designed for structural production applications. This resin, brand name PR2032, laminates very easily and wets out fiberglass, carbon, and aramid fibers readily. As recommended by the manufacturer, PR2032 mixed with PH3665 hardener cures at 65°C in 4 h. The authors have used this epoxy system to manufacture both FRPs (Al-Haik et al., 2004, 2006) and nanocomposites based on SWCNT (Garmestani et al., 2003) and MWCNT (Tehrani et al., 2011). The typical properties of this epoxy system are listed in Table 5.1.

Short MWCNTs (0.5–2.0 μm) from Cheaptubes, Inc., with a purity of 95 wt% and an average diameter of 8 nm were utilized as the nanofillers, Fig. 5.1.

The MWCNTs accounted for 2.0% of the total weight of the matrix. The CNTs were dispersed in the PH3665 epoxy hardener by an IKA Ultra-Turrax T-18 basic shear mixer-dispersion unit. The unit was set to the maximum speed (24,000 rpm) and run for 30 min. Next, the CNT-hardener mixture was homogenized via a Vibra-Cell VCX 500 tip ultrasonic processor for a total of 30 min at amplitude of 40% and a 2-s on/off pulse (the elapsed time was 1 h). The PR2032 epoxy resin was then added to the mixture and was briefly stirred manually. The slurry was mixed for 1 min with the dispersing unit followed by 1 min of mixing with the ultrasonic processor. This mixture constituted the matrix used to fabricate CFRPs with CNT-enhanced epoxy matrix. Without introducing the CNTs, the hardener and resin were subjected to the same procedure from which the reference samples with neat epoxy matrix were produced. To assess the degree of CNT dispersion, a small amount of the hardener-MWCNT-resin mixture was poured into a plastic container and cured overnight at room temperature. Dispersion and homogenization of the MWCNTs through extensive shear mixing and sonication, respectively, resulted in a satisfactory

Table 5.1 Properties of the aeropoxy system

Component	PR2032 epoxy resin	PH3660 curing agent
Viscosity of the mix	0.90–0.95 N s/m^2	
Mix ratio by weight	100	27
Density of the mix	11.09 kg/L	
Tensile strength	67.76 MPa	
Tensile modulus	2.885 GPa	

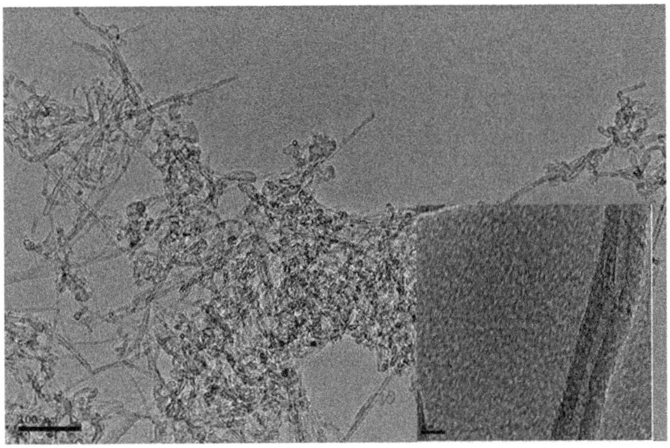

Figure 5.1 TEM image of MWCNTs bundles.nset: high-resolution TEM image showing an MWCNT with an average diameter of 20 nm. The scale bar reads 10 nm.

global dispersion, while slight local agglomerations were observed (see Fig. 5.2(a) and (b)). The fractured surface shown in Fig. 5.2(b) reveals the individualization of MWCNTs bundles. Cross-sectional micrographs of the CFRPs with and without the MWCNTs are shown in Fig. 5.3. It is observed that the matrix entirely infused in between the fibers and impregnated them. However, compared to the referenced neat epoxy sample, few voids exist in the MWCNT-enriched matrix. This is due to the high MWCNTs loadings (ie, more than 2.0 wt%) that lead to an excessively viscous mixture that does not wet and/or impregnate the fibers and, hence, when subjecting the composite to in-plane displacements, stresses will not be effectively transferred to the fibers, and the overall structural performance of the composite will be compromised.

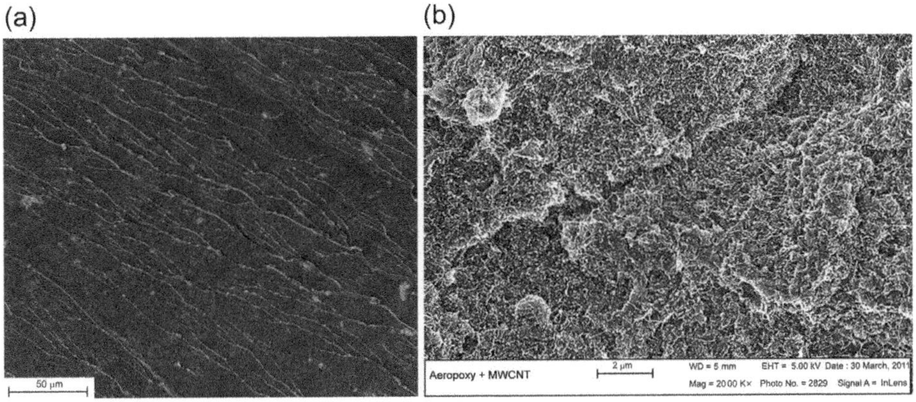

Figure 5.2 SEM micrographs for the fracture surface of MWCNT-epoxy composite sample: (a) global dispersion and (b) local dispersion (Tehrani et al., 2013a).

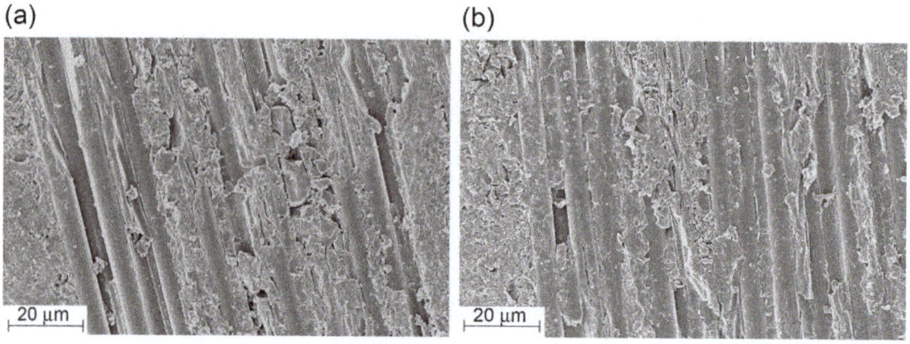

Figure 5.3 SEM micrographs for the cross-section of the CFRP containing MWCNTs (a) and CFRP based on neat epoxy (b) (Tehrani et al., 2013a).

For the purpose of mechanical testing, a set of flat, five-layer laminates of 12.5 × 12.5 cm were manufactured using vacuum- and press-assisted hand lay-up process. After 4 h curing at 65°C the composites panels were removed out of the mold and were left at room temperature overnight for full curing. Abraded G-10 tabs were bonded to the ends of the tensile specimens using the same resin (ie, Aeropoxy). The tensile test coupons of 12.5 × 1.3 cm were cut using a saw. Without any further tailoring, other panels were left for impact testing, whereas cut squares of 4.0 × 4.0 cm were used for the shear punch test. The tensile strength, Young's modulus, and the strain to failure for the specimens were measured from the tension tests utilizing an Instron® 4400R frame. The tensile tests were performed under constant crosshead speed of 2.0 mm/min until failure occurred. Strains up to the failure were recorded via an extensometer. To determine the tensile properties, a total of 10 samples for each configuration were tested. As shown in Fig. 5.4, both composite

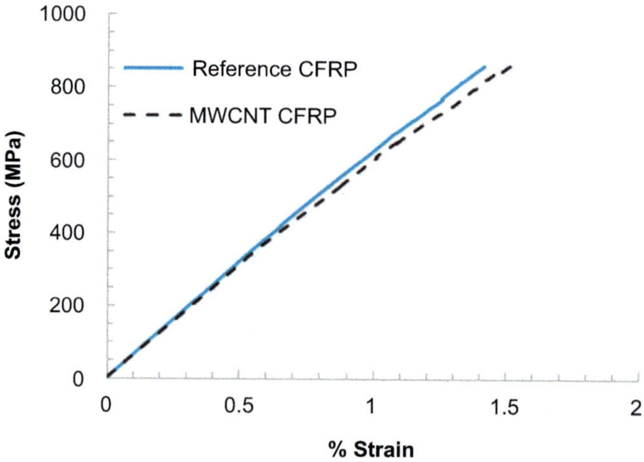

Figure 5.4 Representative stress–strain curves for the composites comprising neat epoxy and CNT-epoxy matrices (Tehrani et al., 2013a).

systems demonstrated a brittle tensile behavior and did not undergo significant plastic deformation. The tensile curve for the composite comprising MWCNTs deviates from the elastic behavior and exhibits a nonlinear deformation. This effect can be attributed to the failure of the MWCNT–polymer interface and the pull-out and rupture of the MWCNTs due to extensive shear stresses exerted on the MWCNTs by the matrix. While a very slight increase in the Young's modulus and the strength were achieved, the strain to failure was improved by 12%. The tensile modulus and strength of the CFRPs are dominated by their fiber phase properties, and hence the improvements in the stiffness and strength of the matrix slightly affect the overall tensile properties. The MWCNTs act as stiff barriers that inhibit cracks within the matrix from progressing, and thus the matrix can carry loads even after cracks are initiated. In addition, the enhanced fiber-matrix bonding promotes the load transfer to the fibers and leads to an impeded fiber-matrix interface failure and, hence, a higher strain to failure is achieved. The enhanced adhesion between the fiber and the matrix was due to the relief of stress concentrations at the fiber-matrix interface due to the existence of CNTs. While the strain to failure for the single dry fibers is reported by the manufacturer to be 1.9%, the reference composite exhibited a 1.38% strain to failure. This value was increased to 1.55% for the samples based on the CNT-epoxy matrix.

To conduct the impact tests, an in-house design of a small single-stage gas gun capable of projecting a 17 g impactor at speeds up to 500 m/s was utilized (Hartman et al., 2010). Each composite panel was impacted using a 1.27-cm alloyed steel ball bearing with a weight of approximately 8.4 g. A 1.30-cm drill bit was used to create holes in a polycarbonate sabot, in which the impactor was placed. As schemed in Fig. 5.5, purified nitrogen gas (*A*) fills a reservoir (*C*) through a hand-operated needle valve (*B*) to a desired pressure. Once the trigger mechanism (*D*) is opened, the projectile is pushed through the barrel (*E*), and the sabot is stripped off via a stripper plate at the muzzle. The impactor impacts the specimen (*F*), and it stops inside a steel containment trap (*G*).

The trajectory of the projectile was captured through a high-speed camera. The trigger for the camera was activated using a laser–photodiode pair. The output of the photodiode is connected to an operational amplifier, then a voltage comparator, so that the photodiode output is converted into a digital signal. The high-speed camera equipped with a wide lens was operated at 30,000 frames per second and a shutter speed of 9.6 μs. The images captured by the camera were subsequently analyzed to calculate the incident and residual velocities of the impactor. While extending the boundaries of the sample away from the center of the impact lessens the role of the boundary condition in the test, a large specimen size is required. However, for testing stiff composite panels at a high impact velocity of 100 m/s, it was observed that the perforation and damage are very local phenomena and are almost not affected by the far-field boundary conditions. Square panels with edge length of 12.5 cm were sandwiched within a frame fixture made of two 2.54-cm-thick steel plates. Using simple bolt calculations and a bolt torque of 80 N m, a clamping pressure of approximately 37.0 MPa was applied to the panel edges. The samples were shot with a steel impactor at velocities ranging from 90 to 110 m/s, and the absorbed kinetic energy during the impact was calculated from the incident and residual velocities. An image processing

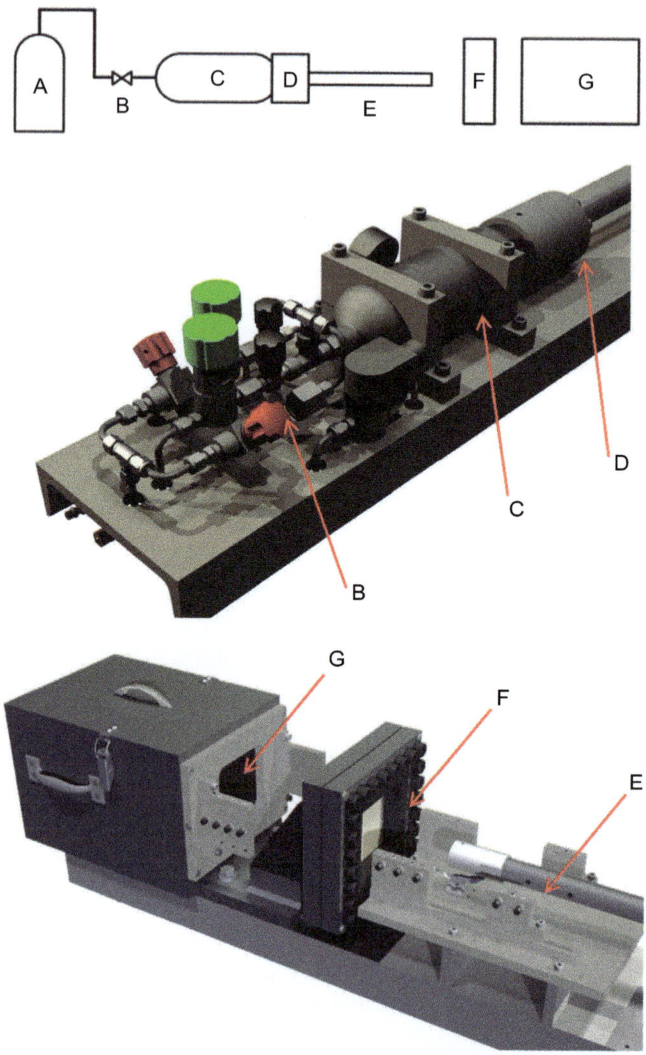

Figure 5.5 Schematic of the gas gun setup.

code was developed in MATLAB® to track the centroid of the impactor during the impact experiment. The image processing allowed for measuring the impactor speed with a high accuracy. As shown in Fig. 5.6, the impactor images acquired at 30,000 frames per second are well defined, and the image processing allowed for accurate tracking of the centroid of the spherical impactor. Four samples for each FRP configuration were tested for the impact energy absorption measurements.

The kinetic energy of the steel impactor was calculated before and after it perforated the composite panels. The average absorbed energies are plotted in Fig. 5.7. These averaged values suggest that the incorporation of MWCNTs in the matrix of a

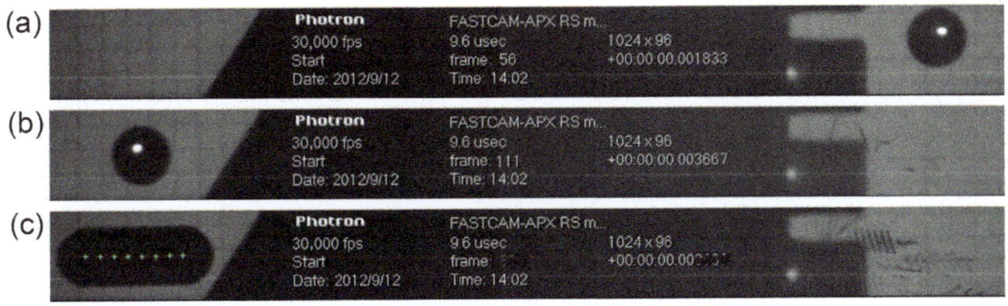

Figure 5.6 The impactor captured at 30,000 frames per second rate (a) before and (b) after the impact. (c) Image-processed cumulative images used for calculating the residual velocity.

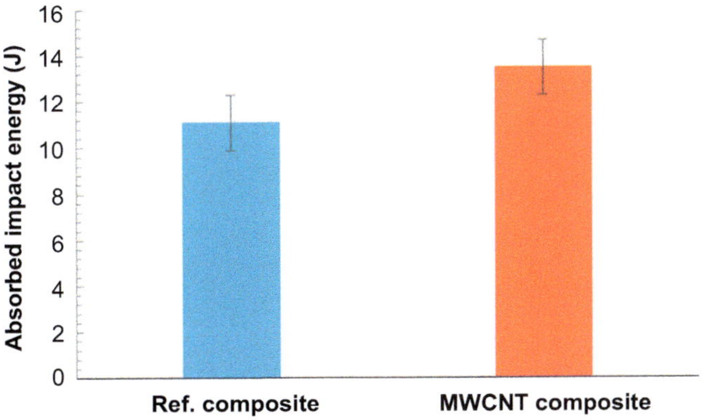

Figure 5.7 Absorbed impact energies for the composites based on neat epoxy and MWCNT-epoxy matrices (Tehrani et al., 2013a).

CFRP enhances the energy absorption capacity of the composites by almost 21.3%. This enhanced impact resistance can be attributed to the increased interlaminar fracture toughness and delamination resistance that alleviates the crushing/buckling strength of the fibers.

In addition, the presence of the MWCNTs delays the delamination of the plies and prevents the expansion of microcracks. Due to the high surface area of the MWCNTs, failure of the MWCNT-epoxy interphases and MWCNTs pull-out dissipate a considerable amount of energy.

Upon establishing the contact of the high-velocity impactor with the composite panel, stress waves propagate through the panel. The amplitude of these waves changes due to the anisotropy and inhomogeneity of the composite at macro-, micro-, and nanoscales. These waves are attenuated when passing through volumes with a high damping property. Impact involves the frictional interactions of different constituents of the composite that generate considerable amounts of heat which does not conduct in time scales of the impact (microseconds in our case). This accumulated heat causes a local temperature rise and gives rise to the viscous response from the constituents of the composite (especially polymer matrix). In addition, the high temperatures can activate the sliding nanotube-matrix energy dissipation mechanisms. Therefore, CNTs promote the attenuation of stress waves and dissipation of energy.

5.3.2 Composites based on carbon fibers sheathed with MWCNTS

Plain-weave, unsized SGP196 carbon fiber (IM7-GP) fabric with 6000-filament counts in a tow, provided by Hexcel Inc., was used as the carbon fiber reinforcement. According to the manufacturer, to attain a thinner, more closed and less porous fabric, PrimeTex™ ZB finishing was performed on the fabric. Square samples of 13.0 × 13.0 cm were cut from the as-received fabric. Different surface treatments were performed on the cut fabrics to delineate the effects of different aspect of the growth process on the carbon fiber. To shield the fabrics with the thermal barrier coating (TBC), a 75-nm-thick layer of SiO_2 was sputtered on some fabrics. For the fabrics to be used as substrates for growth of the CNTs, a 2-nm-thick film of nickel, the catalyst metal, was sputtered on these fabrics. To prepare samples with patterned growth of CNTs, for example, a checkerboard pattern, a polyester meshed mask with a thread size of 76 μm, an opening size of 140 μm and subsequently 42% open area was clamped on top of the fabrics during the nickel sputtering process. A magnetron, high-vacuum sputtering system, ATC Orion from AJA International, Inc., was utilized to deposit the SiO_2 thin films at 3 millitorr pressure under inert (argon) environment using 300 Watts of DC sand RF power supplies.

All the cut carbon fabric samples, except the reference sample, underwent the thermal and gas exposures associated with the GSD growth process. The GSD protocol (Luhrs et al., 2009; Tehrani et al., 2013c) was carried out using a quartz tube reactor, equipped with three-input gas mass flow controllers and a thermal controller. The CNT growth followed the following three steps: (1) a reduction process was carried out, under N_2/H_2 environment and at 550°C for 120 min, to reduce nickel oxides and break the thin, sputtered layer of nickel into nanometer-sized particles, (2) the furnace was flushed by the inert N_2 gas to remove any residuals of the previous process, and (3) a mixture of $N_2/H_2/C_2H_4$ gasses was flown into the reactor to deposit the CNTs on the fibers, maintaining the 550°C temperature for 30 min. At the end of the GSD process, CNTs were grown wherever the catalyst (Ni) had been presputtered. The SEM/image in Fig. 5.8(a) shows a uniform growth of MWCNTs utilizing the GSD technique on the surface of PAN-based carbon fibers where the nickel catalytic particles were deposited. Unlike other methods for deposition of metal catalysts on the fabric, physical vapor deposition ensures a uniform deposition of an ultra-thin film of nickel. The capability to uniformly sputter a 2-nm-thick nickel film reflects in the persistent diameter of the CNTs grown via GSD, and it results in a small variation in the diameter of the grown CNTs. The structure of the MWCNTs along with their parallel side walls are shown in the TEM images, Fig. 5.8(b) and (c). The GSD-synthesized nanotubes have a diameter of less than 50 nm.

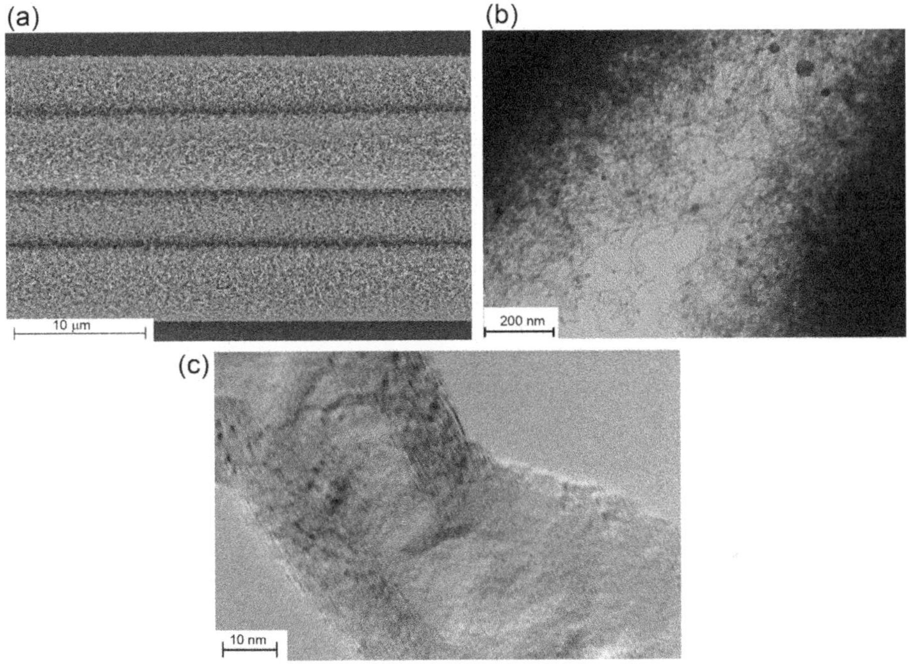

Figure 5.8 (a) SEM micrograph of a carbon fiber grafted with surface-grown CNTs utilizing the GSD technique. (b and c) TEM micrograph of MWCNTs grown via GSD.

Raman spectra were obtained for the different sheathed fiber architectures utilizing a ProSeek Raman system from Raman System, Inc. These spectra were obtained with a confocal Raman microscope, using a 5-mW, 785-nm excitation wavelength laser beam focused on the sample with a 50× objective. Spectra were obtained as the sum of 30 s integration time. Raman peaks, Fig. 5.9, for the samples without CNTs are very weak and do not exhibit the presence of significant crystalline form of carbon. It is well known that PAN-based carbon fibers do not exhibit the G band (unlike graphitic pitch-based fibers); rather, they exhibit spectroscopic appearance of the D line, which corresponds to the structural disorder caused by the existence of the sp^3 bonds (Fitzer, 1989). The CNTs grown on the surfaces of the PAN-based carbon fiber fabric via GSD demonstrate the D-band center value at 1350 cm^{-1} and the G-band at 1595 cm^{-1}, respectively. Although these peaks are also observed for graphite (Jorio et al., 2003), it is evident from the SEM and TEM images of the uniformly distributed filaments that the Raman peaks cannot be from graphite.

Four different fiber configurations were utilized to make the different composites, namely: R (based on fiber fabrics as received), HT (based on fabrics without any catalyst or TBC, underwent the GSD process), G (based on fabrics with uniformly surface-grown CNTs), and PG (based on fabrics with patterned surface-grown CNTs). Laminated composites were fabricated using the abovementioned fabrics, utilizing a vacuum- and pressure-assisted hand lay-up technique. Laminates prepared

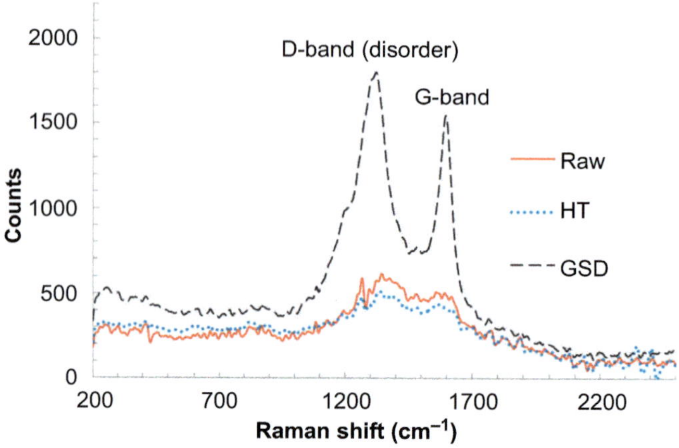

Figure 5.9 Raman spectra of the surface of different processed PAN-carbon fibers. Specimens are based on raw PAN-based carbon fabric (Raw), heat-treated fabric (HT), and the samples with CNTs grown on their surfaces via GSD.

for tensile tests comprised three layers of fabrics, while the impact and spherical punch tests utilized five-layer samples. The matrix was Aeroepoxy™ by PTM&W Industries, Inc. A vacuum- and pressure-assisted hand lay-up was utilized to manufacture the composite panels (Tehrani et al., 2013b,c). Each panel was cut into 12.5 × 125.0 mm strips to make the tensile samples following the ASTM standards: ASTM D3039/D3039M-08 (Standard, 2005). The tensile tests were carried out utilizing an Instron® 4400R frame under constant crosshead speed of 1.0 mm/min until failure occurred. The strain was recorded using an MTS extensometer. Tensile strength, elastic modulus (tangent modulus in the case of off-axis tests) and strain to failure were measured for different specimens. A total of 10 samples—for each configuration—were tested to determine the tensile properties. The manufactured composite laminates retained 62.0% ± 2.0 volume fractions. Results of the tensile test are shown in Figs. 5.10 and 5.11.

Because the elastic modulus for the on-axis tensile test is dominated by the stiffness of carbon fibers (Daniel and Ishai, 2005), the fibers' surface treatments are not expected to affect the elastic modulus for the on-axis loading.

In comparison to the sample based on raw fibers (R), the heat treatment (HT) reduced the ultimate strength by 3%, but the presence of surface-grown CNTs (G) not only compensated for this reduction, but it also significantly increased the final strain to failure (22%). This was because the CNTs intertwine with the polymer matrix and thus enhance the shear strength of the matrix/fiber interface. This in return improved the shear lag capability of the matrix.

The checkerboard patterning of CNTs is envisioned to mechanically engage the matrix around the fiber into interlocking/frictional mechanisms to carry more tensile load and, hence, increase the strength. To study the effect of the growth pattern solely, the samples with patterned growth were not precoated with SiO_2. For the on-axis case, the composite samples based on the fine mesh−patterned (PG) exhibited higher

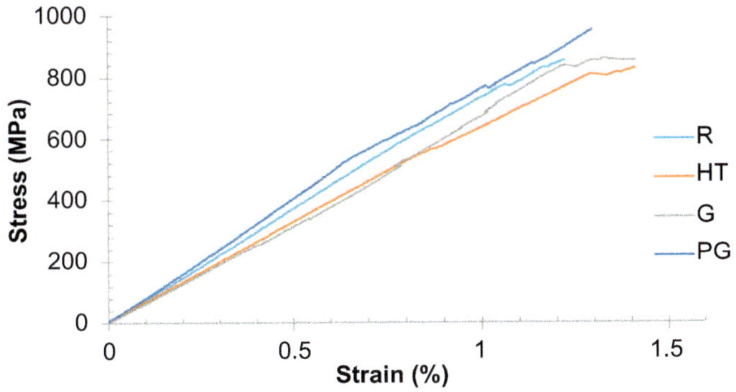

Figure 5.10 Representative stress–strain curves of the CFRPs tensile samples based on raw (R), heat-treated (HT), uniformly CNT grafted (G), and patterned CNT grafted (PG) graphite fiber fabrics.

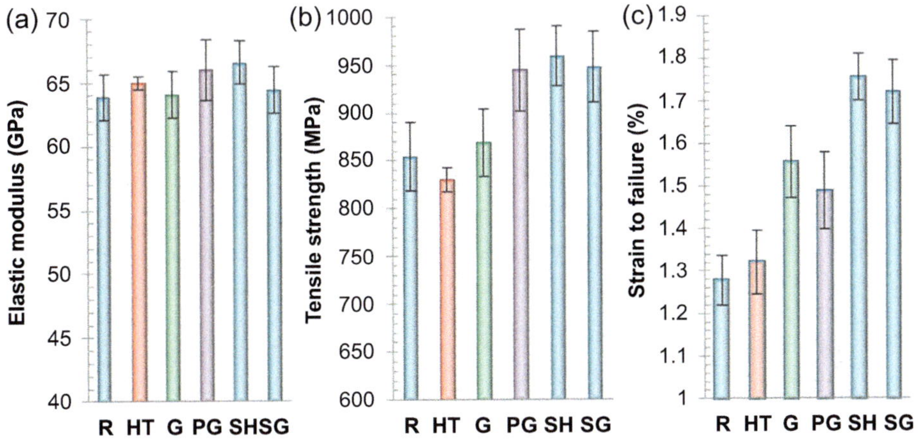

Figure 5.11 Tensile tests results for all the CFRPs: (a) elastic modulus, (b) tensile strength, and (c) strain to failure.

strength (~10%) and modulus compared to the samples with uniformly grown CNTs (Fig. 5.10). This demonstrates the importance of the spacing in between the grown CNT patches. The sample with fine-pattern growth of CNTs (PG) attained the highest stiffness and strength along the fiber direction, without any presputtered TBC. Furthermore, the ductility of the fine-patterned sample was 16% higher than the raw sample's.

The five-layered laminates for impact were plates of 130.0 × 130.0 mm. For each of the four fabric configurations (R, HT, G, and PG) a total number of four samples were tested in impact. For the impact tests, composite panels were subjected to an 8.4-g alloy steel ball impactor with a diameter of 12.7 mm, at a speed of around 104.0 ± 6.0 m/s. Fig. 5.12 demonstrates the absorbed impact energies for the four different composite configurations. Eventually, surface treatment of the PAN fiber

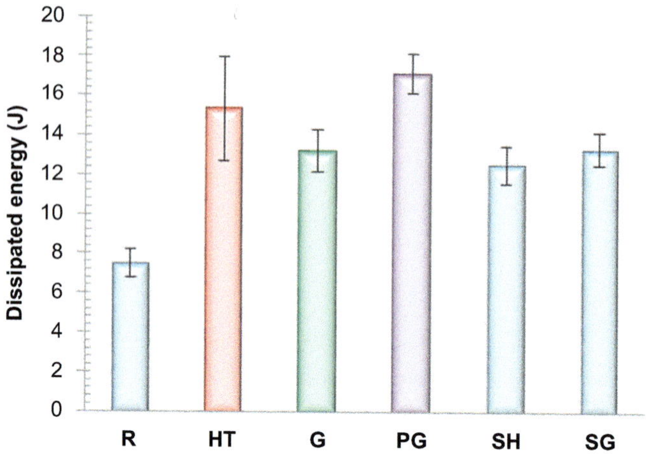

Figure 5.12 Absorbed impact energies for the designed CFRPs examined in high-velocity impact.

either by heat treatment or by growing MWCNTs enhances the energy absorption significantly, where the fine-patterned growth yielded the highest improvement of almost 127% over the composites based on raw carbon fibers.

To put these results in perspective, some damage analysis is needed. The postimpact analysis (Fig. 5.13) revealed that the samples based on the R, HT, and G configurations have shown large delamination zones: greater than 20 mm in length. Revisiting the energy dissipating mechanisms, delamination is the primary mechanism by which the CFRP dissipates energy during impact. Stronger interlaminar bonds require more energy to overcome the friction force between the laminates during delamination; this is evident from the lower absorbed impact energies in the R samples compared to the HT, G, and PG samples. The second form of energy dissipation is via fiber breakage. For more fibers to break, they will need more energy. Under impact, high strain rate loading the epoxy matrix exhibits more brittle behavior than in a quasistatic loading. Hence, the number of matrix cracks initiated in the matrix increase at higher strain rates. As a result, the large number of cracks leads to more fiber

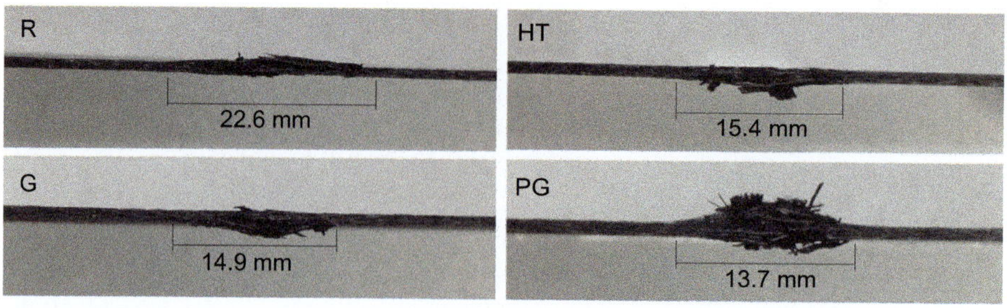

Figure 5.13 Delaminated cross-sections of the impacted CFRPs based on raw (R), heat-treated (HT), uniformly grafted CNT (G), and patterned CNT grafted (PG).

breakages. This is more pronounced for the case where a crack-stopping mechanism (ie, CNT forests) is lacking at the fiber/matrix interface. The other mechanism that dissipates energy during impact is fiber pull-out. A stronger fiber/matrix interface needs more energy to pull a fiber out. That was the mechanism by which the PG sample dissipated more energy than both the G and HT configurations. The exposed interspacing between the CNT batches on the surface of the fibers provides regions for matrix cracks to attack the fibers. The PG configuration utilizes this feature, to have more fiber breakages during the impact than during the punch test and, consequently, dissipates more energy in the impact than in the punch test. The unique features of the PG sample (having the strongest matrix/fiber bond and higher energy dissipation due to fiber breakage) made this configuration exhibit more pronounced improvement in the impact energy dissipation by 127% compared to the reference sample.

5.4 Comments and future trends

It was shown that the quasistatic tensile properties (stiffness and strength) do not change along the axial directions of the fibers in a CFRP with incorporated MWCNTs. Compared to the reference CFRPs, the MWCNT-enriched CFRP exhibited a higher absorbed energy under high strain rate (ie, impact) loading. These enhancements are due to the improved fiber-matrix adhesion and CNT-matrix interactions. Among the different energy dissipation mechanisms, the impact energy absorption is to a lesser extent controlled by delamination and matrix cracking. To predict the impact behavior of composites, it is crucial to take into account their strain rate–dependent behavior: stress wave accumulation. Alternative to mixing CNTs with the matrix prior to FRPs fabrication, they could be surface grown on the structural carbon fibers to form two-scale reinforcement. Growing CNTs directly over the carbon fibers improved the fiber/matrix adhesive bond and the interlaminar strength of the CFRPs, resulting in enhanced in-plane strength and impact energy dissipation, while retaining smaller delamination area. Patterned growth of CNT over the carbon fibers resulted in tremendous improvements on the in-plane performance and impact energy dissipation, with minimal delamination area.

It is unlikely that CFRPs can achieve acceptable energy absorption on their own; however, they can be utilized as sacrificial layer to enforce other structures such as metals or cementitious structures or sandwiched ceramic structures. The negative effects of the growth techniques on the substrate carbon fibers, if avoided, could lead to better energy absorption of CNT-enhanced fibrous reinforcements. The use of TBCs (SiC, SiO_2, etc.) thin films and patterned growth rather than dense growth of the CNTs can lessen the side effects of the growth techniques on the substrate fibers. Other nano reinforcements also could yield better energy absorption. For example, utilizing zinc oxide (ZnO) or barium titanate ($BaTiO_3$) nanowires could add piezoelectric functionality, and thorough electromechanical coupling might dissipate a portion of the blast energies into electric power.

Further reading sources

There are several articles that treat the subject of energy absorption in CFRPs, for example, the reviews by Pichandi et al. (2013) and Langdon et al. (2014). There are numerous classical references for the analysis of impact (stress) wave propagation in anisotropic materials like the books by Kolsky (1963), Zukas et al. (1982), and Jones (1989) and more recent ones like the book by Li-Li (2007). The inclusion of CNTs in polymeric matrices was reviewed extensively through several articles (Thostenson et al., 2001; Liu and Kumar, 2014; Xie et al., 2005). Also, the hybrid composites (sometimes referred to as hierarchal composites) incorporating both micro- and nanoscale carbon fiber were investigated thoroughly (Qian et al., 2010; Diez-Pascual et al., 2014; Sathishkumar et al., 2014).

References

Abrate, S., 1994. Impact on laminated composites: recent advances. Applied Mechanics Review 47, 517–544. http://dx.doi.org/10.1115/1.3111065.

Abrate, S., 1998. Impact on Composite Structures. Cambridge University Press, Cambridge, UK.

Advani, S.G., Fan, Z., 2007. Dispersion, bonding and orientation of carbon nanotubes in polymer matrices. In: Processing and Properties of Nanocomposites. 2nd Ch. World Scientific Publishing Co. Pte Ltd, Hackensack, NJ, pp. 61–85.

Ajayan, P.M., Schadler, L.S., Braun, P.V., 2003. Nanocomposite Science and Technology. Wiley-VCH Verlag Co, Weinheim.

Aksoy, A., Carlsson, L.A., 1992. Interlaminar shear fracture of interleaved graphite epoxy composites. Composites Science and Technology 43, 55–69. http://dx.doi.org/10.1016/0266-3538(92)90133-N.

Al-Haik, M., Luhrs, C., Reda Taha, M., Roy, A., Dai, L., Phillips, J., Doorn, S., 2010. Hybrid carbon fibers/carbon nanotubes structures for next generation polymeric composites. Journal of Nanotechnology 2010. http://dx.doi.org/10.1155/2010/860178.

Al-Haik, M.S., Garmestani, H., Savran, A., 2004. Explicit and implicit viscoplastic models for polymeric composite. International Journal of Plasticity 20, 1875–1907. http://dx.doi.org/10.1016/j.ijplas.2003.11.017.

Al-Haik, M.S., Hussaini, M.Y., Garmestani, H., 2006. Prediction of nonlinear viscoelastic behavior of polymeric composites using an artificial neural network. International Journal of Plasticity 22, 1367–1392. http://dx.doi.org/10.1016/j.ijplas.2005.09.002.

An, Q., Rider, A.N., Thostenson, E.T., 2012. Electrophoretic deposition of carbon nanotubes onto carbon-fiber fabric for production of carbon/epoxy composites with improved mechanical properties. Carbon 50, 4130–4143. http://dx.doi.org/10.1016/j.carbon.2012.04.061.

Askari, D., Ghasemi-Nejhad, M.N., 2011. Inter-laminar mechanical properties improvements in carbon nanotubes reinforced laminated nanocomposites. ASME. http://dx.doi.org/10.1115/IMECE2011-64283.

Batra, R.C., Hassan, N.M., 2008. Blast resistance of unidirectional fiber reinforced composites. Composites Part B: Engineering 39, 513–536. http://dx.doi.org/10.1016/j.compositesb.2007.03.002.

Bernholc, J., Brenner, D., Nardelli, M.B., Meunier, V., Roland, C., 2002. Mechanical and electrical properties of nanotubes. Annual Reviews of Materials Research 32, 347–375. http://dx.doi.org/10.1146/annurev.matsci.32.112601.134925.

Bibo, G.A., Hogg, P.J., 1996. Review: the role of reinforcement architecture on impact damage mechanisms and post-impact compression behavior. Journal of Materials Science 31. http://dx.doi.org/10.1007/BF00353091.

Boroujeni, A.Y., Tehrani, M., Nelson, A.J., Al-Haik, M.S., 2014. Hybrid carbon nanotube-carbon fiber composites with improved in-plane mechanical properties composites. Composites Part B: Engineering 66, 475–483. http://dx.doi.org/10.1016/j.compositesb.2014.06.010.

Boskovic, B., Golovko, V., Cantoro, M., Kleinsorge, B., Chuang, A., Ducati, C., Hofmann, S., Robertson, J., Johnson, B., 2005. Low temperature synthesis of carbon nanofibres on carbon fibre matrices. Carbon 43, 2643–2648. http://dx.doi.org/10.1016/j.carbon.2005.04.034.

Chen, X., Saito, T., Kusunoki, M., Motojima, S., 1999. Three-dimensional vapor growth mechanism of carbon microcoils. Journal of Materials Research 14, 4329–4336. http://dx.doi.org/10.1557/JMR.1999.0586.

Chung, D.D.L., 2003. Structural composite materials tailored for damping. Journal of Alloys and Compounds 355, 216–223. http://dx.doi.org/10.1016/S0925-8388(03)00233-0.

Chung, D.L., 1994. Carbon Fiber Composites. Butterworth-Heinemann, Newton, MA.

Dalton, A.B., Collins, S., Emuñoz, E.A., 2003. Super-tough carbon-nanotube fibres. Nature 423, 703.

Daniel, I.M., Ishai, O., 2005. Engineering Mechanics of Composite Materials. Oxford University Press, NY.

Diez-Pascual, A.M., Naffakh, M., Marco, C., Gomez-Fatou, M.A., Ellis, G.J., 2014. Multiscale fiber-reinforced thermoplastic composites incorporating carbon nanotubes: a review. Current Opinion in Solid State and Materials Science 18, 62–80. http://dx.doi.org/10.1016/j.cossms.2013.06.003.

Donadon, M.V., Lannucci, L., Falzon, B.G., Hodgkinson, J.M., De Almeida, S.F., 2008. A progressive failure model for composite laminates subjected to low velocity impact damage. Computers and Structures 86, 1232–1252. http://dx.doi.org/10.1016/j.compstruc.2007.11.004.

Dresselhaus, M., Dresselhaus, G., Avouris, P., 2001. Carbon Nanotubes Synthesis, Structure, Properties, and Applications. Springer-Verlag, Heidelberg.

Endo, M., Iijima, S., Dresselhaus, M.S., 1997. Carbon Nanotubes. Elsevier.

Fitzer, E., 1989. PAN-based carbon fibers-present state and trend of the technology from the viewpoint of possibilities and limits to influence and to control the fiber properties by the process parameters. Carbon 27, 621–645. http://dx.doi.org/10.1016/0008-6223(89)90197-8.

Fleck, N.A., Deshpande, V.S., 2004. The resistance of clamped sandwich beams to shock loading. Journal of Applied Mechanics 71, 386–401. http://dx.doi.org/10.1115/1.1629109.

Garcia, E.J., Wardle, B.L., Guzman Devilloria, R., Wicks, S., Ishiguro, K., Yamamoto, N., Hart, A.J., 2008. Aligned carbon nanotube reinforcement of advanced composite ply interfaces. In: 49th AIAA/ASME/ASCE/AHS/ASC Structures, Structural Dynamics, and Materials Conference, pp. 7–10. http://dx.doi.org/10.2514/6.2008-1768.

Garmestani, H., Al-Haik, M.S., Dahmen, K., Tannenbaum, R., Li, D.S., Sablin, S.S., Hussaini, M.Y., 2003. Polymer-mediated alignment of carbon nanotubes under high magnetic fields. Advanced Materials 15, 1918–1921. http://dx.doi.org/10.1002/adma.200304932.

Gibson, R.F., 2010. A review of recent research on mechanics of multifunctional composite materials and structures. Composite Structures 92, 2793–2810. http://dx.doi.org/10.1016/j.compstruct.2010.05.003.

Gojny, F.H., Wichmann, M.H.G., Köpke, U., Fiedler, B., Schulte, K., 2004. Carbon nanotube-reinforced epoxy-composites: enhanced stiffness and fracture toughness at low nanotube content. Composites Science and Technology 64, 2363–2371. http://dx.doi.org/10.1016/j.compscitech.2004.04.002.

Gorga, R.E., Cohen, R.E., 2004. Toughness enhancements in poly(methyl methacrylate) by addition of oriented multiwall carbon nanotubes. Journal of Polymer Science: Part B: Polymer Physics 42, 2690–2702. http://dx.doi.org/10.1002/polb.20126.

Grujicic, M., Bell, W.C., Biggers, S.B., Koudela, K.L., Tarter, J.F., Cheeseman, B.A., 2008a. The effect of a carbon-nanotube forest-mat strike face on the ballistic-protection performance of E-glass reinforced poly-vinyl-ester-epoxy composite armour. Proceedings of the Institution of Mechanical Engineers Part L – Journal of Materials-Design and Applications 222, 15–27. http://dx.doi.org/10.1243/14644207JMDA136.

Grujicic, M., Bell, W.C., Thompson, L.L., Koudela, K.L., Cheeseman, B.A., 2008b. Ballistic-protection performance of carbon-nanotube-doped poly-vinyl-ester-epoxy matrix composite armor reinforced with E-glass fiber mats. Materials Science and Engineering: A – Structural Materials Properties Microstructure and Processing 479, 10–22. http://dx.doi.org/10.1016/j.msea.2007.06.013.

Gu, B.H., Li, Y.L., 2005. Ballistic perforation of conically cylindrical steel projectile into three-dimensional braided composites. AIAA Journal 43, 426–434. http://dx.doi.org/10.2514/1.11729.

Hartman, T.B., Allen, D.W., Case, S.W., 2010. Failure of spectra shield UHMWPE composite due to low velocity hard body impact. In: SAMPE Fall Technical Conference, Salt Lake City, UT.

Hudnut, S.W., Chung, D.D.L., 1995. Use of submicron diameter carbon filaments for reinforcement between continuous carbon-fiber layers in a polymer-matrix composite. Carbon 33, 1627–1631. http://dx.doi.org/10.1016/0008-6223(95)00148-7.

Inam, F., Wong, D.Y., Kuwata, M., Peijs, T., 2010. Multiscale hybrid micro-nanocomposites based on carbon nanotubes and carbon fibers. Journal of Nanomaterials. http://dx.doi.org/10.1155/2010/453420.

Jones, N., 1989. Structural Impact. Cambridge University Press, Cambridge, UK.

Jorio, A., Pimenta, M.A., Filho, A.G.S., Saito, R., Dresselhaus, G., Dresselhaus, M.S., 2003. Characterizing carbon nanotube samples with resonance Raman scattering. New Journal of Physics 5, 139. http://dx.doi.org/10.1088/1367-2630/5/1/139.

Joshi, S.C., Dikshit, V., 2012. Enhancing interlaminar fracture characteristics of woven CFRP prepreg composites through CNT dispersion. Journal of Composite Materials 46, 665–675. http://dx.doi.org/10.1177/0021998311410472.

Kang, H.S., 2004. Organic functionalization of sidewall of carbon nanotubes. The Journal of Chemical Physics 121, 7946–7954. http://dx.doi.org/10.1063/1.1775783.

Kau, H.T., 1990. A study of the impact behavior of chopped fiber reinforced composite. Polymer Composites 11, 253–264. http://dx.doi.org/10.1002/pc.750110502.

Kelly, A., 1973. Strong Solids. Clarendon Press, Oxford, UK.

Khan, S.U., Kim, J.-K., 2011. Impact and delamination failure of multiscale carbon nanotube-fiber reinforced polymer composites: a review. International Journal of Aeronautical and Space Sciences 12, 115–133. http://dx.doi.org/10.5139/IJASS.2011.12.2.115.

Kolsky, H., 1963. Stress Waves in Solids. Dover Publication, New York, NY.

Kostopoulos, V., Baltopoulos, A., Karapappas, P., Vavouliotis, A., Paipetis, A., 2010. Impact and after-impact properties of carbon fibre reinforced composites enhanced with multi-wall

carbon nanotubes. Composites Science and Technology 70, 553–563. http://dx.doi.org/10.1016/j.compscitech.2009.11.023.
Langdon, G.S., Cantwell, W.J., Guan, Z.W., Nurick, G.N., 2014. The response of polymeric composite structures to air-blast loading: a state-of-the-art. International Materials Reviews 59, 159–177. http://dx.doi.org/10.1179/1743280413Y.0000000028.
Langdon, G.S., Nurick, Lemanski, S.L., Simmons, M.C., Cantwell, W.J., Schleyer, G.K., 2007. Failure characterisation of blast-loaded fibre-metal laminate panels based on aluminium and glass-fibre reinforced polypropylene. Composites Science and Technology 67, 1385–1405. http://dx.doi.org/10.1016/j.compscitech.2006.09.010.
Li-Li, W., 2007. Foundations of Stress Waves. Elsevier, Oxford, UK.
Liu, Y.D., Kumar, S., 2014. Polymer/carbon nanotube nano composite fibers-a review. ACS Applied Materials and Interfaces 6, 6069–6087. http://dx.doi.org/10.1021/am405136s.
Lopresto, V., Melito, V., Leone, C., Caprino, G., 2006. Effect of stitches on the impact behaviour of graphite/epoxy composites. Composites Science and Technology 66, 206–214. http://dx.doi.org/10.1016/j.compscitech.2005.04.029.
Luhrs, C.C., Garcia, D., Tehrani, M., Al-Haik, M., Taha, M.R., Phillips, J., 2009. Generation of carbon nanofilaments on carbon fibers at 550 C. Carbon 47, 3071–3078. http://dx.doi.org/10.1016/j.carbon.2009.07.019.
Madhukar, M.S., Drzal, L.T., 1992a. Fiber-matrix adhesion and its effect on composite mechanical properties, III. Longitudinal (0-degrees) compressive properties of graphite epoxy composites. Journal of Composite Materials 26, 310–333. http://dx.doi.org/10.1177/002199839202600301.
Madhukar, M.S., Drzal, L.T., 1992b. Fiber-matrix adhesion and its effect on composite mechanical properties, IV. Mode I and Mode II fracture toughness of graphite/epoxy composites. Journal of Composite Materials 26, 936–968. http://dx.doi.org/10.1177/002199839202600701.
Miyagawa, H., Drzal, L.T., 2004. Thermo-physical and impact properties of epoxy nanocomposites reinforced by single-wall carbon nanotubes. Polymer 45, 5163–5170. http://dx.doi.org/10.1016/j.polymer.2004.05.036.
Morozov, E.V., Sylantiev, S.A., Evseev, E.G., 2003. Impact damage tolerance of laminated composite helicopter blades. Composite Structures 62, 367–371. http://dx.doi.org/10.1016/j.compstruct.2003.09.034.
Papakonstantopoulos, G.J., Yoshimoto, K., Doxastakis, M., Al, E., 2005. Local mechanical properties of polymeric nanocomposites. Physical Review E 72, 031801. http://dx.doi.org/10.1103/PhysRevE.72.031801.
Phillips, J., Luhrs, C., Leseman, Z.C., Al-Haik, M., 2007. Novel graphitic structures by design. In: Proceedings of ASME International Mechanical Engineering Congress and Exposition IMECE07. Seattle, Washington, USA. http://dx.doi.org/10.1115/IMECE2007-42977.
Pichandi, S., Rana, S., Oliveira, D., Fangueiro, R., 2013. Fibrous and composite materials for blast protection of structural elements-state-of-the-art review. Journal of Reinforced Plastics and Composites 32, 1477–1500. http://dx.doi.org/10.1177/0731684413498297.
Qian, H., Greenhalgh, E.S., Shaffer, M.S.P., Bismarck, A., 2010. Carbon nanotube-based hierarchical composites: a review. Journal of Materials Chemistry 20, 4751–4762. http://dx.doi.org/10.1039/C000041H.
Safdari, M., Al-Haik, M.S., 2012. Optimization of stress wave propagation in a multilayered elastic/viscoelastic hybrid composite based on carbon fibers/carbon nanotubes. Polymer Composites 33, 196–206. http://dx.doi.org/10.1002/pc.22137.

Sathishkumar, T.P., Naveen, J., Satheeshkumar, S., 2014. Hybrid fiber reinforced polymer composites – a review. Journal of Reinforced Plastics and Composites 33, 454–471. http://dx.doi.org/10.1177/0731684413516393.

Shyr, T.-W., Pan, Y.-H., 2003. Impact resistance and damage characteristics of composite laminates. Composite Structures 62, 193–203. http://dx.doi.org/10.1016/S0263-8223(03)00114-4.

Sinnott, S.B., Andrews, R., 2001. Carbon nanotubes: synthesis, properties, and applications. Critical Reviews in Solid State and Materials Sciences 26, 145–249. http://dx.doi.org/10.1080/20014091104189.

Soliman, E., Al-Haik, M., Taha, M.R., 2012a. On and off-axis tension behavior of fiber reinforced polymer (FRP) composites incorporating multi-walled carbon nanotubes. Journal of Composite Materials 46, 1661–1675. http://dx.doi.org/10.1177/0021998311422456.

Soliman, E.M., Sheyka, M.P., Taha, M.R., 2012b. Low-velocity impact of thin woven carbon fabric composites incorporating multi-walled carbon nanotubes. International Journal of Impact Engineering 47, 39–47. http://dx.doi.org/10.1016/j.ijimpeng.2012.03.002.

Standard, A., 2005. D3039/D3039M, Standard Test Method for Tensile Properties of Polymer Matrix Composite Materials. ASTM Int., West Conshohocken, PA.

Stephan, C., Nguyen, T.P., Chapelle, M.L., Lefrant, S., Journet, C., Bernie, P., 2000. Characterization of singlewalled carbon nanotubes-PMMA composites. Synthetic Metals 108, 139–149. http://dx.doi.org/10.1016/S0379-6779(99)00259-3.

Treacy, M.M., Ebbesen, T.W., Gibson, J.M., 1996. Exceptional high Young's modulus observed for individual carbon nanotubes. Nature 381, 678–680. http://dx.doi.org/10.1038/381678a0.

Taha, M.M.R., Shrive, N.G., 2001. Enhancing fracture toughness of high-performance carbon fiber cement composites. ACI Materials Journal 98, 168–179. http://dx.doi.org/10.14359/10200.

Tehrani, M., Boroujeni, A.Y., Hartman, T.B., Haugh, T.P., Case, S.W., Al-Haik, M.S., 2013a. Mechanical characterization and impact damage assessment of a woven carbon fiber reinforced carbon nanotube-epoxy composite. Composites Science and Technology 75, 42–48. http://dx.doi.org/10.1016/j.compscitech.2012.12.005.

Tehrani, M., Boroujeni, A.Y., Hartman, T.B., Haugh, T.P., Case, S.W., Al-Haik, M.S., 2013b. Mechanical characterization and impact damage assessment of a woven carbon fiber reinforced carbon nanotube–epoxy composite. Composites Science and Technology 75, 42–48. http://dx.doi.org/10.1016/j.compscitech.2012.12.005.

Tehrani, M., Boroujeni, A.Y., Luhrs, C.C., Phillips, J., Al-Haik, M., 2014. Hybrid composites based on carbon fibers/carbon nanofilament reinforcement. Materials 7, 4182–4195. http://dx.doi.org/10.3390/ma7064182.

Tehrani, M., Safdari, M., Al-Haik, M.S., 2011. Nanocharacterization of creep behavior of multiwall carbon nanotubes/epoxy nanocomposite. International Journal of Plasticity 27, 887–901. http://dx.doi.org/10.1016/j.ijplas.2010.10.005.

Tehrani, M., Safdari, M., Boroujeni, A., Razavi, Z., Case, S., Dahmen, K., Garmestani, H., Al-Haik, M., 2013c. Hybrid carbon fiber/carbon nanotube composites for structural damping applications. Nanotechnology 24, 155704. http://dx.doi.org/10.1088/0957-4484/24/15/155704.

Tekalur, S.A., Shukla, A., Shivakumar, K., 2008. Blast resistance of polyurea based layered composite materials. Composite Structures 84, 271–281. http://dx.doi.org/10.1016/j.compstruct.2007.08.008.

Thostenson, E.T., Chou, T.-W., 2002. Aligned multi-walled carbon nanotube-reinforced composites: processing and mechanical characterization. Journal of Physics D: Applied Physics 35, L77. http://dx.doi.org/10.1088/0022-3727/35/16/103.

Thostenson, E.T., Ren, Z.F., Chou, T.W., 2001. Advances in the science and technology of carbon nanotubes and their composites: a review. Composites Science and Technology 61, 1899–1912. http://dx.doi.org/10.1016/S0266-3538(01)00094-X.

Trabelsi, W., Michel, L., Othomene, R., 2010. Effects of stitching on delamination of satin weave carbon-epoxy laminates under mode I, mode II and mixed-mode I/II loadings. Applied Composite Materials 17, 575–595. http://dx.doi.org/10.1007/s10443-010-9128-0.

Tsai, L., Prakash, V., 2005. Structure of weak shock waves in 2-D layered material systems. International Journal of Solids and Structures 42, 727–750. http://dx.doi.org/10.1016/j.ijsolstr.2004.06.030.

Veedu, V.P., Cao, A., Li, X., Ma, K., Soldano, C., Kar, S., Ajayan, P.M., Ghasemi-Nejhad, M.N., 2006. Multifunctional composites using reinforced laminae with carbon-nanotube forests. Nature Materials 5, 457–462. http://dx.doi.org/10.1038/nmat1650.

Vlasveld, D.P.N., Bersee, H.E.N., Picken, S.J., 2005. Nanocomposite matrix for increased fibre composite strength. Polymer 46, 10269–10278. http://dx.doi.org/10.1016/j.polymer.2005.08.003.

Warrier, A., Godara, A., Rochez, O., Mezzo, L., Luizi, F., Gorbatikh, L., Lomov, S.V., Vanvuure, A.W., Verpoest, I., 2010. The effect of adding carbon nanotubes to glass/epoxy composites in the fibre sizing and/or the matrix. Composites Part A – Applied Science and Manufacturing 41, 532–538. http://dx.doi.org/10.1016/j.compositesa.2010.01.001.

Westwood, M., Webster, J., Day, R., Hayes, F., Taylor, R., 1996. Oxidation protection for carbon fibre composites. Journal of Materials Science 31, 1389–1397. http://dx.doi.org/10.1007/BF00357844.

Xie, X.L., Mai, Y.W., Zhou, X.P., 2005. Dispersion and alignment of carbon nanotubes in polymer matrix: a review. Materials Science and Engineering R-Reports 49, 89–112. http://dx.doi.org/10.1016/j.mser.2005.04.002.

Yamamoto, N., Wardle, B.L., 2008. Electrical and thermal properties of hybrid woven composites reinforced with aligned carbon nanotubes. In: 49th AIAA/ASME/ASCE/AHS/ASC Structures, Structural Dynamics, and Materials Conference. American Institute of Aeronautics and Astronautics. http://dx.doi.org/10.2514/6.2008-1857.

Zhandarov, S.F., Mader, E., Yurkevich, O.R., 2002. Indirect estimation of fiber polymer bond strength and interfacial friction from maximum load values recorded in the microbond and pull-out tests. Part 1: local bond strength. Journal of Adhesion Science and Technology 16, 1171–1200. http://dx.doi.org/10.1163/156856102320256837.

Zhang, Q., Liu, J., Sager, R., Dai, L., Baur, J., 2009. Hierarchical composites of carbon nanotubes on carbon fiber: influence of growth condition on fiber tensile properties. Composites Science and Technology 69, 594–601. http://dx.doi.org/10.1016/j.compscitech.2008.12.002.

Zhu, S., Su, C.-H., Lehoczky, S., Muntele, I., Ila, D., 2003. Carbon nanotube growth on carbon fibers. Diamond and Related Materials 12, 1825–1828. http://dx.doi.org/10.1016/S0925-9635(03)00205-X.

Zukas, J.A., Nicholas, T., Swift, H., Greszxzuk, L.B., Curran, D.R., 1982. Impact Dynamics. John Wiley & Sons, Inc., New York, NY.

Modelling of 3D woven fabrics for ballistic protection

6

Bohong Gu
Donghua University, Shanghai, China

6.1 Introduction

Woven fabrics manufactured from high-performance fibres (for example, Kevlar®, Twaron® and Dyneema®) have been regarded as effective materials in protective applications. Such a kind of 3D woven fabrics and composites have great potential applications to the higher impact damage tolerance area, such as lightweight structures, protective structures and aircrafts. Three-dimensional (3D) woven fabrics have been widely applied to ballistic protection in the form of 3D woven composite materials. The first step to characterize the ballistic penetration damage of 3D woven composite materials is to investigate the ballistic impact behaviours of 3D woven preform which has integrated construction along the thickness direction. Generally, there are two types of 3D woven fabric: 3D orthogonal and 3D angle-interlock woven fabric, as shown in Fig. 6.1.

Investigations on the ballistic impact damage of woven fabrics have been extensively reported. Since the pioneering works of Smith and coworkers [1–9], Petterson et al. [10,11] on the impact of textile yarns, Susich et al. [12] on the damage morphology of 2 × 2 basket weave nylon fabric, Roylance et al. [13] on a dynamic finite element computer technique to model the ballistic impact of fabric panel, Wilde et al. [14,15] on the high-speed photography observations of transient responses of the fabric under ballistic impact and Vinson et al. [16] on an analytical model of cone deformation of fabric, the ballistic impact damage and ballistic limit prediction methods of 2D planar woven fabric have been investigated from the three approaches of experimental, analytical and finite element modelling thoroughly. In experimental modelling, Cunniff [17,18] presented systematic ballistic performance results of fabric-based armour systems. The system effects that occur during the ballistic impact of woven fabric body armour materials were discussed from a conceptual framework developed to relate single yarn impact mechanics to fabric impact mechanics. A parametric model has been developed to predict the performance of body armour systems under impact by a chunky steel projectile of arbitrary mass, shape, impact velocity and impact obliquity. For analytical models, Vinson and Zukas [16], Taylor and Vinson [19], Parga-Landa and Hernandez-Olivares [20], Chocron-Benloulo et al. [21], Navarro [22], Billon [23], Lim et al. [24] and Gu [25] analysed the multilayered fabric target penetrated by a projectile. For finite element modelling, Leech et al. [26], Hearle et al. [27], Roylance et al. [13,28,29], Shim et al. [30], Lim et al. [31], Tan et al. [32] and Gu [33] put forward numerical models to analyse a single-layer fabric target under ballistic impact.

Advanced Fibrous Composite Materials for Ballistic Protection. http://dx.doi.org/10.1016/B978-1-78242-461-1.00006-6

(a)

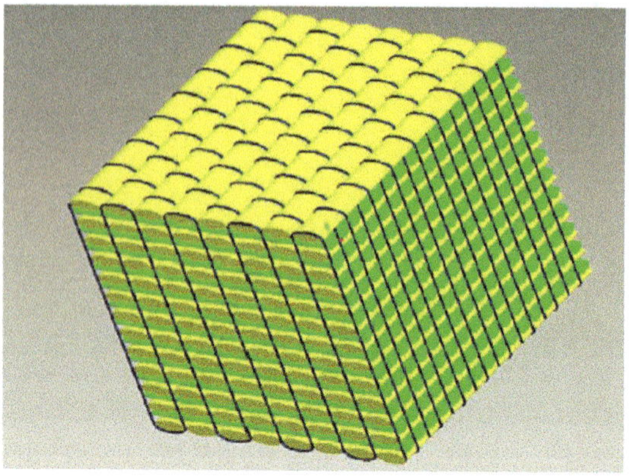

(b)

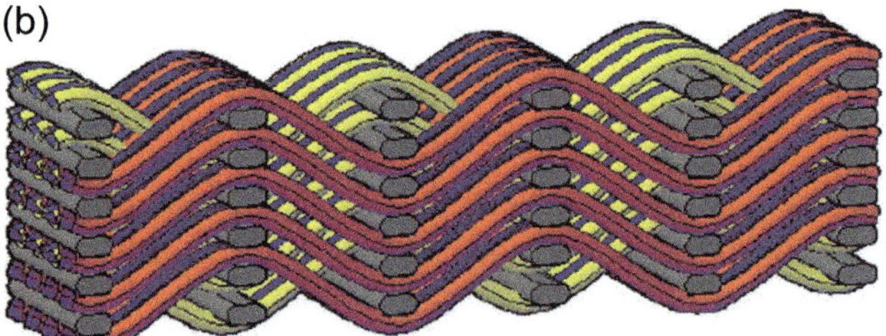

Figure 6.1 Diagram of 3D woven fabric construction: (a) 3D orthogonal woven fabric; (b) 3D angle-interlock woven fabric.

More recently, Phoenix and Porwal [34] developed a membrane model for predicting ballistic impact response. But its sophisticated formulation limits the application in designing. In this lengthy paper, Phoenix and Porwal [34] reviewed almost all published papers on ballistic penetration of woven fabrics to 2003. After 2003, many studies focused on the influence of yarn interlacing points (ie, stress wave reflection), yarn crimp, yarn friction, boundary condition, bullet shape and fabric structure on fabric ballistic impact behaviour. For example, Duan et al. [35–38], Tan and Ching [39] and Rao et al. [40] showed crimp and friction of yarns are needed to examine ballistic performance of high-strength fabric structures from finite element results. Novotny et al. [41] modelled ballistic efficiency of woven fabric targets when either the number of layers in the panel or the yarn denier was increased. Talebi et al. [42] developed finite element evaluation of projectile nose angle effects in ballistic perforation of high-strength fabric. It was showed by Gu [33] and Ivanov and Tabiei [43] that strain rate was an important factor in a finite element model. Influence of boundary conditions was reported by Zeng et al. [44]. Karahan et al. [45,46] and Bilisik and Turhan [47] presented experimental investigation that stitching type had significant effects on ballistic properties. Cork and Foster [48] and Barauskas and Abraitienė [49] investigated the ballistic impact damage of plain-woven Kevlar® fabric from finite element modelling. Zohdi and Powell [50] presented a multiscale model of woven fabric for ballistic impact damage simulation. Furthermore, Joo and Kang [51] employed the energy analysis method for understanding the impact behaviour of multi-ply plain-woven fabric, such an effort also by Zhang et al. [52]. Nasr-Isfahani et al. [53] calculated the residual velocity of projectile-penetrating targets composed of planar plain-woven fabric and simulated the response of plain-woven fabric under high-speed projectile impact. Wang et al. [54] developed an explicit digital element method for numerical simulation of ballistic impact and penetration of textile fabrics. Each yarn is digitized as an assembly of digital fibres. Parsons et al. [55] introduced a mesostructure-based continuum model that permits yarn slip and use the model to simulate the ballistic impact of woven fabric. It was found that simulating yarn slip significantly improves finite element predictions of the ballistic impact of a Kevlar® woven fabric, in particular by increasing the energy absorbed at high initial projectile velocities. Further simulations elucidate the micromechanisms of deformation of ballistic impact of woven fabric with yarn slip. However, all the investigations mentioned earlier are focused on the 2D planar woven fabrics. As for 3D woven fabric, little information reports the ballistic impact behaviour either in experimental or in theoretical modelling.

The main absorption mechanisms of kinetic energy of a projectile by 2D woven fabric targets are deformation of fabric structure (yarn crimp, fabric construction), failure and deformation of fibres and yarns. The high absorption capacity of kinetic energy of a projectile could be obtained by manufacturing fabrics with fibres of high modulus, high failure stress and high failure strain or by increasing the deformation of fabric structure during impact. Multilayered fabrics are often used to increase the ballistic protection capability by increasing the number of damaged fibres and yarns and the deformation zone of fabrics. In an optimum design of a bullet-proof fabric target, various models have been established to explain the influence of many factors, such as yarn elastic modulus, yarn tensile strength, yarn failure strain, yarn linear density, fabric construction, area density of fabric target, etc., on the fabric target deformation, damage, absorption of kinetic energy of projectile and ballistic limit.

Three-dimensional woven fabrics have higher interlaminar strength than a multilayered 2D woven fabric system. Such a structure feature will lead to higher ballistic impact energy absorption. However, all the investigations mentioned earlier are focused on the 2D planar woven fabrics. As for 3D woven fabric, little information reports the ballistic impact behaviour either in experimental or in theoretical. Gama et al. [56] performed numerical simulation of ballistic impact, damage and penetration of a single-layer 3D orthogonal weave fabric composite from a unit cell model of the composite. Nilakantan et al. [57] investigated the ballistic impact behaviour of 3D woven fabrics with different inclination angles of Z-tows from finite element simulation. A finite element (FE) model that incorporates the actual geometry and undulations of the warp, fill and Z-tows obtained from micrographs was presented. The results obtained were compared against an equivalent three-layer 2D plain-weave fabric system, both with and without through-thickness stitching that holds the layers together. Chen and Yang [58,59] reported an investigation on the ballistic performance of angle-interlock fabrics and found that the angle-interlock fabric is not less than the conventional fabric constructions used for body armour against ballistic impact.

The modelling strategies of 3D woven fabrics could be generally classified as analytical modelling and numerical modelling. Both methods are based on the fabric construction and the mechanical behaviours of fibres and yarns and the interactions of fibres and yarns. The analytical model employed an analytical method and derived the fabric energy absorption from analytical formulas. More specifically, the friction among yarns and the strain rate effect of mechanical behaviours of fibres would be considered in calculating the ballistic impact damages. The maximum stress or strain failure criteria are usually used to find the starting point of damage initiation and propagation. The numerical modelling methods are an effective way to unveil the ballistic impact damage propagation and visually reveal fabric construction deformation. Finite element methods (FEM) or finite difference methods are usually used in modelling the impact damage of fabric at the unit cell level and microstructure level. At the unit cell level, the woven fabric is assumed as a continuum or a 2D membrane material. The constitutive equations of the woven fabric are derived from the constitutive equations of fibre tows and the assumptions of simplifying the woven fabric into a 2D continuum. At the microstructure level, the geometrical model of the woven fabric is established based on the structure parameters of woven fabric. Some geometrical features, such as cross-sections of yarns, are assumed as elliptical in shape. The yarns are considered as a continuum which is defined from the mechanical behaviours of fibre tows. Combined with the constitutive equations of yarns under high strain rates and explicit form algorithms of impact dynamics, the ballistic penetration damage of the woven fabric could be calculated and numerically simulated.

In this chapter, the ballistic impact damages of 3D woven fabrics will be reported both from the analytical and numerical methods. The 3D woven fabrics include the 3D orthogonal woven fabric (3DOWF) and 3D angle-interlock woven fabric (3DAWF). The impact energy absorption and failure mechanisms will be analysed. The ballistic impact damage of 2D woven fabric will also be given for comparisons.

6.2 Numerical modelling of ballistic impact simulation

6.2.1 Geometrical modelling of 3D woven fabric at microstructure level

The principle of 3D orthogonal woven fabric architecture is to bind straight warp yarns and weft yarns together by using Z-yarns which run through the whole thickness of the fabric. The warp and weft yarns provide high in-plane stiffness and strength, and the binder yarns run through the thickness direction to stabilize the woven structure.

As shown in Figs. 6.2 and 6.3, the 3DOWF is composed of three kinds of yarns which are perpendicular to each other, named warp yarn, weft yarn and Z-yarn, respectively. All the warp and weft yarns are cross-plied and not interwoven. Such a structure feature will lead to the highest stress wave propagation speed in the warp and weft yarns. The impact energy can spread to the largest fabric area.

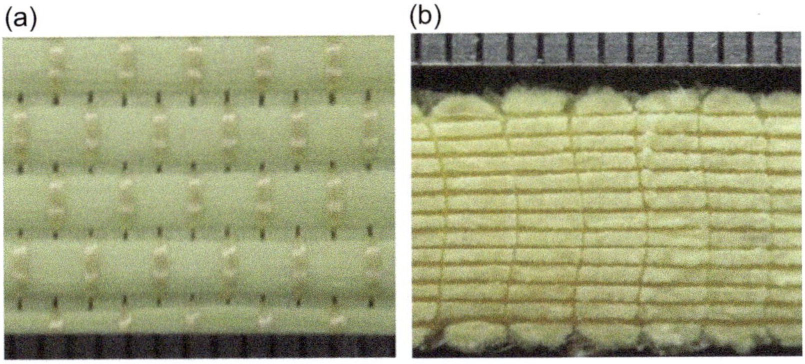

Figure 6.2 Photographs of 3D orthogonal woven fabric: (a) top view; (b) cross-section.

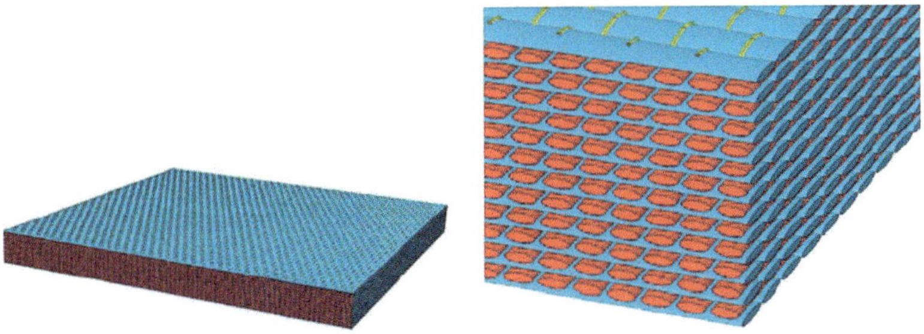

Figure 6.3 Geometrical model of 3D orthogonal woven fabric.

The microstructure model of the 3DOWF was established based on the aforementioned structures parameters and shown in Fig. 6.3. In Fig. 6.3, all the cross-sections of fibre tows were simplified as an elliptical form.

The geometrical microstructure models of the 3DAWF were established as shown in Fig. 6.4. In Fig. 6.4(a) and (b), the cross-section of the warp and weft yarns is assumed as an ellipse form in the microstructure model. This assumption was confirmed by observing the yarn cross-sections of the fabric shown in Fig. 6.5. The whole geometrical microstructure model of the 3DAWF is shown in Fig. 6.4(c).

6.2.2 Algorithms

The interaction between a rigid projectile and a 3DOWF target involves the sliding, erosion and deformation at the interface. Interfaces can be defined in three dimensions by listing in arbitrary order all triangle and quadrilateral segments that comprise each side of the interface. One side of the interface is designated as the slave side, and the other is designated as the master side. Nodes lying in those surfaces are referred to as slave and master nodes, respectively [60].

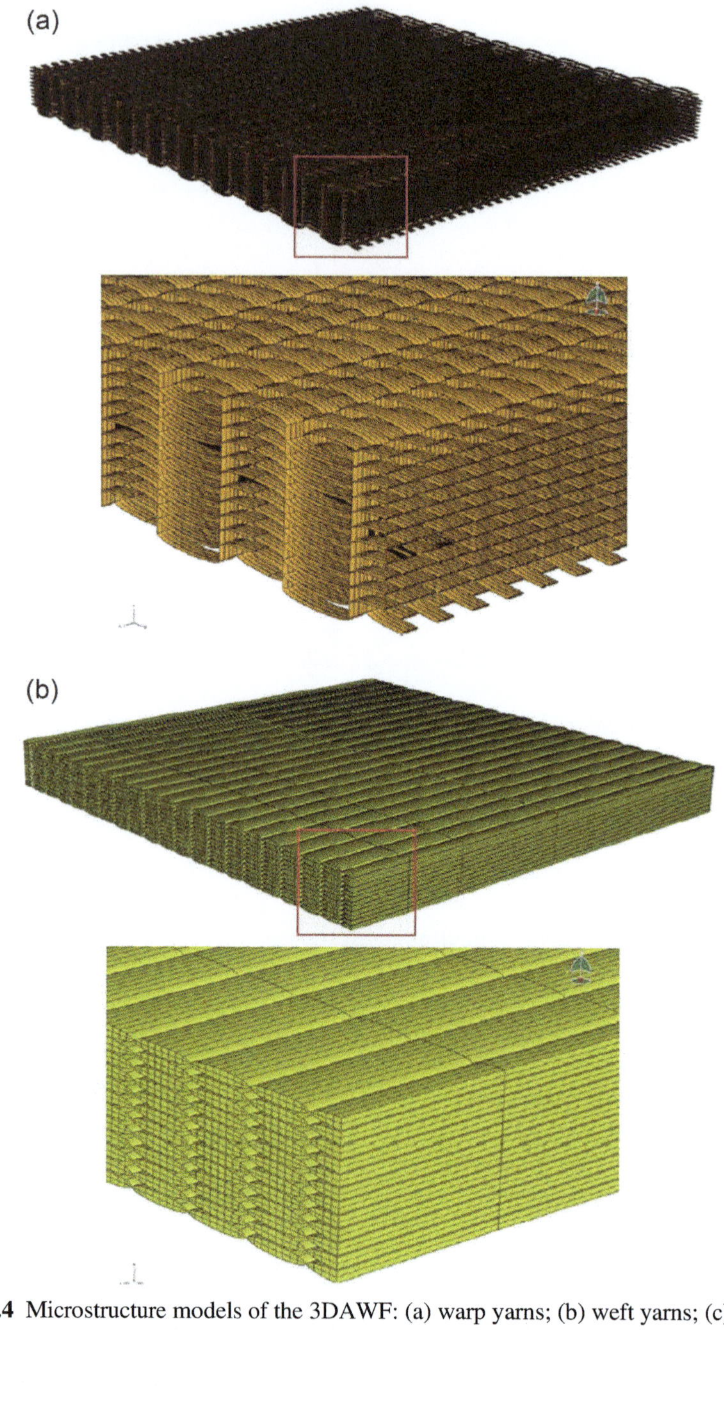

Figure 6.4 Microstructure models of the 3DAWF: (a) warp yarns; (b) weft yarns; (c) 3DAWF.

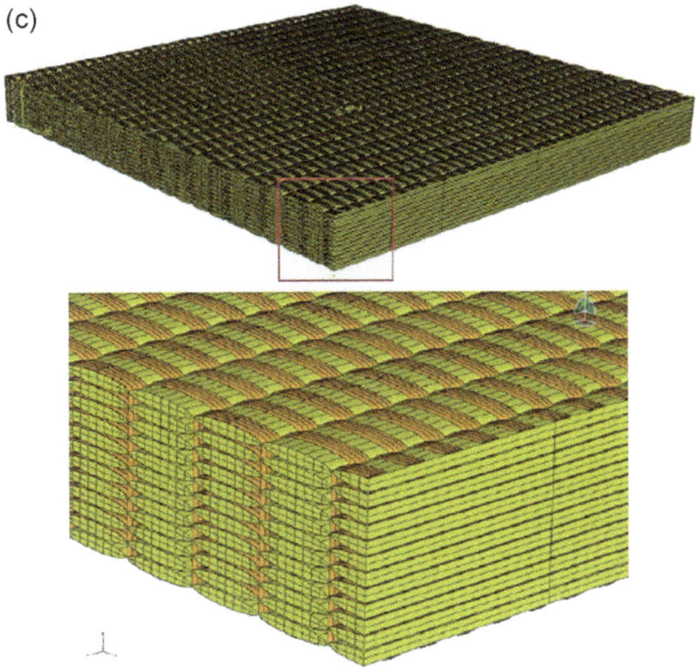

Figure 6.4 Continued.

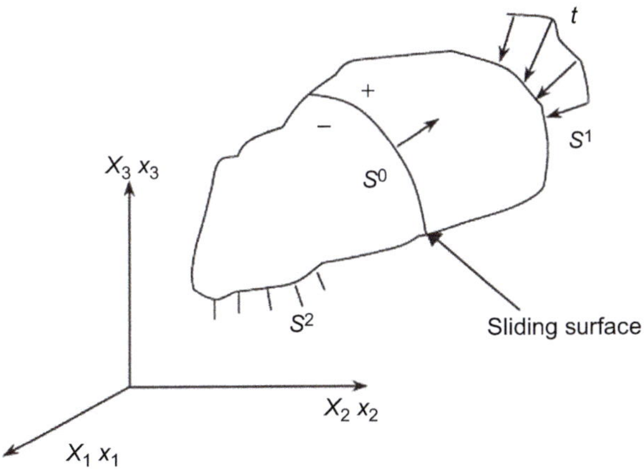

Figure 6.5 Contact notations.

The impact dynamics and concerned FE algorithms have been reported in many books intensively in great length, such as Hallquist [60], Belytschko et al. [61] and Zukas et al. [62]. The algorithms can be simplified and described concisely as follows.

Fig. 6.5 shows the impact of two bodies and the sliding surface between them. In a fixed rectangular Cartesian coordinate system, the position of an arbitrary point at time $t = 0$ is X_i ($I = 1,2,3$). At time t the new position is x_i ($I = 1,2,3$). The equation for the movement of the point is shown in Eq. [6.1]:

$$x_i = x_i(X_j, t) \quad i,j = 1, 2, 3 \tag{6.1}$$

At time $t = 0$, the following is given:

$$\begin{aligned} x_i(X_j, 0) &= X_i \\ \dot{x}_i(X_j, 0) &= V_i(X_j, 0) \end{aligned} \tag{6.2}$$

where V_i is the initial velocity.

6.2.2.1 Governing equations

We seek a solution to the momentum equation as follows:

$$\sigma_{ij,j} + \rho f_i = \rho \ddot{x}_i \tag{6.3}$$

that satisfies the contact discontinuity along an interior boundary $x_i^+ = x_i^-$:

$$\left(\sigma_{ij}^+ - \sigma_{ij}^-\right) n_i = 0 \tag{6.4}$$

Here σ_{ij} is the Cauchy stress, ρ is the current density, f is the body force density, $\ddot{x}$ is the acceleration, the comma denotes covariant differentiation and n_i is a unit outward normal to a boundary element.

Mass conservation is trivially stated as follows:

$$\rho V = \rho_0 \tag{6.5}$$

where V is the relative volume, that is, the determinant of the deformation gradient matrix, F_{ij}, and ρ_0 is the reference density.

$$F_{ij} = \frac{\partial x_i}{\partial X_j} \tag{6.6}$$

The energy equation, shown in Eq. [6.7], is integrated in time and is used for equation of state evaluations and a global energy balance.

$$\dot{E} = V S_{ij} \dot{\varepsilon}_{ij} - (p + q)\dot{V} \tag{6.7}$$

In Eq. [6.7], S_{ij} and p represent the deviatoric stresses and pressure, respectively, q is the bulk viscosity, and $\dot{\varepsilon}_{ij}$ is the strain rate tensor.

$$S_{ij} = \sigma_{ij} + (p + q)\delta_{ij}$$
$$p = -\frac{1}{3}\sigma_{ij}\delta_{ij} - q = -\frac{1}{3}\sigma_{kk} - q \qquad [6.8]$$

where δ_{ij} is the Kronecker delta. The discussion on the advantage of bulk viscosity in the numerical calculation of impact dynamics can be found in Refs [61] and [62].

We can write the following:

$$\int_V (\rho\ddot{x}_i - \sigma_{ij,j} - \rho f_i)\delta x_i \, dV + \int_{S^0} \left(\sigma_{ij}^{+} - \sigma_{ij}^{-}\right) n_j \delta x_i \, dS + \int_{S^1} (\sigma_{ij} n_j - t_i)\delta x_i \, dS = 0 \qquad [6.9]$$

where δx_i satisfies all boundary conditions on S^2, and the integrations are over the current geometry. Applications of the divergence theorem gives Eq. [6.10]:

$$\int_V (\sigma_{ij}\delta x_i)_{,j} \, dV = \int_{S^1} \sigma_{ij} n_j \delta x_i \, dS + \int_{S^0} \left(\sigma_{ij}^{+} - \sigma_{ij}^{-}\right) n_j \delta x_i \, dS \qquad [6.10]$$

where Eq. [6.11] is noted:

$$(\sigma_{ij}\delta x_i)_{,j} - \sigma_{ij,j}\delta x_i = \sigma_{ij}\delta x_{i,j} \qquad [6.11]$$

This leads to the weak form of the equilibrium equations:

$$\delta\pi = \int_V \rho\ddot{x}_i \delta x_i \, dV + \int_V \sigma_{ij}\delta x_{i,j} \, dV - \int_V \rho f_i \delta x_i \, dV - \int_{S^1} t_i \delta x_i \, dS = 0 \qquad [6.12]$$

Eq. [6.12] is a statement of the principles of virtual work.

6.2.2.2 FE discretization

We superimpose a mesh of FE interconnected at nodal points on a reference configuration and track particles through time:

$$x_i(X_j, t) = x_i(X_j(\xi, \eta, \zeta), t) = \sum_{j=1}^{k} \phi_j(\xi, \eta, \zeta) x_i^j(t) \qquad [6.13]$$

where ϕ_j are the shape functions of the parametric coordinates (ξ, η, ζ), k is the number of nodal points defining the element and x_i^j is the nodal coordinate of the jth node in the ith direction.

Summing the n elements, we may approximate $\delta\pi$ with the following:

$$\delta\pi = \sum_{m=1}^{n} \delta\pi_m = 0 \qquad [6.14]$$

And, we may write Eq. [6.15]:

$$\sum_{m=1}^{n}\left\{\int_{V_m} \rho\ddot{x}_i\mathbf{\Phi}_i^m\, dV + \int_{V_m} \sigma_{ij}^m\mathbf{\Phi}_{i,j}^m\, dV - \int_{V_m} \rho f_i\mathbf{\Phi}_i^m\, dV - \int_{S^1} t_i\mathbf{\Phi}_i^m\, dS\right\} = 0 \qquad [6.15]$$

where the following is true:

$$\mathbf{\Phi}_i^m = (\phi_1, \phi_2, \cdots, \phi_k)_i^m \qquad [6.16]$$

In matrix notation, Eq. [6.15] becomes the following:

$$\sum_{m=1}^{n}\left\{\int_{V_m} \rho\mathbf{N}^t\mathbf{N}\mathbf{a}\, dV + \int_{V_m} \mathbf{B}^t\boldsymbol{\sigma}\, dV - \int_{V_m} \rho\mathbf{N}^t\mathbf{b}\, dV - \int_{S^1} \mathbf{N}^t\mathbf{t}\, dS\right\}^m = 0 \qquad [6.17]$$

where $\mathbf{N}$ is the interpolation matrix, $\boldsymbol{\sigma}$ is the stress vector and the following is true:

$$\boldsymbol{\sigma}^t = (\sigma_{xx}, \sigma_{yy}, \sigma_{zz}, \sigma_{xy}, \sigma_{yz}, \sigma_{zx})$$

$\mathbf{B}$ is the strain-displacement matrix, $\mathbf{a}$ is the nodal acceleration vector:

$$\begin{bmatrix} \ddot{x}_1 \\ \ddot{x}_2 \\ \ddot{x}_3 \end{bmatrix} = \mathbf{N}\begin{bmatrix} a_{x_1} \\ a_{y_1} \\ \vdots \\ a_{y_k} \\ a_{z_k} \end{bmatrix} = \mathbf{N}\mathbf{a} \qquad [6.18]$$

$\mathbf{b}$ is the body force load vector, and $\mathbf{t}$ are applied traction loads:

$$\mathbf{b} = \begin{bmatrix} f_x \\ f_y \\ f_z \end{bmatrix}, \quad \mathbf{t} = \begin{bmatrix} t_x \\ t_y \\ t_z \end{bmatrix}$$

After the completion of element calculation and assembly, Eq. [6.17] could be written as follows:

$$\mathbf{M}\ddot{\mathbf{x}}(t) = \mathbf{P}(\mathbf{x}, t) - \mathbf{F}(\mathbf{x}, \dot{\mathbf{x}}) \qquad [6.19]$$

where $\mathbf{M}$ is the global mass matrix, $\ddot{\mathbf{x}}(t)$ is the global nodal acceleration vector, $\mathbf{P}$ is the global load vector and the following:

$$\mathbf{F} = \sum_{m=1}^{n} \int_{V_m} \mathbf{B}^t \boldsymbol{\sigma} \, dV$$

6.2.2.3 Geometrical FE modelling

The microstructure model of the 3DOWF was meshed with hexahedral solid elements. Fig. 6.6(a) is the mesh scheme of the whole 3DOWF fabric structure. And the mesh schemes of the weft yarns, warp yarns and Z-yarns are shown in Fig. 6.6(b–d), respectively. Fig. 6.7 is the FE model of a fabric-projectile system. Fig. 6.8 is the mesh scheme for the 3DAWF and projectile system.

The calculation time required to finish the simulation depends on the mesh density. For an eight-node solid element, the critical time step size Δt_e is as follows [60]:

$$\Delta t_e = \frac{L_e}{Q + (Q^2 + C^2)^{1/2}} \qquad [6.20]$$

where Q is a function of bulk viscosity coefficients C_0 and C_1:

$$Q = \begin{cases} C_1 c + C_0 L_e |\dot{\varepsilon}_{kk}| & \text{for} \quad \dot{\varepsilon}_{kk} < 0 \\ 0 & \text{for} \quad \dot{\varepsilon}_{kk} \geq 0 \end{cases} \qquad [6.21]$$

L_e is characteristic length, and $L_e = v_e/A_{e\max}$. v_e is the element volume, $A_{e\max}$ is the area of the largest side and c is the adiabatic sound speed of element material.

For a material with m elements, the time step Δt at next calculating step is as follows:

$$\Delta t^{n+1} = \alpha \cdot \min(\Delta t_{e1}, \Delta t_{e2}, \ldots, \Delta t_{em}) \qquad [6.22]$$

The scale factor α in Eq. [6.22] is 0.9 or a small value for stability.

From Eq. [6.3], it is shown that the time required to finish calculation will become longer when the element density is higher, while the calculating results will be closer to the actual situation. The element sizes were different according to different regions, and the contact area between the bullet and the composite was meshed in finer sizes.

To find the precise results of numerical simulation, the finite element analysis (FEA) solution should be checked by means of mesh dependency. To test the mesh

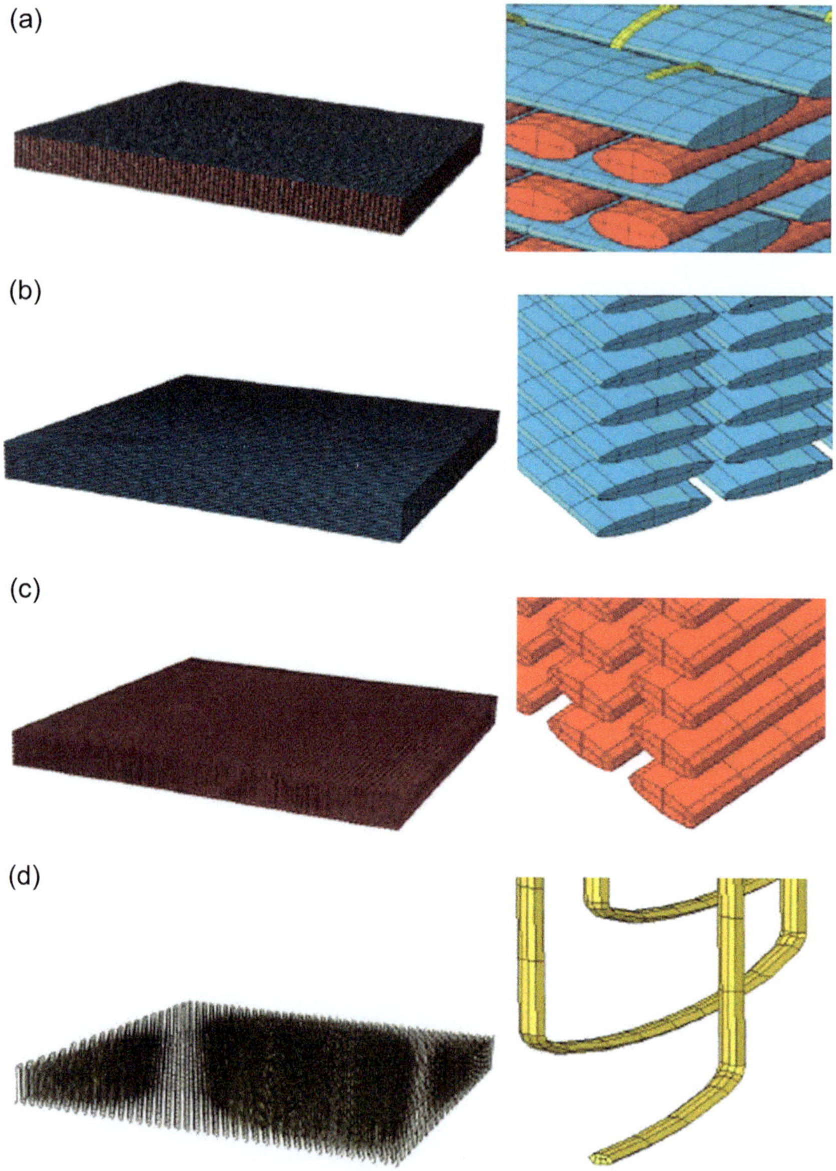

Figure 6.6 Mesh scheme of (a) 3D orthogonal woven fabric; (b) weft yarn; (c) warp yarn; (d) Z-yarn.
Jia XW, Sun BZ, Gu BH. A numerical simulation on ballistic damage of 3-D orthogonal woven fabric. International Journal of Damage Mechanics 2012;21(2):237–66.

Figure 6.7 Finite element model of projectile-fabric system.

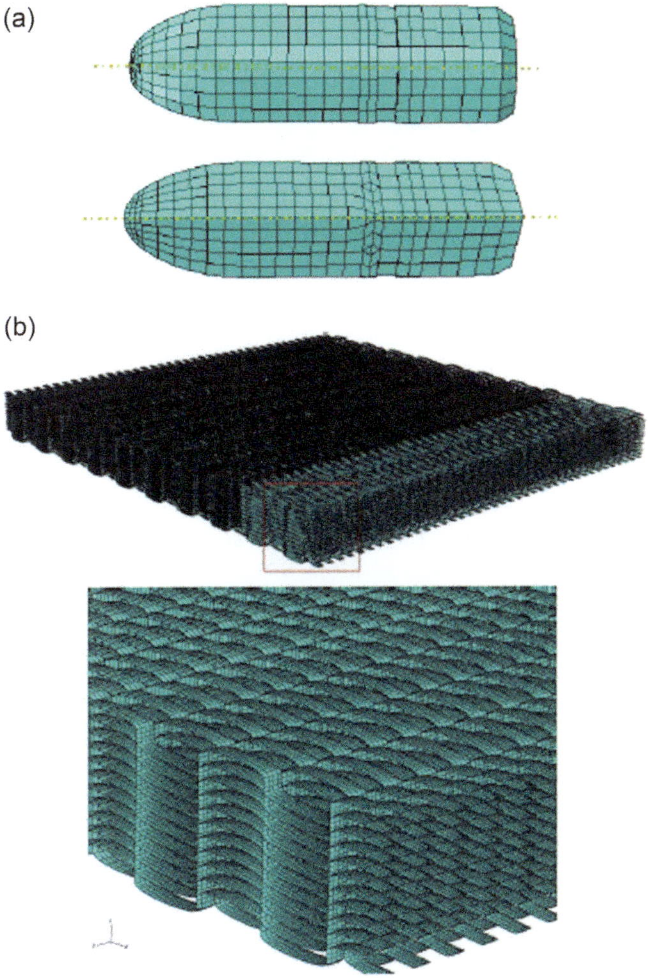

Figure 6.8 Mesh scheme of the projectile (a) and each component of the 3DAWF, (b) warp yarn; (c) weft yarn.
Jin LM, Sun BZ, Gu BH. Finite element simulation of three-dimensional angle-interlock woven fabric undergoing ballistic impact. Journal of the Textile Institute 2011;102(11):982–93.

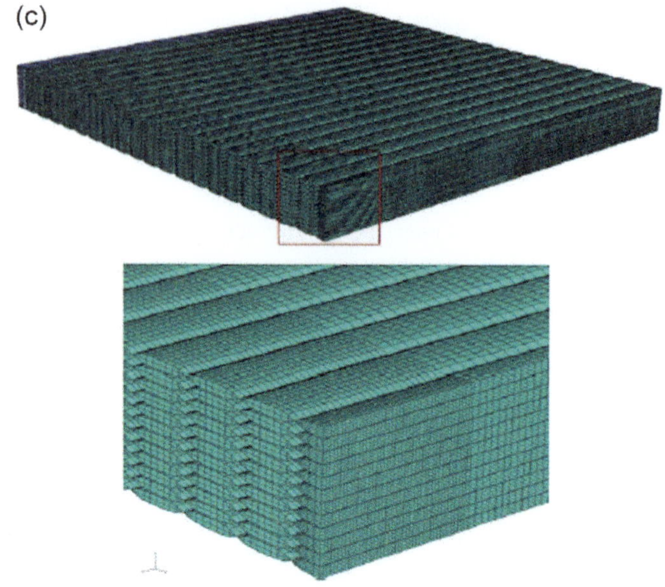

Figure 6.8 Continued.

convergence, one needs to have at least three sets of different meshes. In FE calculation, the maximum stress criteria were used for the comparison of critical failure strength and the determination of the element delete.

In the FEM model, the projectile was defined as a rigid body because the projectile was not deformed after ballistic penetration. The top centre of the projectile was selected as the reference point for the rigid body. The initial strike velocity was loaded at this reference point. The density of projectile is 7.81 g/cm^3, Young's modulus is defined as 200 GPa and Poisson's ratio is 0.292. The friction coefficient of contact between the projectile and the composite is 0.2.

6.2.3 Ballistic impact damage

Fig. 6.9 displays the impact deformation and damage of the 3DOWF target at the different time steps. In each time step, the impact damages viewed from the incident surface, rear surface and cross-section are presented. The stress wave propagated along the fibre tows and spread to the neighbouring fibre tows. In Fig. 6.9(d), the extension of the Z-yarns was clearly shown. The weft and warp yarns were bonded with the Z-yarns (Fig. 6.9(c)). When the breakage of the Z-yarns occurrs (Fig. 6.9(d) and (e)), the warp or weft yarns will be in delamination. Therefore, the higher modulus and strength Z-yarns should be used in the preparation of the 3DOWF.

Fig. 6.10(a) and (b) compares the final stage of the fabric damage with the original fabric stage. The breakage of the Z-yarns initiates the final collapse of the 3DOWF structure. We chose a node No.1041 for analysing the deflection versus time history. The deflection curves of this node under the different striking velocities were shown in Fig. 6.10(c). The maximum deflection for all the strike velocities is almost the same, while the slope of the curves varied with the strike velocities. Under higher strike velocities, the time for reaching the maximum deflection will be shorter.

Fig. 6.11 shows the ballistic impact damage of 3DAWF from FEM simulation. Three times (20, 40 and 65 μs) were chosen for illustrating the impact damage evolution. It can be seen that the stress waves in fabric were mainly propagated along the warp and weft directions. And it was found that the yarns tensile breakage was the main impact damage modus. As the ballistic penetration process goes on, the

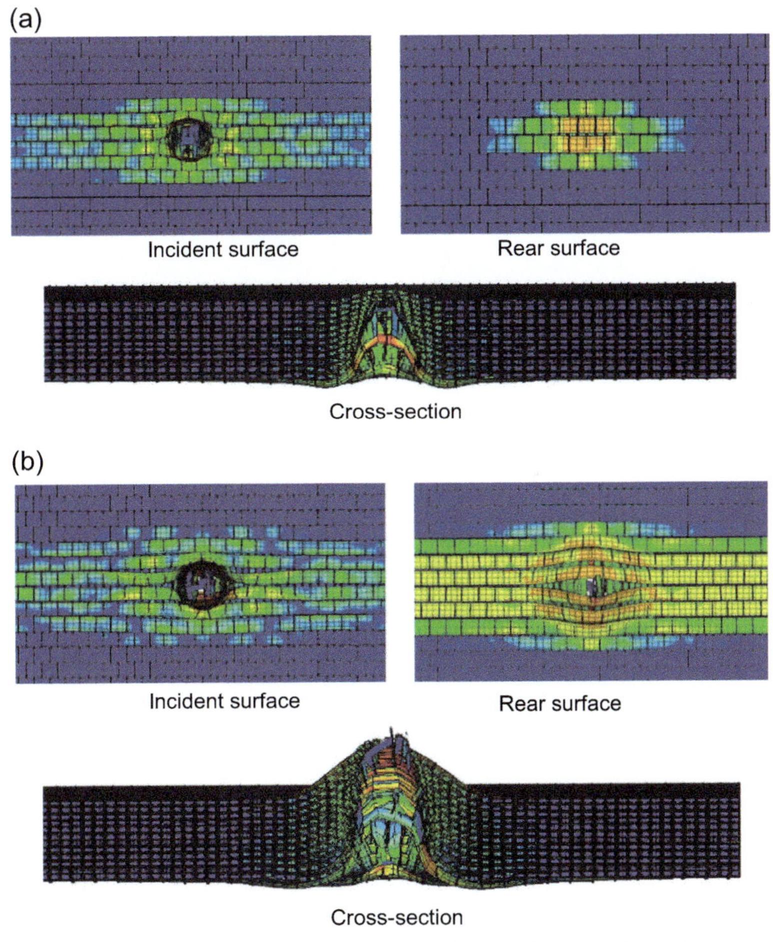

Figure 6.9 Impact deformation evolution of the 3DOWF under the strike velocity of 633 m/s. (a) $t = 13$ μs; (b) $t = 26$ μs; (c) $t = 39$ μs; (d) $t = 52$ μs; (e) $t = 65$ μs.
Jia XW, Sun BZ, Gu BH. A numerical simulation on ballistic damage of 3-D orthogonal woven fabric. International Journal of Damage Mechanics 2012;21(2):237−66.

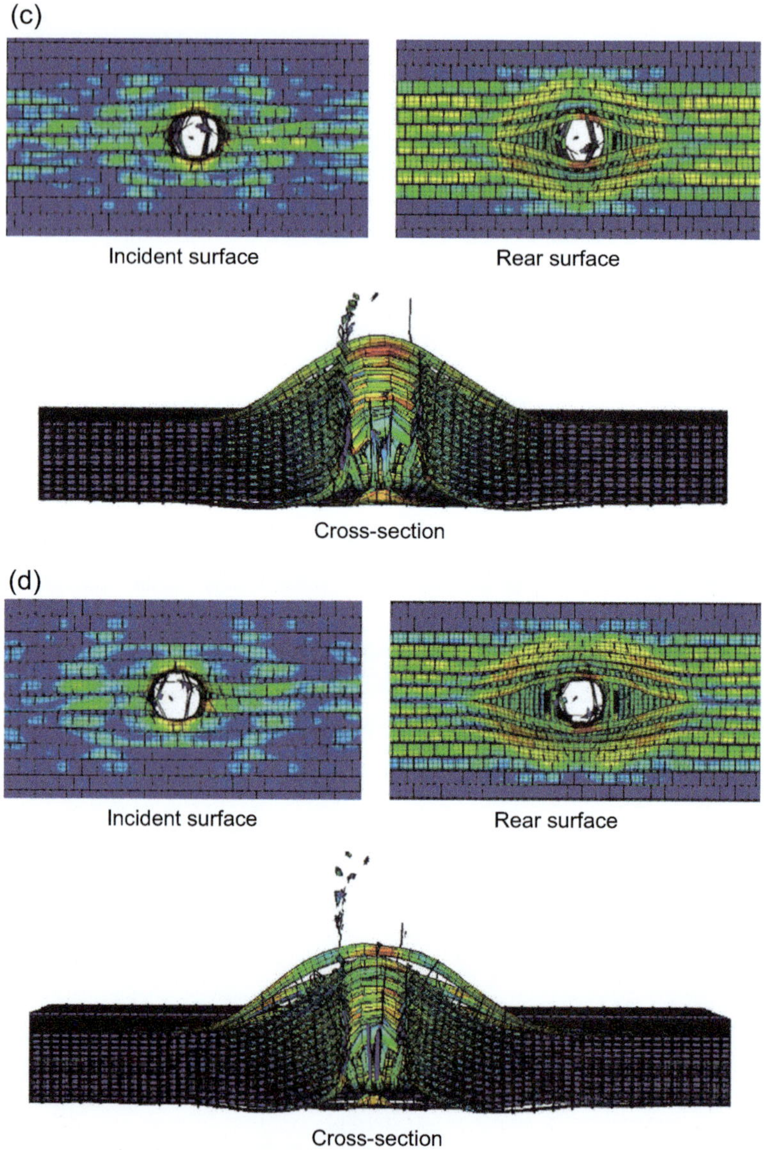

Figure 6.9 Continued.

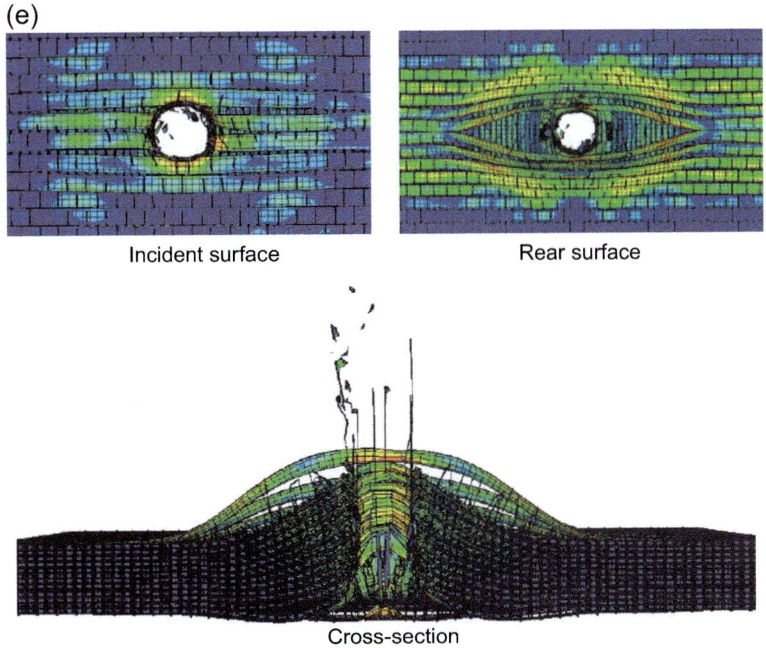

Figure 6.9 Continued.

damage area developed rapidly until the generation of ballistic hole. Fig. 6.12 depicts the fabric target, warp and weft yarn-system damage pattern at the time of 65 μs. The fracture of yarns can be found at the ballistic penetration area in Figs 6.18–6.20.

Figs 6.13 and 6.14 show the comparisons of the impact damage pattern between FEM-simulated results and experimental ones at the strike velocities of 597 and 248 m/s. Good agreements were obtained. It can be clearly observed that there was fracture of the yarns, and the yarns were pulled out of the surface of the fabric target. For the ballistic impact of 597 m/s, the result from FEM is from 0 μs (the process of ballistic penetration begins) to 65 μs (the process of ballistic penetration ends). In fact, the recovery of the elastic deformation of the fabric will occur after 65 μs. Fig. 6.13 shows the phenomenon of ballistic impact at 65 μs but not the final state. Therefore, there were few differences between the FEM result and experimental. And also for the difference of energy absorption, the main reason is that the damage mode for FEA is element deletion; while for experimental, it is the fibre breakage at the different penetration time. The most interesting and valuable fact is that the projectile was captured by the target at the velocity of 248 m/s. This means that all the energy of the projectile was absorbed by the target. It is very meaningful for lightweight armour design.

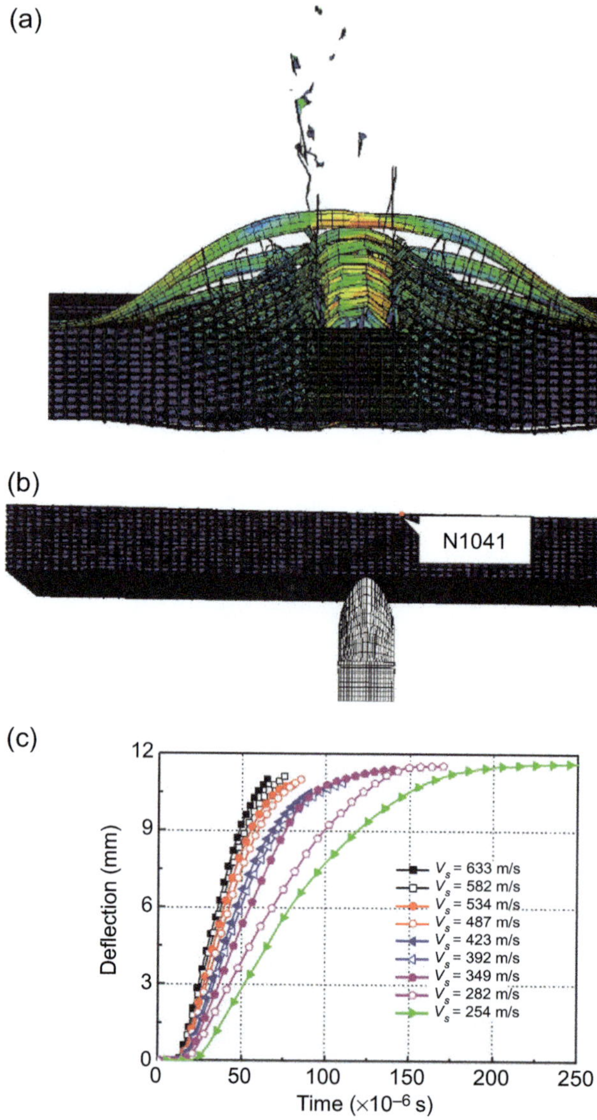

Figure 6.10 Deflection distribution of the 3DOWF under the strike velocity of 633 m/s. (a) Impact deflection; (b) location of a specific node No. 1041 in weft; (c) deflection of the node No. 1041 under different impact velocities.
Jia XW, Sun BZ, Gu BH. A numerical simulation on ballistic damage of 3-D orthogonal woven fabric. International Journal of Damage Mechanics 2012;21(2):237–66.

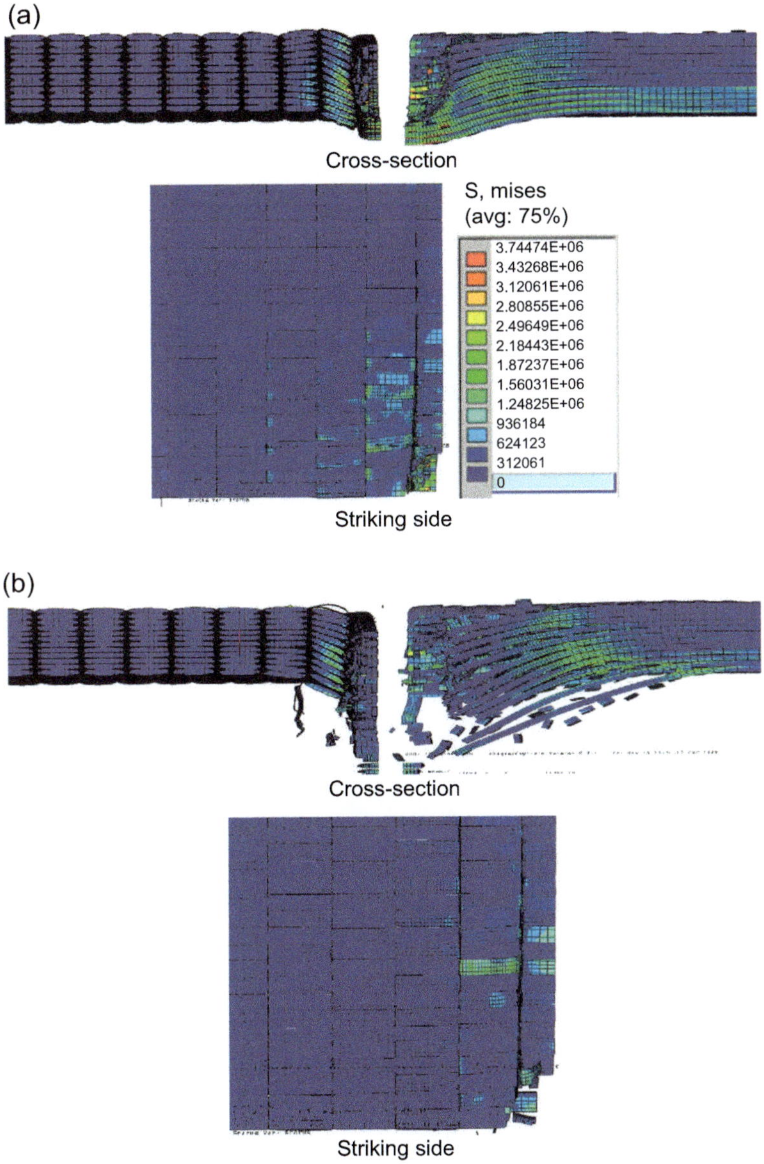

Figure 6.11 Damage evolution of the fabric under ballistic perforation ($v_s = 597$ m/s). (a) $t = 20$ μs; (b) $t = 40$ μs; (c) $t = 65$ μs.
Jin LM, Sun BZ, Gu BH. Finite element simulation of three-dimensional angle-interlock woven fabric undergoing ballistic impact. Journal of the Textile Institute 2011;102(11):982–93.

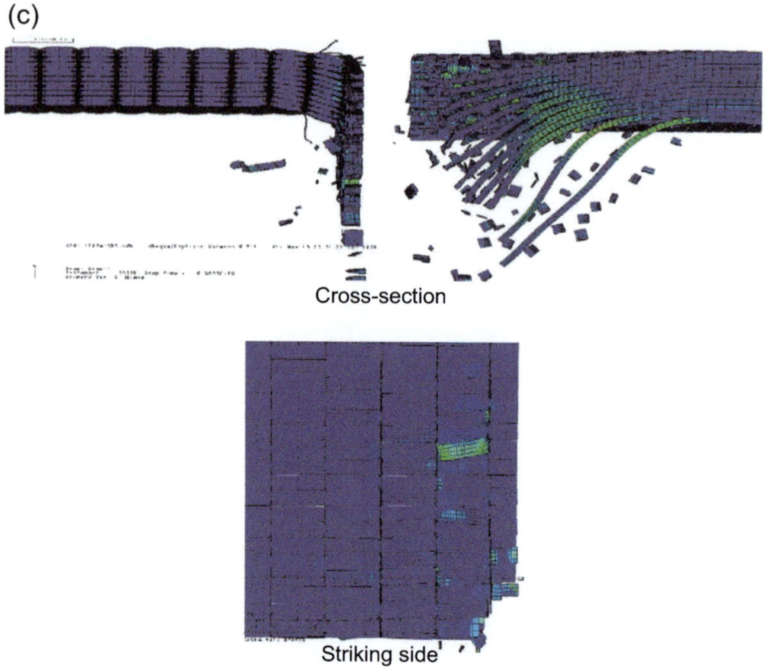

Figure 6.11 Continued.

6.3 Analytical modelling and optimization

6.3.1 3DOWF

6.3.1.1 A simple deformation model of the 3DOWF under normal ballistic impact

As shown in Fig. 6.1, the 3DOWF is composed of three kinds of yarns which are perpendicular to each other: named warp yarn, weft yarn and Z-yarn, respectively. As shown in the fabric architecture diagram (Fig. 6.1), all the warp and weft yarns are cross-plied and not interwoven. Therefore the impact deformation of the fabric can be regarded as a pyramid shape as shown in Figs 6.15 and 6.16. This kind of deformation of woven fabric has also been described by other researchers, such as Refs [63] and [64]. In Fig. 6.16, there are two stress wave regions along the warp and weft yarns. One is a longitudinal wave region, and the other is a transverse wave region. The boundary between the two regions is a rhombus, as shown in Fig. 6.16. The rhombus region increases with time until the projectile perforates through the fabric or is arrested by the fabric. And also, the fabric in the transverse wave region will move forward with the projectile.

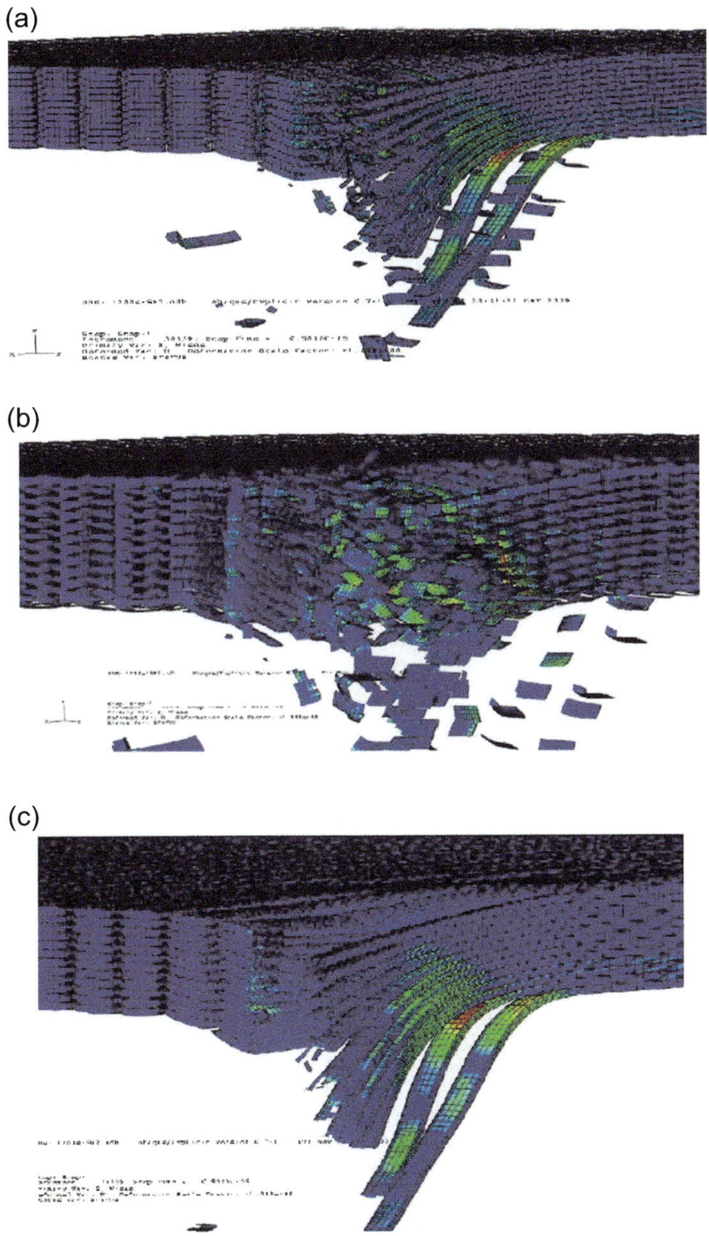

Figure 6.12 Fabric target and yarn-system impact damage at the time of 65 μs ($v_s = 597$ m/s). (a) Fabric target; (b) warp yarn-system; (c) weft yarn-system.
Jin LM, Sun BZ, Gu BH. Finite element simulation of three-dimensional angle-interlock woven fabric undergoing ballistic impact. Journal of the Textile Institute 2011;102(11):982–93.

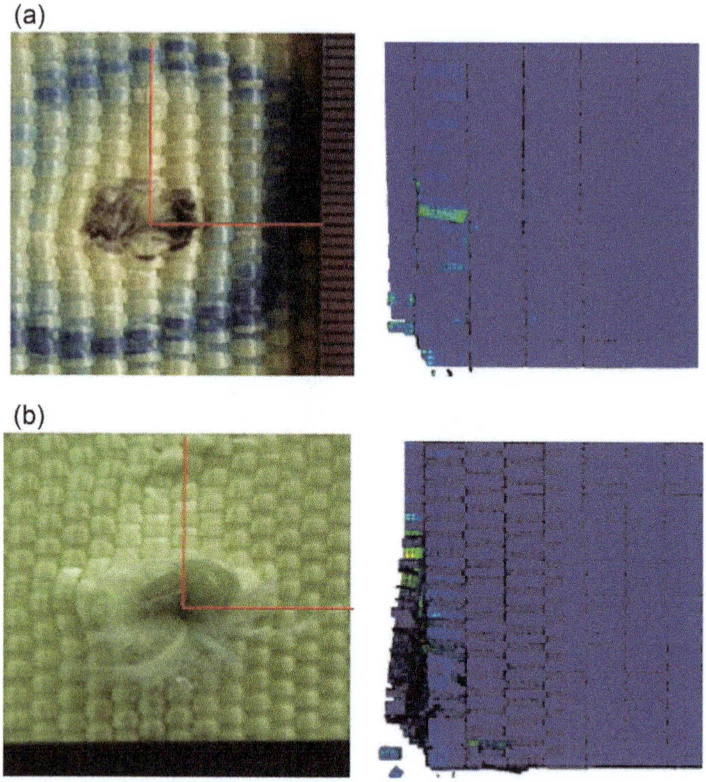

Figure 6.13 Comparison of the fabric damage between FEA and experimental ($v_s = 597$ m/s). (a) Striking side; (b) distal side.

Assuming that the heat generated during ballistic penetration can be neglected, then the loss of the kinetic energy of the projectile equals the energy absorbed by the 3DOWF target. By calculating the absorbed energy from fabric architecture, we can get the residual velocity of the projectile.

6.3.1.2 Energy absorption of principal yarns

As shown in Fig. 6.16, the warp or the weft yarns are defined as the principal yarns as they directly contact with the projectile. Owing to the noncrimp and noninterwoven features of warp and weft yarns in the 3DOWF, the stress wave propagation in the warp and weft yarns is similar with that in a constrained yarn. The stress wave propagation theory in a constrained yarn was fairly well developed by Smith [2–9], more recently by Daniel et al. [65] and also validated by high-speed photography and FEM modelling [13]. As shown in Fig. 6.17, at a given striking velocity v_s, the stress wave propagation can be divided into five periods.

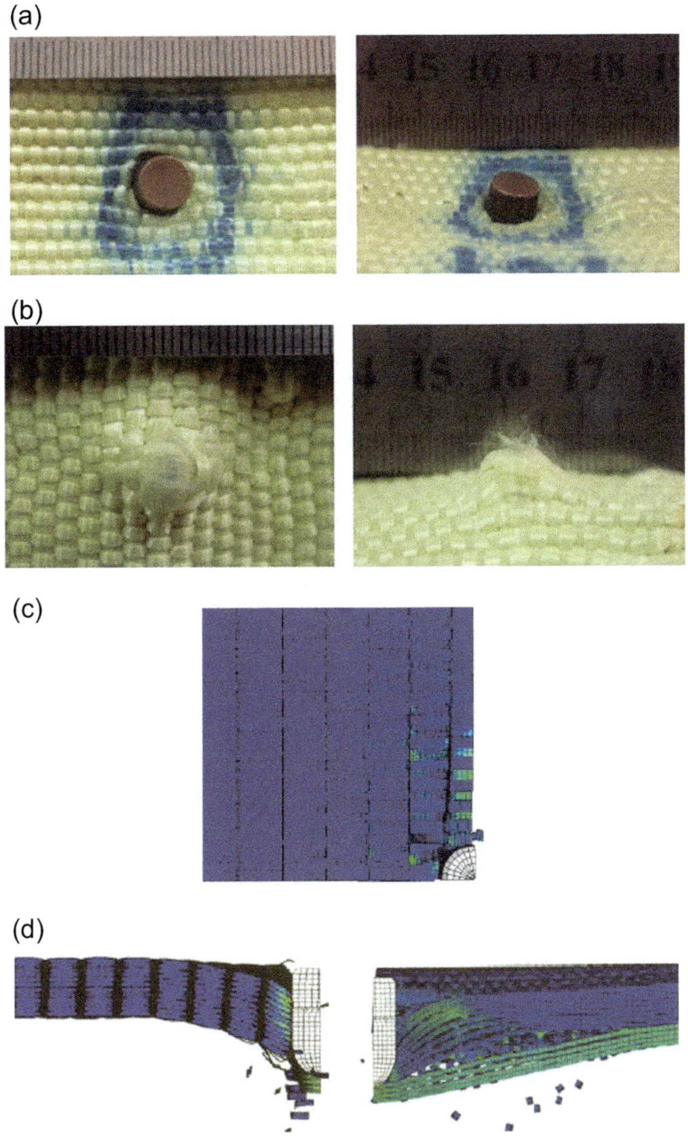

Figure 6.14 Comparison of the fabric damage between FEA and experimental ($v_s = 248$ m/s). (a) Striking side; (b) distal side; (c) striking side; (d) cross-section.

Assuming that the yarn is linearly elastic, ahead of the elastic wave front, the strain is zero, and behind, the strain ε_{ins} (generated in yarn at the instance when the projectile contacts the yarn) is constant and can be calculated as follows [4]:

$$v_s = c\sqrt{2\varepsilon_{\text{ins}}(1 + \varepsilon_{\text{ins}}) - \varepsilon_{\text{ins}}^2} \qquad [6.23]$$

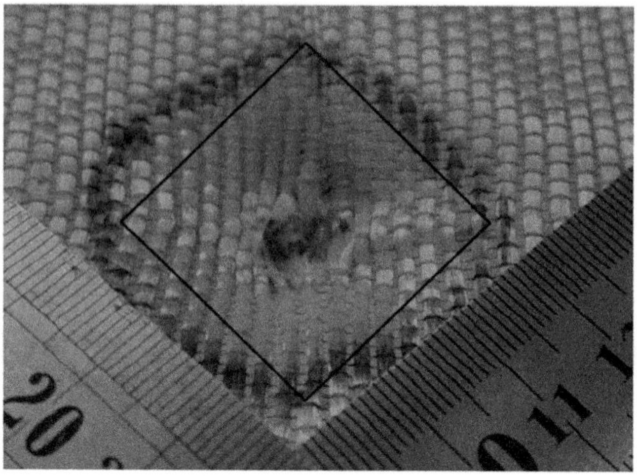

Figure 6.15 Photographs of 3DOWF (surface) after ballistic perforation.
Shi WF, Hu H, Sun BZ, Gu BH. Energy absorption of 3D orthogonal woven fabric under ballistic penetration of hemispherical-cylindrical projectile. Journal of the Textile Institute 2011; 102(10):875–89.

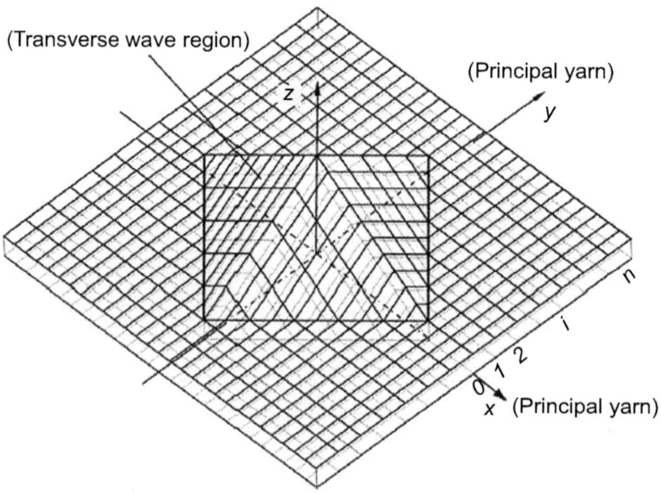

Figure 6.16 Deformation sketch of 3DOWF under ballistic impact.
Shi WF, Hu H, Sun BZ, Gu BH. Energy absorption of 3D orthogonal woven fabric under ballistic penetration of hemispherical-cylindrical projectile. Journal of the Textile Institute 2011; 102(10):875–89.

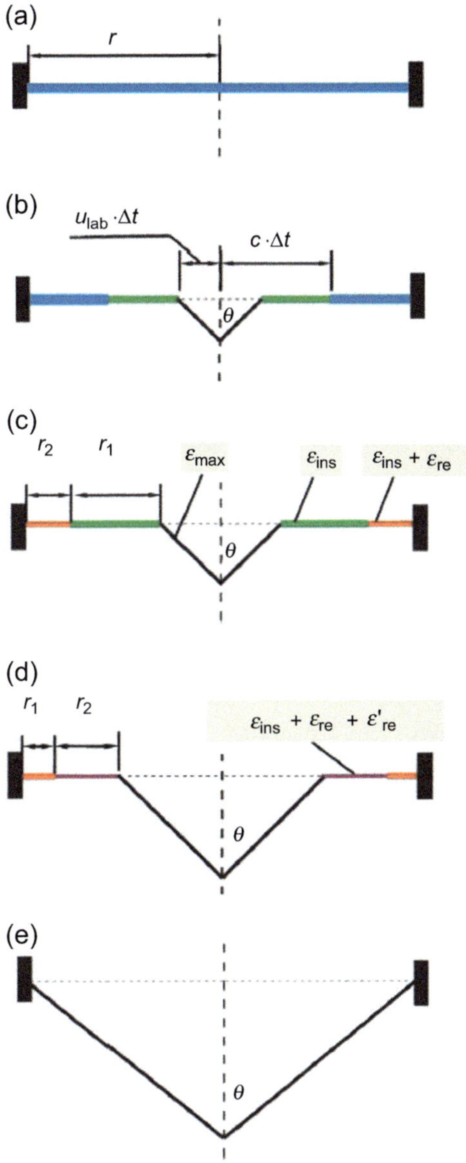

Figure 6.17 Deformation of a constrained yarn under normal impact. (a) $t = 0$; (b) $c\ \Delta t \leq r$; (c) strain wave reflects on the clamping point; (d) strain wave reflection on the impact point; (e) $u_{lab}\ \Delta t > r$.
Shi WF, Hu H, Sun BZ, Gu BH. Energy absorption of 3D orthogonal woven fabric under ballistic penetration of hemispherical-cylindrical projectile. Journal of the Textile Institute 2011;102(10):875–89.

where v_s is the striking velocity of the projectile, c is the longitudinal wave velocity in the yarn and $c = \sqrt{E/\rho}$ (E is the Young's modulus at high strain rate, and ρ is the volume density of the yarn).

$$u_{\text{lab}} = c\left(\sqrt{\varepsilon_{\text{ins}}(1+\varepsilon_{\text{ins}})} - \varepsilon_{\text{ins}}\right) \tag{6.24}$$

$$u = c\sqrt{\varepsilon_{\text{ins}}/(1+\varepsilon_{\text{ins}})} \tag{6.25}$$

u_{lab} is the transverse wave velocity with respect to the laboratory, and u is the velocity of transverse wave front relative to points on the strained yarn (ie, u is the Lagrangian wave speed). And the angle θ between the line of impact and the yarn can be given as follows:

$$\cos\theta = \frac{1}{1+\varepsilon_{\text{ins}}}\frac{v_s}{u} \tag{6.26}$$

Because the yarn is a flexible material, the ballistic impact damage of the yarn can be attributed to tensile failure. The tensile strain energy stored in per cubic metre yarn is $\int_0^{\varepsilon_{\max}} \sigma(\varepsilon)\,d\varepsilon$. Then the strain energy of the principal yarn under ballistic impact after time Δt is calculated next.

When $c\Delta t \le r$, in Fig. 6.17(b), the strain energy of the principal yarn can be written as follows:

$$W_{sy} = 2c\Delta tS\int_0^{\varepsilon_{\max}} \sigma(\varepsilon)\,d\varepsilon + 2LS\int_{\varepsilon_{\text{ins}}}^{\varepsilon_{\max}} \sigma(\varepsilon)\,d\varepsilon \quad 0 < \Delta t \le \frac{r}{c} \tag{6.27}$$

where $L = \frac{u_{\text{lab}}\Delta t}{(1+\varepsilon_{\max})\sin\theta}$ is the original length of half the principal yarn in the transverse region.

When $c\Delta t > r$, in Fig. 6.17(c), and $u_{\text{lab}}\Delta t \le r$, the longitudinal wave reaches the fix boundary and then reflects from two boundaries. The strain of the yarn outside of the transverse wave is the summation of ε_{ins} and the reflection strain ε_{re} (ε_{ins} is equal to ε_{re} without considering energy reduction), and the strain of yarn within the transverse region would be increased from ε_{ins} to $\varepsilon_{\max}$, where $\varepsilon_{\max}$ is the failure strain of the yarn.

In Fig. 6.17(c) and (d), the principal yarn is divided into three sections: the transverse wave region, the longitudinal wave region and the reflective wave region (different colours and different thickness of the lines symbolize the different strain values).

$$n = \text{trunc}\left(\frac{c\Delta t}{r}\right), \quad L = \frac{u_{\text{lab}}\Delta t}{(1+\varepsilon_{\max})\sin\theta}$$

where n is integral part of $\frac{c \cdot \Delta t}{r}$, and when the longitudinal wave is odd-reflection, the strain of the principal yarn can be shown in Fig. 6.17(c). On the contrary, it can be shown in Fig. 6.17(d).

$$\begin{cases} r_2 = c\Delta t - nr; & r_1 = r - r_2 - L \quad (n = 2N - 1) \\ r_2 = c\Delta t - nr - L; & r_1 = r - r_2 - L \quad (n = 2N) \end{cases} \qquad [6.28]$$

where N is natural number. Then strain energy of the principal yarn could be written as follows:

$$W_{sy} = \begin{cases} 2r_2 S \int_0^{2^n \varepsilon_{ins}} \sigma(\varepsilon)\, d\varepsilon + 2r_1 S \int_0^{2^{n-1} \varepsilon_{ins}} \sigma(\varepsilon)\, d\varepsilon + 2LS \int_0^{\varepsilon_{max}} \sigma(\varepsilon)\, d\varepsilon \ (2^n \varepsilon_{ins} \le \varepsilon_{max}) \\ 2r_2 S \int_0^{\varepsilon_{max}} \sigma(\varepsilon)\, d\varepsilon + 2r_1 S \int_0^{2^p \varepsilon_{ins}} \sigma(\varepsilon)\, d\varepsilon + 2LS \int_0^{\varepsilon_{max}} \sigma(\varepsilon)\, d\varepsilon \ (2^n \varepsilon_{ins} > \varepsilon_{max}) \end{cases} \qquad [6.29]$$

where p is the biggest natural number, $p \in \{n\}$, and $2^p \varepsilon_{ins} \le \varepsilon_{max}$.

In Fig. 6.17(d), when $u_{lab}\Delta t > r$, the transverse wave reaches the clamping point, and the whole yarn is under the tensile state with the strain ε_{max}.

$$W_{sy} = 2rS \int_0^{\varepsilon_{max}} \sigma(\varepsilon)\, d\varepsilon, \quad \frac{r}{u_{lab}} < \Delta t \qquad [6.30]$$

Because only the principal yarn behind the transverse wave front moves forward with the projectile, the principal yarn has the highest velocity v at the striking point and the minimum value of zero at the transverse wave front boundary. The velocity decreases linearly from the impacting point v to zero along the principal yarn; it is assumed $\frac{v}{2}$ in average as the velocity of the principal yarn for calculation. The kinetic energy of the principal yarn at different time intervals can be written as follows:

$$W_{ky} = \frac{L \cdot \mathrm{Tex}}{10^6} \left(\frac{v}{2}\right)^2, \quad 0 < \Delta t \le \frac{r}{u_{lab}} \qquad [6.31]$$

$$W_{ky} = \frac{r \cdot \mathrm{Tex}}{10^6} \left(\frac{v}{2}\right)^2, \quad \Delta t > \frac{r}{u_{lab}} \qquad [6.32]$$

6.3.1.3 Energy absorption of slave yarns of warp and weft

In Fig. 6.16, the slave yarns are defined as the yarns in the 3DOWF except the principal yarns. To simplify the model, only one layer of 3DOWF without Z-yarn under ballistic impact is analysed first, as shown in Fig. 6.16. The deformation region shows that points on the principal yarns within the transverse wave region appear to be moving faster than adjacent points on other parallel yarns in the impacting direction. In Fig. 6.18, it shows strain distribution in different yarns: the strain of the ith slave yarn along the warp or weft direction can be calculated by linearly diminishing from the principal yarn strain to zero. At the same time, each slave yarn can be calculated by linear interpolation from the strain of maximum value to zero. The same rule can also be applied to the velocity distribution of the salve yarns. Therefore the strain energy and the kinetic energy of slave yarns can be written as follows:

$$n = \mathrm{round}(u_{\mathrm{lab}} \cdot \Delta t \cdot n_X) \quad \text{or} \quad n = \mathrm{round}(u_{\mathrm{lab}} \cdot \Delta t \cdot n_Y) \tag{6.33}$$

where n (n_x or n_y) is the round-off number of the warp or weft slave yarns in half of the rhombus region;

$$l_i = L\left(1 - \frac{i}{n}\right) \tag{6.34}$$

where l_i is the original length of half of the ith slave yarn.

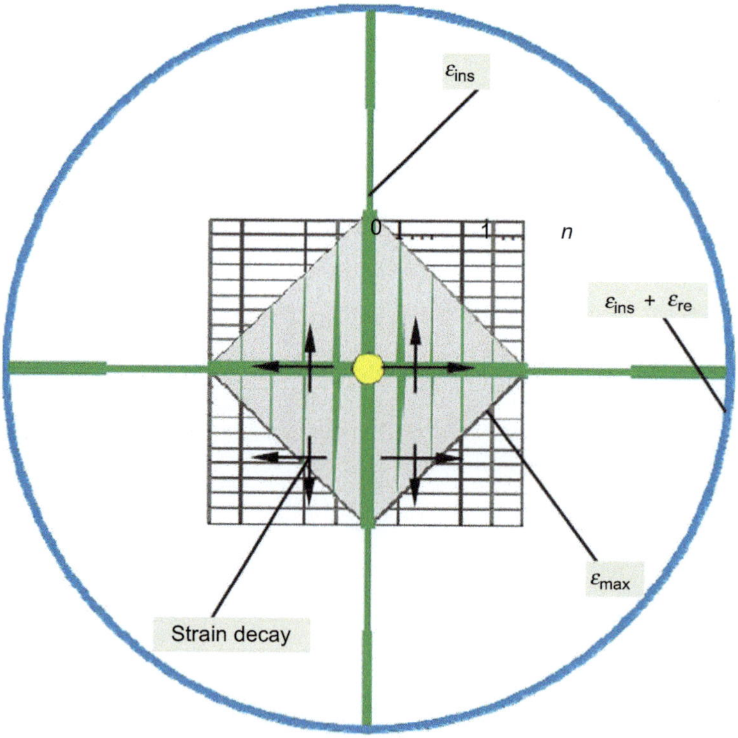

Figure 6.18 Strain propagation in 3DOWF during normal ballistic impact.
Shi WF, Hu H, Sun BZ, Gu BH. Energy absorption of 3D orthogonal woven fabric under ballistic penetration of hemispherical-cylindrical projectile. Journal of the Textile Institute 2011; 102(10):875–89.

$$\varepsilon_{ix} = \left(1 - \frac{i}{n}\right) \cdot \varepsilon_{\max} \cdot x / l_i \quad [6.35]$$

where ε_{ix} (ε_{ixx}, ε_{iyx} for warp and weft, respectively) is the strain of the ith slave yarn in the x position. Then the strain energy of the ith slave yarn is shown in Eq. [6.36]:

$$W_i = 2 l_i S \int_0^{l_i} \left(\int_0^{\varepsilon_{ix}} \sigma(\varepsilon)\, d\varepsilon \right) dx \quad [6.36]$$

Therefore, we arrive at the following:

$$W_{ix} = 2 l_{ix} S_x \int_0^{l_{ix}} \left(\int_0^{\varepsilon_{ixx}} \sigma_x(\varepsilon)\, d\varepsilon \right) dx \quad [6.37]$$

$$W_{iy} = 2 l_{iy} S_y \int_0^{l_{iy}} \left(\int_0^{\varepsilon_{iyx}} \sigma_y(\varepsilon)\, d\varepsilon \right) dx \quad [6.38]$$

The whole strain energy of the rhombus region of one layer can be written as Eq. [6.39]:

$$W_s = W_{syx} + W_{syy} + 2 \sum_{i=1}^{n_x} W_{ix} + 2 \sum_{i=1}^{n_y} W_{iy} \quad [6.39]$$

The sum of kinetic energy of warp yarns from 1 to n is as follows:

$$W_{kfx} = 4 \sum_{i=1}^{n_x} \frac{1}{2} \frac{l_{ix} T_x}{10^6} \left(\frac{v}{2} \left(1 - \frac{i}{n_x} \right) \right)^2 = 2 \sum_{i=1}^{n_x} L \frac{\left(1 - \frac{i}{n_x}\right)^3 T_x}{10^6} \left(\frac{v}{2}\right)^2 \quad [6.40]$$

In the same way, the sum of kinetic energy of weft yarns from 1 to n is obtained next:

$$W_{kfy} = 2 \sum_{i=1}^{n_y} L \frac{\left(1 - \frac{i}{n_y}\right)^3 T_y}{10^6} \left(\frac{v}{2}\right)^2 \quad [6.41]$$

6.3.1.4 Z-yarns analysis

When 3DOWF is impacted by a projectile, Z-yarns bind the noninterwoven warp and weft yarns together and prevent delamination. Because Z-yarns run through the

3DOWF, it is assumed only the Z-yarns in the rhombus region move forward along impacting direction and only the ones contacting the projectile (which are called the principal Z-yarns) will break. The following hypotheses were made to simplify the analysis of Z-yarn deformation.

1. The strain of the principal Z-yarns linearly decreases from the impacting point ε_{max} to zero in the warp direction.
2. The strain of the slave Z-yarns parallel to the principal Z-yarns along the weft direction can be calculated by linear interpolation.
3. The same rule can also be applied to the velocity distribution of the Z-yarns.

In Fig. 6.19, the black arrow shows the penetration direction, and the line represents the strain in the principal Z-yarns. Higher thickness corresponds to lower strain value. And, the strain energy and kinetic energy of the principal Z-yarns can be calculated as follows:

$$n_z = \text{round}(u_{\text{lab}} \cdot \Delta t \cdot n_Z) \tag{6.42}$$

where n_z is the round-off number of the slave Z-yarns in half of the rhombus region;

$$L_z = u_{\text{lab}}\Delta t + u_{\text{lab}}\Delta t n_z h_z \tag{6.43}$$

where L_z is half of the original length of the principal Z-yarn in the transverse wave region, and h_z is the thickness of fabric. Then the strain energy of the principal Z-yarn can be written as follows:

$$W_{syz} = 2S_z \int_0^{L_z} \left(\int_0^{\frac{\varepsilon_{max}}{L_z}x} \sigma_z(\varepsilon)\, d\varepsilon \right) dx \tag{6.44}$$

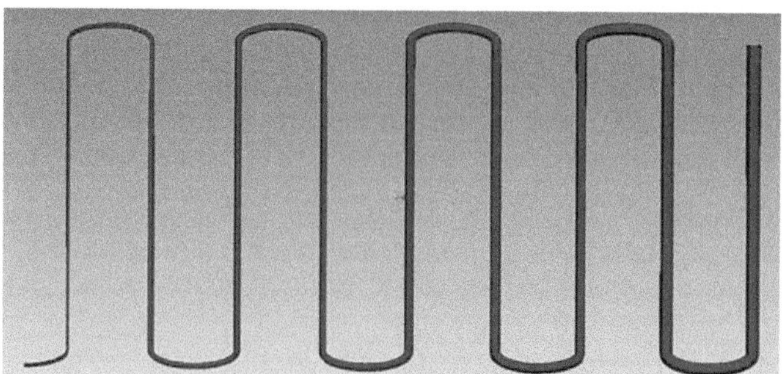

Figure 6.19 Z-yarn penetrated by the projectile (the thinner, the larger strain).
Shi WF, Hu H, Sun BZ, Gu BH. Energy absorption of 3D orthogonal woven fabric under ballistic penetration of hemispherical-cylindrical projectile. Journal of the Textile Institute 2011; 102(10):875–89.

And the kinetic energy of the principal Z-yarn is shown next:

$$W_{kyz} = \frac{L_z T_z}{10^6}\left(\frac{v}{2}\right)^2 \quad [6.45]$$

The strain energy and kinetic energy of the slave Z-yarns can be calculated as follows:

$$l_{iz} = L_z\left(1 - \frac{i}{n_z}\right) \quad [6.46]$$

where l_{iz} is the half length of the ith Z-yarn in the transverse wave region;

$$\varepsilon_{izx} = \frac{\left(1 - \frac{i}{n_z}\right)\cdot \varepsilon_{\max z}}{l_{iz}} x \quad [6.47]$$

where ε_{izx} is the strain of the ith slave Z-yarn in the x position. Then the strain energy of the ith slave yarn is as follows:

$$W_{sfz} = 4\sum_{i=1}^{n_z}\int_0^{l_{iz}}\left(\int_0^{\varepsilon_{izx}} \sigma_z(\varepsilon)d\varepsilon\right)dx \quad [6.48]$$

$$W_{skz} = 4\sum_{i=1}^{n_z}\frac{1}{2}\frac{l_{iz}T_z}{10^6}\left(\frac{v}{2}\left(1 - \frac{i}{n_z}\right)\right)^2 = 2\sum_{i=1}^{n_z} L\frac{\left(1 - \frac{i}{n_z}\right)^3 T_z}{10^6}\left(\frac{v}{2}\right)^2 \quad [6.49]$$

6.3.1.5 Energy conservation function

As for 3DOWF, the friction between different layers of warp and weft yarn can be neglected because there is no interlacing between different layers. In the analytical model, it is assumed that every layer can absorb the same energy and reach the maximum strain rate at the same time. The residual velocity of projectile v_r can replace v in the earlier equations, and then the kinetic energy absorbed by the fabric could be obtained as follows:

$$\begin{aligned} W = {} & N_x(W_{syx} + W_{skx} + W_{sfx} + W_{skx}) + N_y(W_{syy} + W_{sky} + W_{sfy} + W_{sky}) \\ & + (W_{syz} + W_{skz} + W_{sfz} + W_{skz}) = \frac{1}{2}m\left(v_s^2 - v_r^2\right) \end{aligned} \quad [6.50]$$

where N_x is the layers of warp yarn, and N_y is the layers of weft yarn.

Then the residual velocity v_r of the projectile can be calculated, and energy distribution in the rhombus region can be obtained.

6.3.2 3DAWF

From impact deformation and damage of the 3DAWF shown in Fig. 6.14, it was found that only the yarns in direct contact with the projectile would be fractured, while the other yarns in the deformation region would not. The yarns in the deformation region can be divided into two categories: (a) principal yarns, which directly contacted with the projectile, and (b) secondary yarns. Fig. 6.20 indicates the number of yarns in direct contact with the projectile, that is, the number of principal yarns. Fig. 6.16 is a

Figure 6.20 Number of the principal yarns (contacted with projectile directly during ballistic penetration).
Shi WF, Hu H, Sun BZ, Gu BH. Energy absorption of 3D orthogonal woven fabric under ballistic penetration of hemispherical-cylindrical projectile. Journal of the Textile Institute 2011; 102(10):875–89.

simplified geometric model of the impact deformation of the 3DAWF, where the number zero represents a principal yarn, and 1 to N are secondary yarns.

Compared with the flexible fibre tows, the projectile of Type 56 (steel core coated with copper skin) can be regarded as a rigid body. The deformation of the projectile in modelling can be neglected, assuming the following:

1. The contacting point between the projectile and the fabric target is located at the interweave point of warp and weft yarns.
2. Friction between warp and weft yarns is ignored under high-velocity impact.
3. Filament tow is a linear-elastic material.

The energy absorptions are from (1) strain energy of the principal yarns, (2) kinetic energy of the principal yarns, (3) strain energy of secondary yarns in a single layer of 3DWAF and (4) kinetic energy of secondary yarns in a single layer of 3DAWF. They can be calculated as those of 3DOWF. Then the energy absorption of the whole fabric during ballistic impact equals the product of absorbed energy of a single-layer fabric and the number of layers. And, the heat energy generated in the whole process of ballistic penetration can be ignored compared with the strain energy and kinetic energy. Based on the energy conservation law, the following equation can be obtained:

$$n\cdot(W_{\mathrm{SP}}+W_{\mathrm{KP}}+W_{\mathrm{SF}}+W_{\mathrm{KF}})=\frac{1}{2}\cdot m\cdot\left(v_s^2-v_r^2\right) \quad [6.51]$$

where n is the number of the fabric layers, W_{SP} is strain energy of a single principal yarn, W_{KP} is kinetic energy of a single principal yarn, W_{SF} is strain energy of secondary yarns in a single layer of 3DWAF, and W_{KF} is the kinetic energy of secondary yarns in a single layer of 3DWAF.

Then the residual velocity of the projectile could be obtained from Eq. [6.51].

As for 3DAWF, the energies absorbed by the warp and weft yarns should be calculated separately because of the different diameters and yarn crimps. All variables and parameters included in the aforementioned algorithms are distinguished with subscripts B and W, which indicate the warp and weft yarns, respectively.

The warp yarns are in crimp form while the weft yarns are in straight form in 3DAWF. The longitudinal wave velocity of the warp principal yarn should be divided by its crimp coefficient [66,67] due to its crimp. Crimp coefficient refers to the ratio between yarn actual length to its horizontal distance in fabric. Through unravelling warp yarns in 3DAWF, we found that the crimp coefficient was 1.08 for the 3DAWF in this paper. Hence, the longitudinal wave velocity of the warp principal yarns is $c_B=\frac{\sqrt{E/\rho_B}}{1.08}$ in actual calculation. Then Eq. [6.51] can be further amended to be the following:

$$n_{\mathrm{B}}(W_{\mathrm{SPB}}+W_{\mathrm{KPB}}+W_{\mathrm{SFB}}+W_{\mathrm{KFB}})+n_{\mathrm{W}}(W_{\mathrm{SPW}}+W_{\mathrm{KPW}}+W_{\mathrm{SFW}}+W_{\mathrm{KFW}})=\frac{1}{2}m\left(v_s^2-v_r^2\right) \quad [6.52]$$

where n_{B} and n_{W} are the number of the warp and weft yarn layers in the 3DAWF.

6.3.3 Ballistic penetration time interval Δt

The time interval Δt, from the beginning of impact to the instant when the 3D woven fabric is perforated or the projectile is arrested, is the important and indispensable parameter to solve the aforementioned equations. In experimental modelling, the time interval Δt can be obtained from high-speed photograph technique. This approach for determining the time interval Δt requires one to perform ballistic tests first and then to calculate the ballistic performance of the fabric target. Obviously, the ballistic performance of the 3D woven fabric cannot be obtained in advance. This is not feasible in a theoretical modelling. Here we employ the concept of strain rate to approximately estimate the time interval Δt:

$$\Delta t = \frac{\Delta \varepsilon}{\dot{\varepsilon}} = \frac{\varepsilon_{\max} - \varepsilon_{\text{ins}}}{\dot{\varepsilon}} \quad [6.53]$$

where $\dot{\varepsilon}$ is the strain rate in ballistic penetration.

Because the fabrics and yarns are in a strain rate range of 500–2000/s under ballistic penetration [68], the penetration time interval Δt can be calculated from the different strain rate values.

6.4 Energy absorption and penetration mechanisms

6.4.1 3DOWF

Fig. 6.21 shows the residual velocity versus strike velocity curves both in FEA and experimental modelling. The residual velocities from FEA agree well with those in experimental although the differences exist at low and high strike velocity ranges. This means that the FEA model agrees well with that in ballistic tests and could be used in the 3DOWF ballistic limit estimation. The difference between the FEA and experimental modelling mainly exists at the low strike velocity range, especially when the strike velocity is lower than 350 m/s. The reason for this difference most probably attributes to the Twaron® fibre tows being regarded as a continuum rod in FEM, while they are composed of thousands of filament in each tow. The unsimultaneous failure of each filament will lead the higher energy absorption in experimental modelling.

Fig. 6.22 is the velocity versus time curves of the projectile at different strike velocities obtained from FEA when the projectile penetrated the 3DOWF target plate. The penetration time is longer when the strike velocity is lower. Therefore, the projectile takes a longer time to perforate the composite target. Fig. 6.23 depicts the projectile acceleration during penetration. The absolute value of acceleration is greater when the strike velocity is higher. The fluctuation of the acceleration means the forces acting on the projectile fluctuated. The nonsimultaneous failures of the fibre tows during penetration lead to the fluctuation of the penetration resistance. This attributes to the higher strike velocity, resulting in the lower penetration time. Then the stress waves at the cross-section points between the fibre tows are more difficult to reach the

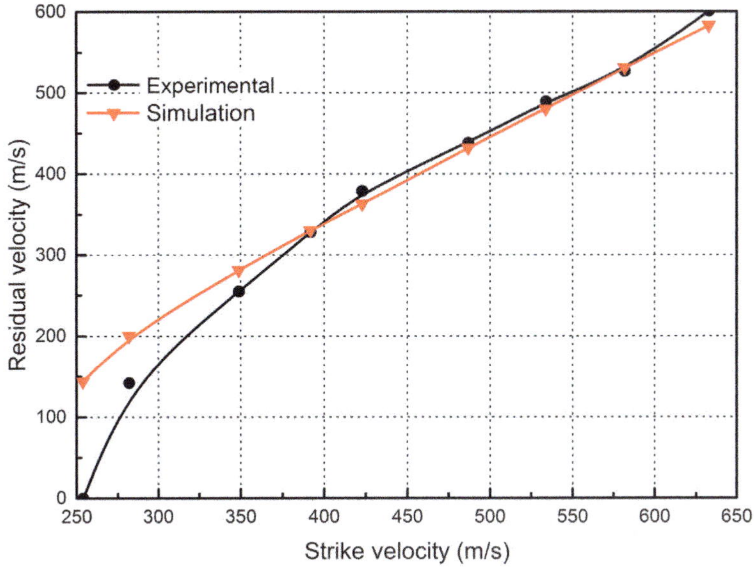

Figure 6.21 Strike velocity versus residual velocity curves.

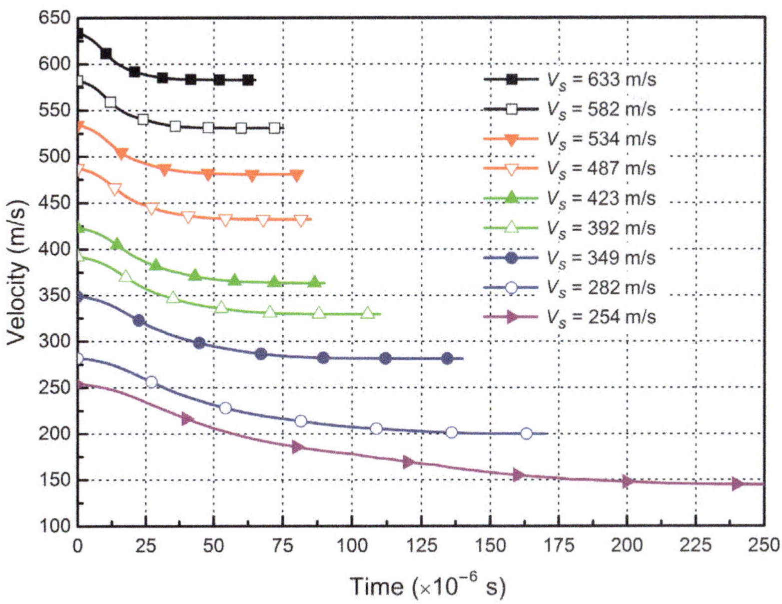

Figure 6.22 Velocity history of projectile at different strike velocities.

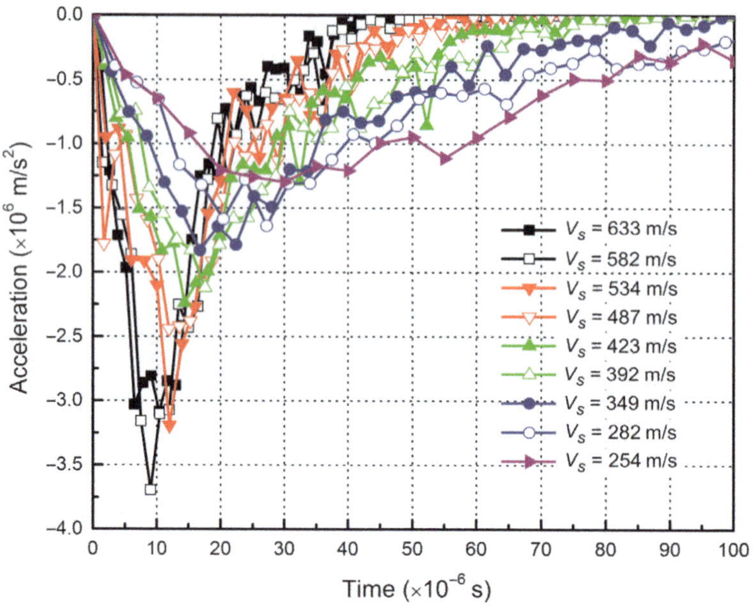

Figure 6.23 Acceleration of projectile under different strike velocities.

equilibrium state than in longer penetration time. The nonequilibrium stress wave would induce the higher stress fluctuation for the breakages of fibre tows and resins in different time steps.

Furthermore, owing to the cylindrical-conical shape of the projectile, the penetration resistance of the 3DOWF acted on the projectile varies the shape of the projectile. When the conical part of the projectile finishes penetrating the fabric, the acceleration reaches a maximum value. The time corresponding to the maximum value will be later when the strike velocity is lower. This phenomenon is obviously shown in Fig. 6.22.

Fig. 6.24 shows the damage patterns under ballistic impact. It is evident to find that the failure modes are tensile and shear failure on the rear side and compressive and shear failure on the incident side. Fig. 6.25 shows the FEA results of the impact damage. To find the failure mechanisms of the 3DOWC under impact, the impact damages of the constituent yarns of the 3DOWF were chosen for further illustration. Fig. 6.26 shows the impact damage of the 3DOWF at the time of 65 μs. Fig. 6.26(a) and (b) illustrate the same deformation mode and failure mechanics of the warp and weft yarns. The compression failure in the incident surface and the tension failure in the rear surface manifest that the high modulus and strength of the yarns will lead the higher ballistic impact energy absorption. Fig. 6.26(c) displays the impact damage of the Z-yarns. The Z-yarns bind the warp and the weft yarns together and also absorb a small fraction of the impact energy. However, the structure stability of the 3DOWF depends on the Z-yarns, and the breakage of Z-yarns will induce the structure collapse of the 3DOWF. So, the higher strength Z-yarns should be used in the 3DOWF.

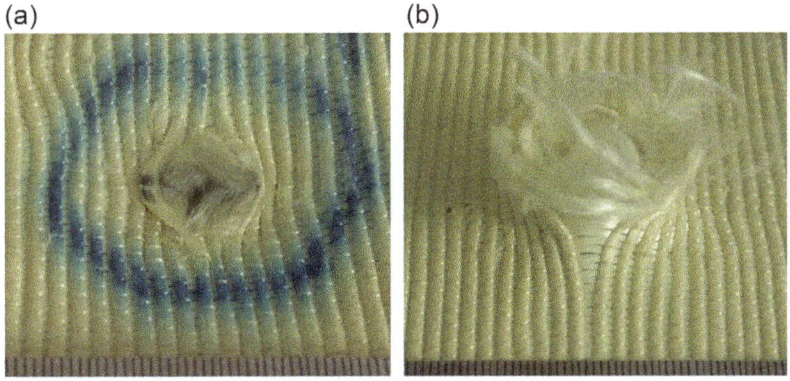

Figure 6.24 Impact damage of the 3DOWF target under the strike velocity of 633 m/s. (a) Top surface; (b) bottom surface.
Jia XW, Sun BZ, Gu BH. A numerical simulation on ballistic damage of 3-D orthogonal woven fabric. International Journal of Damage Mechanics 2012;21(2):237–66.

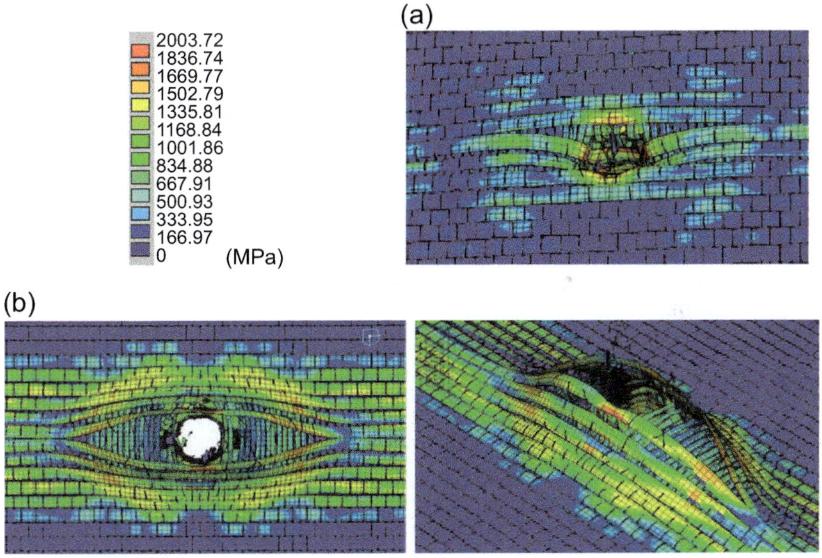

Figure 6.25 Impact damage simulation results. (a) Top surface; (b) bottom surface.
Jia XW, Sun BZ, Gu BH. A numerical simulation on ballistic damage of 3-D orthogonal woven fabric. International Journal of Damage Mechanics 2012;21(2):237–66.

6.4.2 3DAWF

Fig. 6.27 depicts the comparisons of strike velocity v_s versus residual velocity v_r curves between theoretical and experimental. Fig. 6.28 displays the velocity history of the projectiles at different strike velocities. It is interesting to note that the projectile was captured by the fabric target at the strike velocity of 248 m/s. Fig. 6.29 compares

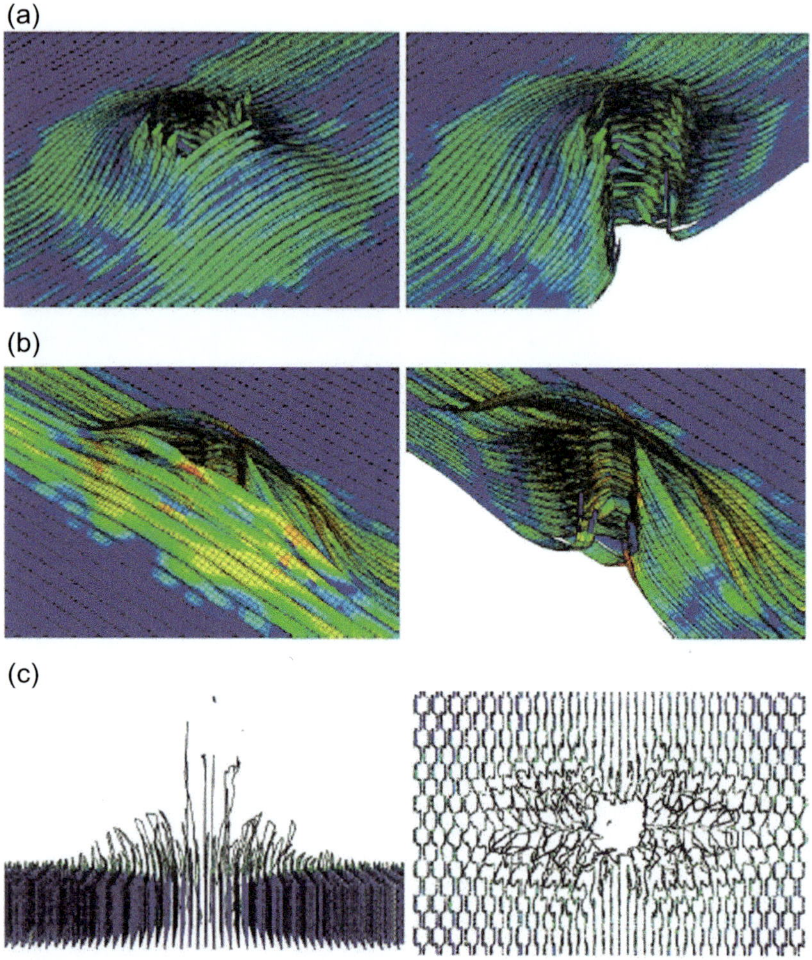

Figure 6.26 Impact damage of constituent yarns in the 3DOWF. (a) Warp yarn; (b) weft yarn; (c) Z-yarn.
Jia XW, Sun BZ, Gu BH. A numerical simulation on ballistic damage of 3-D orthogonal woven fabric. International Journal of Damage Mechanics 2012;21(2):237–66.

projectile acceleration curves during penetration. The fluctuation of the acceleration means that the forces acted on the projectile fluctuated. The reason is that as the contact area between the projectile and the fabric during ballistic penetration increased, the number of yarns which contacted with the projectile increased. And the yarns were broken at different times. These factors resulted in the fluctuation of forces acted on the projectile.

As shown in Fig. 6.30, a node N was selected for analysis of the fabric target deformation under ballistic impact. The deflection–time curves at this node under different striking velocities are shown in Fig. 6.31. Compared with the higher impact velocities,

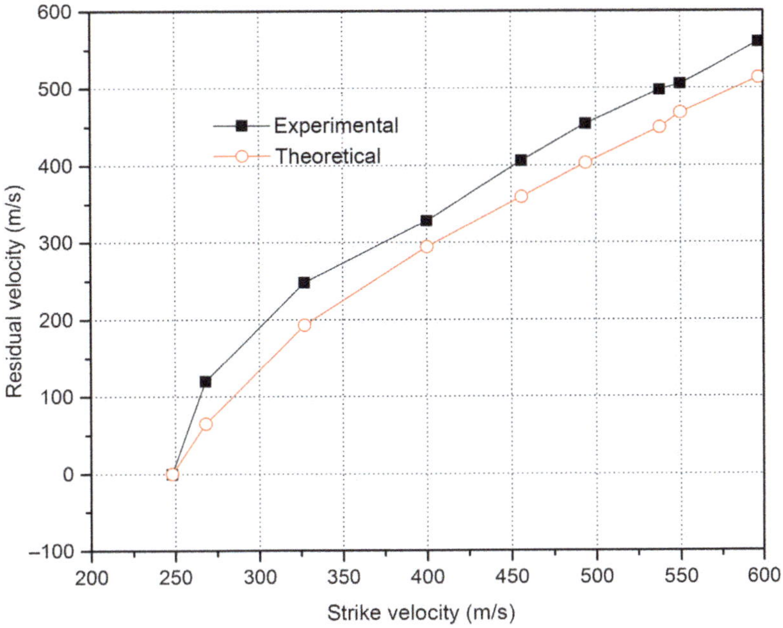

Figure 6.27 Strike velocity versus residual velocity curves.

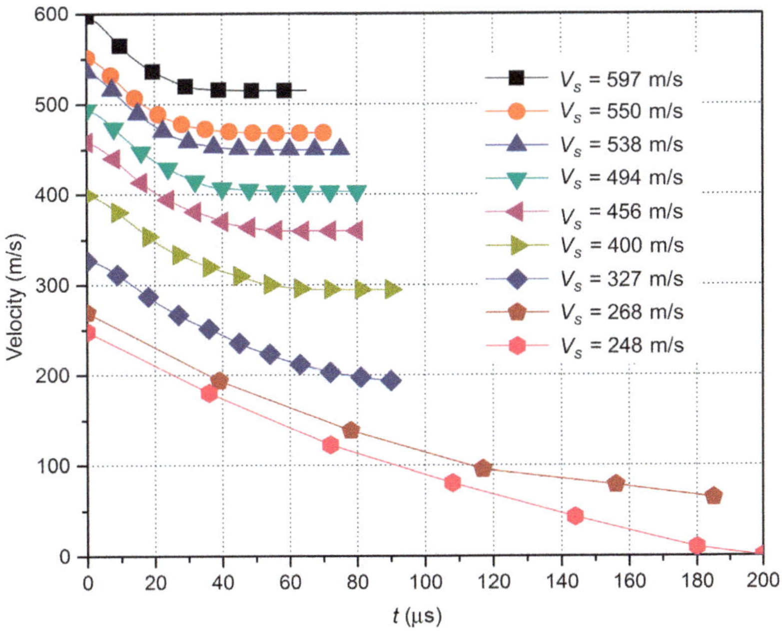

Figure 6.28 FEM results of velocity history of projectile at different strike velocities.

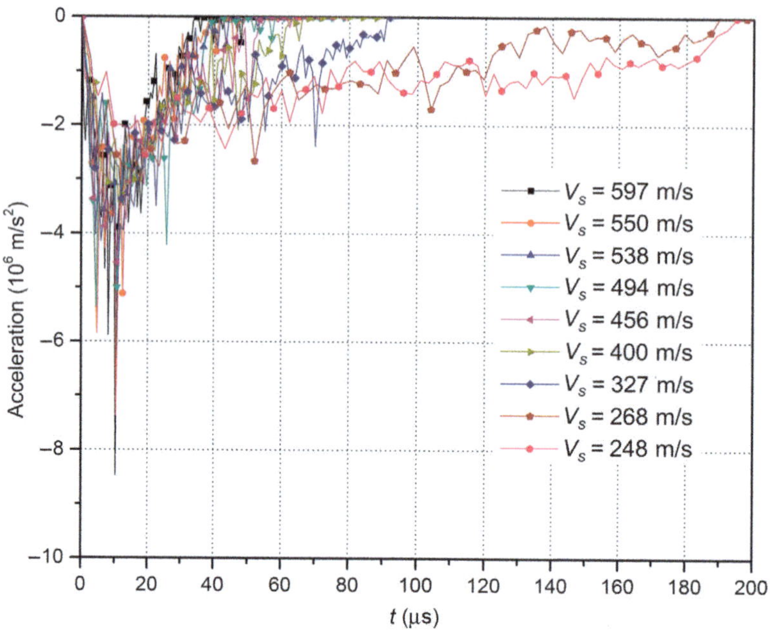

Figure 6.29 FEM results of acceleration history of projectile at different strike velocities.

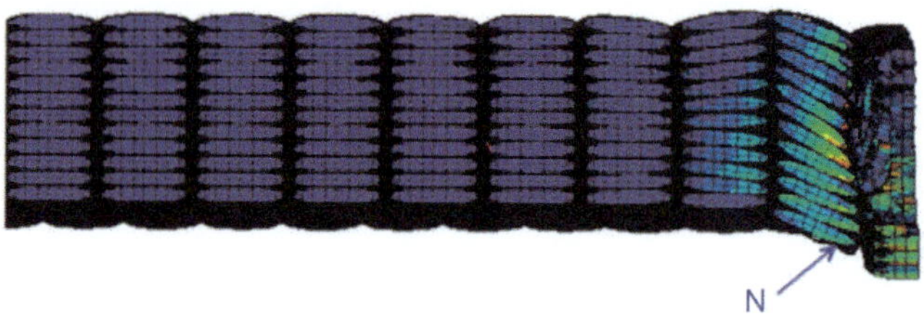

Figure 6.30 Location of node N in the model.
Jin LM, Sun BZ, Gu BH. Finite element simulation of three-dimensional angle-interlock woven fabric undergoing ballistic impact. Journal of the Textile Institute 2011;102(11):982–93.

the lower impact velocities can provide the fabric with more time to deform, so the maximum deflection values increase gradually with the decrease of the strike velocities. The tendency of the curves is the result of the change of fabric resistance force to the projectile, considering that the conical part of the projectile perforated the fabric first and then the cylindrical part. The resistance force to the projectile reaches to a maximum value until the end of the conical part. After then, the resistance force decreases gradually. The resistance force reflects the deformation of the fabric which leads to the tendency of the curves. Moreover, the recovery of the elastic deformation of the fabric target also contributes to this process.

Six nodes (nodes A, B, C, D, E and F) as shown in Fig. 6.32 were chosen for the discussion of von Mises stress wave propagation. Nodes A and B are located at

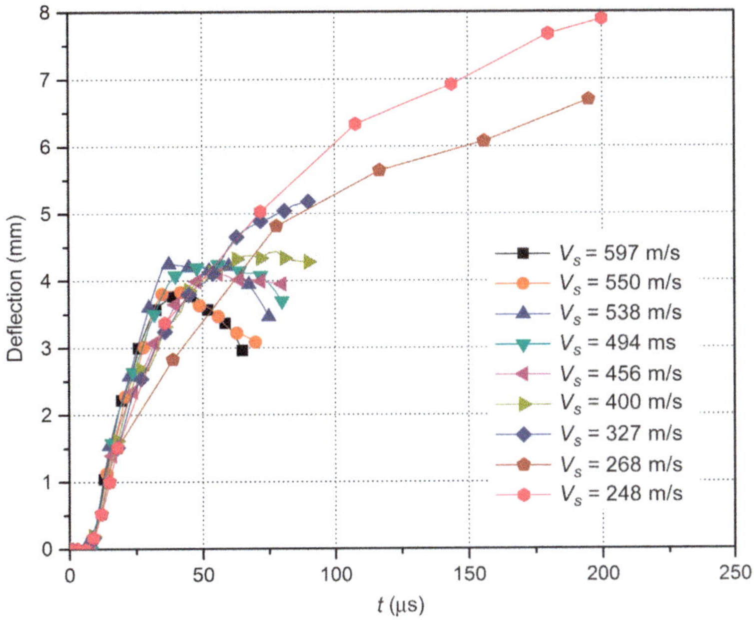

Figure 6.31 Deflection history of node N at different strike velocities.

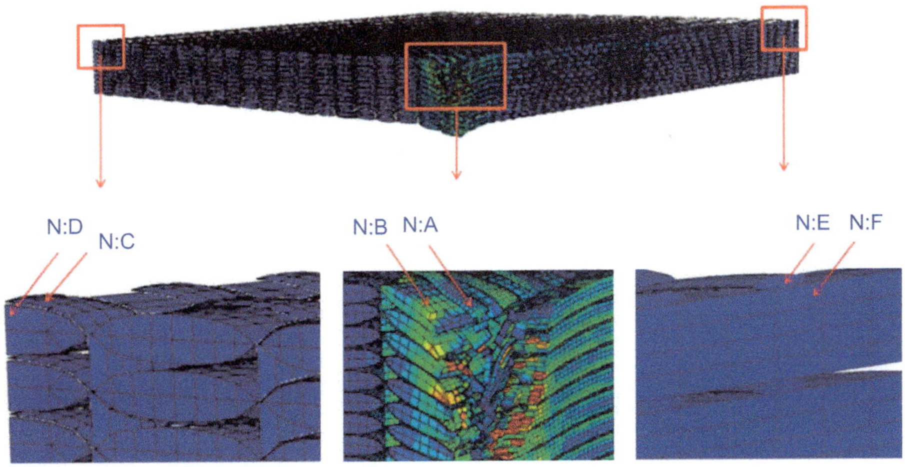

Figure 6.32 Locations of specific nodes.
Jin LM, Sun BZ, Gu BH. Finite element simulation of three-dimensional angle-interlock woven fabric undergoing ballistic impact. Journal of the Textile Institute 2011;102(11):982–93.

the near-impact area, while nodes C, D, E and F are located at the edge region of the fabric target. In addition, nodes A, C and E belong to warp yarns, and nodes B, D and F belong to weft yarns. The stress wave histories of the nodes under impact velocity of 597 m/s are shown in Fig. 6.33. For the nodes A and B, as the time goes on, the amplitude of the curves decays gradually because of the impact energy absorbed by the fabric. For the nodes C, D, E and F, the stress is zero at the beginning of penetration process due to the time consumed for stress to arrive at these nodes. Because the stress at node E is primarily from the adjacent weft yarns, its stress values are much smaller than those of node F. Similarly, the stress at node D is primarily from the adjacent warp yarns; however, the stress value is smaller than that of

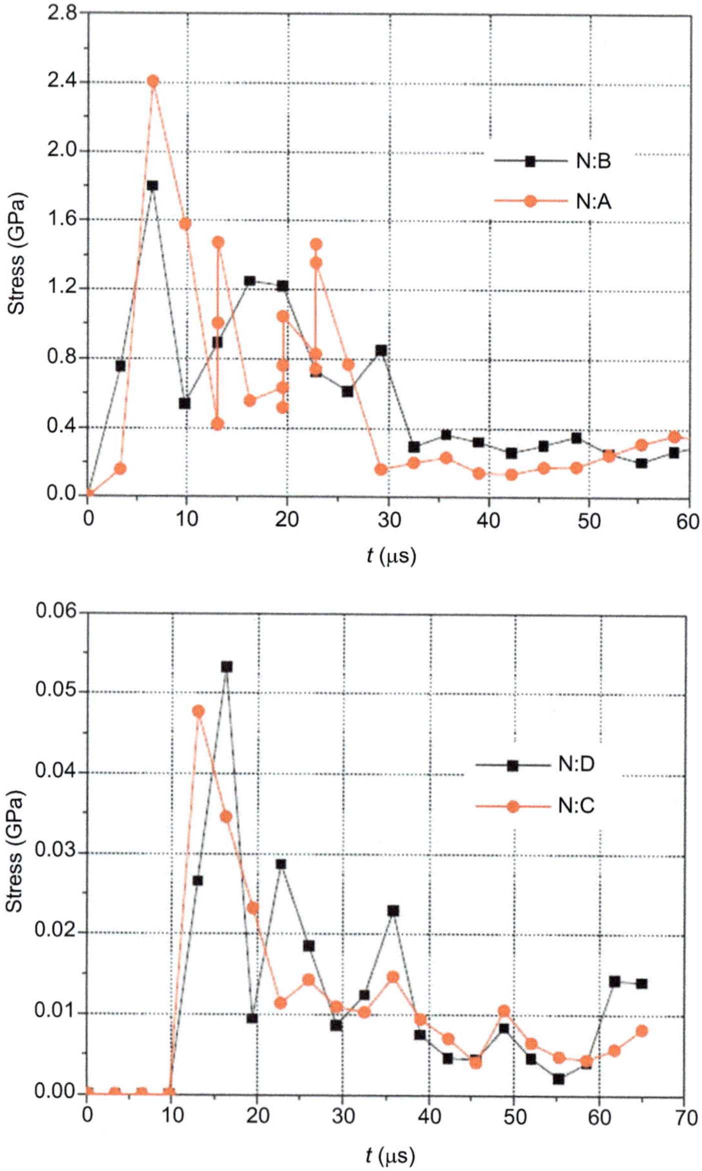

Figure 6.33 Stress-time history at the specific nodes.

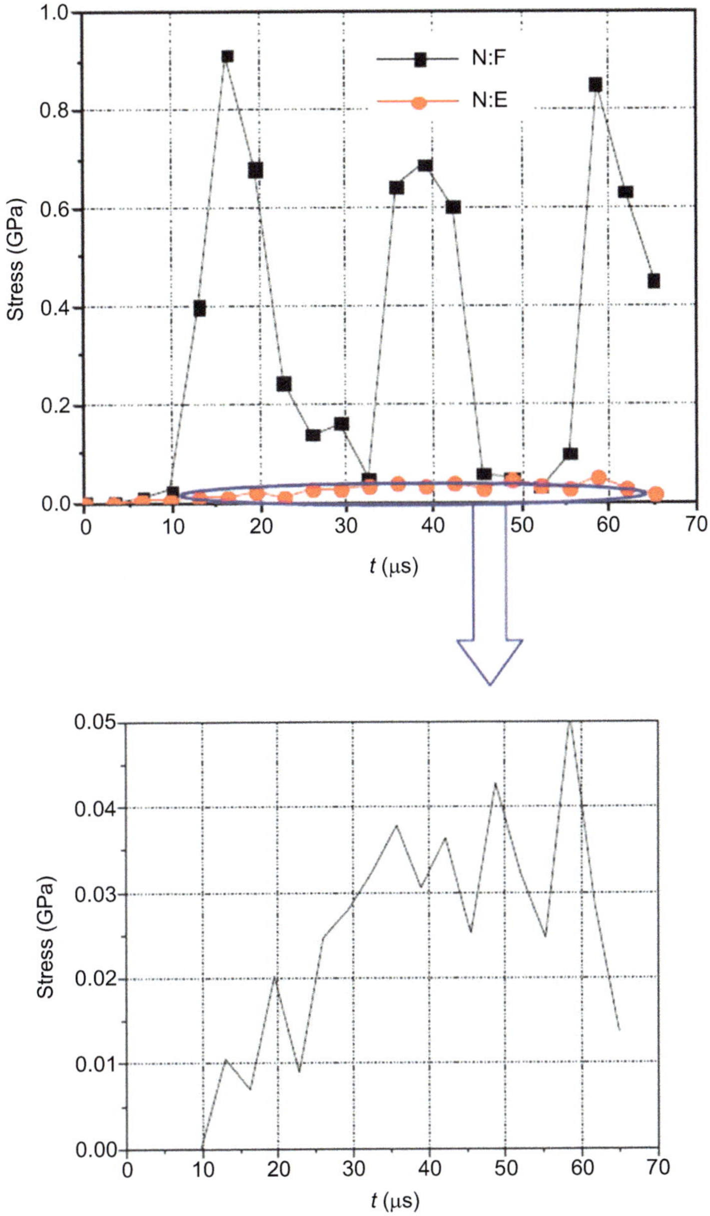

Figure 6.33 Continued.

node C. This means that the warp yarns play an important role in transferring the impact energy, and the weft yarns do likewise in absorbing the impact energy. With such a kind of structure, the 3DAWF has a significant effectiveness in the ballistic protection.

6.5 Design of 3D woven fabrics for ballistic protection

6.5.1 3DOWF

In the analytical model, strain rate can be described as an important factor which affects the accuracy of the results. The fitting method of different strain rates versus different striking velocities shows very good agreement between the calculated and experimental results. The strain rates of yarns are different under the different striking velocities. Therefore, accurate strain rate estimation is the basis of the calculation. Through numerical calculation, an interesting phenomenon was found that when the higher strain rate was used for calculation, a smaller kinetic energy was absorbed by the fabric. This means if we can reduce the strain rate significantly when the fabric is under ballistic impact, a higher bulletproof performance could be obtained. And also, the fitting method of different strain rates versus different striking velocities will provide a way to predict the strain rate of the 3DOWF under ballistic impact. This method is also referred as a good method with a new technique [69] to measure yarn strain in a fabric subjected to ballistic impact.

Fig. 6.34 shows the absorbed energy distribution from the analytical model.

As shown in Fig. 6.34(a), the warp and weft yarns absorb 99% energy, while the Z-yarns only absorb 1% of the total energy. The warp and weft yarns play an important

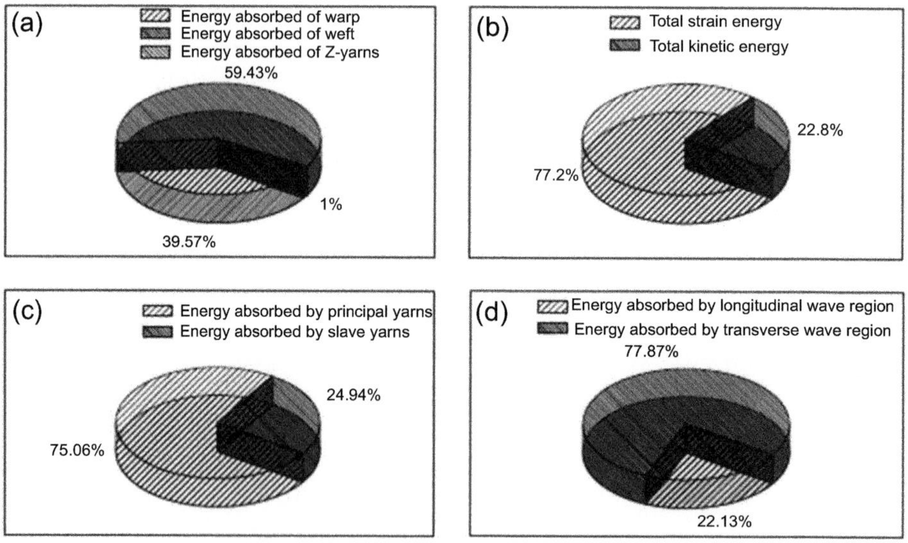

Figure 6.34 Energy proportion when striking velocity is 534 m/s (strain rate at 1800/s). (a) Energy absorbed by warp, weft and Z-yarns; (b) total strain energy and total kinetics energy; (c) energy absorbed by principal yarn and slave yarn; (d) energy absorbed by longitudinal wave region and transverse wave region.
Shi WF, Hu H, Sun BZ, Gu BH. Energy absorption of 3D orthogonal woven fabric under ballistic penetration of hemispherical-cylindrical projectile. Journal of the Textile Institute 2011; 102(10):875–89.

role in ballistic performance. In the 3DOWF armour design, the higher linear density of warp and weft yarns should be used. To weave the compact fabric structure, the lower linear density Z-yarns could be used for binding the noncrimp warp and weft yarns together to form a compact fabric structure.

Fig. 6.34(b) shows that 77.2% of the total energy absorbed by the 3DOWF was strain energy in the fabric. This means that fabric woven from higher strength, higher modulus and higher failure strain yarns will absorb higher energy.

As shown in Fig. 6.34(c), the principal yarns absorbed 75.06% of the total energy, while the slave yarns absorbed 24.94%. This reveals that the breakage of the principal yarns is the main energy absorbed mechanism of 3DOWF. In the design of the 3DOWF body armour, the higher weaving density is necessary for increasing the number of the principal yarns which could be involved in energy absorption.

Fig. 6.34(d) shows that the energy absorbed by the transverse wave region (77.9%) is much higher than the energy absorbed by the longitudinal wave region (22.1%). This means fabric deformation is also an important factor for armour design, and the higher failure strain fibre tows should be used for 3DOWF body armour. The warp and weft yarns could have higher linear density while the Z-yarns should have much lower linear density. Increasing the weaving density of fabric structure is also an important factor for design.

6.5.2 3DAWF

Taking the strike velocity of 536 m/s as an example, Fig. 6.35 shows the energy absorption proportion of principal and secondary yarns and the energy absorption mechanism of 3DAWFs.

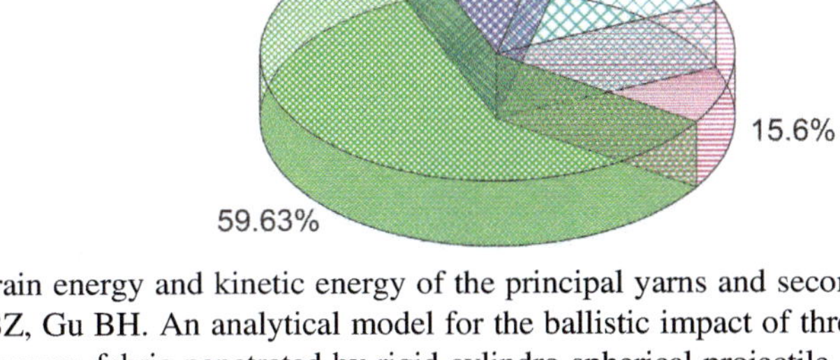

Figure 6.35 Strain energy and kinetic energy of the principal yarns and secondary yarns. Hou YQ, Sun BZ, Gu BH. An analytical model for the ballistic impact of three-dimensional angle-interlock woven fabric penetrated by rigid cylindro-spherical projectile. Textile Research Journal 2011;81(12):1287–303.

1. 69.7% of the total energy was absorbed by the principal yarns, while the rest of the energy was absorbed by the secondary yarns. The principal yarns are the key factor for energy absorption. Even so, the energy absorption by secondary yarns cannot be neglected. In fabric armour design, it is suggested to use fibres with high tenacity, modulus and fracture strain for increasing the energy absorption by principal yarns and the deformation of the secondary yarns.
2. 35.4% of the total energy was absorbed by warp yarns, while the weft yarns absorb the rest of the total energy. Because the weft yarns are in straight line in the 3DAWF structure, the weft yarns will absorb much of the total energy owing to the stress wave that will spread to a larger area than that in crimped warp yarns. In fabric armour design, the straight yarns should be used as much as possible; if necessary, the straight stuffer yarns could be inserted in the fabric structure.
3. 74.3% of the total energy was absorbed by the strain energy of the 3DAWF, while the rest is by fabric kinetic energy. This manifests that the energy of the projectile was mainly absorbed by the fabric deformation. This is the main mechanism for ballistic impact energy absorption. In the total strain energy of the fabric, the principal yarns absorb 59.6% of the energy. Therefore the fibre tows with high extension rate or failure strain should be used for fabric armour design.

6.5.3 Comparison among different fabric constructions

The comparison among multilayered plain-woven fabrics, 3D orthogonal woven fabrics (3DOWFs) and 3D angle-interlock woven fabrics was made in terms of energy absorption capabilities and the projectile residual velocities. Relevant works [25,33] on multilayered plain-woven fabrics have already been done, and good results were obtained. Coincidently, these three structural fabrics are all manufactured with Twaron® filament tows, and their ballistic penetration experiments results are separately listed in Table 6.1. As we can see, for the same rigid projectiles and almost the same strike velocities, the multilayered plain-woven fabric absorbs the highest energy, the 3D orthogonal woven fabrics the second most and the 3DAWFs absorbs the lowest energy. It is obvious that the 3DAWFs and 3DOWFs have better delamination resistance as an integrated structure. Also, we can find that 3DOWFs can absorb more energy than 3DAWFs in similar conditions, which can be explained from their structural differences. The three yarn systems in 3DOWFs are all straight with no crimps, so the stress wave in them can propagate faster to dissipate energy in lager areas; that is, more deformation will occur in 3DOWFs. This also accounts for the crimp effect which plays an important role in the ballistic performance of fabrics. Therefore we can achieve this goal by designing the binding pattern of the warp yarns in 3DAWFs to obtain smaller crimp angle in warp yarns.

When the energy absorption of the fabrics was compared according to the areal density in Table 6.1, the multilayered plain-woven fabric has the highest energy absorption capacity. And the 3DOWFs absorbs an amount of kinetic energy just less than the plain-woven fabric. This is mainly attributed to the Z-yarns in the 3DOWFs not being perpendicular to the penetration direction and having less contribution to the energy absorption compared with the warp and weft yarns. For the 3DAWFs, the crimp of the weft yarns will lead to the lower stress wave spread velocity and also the lower kinetic energy absorption. Compared with the plain-woven fabric, the crimp of the weft yarns in the 3DAWFs is much higher, and also, the energy absorption is much lower.

Table 6.1 Comparison of the energy absorption capabilities among multilayered plain-woven fabrics, 3D orthogonal woven fabrics and 3D angle-interlock woven fabrics

Target type	Target layers	Areal density (g/cm^2)	Strike velocity (m/s)	Residual velocity (m/s)	Energy absorbed (J)	Energy/areal density ($J/g/cm^2$)
Plain weaving	10 (11-layer warp + 12-layer weft)	0.48	362	334	77.5	161.46
			404	382	68.7	143.12
						Average: 152.29
3D orthogonal weaving	1 (12-layer warp + 13-layer weft)	0.91	392	328	128.8	141.54
			487	438	126.7	139.23
			534	489	128.7	141.43
						Average: 140.73
3D angle-interlock weaving	1 (11-layer warp + 12-layer weft)	0.82	399	336	129.4	157.80
			494	463	82.9	101.10
			536	508	81.7	99.63
						Average: 119.51

6.5.4 Applications to flexible armour design

From the discussions in Section 6.5.2, it is suggested that the higher weaving density of the 3DAWF should be used to increase the number of principal yarns because the principal yarns absorb much of the total energy. And also, the flexibility of the 3DAWF could be designed by adjusting fabric structure parameters. Because the strain energy of the principal yarns is the main factor in energy absorption, the yarns with high strength, high modulus and high fracture strain should be used in fabric armour design to increase the failure strain energy absorption.

In addition, the stress wave will spread to a larger fabric area when the stress wave velocity increases. Then the fabric deformation area will also be increased, and the energy absorption capacity will be enlarged. From Eqs [6.23]–[6.25], the higher Young's modulus and lower crimp fibre tow should be used in fabric architecture design.

6.6 Future trends

The inventions of high-performance fibres will provide great potentials for designing flexible ballistic protection armours and materials. Combined with analytical models and numerical models, the ballistic impact damage mechanisms could be revealed for optimizations of fabric constructions. For example, for 3DOWF, it was found that the ballistic impact deformation of the weft and warp yarns were the same; however, the Z-yarns play an important role, although the energy absorption of the Z-yarns was much less than that of weft and warp yarns. The Z-yarns bind the weft and warp yarns together and maintain the fabric structure stability. The higher strength of Z-yarns should be used for obtaining a higher ballistic impact–resistant fabric. For 3DAWF, it was found that only the yarns in contact with the projectile (named as principal yarns) were in breakage, and other yarns (named as secondary yarns) were in a cone deformation region. The yarn crimp and weaving density are the key factors in the 3DAWF armour design. By employing the high-modulus, high failure strain and high-strength fibres and combining the fabric architecture design, the fabrics with high ballistic performance could be manufactured. Furthermore, one important parameter in the design of fabric armours is the hybrid armour structures, for example, the hybrid of the 3DAWF and 3DOWFs. This can be recommended for the armour design due to increasing the energy absorption capacity of the system.

Sources of further information and advice

In this chapter, we have provided references concerned with the modelling of ballistic protection of 3D woven fabrics. For more information on this topic, the following resources or books are recommended for further reading.

Book

Gu B, Sun B. Impact dynamics of textile structural composite materials. Beijing: Science Press; 2012 [in Chinese].

Articles

(Part of author and coworker's papers on ballistic impact damage of woven fabrics and composites in last five years):

[1] Luan K, Sun BZ, Gu BH. Ballistic impact damages of 3-D angle-interlock woven composites based on high strain rate constitutive equation of fiber tows. International Journal of Impact Engineering July 2013;57:145–58.
[2] Zhu LT, Sun BZ, Hu H, Gu BH. Ballistic impact damage of biaxial multilayer knitted composite. Journal of Composite Materials 2012;46(5):527–47.
[3] Hou YQ, Sun BZ, Gu BH. An analytical model for the ballistic impact of three dimensional angle-interlock woven fabric penetrated by rigid cylindro-spherical projectile. Textile Research Journal 2011;81(12):1287–303.
[4] Jia XW, Sun BZ, Gu BH. A numerical simulation on ballistic damage of 3-D orthogonal woven fabric. International Journal of Damage Mechanics 2012;21(2):237–66.
[5] Jia XW, Sun BZ, Gu BH. Ballistic penetration of conically cylindrical steel projectile into 3-D orthogonal woven composite: a finite element study. Journal of Composite Materials 2011;45(9):965–87.
[6] Shi WF, Hu H, Sun BZ, Gu BH. Energy absorption of 3D orthogonal woven fabric under ballistic penetration of hemispherical-cylindrical projectile. Journal of the Textile Institute 2011;102(10):875–89.
[7] Jin LM, Sun BZ, Gu BH. Finite element simulation of three-dimensional angle-interlock woven fabric undergoing ballistic impact. Journal of the Textile Institute 2011;102(11): 982–93.
[8] Luan K, Sun BZ, Gu BH. A multi-scale geometrical model for finite element analyses of three-dimensional angle-interlock woven composite under ballistic penetration. CMES: Computer Modeling in Engineering & Sciences 2011;79(1):31–62.
[9] Li ZJ, Sun BZ, Gu BH. FEM simulation of 3D angle interlock woven composite under ballistic impact from unit cell approach. Computational Material Science 2010;49(1): 171–83.
[10] Jin LM, Hu H, Sun BZ, Gu BH. A simplified microstructure model of bi-axial warp-knitted composite for ballistic impact simulation. Composites Part B 2010;41(4):337–53.

References

[1] McCrackin FL, Schiefer HF, Smith JC, Stone WK. Stress-strain relationships in yarns subjected to rapid impact loading: Part II. Breaking velocities, strain energies, and theory neglecting wave propagation. Textile Research Journal 1955;25(6):529–34.
[2] Smith JC, McCrackin FL, Schiefer HF. Stress-strain relationships in yarns subjected to rapid impact loading: Part III. Effect of wave propagation. Textile Research Journal 1955; 25(8):701–8.
[3] Smith JC, McCrackin FL, Schiefer HF, Stone WK, Towne KM. Stress-strain relationships in yarns subjected to rapid impact loading: Part IV: transverse impact tests. Textile Research Journal 1956;26(11):821–8.
[4] Smith JC, McCrackin FL, Schiefer HF. Stress-strain relationships in yarns subjected to rapid impact loading: Part V: wave propagation in long textile yarns impacted transversely. Textile Research Journal 1958;28(4):288–302.
[5] Smith JC, Blandford JM, Schiefer HF. Stress-strain relationships in yarns subjected to rapid impact loading: Part VI: velocities of strain waves resulting from impact. Textile Research Journal 1960;30(10):752–60.

[6] Smith JC, Shouse PJ, Blandford JM, Towne KM. Stress-strain relationships in yarns subjected to rapid impact loading: Part VII: stress-strain curves and breaking-energy data for textile yarns. Textile Research Journal 1961;31(8):721–34.

[7] Smith JC, Blandford JM, Towne KM. Stress-strain relationships in yarns subjected to rapid impact loading: Part VIII: shock waves, limiting breaking velocities, and critical velocities. Textile Research Journal 1962;32(1):67–76.

[8] Smith JC, Blandford JM, Shouse PJ, Towne KM. Stress-strain relationships in yarns subjected to rapid impact loading: Part IX: effect of yarn structure. Textile Research Journal 1962;32(6):472–80.

[9] Smith JC, Fenstermaker CA, Shouse PJ. Stress-strain relationships in yarns subjected to rapid impact loading: Part XI: strain distributions resulting from rifle bullet impact. Textile Research Journal 1965;35(8):743–57.

[10] Petterson DR, Stewart GM, Odell FA, Maheux RC. Dynamic distribution of strain in textile materials under high-speed impact: Part I: experimental methods and preliminary results on single yarns. Textile Research Journal 1960;30(6):411–21.

[11] Petterson DR, Stewart GM. Dynamic distribution of strain in textile materials under high-speed impact: Part II: stress-strain curves from strain-position distributions. Textile Research Journal 1960;30(6):422–31.

[12] Susich G, Dogliotti LM, Wrigley AS. Microscopical study of a multilayer nylon body armor panel after impact. Textile Research Journal 1958;28(5):361–77.

[13] Roylance D, Wilde A, Tocci G. Ballistic impact of textile structures. Textile Research Journal 1973;43(1):34–41.

[14] Wilde AF, Roylance DK, Rogers JM. Photographic investigation of high-speed missile impact upon nylon fabric: Part I: energy absorption and cone radial velocity in fabric. Textile Research Journal 1973;43(12):753–61.

[15] Wilde AF. Photographic investigation of high-speed missile impact upon nylon fabric: Part II: retarding force on missile and transverse critical velocity. Textile Research Journal 1974;44(10):772–8.

[16] Vinson JR, Zukas JA. On the ballistic impact of textile body armor. Journal of Applied Mechanics 1975;42(6):263–8.

[17] Cunniff PM. An analysis of the system effects in woven fabrics under ballistic impact. Textile Research Journal 1992;62(9):495–509.

[18] Cunniff PM. A semiempirical model for the ballistic impact performance of textile-based personnel armor. Textile Research Journal 1996;66(1):45–58.

[19] Taylor WJ, Vinson JR. Modeling ballistic impact into flexible materials. AIAA Journal 1990;28(2):2098–103.

[20] Parga-Landa B, Hernandez-Olivares F. An analytical model to predict impact behaviour of soft armours. International Journal of Impact Engineering 1995;16(3):455–66.

[21] Chocron-Benloulo IS, Rodriguez J, Sauchez-Galvez V. A simple analytical model to simulate textile fabric ballistic behavior. Textile Research Journal 1997;67(7):520–8.

[22] Navarro C. Simplified modeling of the ballistic behavior of fabrics and fiber-reinforced polymeric matrix composites. Key Engineering Materials 1998;141–143:383–400.

[23] Billon HH, Robinson DJ. Models for the ballistic impact of fabric armor. International Journal of Impact Engineering 2001;25(4):411–22.

[24] Lim CT, Tan VBC, Cheong CH. Perforation of high-strength double-ply fabric system by varying shaped projectiles. International Journal of Impact Engineering 2002;27(6):577–91.

[25] Gu BH. Analytical modeling for the ballistic perforation of planar plain-woven fabric target by projectile. Composites Part B 2003;34(4):361–71.

[26] Leech C, Hearle JWS, Mansell J. A variational model for the arrest of projectiles by woven cloth and nets. Journal of the Textile Institute 1979;70(11):469−78.
[27] Hearle JWS, Leech CM, Cork CR. Ballistic impact resistance of multi-layer textile fabrics. AD-A127641 (1981.10).
[28] Roylance D, Wang SS. Penetration mechanics of textile structure. In: Laible RC, editor. Ballistic materials and penetration mechanics. Amsterdam: Elsevier Scientific Publishing Company; 1980. p. 272−92.
[29] Roylance D, Chammas P, Ting J, Chi H, scott B. Numerical modeling of fabric impact. In: Proceedings of the national meeting of the American Society of Mechanical Engineers (ASME), San Francisco; October 1995.
[30] Shim VPW, Tan VBC, Tay TE. Modeling deformation and damage characteristics of woven fabric under small projectile impact. International Journal of Impact Engineering 1995;16(4):585−605.
[31] Lim CT, Shim VPW, Ng YH. Finite-element modeling of the ballistic impact of fabric armor. International Journal of Impact Engineering 2003;28(1):13−31.
[32] Tan VBC, Lim CT, Cheong CH. Perforation of high-strength fabric by projectiles of different geometry. International Journal of Impact Engineering 2003;28(2): 207−22.
[33] Gu BH. Ballistic penetration of conically cylindrical steel projectile into plain-woven fabric target − a finite element simulation. Journal of Composite Materials 2004;38(22): 2049−74.
[34] Phoenix SL, Porwal PK. A new membrane model for the ballistic impact response and V-50 performance of multi-ply fibrous systems. International Journal of Solids and Structures 2003;40(24):6723−65.
[35] Duan Y, Keefe M, Bogetti TA, Cheeseman BA. Modeling friction effects on the ballistic impact behavior of a single-ply high-strength fabric. International Journal of Impact Engineering 2005;31(8):996−1012.
[36] Duan YP, Keefe M, Wetzel ED, Bogetti TA, Powers B, Kirkwood JE, Kirkwood KM. Effects of friction on the ballistic performance of a high-strength fabric structure. In: International conference on impact loading of lightweight structure 2005, May 8−12, Forianopolis, Brazil; 2005.
[37] Duan Y, Keefe M, Bogetti TA, Cheeseman BA. Modeling the role of friction during ballistic impact of a high-strength plain-weave fabric. Composite Structures 2005;68(3): 331−7.
[38] Duan Y, Keefe M, Bogetti TA, Cheeseman BA, Powers B. A numerical investigation of the influence of friction on energy absorption by a high-strength fabric subjected to ballistic impact. International Journal of Impact Engineering 2006;32(8):1299−312.
[39] Tan VBC, Ching TW. Computational simulation of fabric armour subjected to ballistic impacts. International Journal of Impact Engineering 2006;32(11):1737−51.
[40] Rao MP, Duan Y, Keefe M, Powers BM, Bogetti TA. Modeling the effects of yarn material properties and friction on the ballistic impact of a plain-weave fabric. Composite Structures 2009;89(4):556−66.
[41] Novotny WR, Cepus E, Shahkarami A, Vaziri R, Poursartip A. Numerical investigation of the ballistic efficiency of multi-ply fabric armours during the early stages of impact. International Journal of Impact Engineering 2007;34(1):71−88.
[42] Talebi H, Wong SV, Hamouda AMS. Finite element evaluation of projectile nose angle effects in ballistic perforation of high strength fabric. Composite Structures 2009;87(4): 314−20.

[43] Ivanov I, Tabiei A. Loosely woven fabric model with viscoelastic crimped fibres for ballistic impact simulations. International Journal for Numerical Methods in Engineering 2004;61(10):1565–83.
[44] Zeng XS, Tan VBC, Shim VPW. Modelling inter-yarn friction in woven fabric armour. International Journal for Numerical Methods in Engineering 2006;66(8):1309–30.
[45] Karahan M. Comparison of ballistic performance and energy absorption capabilities of woven and unidirectional aramid fabrics. Textile Research Journal 2008;78(8):718–30.
[46] Karahan M, Kus A, Eren R. An investigation into ballistic performance and energy absorption capabilities of woven aramid fabrics. International Journal of Impact Engineering 2008;35(6):499–510.
[47] Bilisik AK, Turhan Y. Multidirectional stitched layered aramid woven fabric structures and their experimental characterization of ballistic performance. Textile Research Journal 2009;79(14):1331–43.
[48] Cork CR, Foster PW. The ballistic performance of narrow fabrics. International Journal of Impact Engineering 2007;34(3):495–508.
[49] Barauskas R, Abraitienė A. Computational analysis of impact of a bullet against the multilayer fabrics in LS-DYNA. International Journal of Impact Engineering 2007;34(7): 1286–305.
[50] Zohdi TI, Powell D. Multiscale construction and large-scale simulation of structural fabric undergoing ballistic impact. Computer Methods in Applied Mechanics and Engineering 2006;195(1–3):94–109.
[51] Joo K, Kang TJ. Numerical analysis of energy absorption mechanism in multi-ply fabric impacts. Textile Research Journal 2008;78(7):561–76.
[52] Zhang GM, Batra RC, Zheng J. Effect of frame size, frame type, and clamping pressure on the ballistic performance of soft body armor. Composites Part B 2008;39(3):476–89.
[53] Nasr-Isfahani M, Amani-Tehran M, Latifi M. Simulation of ballistic impact on fabric armour using finite-element method. Journal of the Textile Institute 2009;100(4):314–8.
[54] Wang YQ, Miao YY, Swenson D, Cheeseman BA, Yen CF, LaMattina B. Digital element approach for simulating impact and penetration of textiles. International Journal of Impact Engineering 2010;37(5):552–60.
[55] Parsons EM, King MJ, Socrate S. Modeling yarn slip in woven fabric at the continuum level: simulations of ballistic impact. Journal of the Mechanics and Physics of Solids 2013; 61(1):265–92.
[56] Gama BA, Bogdanovich AE, Coffelt RA, Haque MJ, Rahman M, Gillespie Jr JW. Ballistic impact damage modeling and experimental validation on a 3-D orthogonal weave fabric composite. In: International SAMPE symposium and exhibition (Proceedings). SAMPE 05: New horizons for materials and processing technologies – conference proceedings, vol. 50; 2005. p. 1161–75.
[57] Nilakantan G, Keefe M, Gillespie Jr JW, Bogetti TA, Adkinson R. A numerical investigation into the effects of 3D architecture on the impact response of flexible fabrics. In: Second world conference on 3D fabrics and their applications; April 6–7, 2009. Greenville, South Carolina, USA.
[58] Chen XG, Yang D. Use of 3D angle-interlock woven fabric for seamless female body armour: Part I: ballistic evaluation. Textile Research Journal 2010;80(15):1581–8.
[59] Chen XG, Yang D. Use of three-dimensional angle-interlock woven fabric for seamless female body armour: Part II: mathematical modeling. Textile Research Journal 2010; 80(15):1589–601.
[60] Hallquist JO. Ls-dyna theoretical manual. Livermore Software Technology Corporation; 1998. 5, pp. 18.1–18.7, 19.1–19.2.

[61] Belytschko T, Liu WK, Moran B. Nonlinear finite elements for continua and structures. New York: John Wiley & Sons; 2000 [Chapter 10].
[62] Zukus JA. Penetration and perforation of solids. In: Zukus JA, et al., editors. Impact dynamics. New York: John Wiley; 1982. p. 155–214, 377–382.
[63] Leech CM, Hearle JWS, Mansell JA. Variational model for the arrest of projectiles by woven cloth and nets. Journal of the Textile Institute 1979;70(11):469–78.
[64] Leech CM. The dynamics of flexible filaments assemblies. In: Hearle JWS, Thwaites JJ, Amirbayat J, editors. Mechanics of flexible fibre assemblies. The Netherlands: Sijthoff & Noordhoff; 1980. p. 343–390.
[65] Drodge DR, Mortimer B, Holland C, Siviour CR. Ballistic impact to access the high-rate behaviour of individual silk fibres. Journal of the Mechanics and Physics of Solids 2012; 60(10):1710–21.
[66] Roylance D, Wang SS. Penetration mechanics of textile structures. In: Laible RC, editor. Ballistic Materials and Penetration Mechanics. Elsevier; 1980. p. 273–293.
[67] Tan VBC, Shim VPW, Zeng X. Modelling crimp in woven fabrics subjected to ballistic impact. International Journal of Impact Engineering 2005;32(1–4):561–74.
[68] Sierakowski RL, Chaturvedi SK. Dynamic loading and characterization of fiber-reinforced composites. John Wiley & Sons, Inc.; 1997. p. 252.
[69] Chocron S, Samant KR, Nicholls AE, Figueroa E, Weiss CE, Walker JD, Anderson CE. Measurement of strain in fabrics under ballistic impact using embedded nichrome wires. Part I: technique. International Journal of Impact Engineering 2009;36(10–11):1296–302.

Measurements of dynamic properties of ballistic yarns using innovative testing devices

7

C. Chevalier, C. Kerisit, A. Klavzar
ISL – French-German Research Institute of Saint-Louis, France

F. Boussu
ENSAIT – GEMTEX Laboratory, France

D. Coutellier
LAMIH, UMR CNRS 8201, University of Valenciennes, France

7.1 Introduction

Many studies (Hu, 2008; Tong et al., 2002; Sheng and Hoa, 2003; Mouritz et al., 1999; Tsai et al., 2000; Baucom and Zirky, 2003; Tanzawa et al., 1999) have revealed the interest in 3D warp interlock structures, as fiber reinforcement of composite material, to provide a better impact response than other not-bonded-in-thickness material, such as 2D laminated fabrics. In fact, during a composite material impact, the dynamic behavior mode of the fiber structure is strongly dependent on the initial velocity of the projectile. When the strain rate of the material is much greater than its ability to respond to a very short load, the fabric provides almost no resistance to the projectile, and failure mode in shear-type thickness occurs, modifying only slightly the initial velocity of the projectile. That limitation of use of the fibrous material in a composite material subjected to an impact is difficult to estimate due to the different number of independent parameters acting on its resilience, such as the shape and velocity of the projectile, type of fibrous structure (geometry), the type and nature of the yarn (raw material, yield, and twist) and the type of impregnation of the composite.

Also, to understand the influence of each parameter on the dynamic impact behavior of the composite material, we have decomposed the complex mode of rupture of a 3D warp interlock fabric at different scales. As part of the different research works of Ha-Minh (2011), Provost (2013), Lefebvre (2011), and also those in progress of Chevalier (2012), we could observe, respectively, the deformation mode of a yarn subjected to dynamic loads approaching real conditions of impact, and then of a 2D fabric and finally of a 3D warp interlock fabric.

Initially, the impact theory on a yarn was examined by Stone in 1955 as mentioned in the work of Chocron et al. (2010). Based on his experimental results of a ballistic test (Smith et al., 1956), Smith et al. (1958) built an analytical model to describe the response of a yarn in a transverse impact. This model considers that the moment

Advanced Fibrous Composite Materials for Ballistic Protection. http://dx.doi.org/10.1016/B978-1-78242-461-1.00007-8

when the projectile hits the yarn with a constant initial velocity V_i, two deformation waves are generated and spread from the impact point: one in the longitudinal direction and the other in the transversal direction of the yarn (Fig. 7.1).

The longitudinal wave, close to yarn edges, provides two wavelets: a so-called elastic one for the deformation $\varepsilon = 0$ with a moving speed C_e and another so-called plastic one for the deformation $\varepsilon = \varepsilon_p$ generated upon impact with a moving speed C_p:

$$C_e = \sqrt{\frac{1}{\rho}\left(\frac{d\sigma}{d\varepsilon}\right)_{\varepsilon=0}} \qquad [7.1]$$

$$C_p = \sqrt{\frac{1}{\rho}\left(\frac{d\sigma}{d\varepsilon}\right)_{\varepsilon=\varepsilon_p}} \qquad [7.2]$$

where σ and ε are, respectively, the tensile stress and strain of the yarn, and ρ is its density.

For a yarn considered as infinite in length, the thread is not deformed to the elastic wavelet. Then, between the elastic and plastic wavelet, the deformation of the yarn

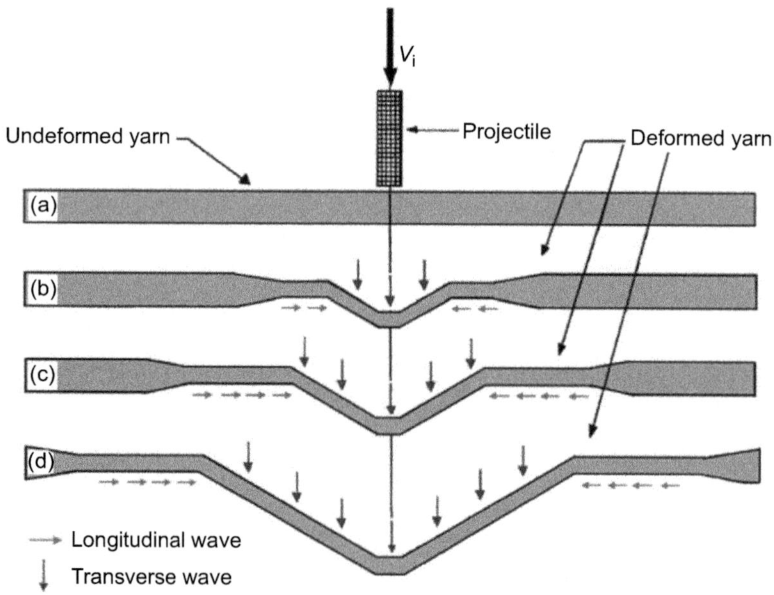

Figure 7.1 Propagation of the two deformation waves in a yarn subjected to a transverse impact.
Reproduced from Smith, J., et al., 1956. Stress–strain relationships in yarns subjected to rapid impact loading: 4 transverse impact tests. Journal of Research of the National Bureau of Standards 57 (2), 83–89.

varies from zero to ε_p, and finally, the yarn deformation remains constant and equal to ε_p just after the start of the plastic wavelet.

When the longitudinal wave reaches a certain position, the material points are directed to the point of impact with a velocity W:

$$W = \int_0^{\varepsilon_p} \sqrt{\frac{1}{\rho}\frac{d\sigma}{d\varepsilon}}\, d\varepsilon \qquad [7.3]$$

This flow is always along the plastic wavelet and hits the transverse wave that propagates at the speed U (Smith et al., 1958):

$$U = \sqrt{\frac{\sigma_p}{\sigma(1+\varepsilon_p)}} \qquad [7.4]$$

where ρ, ε_p, σ_p are, respectively, the density, the elastic limits of strain, and the tensile stress of the yarn. The displacement of this wave led to the development of a pyramid shape (Fig. 7.2). In this pyramid, material points move only in the transverse direction, and they have, consequently, the same speed V_i as the projectile.

The material in the limited area of the transverse wave (Fig. 7.2 pyramid) moves with the speed V_i of the projectile:

$$V_i = \sqrt{(1+\varepsilon_p)^2 U^2 - [(1+\varepsilon_p)U - W]^2} \qquad [7.5]$$

In the particular case where the elastic yarn is a material (assumed as continuous, homogeneous, and isotropic) characterized by a Young's modulus E, the elastic and plastic waves propagate with the same velocity C (Smith et al., 1956):

$$C = \sqrt{\frac{E}{\rho}} \qquad [7.6]$$

The value of Young's modulus appears as a decisive factor in determining the speed of wave propagation. However, to better reflect the dynamic yarn behavior during a

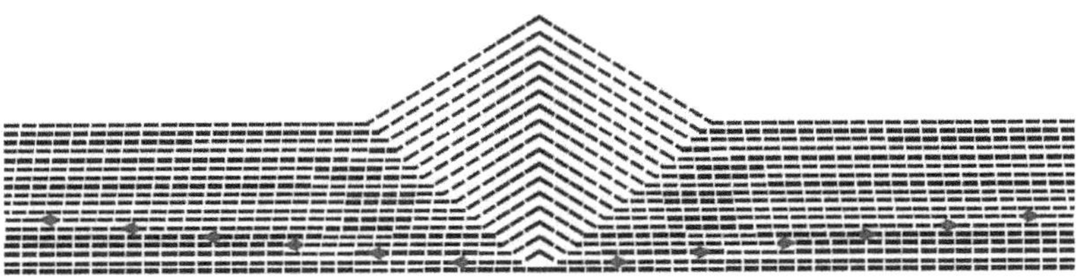

Figure 7.2 Configuration at each 40×10^{-6} s of the elastic yarn after a transverse impact of 180 m/s with the propagation of the longitudinal wave and displacement of material points. Reproduced from Smith, J., McCrackin, F., Schniefer, H., 1958. Stress–strain relationships in yarns subjected to rapid impact loading. Part V: wave propagation in long textile yarns impacted transversely. Textile Research Journal 28 (4), 288–302.

ballistic impact, it can be fruitful to measure this value dynamically. Thus, we have been able to conduct various research projects to identify the dynamic value of Young's modulus of a wire, considered to be an elastic material.

7.2 Testing devices adapted to dynamic properties of yarn

The Hopkinson bars device was introduced in 1914 by Hopkinson and improved in 1949 by Kolsky. This device allows testing materials in dynamic load by the mean of wave transmissions inside the specimen tested. The Hopkinson bars have been developed in several ways, primarily to perform tests on materials in compaction and in tension (Ramesh, 2008).

As part of the work of Rodriguez et al. (1996), the measurement of the dynamic yarn elongation using Hopkinson Tensile Bars was validated for para-aramid and polyethylene fabric composites. The dynamic tensile stress—strain curve appears more linear than that measured statically and shows a greater decrease in the elongation of the polyethylene yarn than the para-aramid yarn, revealing clearly different rupture behavior.

The clamping device proposed by Rodriguez to adapt the maintenance of fabric composites in dynamic tension has made possible the realization of dynamic tensile tests using Hopkinson Tensile Bars on yarns (Zhu et al., 2012; Tan et al., 2008) and on filaments (Lim et al., 2011; Hudspeth et al., 2012; Cheng et al., 2005).

In addition, in the study of Dabboussi et al. (2003), a specific gripping device of the multifilament yarn was used and applied to para-aramid Kevlar® KM2 yarn type to ensure accurate and safe dynamic elongation measurements. The tested yarn was impregnated in a half dog bone shape with epoxy resin to give repeatable results in the specimen gage section. These dynamic values have been compared to quasistatic tests done on a hydraulic cylinder to highlight the influence of the dynamic deformation rate on multifilament yarn strength.

Thus, we have performed (Guillot, 2010) tensile tests using a fast Schenk cylinder of 50 kN, allowing movement at a maximum speed of 1 m/s. A set of 10 tests for each different speed was achieved on a high-modulus polyethylene Dyneema® SK75 yarn (1760 dTex, untwisted) at 5 mm/min (0.083×10^{-3} m/s), 500 mm/min (8.33×10^{-3} m/s) and 1 m/s to measure the tensile stress—strain curves as given in Table 7.1. This range of speeds corresponds to the application of low strain rates on yarns, which are quasistatic loads. Strain rates from 10^2 s^{-1} are considered to be dynamic loads (Ramesh, 2008).

The obtained results reveal an increase of the Young's modulus of the polyethylene Dyneema® SK75 yarn for higher tensile speeds. But the clamping device is not adapted to test untwisted multifilament yarn in this range of speeds because it gives way to some yarn slippages between jaws. Furthermore, vibrations generated during the dynamic movement of the Schenk cylinder have provided many disturbances in the measurement system, which leads us to not continue with this type of material that is not suitable at a yarn scale.

Table 7.1 Tensile stress–strain measurements at different speeds applied to untwisted high-modulus polyethylene Dyneema® SK75 yarn (1760 dTex)

Speed (m/s)	Maximum stress (GPa)	Failure strain (%)	Young's modulus (GPa)
0.083×10^{-3}	2635	8.3	61.8
8.33×10^{-3}	2679	9	69.7
1	2944	6.6	71

To measure the tensile yarn behavior at a high strain rate, we have developed a first prototype of dynamic measurement in the work of Ha-Minh (2011) at the same time as that proposed by Lopatnikov et al. (2012). The principle of measurement is based on a dynamic tool called the Split Flying Mass (SFM), comprising a sabot (Fig. 7.3, part 2) on which one end of the yarn (Fig. 7.3, part 3) is fixed and a projectile with a smaller diameter than the associated sabot (Fig. 7.3, part 4) on which the other end of the yarn is attached (Fig. 7.3, part 1).

The SFM device is inserted into the barrel of the pneumatic tube, and the sabot is ejected at an initial speed (Fig. 7.4(a)) under the effect of the incident pressure. Then, the support hits the fixed inner parts of the tube (Fig. 7.4(b)) to eject the projectile and provide a dynamic tension of the yarn. A device for the dynamic measurement of the displacement of the projectile to yarn rupture is ensured by a high-speed camera coupled to laser action (Nussbaum et al., 2012).

In parallel, tensile tests achieved on a para-aramid Kevlar® 129 yarn (3300 dTex) at a maximum quasistatic speed of 500 mm/min = (8.33×10^{-3} m/s) have been conducted according to ASTM standard D885-03 on an Instron 8501 machine (Ha-Minh, 2011).

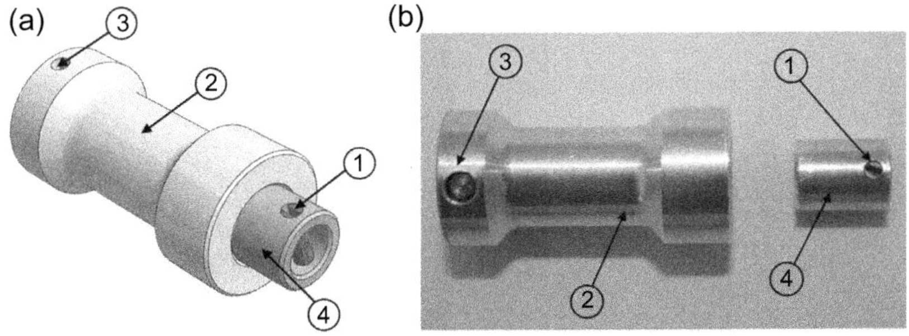

Part ① : screw for attachment of one side of the yarn

Part ② : sabot

Part ③ : screw for attachment of other side of the yarn

Part ④ : projectile

Figure 7.3 Description of the dynamic measurement device adapted to multifilament yarn Split Flying Mass (SFM).

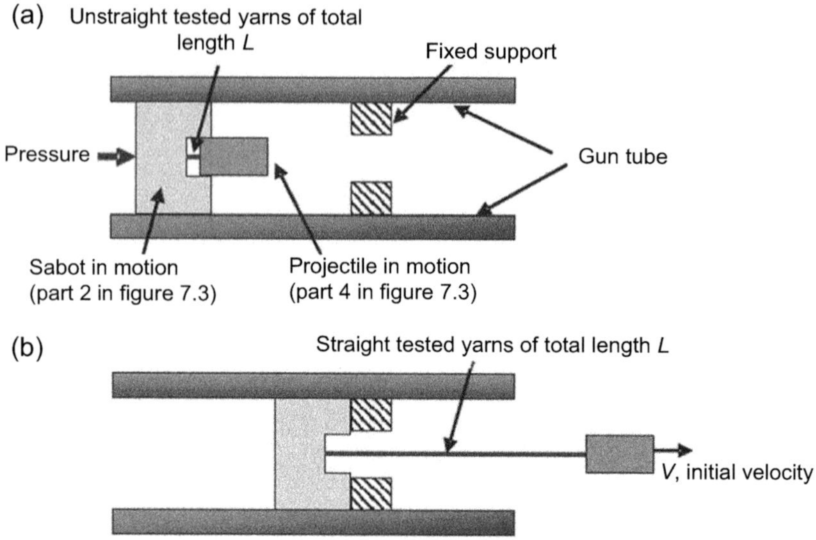

Figure 7.4 Operating steps of the Split Flying Mass (SFM).

In static load, the stress–strain curve shapes of para-aramid yarns tested brings out an elastic linear behavior characterized by a Young's modulus (E), a failure stress (σ_R) and a failure strain (ε_R).

By assuming an elastic linear behavior of yarns tested in dynamic load, the Young's modulus can be determined with the equation of energy conservation:

$$\frac{1}{2}m_p\left(v_i^2 - v_r^2\right) = \frac{1}{2}V_f E(\varepsilon_R)^2$$

where m_p, v_i and v_r, V_f, E, and ε_R are, respectively, the projectile mass, its initial and residual velocities, the yarn volume, and the Young's modulus and the failure strain of the yarn. The measure of the yarn failure strain allows calculating the Young's modulus (E) and the strain energy at break (W), presented in Table 7.2.

Table 7.2 Comparison of static and dynamic tensile tests on para-aramid Kevlar® 129 yarn (3300 dTex)

Mode (actuation frequency $\dot{\varepsilon}$)	Longitudinal modulus (E) in GPa	Failure strain (ε_R) in %	Specific absorbed energy (W) in J/cm^3
Static (0.03 s^{-1})	69.1 (±3.8%)	3.7 (±1.4%)	47.3
Dynamic (225 s^{-1})	214.6 (±28%)	2.2 (±14%)	51.9

In dynamic mode, the yarn rupture is faster due to the significant increase in modulus compared to the measured value in static mode for the same strain energy-to-break consumption value both in the static and dynamic cases.

These first dynamic tensile measurements on yarns allowed us to use the resulting dynamic value of the Young's modulus in ballistic impact simulations of a fabric and thereby improve the accuracy of the simulated fabric behavior as compared with the results observed on real fabric (Ha-Minh et al., 2011).

However, this experimental device of dynamic tensile test on yarn (SFM) reveals several drawbacks during the testing of the yarn. At first, physical limitations cannot yet afford to reach a tensile speed of the yarn significantly closer to the actual conditions of stress during a ballistic impact on yarn. Second, the roll-out of the not taut yarn during the tension of the projectile also avoids applying a uniaxial tension on the yarn. At the end, the clamping system initiates the yarn rupture by transverse shear (Nussbaum et al., 2012).

Therefore, as part of the doctoral research work of Chevalier (2012), improvements have been provided to the dynamic experimental device SFM (Chevalier et al., 2013) (Fig. 7.5).

A new design of the support and projectile has been achieved to improve the measurement accuracy and ensure repeatable tests on yarns. Stress concentrations on yarn have to be reduced to avoid shearing breakage in the transverse direction. The yarn slippage at contact points inside the support and the projectile needs also to be avoided. To respect these two conditions, two different clamping systems have been used. The first one, implemented at the contact between the yarn and the support, is based on the clamping jaws system developed by Chocron Benloulo et al. (1997) to test fabrics and adapted by Shim et al. (2001) to test yarns. The other one, located at the contact point between the yarn and the projectile, consists of unrolling the yarn around a cylinder and fixing it into an epoxy polymer.

At last, the projectile mass has to be high enough to transmit to the yarn a kinetic energy superior to its rupture energy while neglecting the air friction forces. Inversely, the projectile has to be light enough to be able to observe the variation of its displacement, which is specific to the yarn deformation and rupture. In our new configuration, the projectile mass is equal to 1.6 g. At the gas gun level, two copper rings are added at the end of the gun for a dumping effect to avoid any support bounce and thus to have a velocity equal to zero at the yarn end held by the support (Fig. 7.6).

Through high-speed camera pictures, the yarn rupture can be observed using the enhanced SFM (Fig. 7.7) (Chevalier et al., 2014).

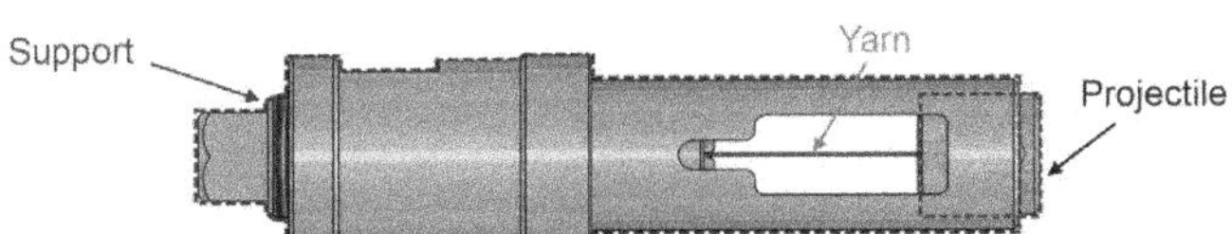

Figure 7.5 Final version of the dynamic tensile measurement device SFM.

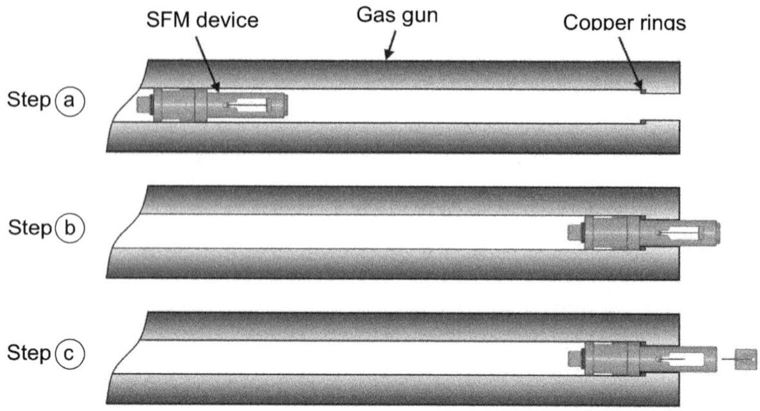

Figure 7.6 Improvement of the experimental setup inside the gas gun.

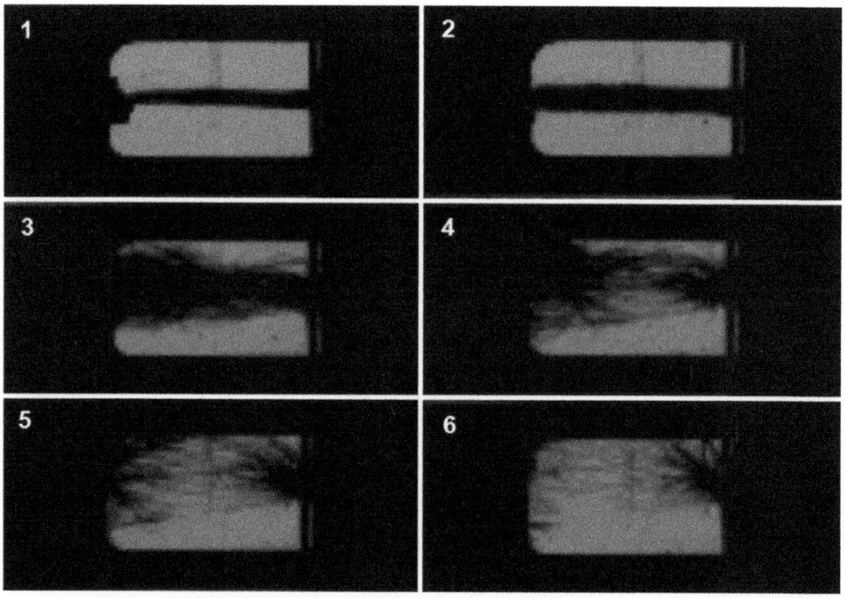

Figure 7.7 High-speed camera observations of the dynamic yarn rupture using the improved Split Flying Mass device (SFM) at 35 m/s.

Initially, the yarn is straight and dynamically elongate in the fiber direction. The wave reflection inside the yarn filaments coming from the contact points of the clamping devices, avoiding any slippery stress concentration in the cross-section direction of the yarn, progressively increases in intensity and leads to decohesion of multifilaments until rupture.

7.3 Optimization of the dynamic tensile device SFM

Due to the different motions of parts during the dynamic test on yarn using the improved SFM device, appropriate and no-contact measurement equipment based on a light barrier produced by laser have been used and calibrated to ensure a safe and accurate measure of the yarn's elongation.

7.3.1 Description of the measurement device

The aim of this characterizing device is to define the mechanical behavior of yarns in dynamic load and to extract their characterized mechanical parameters. To reach that goal, the velocity variation of the projectile, which is due to the characteristic deformation and rupture of the yarn, needs to be determined. This velocity variation is obtained thanks to the light intensity variation measurement performed by a double-laser measurement device (Fig. 7.8) (Nussbaum et al., 2012).

This measurement device is composed of a continuous laser beam (part 2 of Fig. 7.8) and a discontinuous laser of five markers (part 4 of Fig. 7.8). The light intensity variation of these two laser beams is received by two photodiodes (part 3 of Fig. 7.8), then recorded by an oscilloscope.

During the test, the projectile crosses the two laser beams. The measurement of the light intensity variation of the laser beam is proportional to the displacement variation of the projectile and allows us to obtain information about yarn deformation and rupture.

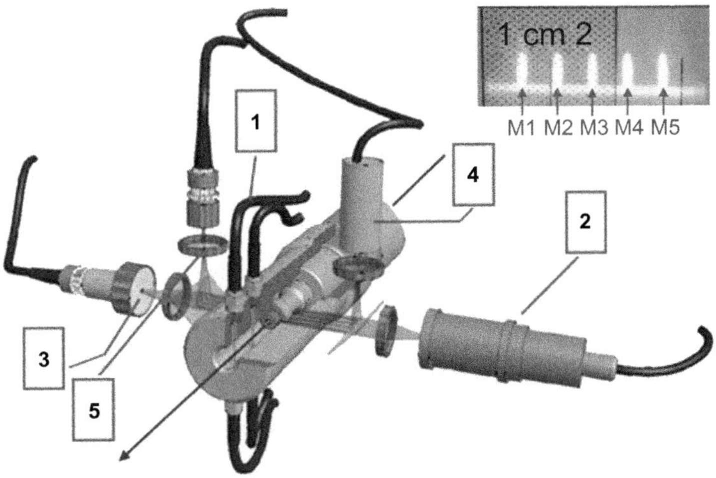

Figure 7.8 Double-laser device: (1) double light barrier, (2) laser line with homogeneous light intensity, (3) optical detection of homogeneous light intensity, and (4) laser markers. Reproduced from Nussbaum, J., et al., September 2–7, 2012. Development of a New Method to Determine Tensile Strength of Ballistic Yarns. Freiburg, Germany.

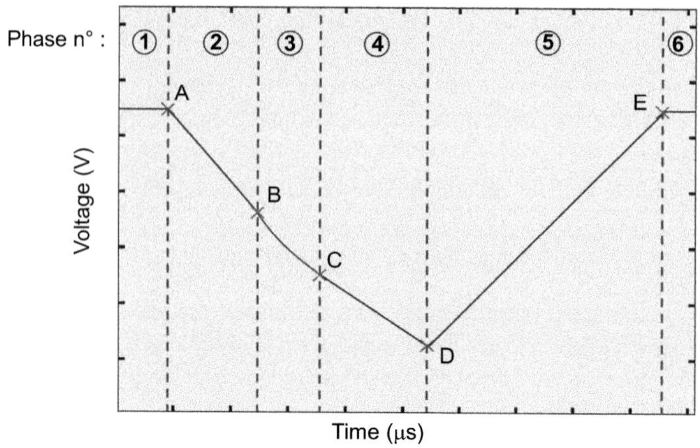

Figure 7.9 Theoretical shape of the light intensity variation of the continuous laser beam.

The theoretical shape of the light intensity variation of the continuous laser beam (Fig. 7.9) and the different steps of the projectile crossing in front of the laser beam (Fig. 7.10), allowing us to identify the different phases of the yarn.

During the dynamic test on yarn, six main phases can be observed, in particular, phases n°2, 3, and 4.

The phase n°2 corresponds to the propulsion of the SFM into the gun. Its initial velocity is calculated from the light intensity variation measured by the double-laser device.

The phase n°3 starts with the stop of the support against the diameter reduction at the end of the gun. In this phase, the projectile keeps being propelled and applies uniaxial tensile force on the yarn, which undergoes an elongation up to its rupture. The velocity variation calculated from double-laser data in this phase is characteristic to the yarn deformation and rupture.

The phase n°4 corresponds to the free flying of the projectile due to the rupture of the yarn at a residual velocity calculated from double-laser data.

We can observe the light intensity variation of phases n°2, n°3, and n°4 during experimental tests on Fig. 7.11.

7.3.2 Calibration of the measurement device

To obtain an accurate measurement of the projectile velocity variation, a calibration is necessary to pass from the light intensity variation to the projectile displacement variation. To reach such a goal, a set of 20 tests have been performed without yarn and at different initial velocities from 26 to 39 m/s. The discontinuous laser of five markers allows measuring accurately these initial velocities by knowing exactly the distance between each marker and the time necessary for the projectile to pass from one marker to the next.

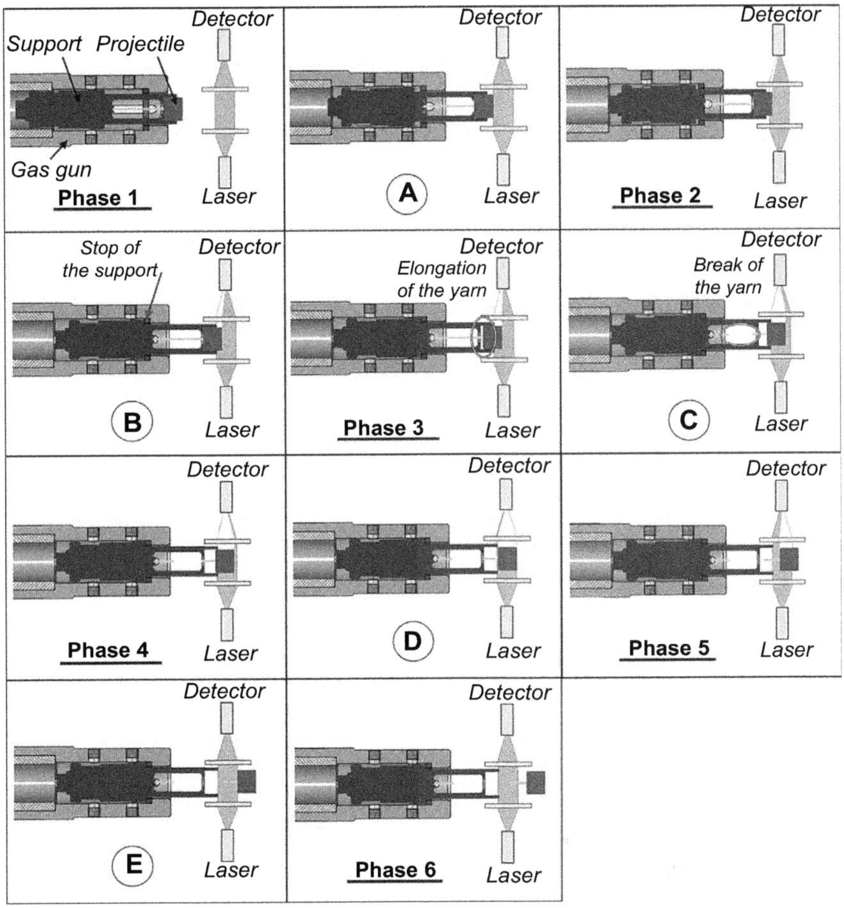

Figure 7.10 Projectile crossing steps in front of the continuous laser beam.

By treatment of the data from experimental tests, it is possible to obtain the projectile displacement variation with the following formula:

$$D = \frac{(V_{\text{full}} - V)}{A} \quad [7.7]$$

where D is the projectile displacement, V_{full} the measured voltage for a maximal light intensity of the continuous laser beam, V the measured voltage of the light intensity during the test, and A is a numerical coefficient.

Thus, we can determine the projectile displacement variation as a function of the measured light intensity variation (Fig. 7.12).

The difference of signal during the projectile's displacement signal versus time allows us to obtain the projectile velocity variation during the test (Fig. 7.13). Because of weak perturbations of the projectile displacement signal, perturbations on the

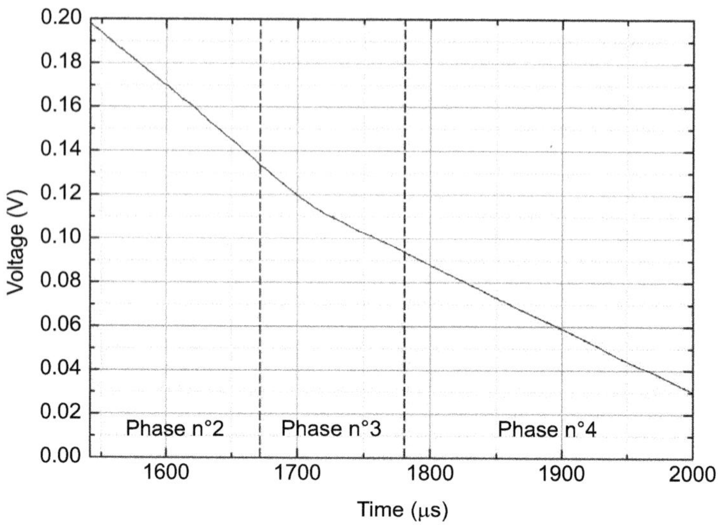

Figure 7.11 Experimental light intensity variation of phase n°2, n°3 and n°4.

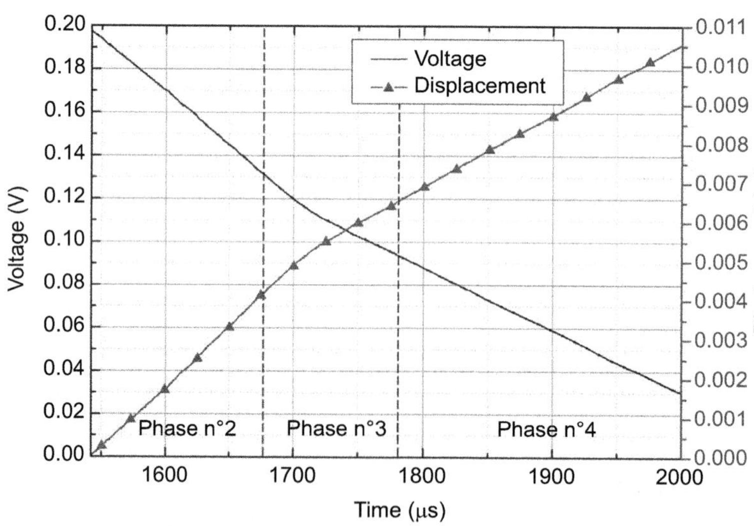

Figure 7.12 Conversion of light intensity curve into projectile displacement variation.

projectile velocity signal are strongly increased. But, thanks to the initial and residual velocities measured with the discontinuous laser of five markers, the initial and residual velocity of the projectile can be accurately recovered and checked. The difference between the two residual velocity measurements is due to a defect of the lens in the double-laser measurement device observed after 2000 μs.

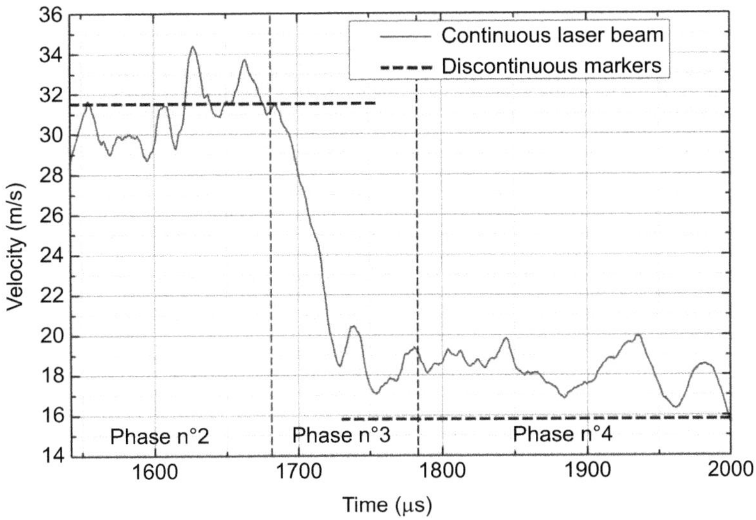

Figure 7.13 Comparison between velocity measurements obtained from the continuous laser beam and from the discontinuous laser of five markers.

7.3.3 Validation of the double-laser measurement device

To confirm these results, a comparison of measurements has been done with those obtained from the PDV (photonic Doppler velocimetry) device chosen as reference. Experimental tests combining our double-laser device and the PDV device have been performed and allow us to confirm the velocity measurement obtained from our measurement device (Fig. 7.14).

Obtained curves from these two measurement devices are quite similar, which allow us to validate our double-laser device measurement procedure.

7.4 Experimental results of dynamic tensile tests on yarn using the optimized SFM

Dynamic tests have been performed with the SFM device on para-aramid Twaron® yarns (336 Tex) of 2-cm length at an initial velocity of 36 m/s, which correspond to a strain rate of 1800 s^{-1}. Static tests have been performed on an MTS quasistatic tensile device on a para-aramid Twaron® (168 Tex) yarn. The same assumptions as those made in the study of Ha-Minh (2011) with the first version of the SFM have been considered. Thus, the measure of the yarn failure strain allows calculating the Young's modulus (E) and the strain energy at break (W) (Table 7.3).

The measurement obtained with the last version of the SFM allows the observation of an increase of the yarn's Young's modulus and an increase of the specific energy absorbed by the yarn. These results are in good agreement with the existence of

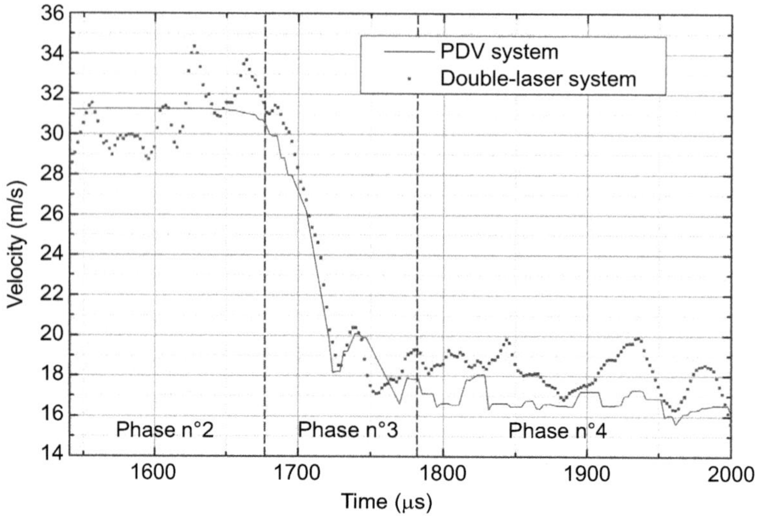

Figure 7.14 Comparison between the velocity variations obtained from the PDV device and our double-laser device.

Table 7.3 Measured values of para-aramid Twaron® yarns at a strain rate of 0.001 s^{-1} and 1800 s^{-1}

Mode (actuation frequency $\dot{\varepsilon}$)	Longitudinal modulus (E) in GPa	Failure strain (ε_R) in %	Specific absorbed energy (W) in J/cm^3
Static (0.001 s^{-1})	62 (±8%)	5 (±5.8%)	77.5
Dynamic (1800 s^{-1})	149	4	119

different failure modes in static and dynamic loads. In static, the failure mode corresponds primarily to the rupture of the secondary bonds between macromolecular chains of each filament. In dynamic, the load on the yarn is too quick, and the rupture of the secondary bond does not have the time to occur. The dynamic load used directly acts on the primary interatomic bonds whom rupture requires a higher specific energy (Carr, 1999; Shim et al., 2001; Billon, 2006; Tan et al., 2008).

7.5 Conclusions

Performing a dynamic tensile test on yarn is quite a challenging task and requires a step-by-step approach to check and ensure safe measurement of the yarn's elongation. An improved version of the dynamic tensile device entitled "Split Flying Mass" has ensured accurate measurements on yarn, especially for multifilament yarn assumed as continuous and heterogeneous. This accuracy on measurements has been obtained by light barrier equipment coming from double-laser devices and helps to safely correlate the measured electric signal to the yarn displacement. The first results obtained on a para-aramid yarn (336 Tex) at a strain rate of 1800 s^{-1} compared with results obtained at a strain rate of 0.001 s^{-1} reveal an increase of the stiffness and of the specific energy absorbed by the yarn due to an increase of the strain rate and, therefore, a modification of the failure mode of the yarn.

Several dynamic tests on yarns are still in progress as of 2015, varying yarn's parameters as raw material, linear density, and number of twists per meter.

Higher strain rates will be proposed in the near future using the dynamic tensile device SFM to characterize the ballistic yarns close to the real conditions during a ballistic impact.

Acknowledgment

The authors of this paper would like to thank the French MoD for providing funds to conduct this research.

References

Baucom, J., Zirky, M., January 1, 2003. Evolution of failure mechanisms in 2D and 3D woven composite systems under quasi-static perforation. Journal of Composite Materials 37 (18), 1651–1674.

Billon, N., 2006. Comportement Mécanique des Polymères. In: Matériaux Pour L'ingénieur. Ed. Mines, Paris, France, p. 153.

Carr, D., 1999. Failure mechanisms of yarns subjected to ballistic impact. Journal of Materials Science Letters 18, 585–588.

Cheng, M., Chen, W., Weerasooriya, T., 2005. Mechanical properties of Kevlar KM2 single fiber. Journal of Engineering Materials and Technology 127, 197–203.

Chevalier, C., 2012. Détermination des caractéristiques dynamiques des structures fibreuses. Internal Presentation, UVHC, Valenciennes, France.

Chevalier, C., et al., May 2014. Dynamic Mechanical Behaviour or Yarn under High Speed Loading. Bursa, Turkey.

Chevalier, C., et al., September 16–20, 2013. Development of a New Device for Experimental Study of the Dynamic Behaviour of Ballistic Yarns. Leuven, Belgium.

Chocron Benloulo, I., Rodriguez, J., Martinez, M., Sanchez Galvez, V., 1997. Dynamic tensile testing of aramid and polyethylene fiber composites. International Journal of Impact Engineering 19 (2), 135–146.

Chocron, S., et al., 2010. Modeling and validation of full fabric targets under ballistic impact. Composites Science and Technology 70 (13), 2012–2022.

Dabboussi, W., Dooraki Babak, F., Nemes, J.A., February 2003. High Strain Rate Testing of Yarns. McGill University, Montréal, Canada.

Guillot, M., 2010. Caractérisation mécanique dynamique de multi-filaments. Master Report, ENSAIT, Lille, France.

Ha-Minh, C., 2011. Comportement mécanique des matériaux tissés soumis à un impact balistique: approches expérimentale, numérique et analytique. PhD Thesis, Université de Lille 1, Lille, France.

Ha-Minh, C., Kanit, T., Boussu, F., Imad, A., 2011. Numerical multi-scale modeling for textile woven fabric against ballistic impact. Computational Materials Science 50 (7), 2172–2184.

Hudspeth, M., Nie, X., Chen, W., 2012. Dynamic failure of Dyneema SK76 single fibers under biaxial shear/tension. Polymer 53 (24), 5568–5574.

Hu, J., 2008. 3D Fibrous Assemblies, Properties Applications and Modelling of Three Dimensional Textile Structure. Woodhead edn. Woodhead Publishing, Hong Kong, China.

Lefebvre, M., 2011. Résistance à l'impact balistique de matériaux composites à renforts Interlocks tissés. Application au blindage de véhicules. PhD Thesis, UVHC, Valenciennes, France.

Lim, J., Zheng, J., Masters, K., Chen, W., 2011. Effects of gage length, loading rates, and damage on the strength of PPTA fibers. International Journal of Impact Engineering 38 (4), 219–227.

Lopatnikov, S., Schevchenko, N., Gillepsie, J., 2012. Device and Method for Investigation of Mechanical Properties of Fibers under High-strain Rate Tensile Load [Online] Available at: http://arxiv./abs/1201.0916 (accessed 10.01.12.).

Mouritz, A., Bannister, M., Falzon, P., Leong, K., 1999. Review of applications for advanced three-dimensional fibre textile composites. Composites Part A: Applied Science and Manufacturing 30, 1445–1461.

Nussbaum, J., et al., September 2–7, 2012. Development of a New Method to Determine Tensile Strength of Ballistic Yarns. Freiburg, Germany.

Provost, B., 2013. Étude et Réalisation d'une solution à renfort tissé interlock pour la protection balistique de véhicule. PhD Thesis, UVHC, Valenciennes, France.

Ramesh, K.T., 2008. High strain rate and impact experiments. In: Sharpe, J.W.N. (Ed.), Handbook of Experimental Solid Mechanics. Springer, CBS Publishers, New Delhi, India, pp. 929–960.

Rodriguez, J., Chocron, I.S., Martinez, M., Sanchez-Galvez, V., 1996. High strain rate properties of aramid and polyethylene woven fabric composites. Composites: Part B 27 (2), 147–154.

Sheng, S., Hoa, S., 2003. Modelling of 3D angle interlock woven fabric composites. Journal of Thermoplastic Composite Materials 16 (1), 45–59.

Shim, V., Lim, C., Foo, K., 2001. Dynamic mechanical properties of fabric armour. International Journal of Impact Engineering 25 (1), 1–15.

Smith, J., et al., 1956. Stress-strain relationships in yarns subjected to rapid impact loading: 4 transverse impact tests. Journal of Research of the National Bureau of Standards 57 (2), 83–89.

Smith, J., McCrackin, F., Schniefer, H., 1958. Stress-strain relationships in yarns subjected to rapid impact loading. Part V: wave propagation in long textile yarns impacted transversely. Textile Research Journal 28 (4), 288–302.

Tan, V., Zeng, X., Shim, V., 2008. Characterization and constitutive modeling of aramid fibers at high strain rates. International Journal of Impact Engineering 35, 1303–1313.

Tanzawa, Y., Watanabe, N., Ishikawa, T., 1999. Interlaminar fracture toughness of 3-D orthogonal interlocked fabric composites. Composites Science and Technology 59 (8), 1261–1270.

Tong, L., Mouritz, A., Bannister, M., 2002. 3D Fibre Reinforced Composite Materials. Elsevier Applied Science, London.
Tsai, K., Chiu, C., Wu, T., 2000. Fatigue behaviour of 3D multi-layer angle interlock woven composite plates. Composites Science and Technology 60 (2), 241–248.
Zhu, D., et al., 2012. Strain rate and gage length effects on tensile behavior of Kevlar 49 single yarn. Composites Part A: Applied Science and Manufacturing 43 (11), 2021–2029.

Analysis of woven fabric composites for ballistic protection

8

N.K. Naik
Indian Institute of Technology Bombay, Mumbai, India

8.1 Introduction

Protection against external high-velocity projectiles is one of the critical requirements for various structural elements. High velocity impact can lead to penetration/perforation of the target by the projectile. Ballistic impact is a high-velocity impact caused by a propelling source, generally of low mass and high velocity. Protection against ballistic impact loads is a critical issue in the design of ballistic grade armors, shock and impact absorbers, and development of high-grade aerospace, marine, automobile, and civil structures. Extensive research has been carried out on ballistic impact performance of polymer matrix composites in recent years. Various parameters such as mass, shape and size, and velocity of the projectile and geometry and mechanical properties of the target influence the ballistic impact performance of composite targets.

When an impact load is applied to a body, instantaneous stresses are produced. But the stresses are not immediately transmitted to all parts of the body. The remote portions of the body remain undisturbed for some time. The stresses progress in all directions through the body in the form of disturbances of different types. In other words, stresses (and their associated deformations or strains) travel through the body at specific velocities. These velocities are functions of the material properties. Regardless of the method of application of impact load, the disturbances generated have identical properties based only on the target material properties.

During an impact event the stress wave propagation takes place in all the directions. Generally, this problem is analyzed using either 1D, 2D, or 3D approaches. In 1D and 2D studies, wave propagation through the thickness direction is not considered. When these approaches are used to analyze structures, isotropic as well as orthotropic, it is assumed that the deformation behavior along the thickness direction of the target is the same along the entire thickness. Such an assumption can be made for targets of lower thickness or, in other words, can be used for thin plates. If the thickness of the plate is increased, the deformation and the induced stress behavior of the plate would be different at different locations along the thickness direction. For such cases, the analysis is based on also considering the wave propagation along the thickness direction.

As technologies of composites manufacturing are advancing and applications of composites in nonaerospace industries are increasing, more and more thick section composites are used in structural designs. Thick composites behave differently compared to thin composites under impact-loading conditions.

Advanced Fibrous Composite Materials for Ballistic Protection. http://dx.doi.org/10.1016/B978-1-78242-461-1.00008-X

Based on the phenomenon of energy transfer between the projectile and the target, energy dissipation, and damage mechanisms, the impact phenomenon is classified into three categories: low-velocity impact, high-velocity impact, and hypervelocity impact. The reason for this classification is that energy transfer between projectile and target, energy dissipation, and damage mechanisms in the target undergo drastic changes as the velocity of the projectile changes. An impact event is considered to be a low-velocity impact if the contact duration of impact is longer than the time period of the lowest vibrational mode of the structure. On the other hand, in high-velocity or ballistic impact the contact duration of impact is much smaller than the time period of the lowest vibrational mode of the structure. Hypervelocity impact involves projectiles moving at extremely high velocities such that the local target materials behave like fluids. Ballistic impact is an impact caused by a propelling source, generally of low mass and high velocity.

Protection against external high-velocity projectiles is one of the critical requirements. Extensive research is being carried out on ballistic impact behavior of polymer matrix composites. In many aerospace and defense applications, composite structures may be subjected to localized projectile impact. Such impacts could result in indentation, partial penetration, or perforation of the composite target depending on the projectile mass, velocity, and shape. To ensure the safety of composite structures against penetration/perforation by high-velocity projectiles, clear and complete understanding of penetration and perforation process in composites is essential. Protection against ballistic impact loads is a critical issue in the design of ballistic-grade armor, shock and impact absorbers, and development of high-grade aerospace, marine, automobile, and civil structures. Instances of impact loading on aerospace and defense structures include bird strikes, hailstones, impact of runaway debris, small and medium caliber bullets, and blast fragments.

When a ballistic impact event takes place, different types of stress waves propagate in the impacted bodies (Meyers, 1994). Types of stress waves are classified based on the direction of motion of the particles of the body relative to the direction of propagation of the waves and boundary conditions. To study ballistic impact behavior of composite structures, a thorough understanding of stress wave propagation is essential.

Several studies are available in the literature on ballistic impact behavior of composite structures. An analytical model of normal impact and perforation of cylindrical-conical projectiles on laminated Kevlar/polyester composites was developed by Zhu et al. (1992b). Potti and Sun (1996, 1997) used a dynamic response model along with critical deflection failure criterion to predict the residual velocity of the projectile in thick graphite−epoxy composites.

Based on the analytical model developed by Smith et al. (1958) for impact on yarns, Chocron Benloulo et al. (1997) proposed an analytical model of impact on textiles and composites together with an energy-based failure criterion. An energy-based model was proposed by Morye et al. (2000) to predict energy absorption by polymer composites upon ballistic impact. They considered conical deformation and kinetic energy of the moving cone in their formulation.

Wen (2000, 2001) developed an analytical model to predict penetration and perforation of monolithic composite laminates impacted transversely by rigid projectiles with different nose shapes. A membrane model was developed by Phoenix and Porwal (2003) to predict ballistic impact performance of multi-ply fibrous body armor materials impacted transversely by a blunt-nosed projectile.

Naik et al. (2005, 2006) and Naik and Shrirao (2004) presented an analytical model for the ballistic impact behavior of composites. They considered conical deformation and kinetic energy carried by the moving cone in their formulation. The ballistic impact behavior of thick, woven fabric composites was investigated by Naik and Doshi (2005, 2008). The method developed was based on in-plane and through-the-thickness wave propagation and energy balance. Shear plugging was considered in their formulation. Udatha et al. (2012) studied ballistic impact behavior of 3D woven composites. Shaktivesh et al. (2015) presented ballistic impact behavior of fabric targets with multiple layers. Pandya et al. (2015) presented a generalized analytical model for the ballistic impact behavior of composite targets.

Lin and Hoo Fatt (2006) proposed three separate impact models based on contact duration, through-the-thickness transit time, and lateral transit time. High-velocity impact response of E-glass/polyester panels was predicted by a 2D wave propagation model. Gama and Gillespie (2008) developed a quasistatic penetration model based on quasistatic punch shear tests.

Gellert et al. (2000), Haijun et al. (2013), Jia et al. (2012), Mines et al. (1999), Pandya et al. (2012a,b, 2013), Shaktivesh et al. (2013), Wu and Chang (1995), Zee and Hsieh (1998), Zhang et al. (2014), and Zhu et al. (1992a) carried out experimental studies on ballistic impact behavior of composites. Naik et al. (2013) presented an energy-based model for ballistic impact analysis of ceramic-composite armors. Hazell and Appleby-Thomas (2012) and Kasano (1999) presented reviews of ballistic impact behavior of composites. Rojek et al. (2013) discussed polymeric composite materials as applied to ballistic shields.

The behavior of composite structures under ballistic impact loading depends on the thickness of the target. The focus of the present study is on a generalized analytical formulation for ballistic impact behavior of 2D woven fabric composites impacted by a rigid cylindrical projectile. The formulation presented is valid for a wide range of laminate thicknesses. The formulation is based on stress wave propagation and energy balance between the projectile and the composite target. The formulation considers both shear plugging and tensile failure during conical deformation on the back face of the target. The projectile is assumed to be rigid, and the energy is absorbed only by the composite target by different damage- and energy-absorbing mechanisms. Experimental validation is performed on the ballistic impact behavior of composites. Typical results on ballistic limit velocity and energy absorbed by various mechanisms are presented.

8.2 Materials for ballistic protection

Protection against ballistic impact is an important consideration in load-bearing structural components. Selection of materials for ballistic protection is primarily based on its energy absorption characteristics. During the last two decades, many high-performance fibers have been developed for ballistic impact protection applications. These fibers are generally lightweight and possess exceptionally higher energy absorption characteristics.

An important component of the composite armor is the fiber. Although carbon fiber and glass fiber are the reinforcements used widely for structural applications, from a ballistic protection application point of view, the widely used fibers are para-aramids, high molecular weight polyethylene (HMPE), Poly-phenylene Benzobis Oxazole (PBO), and poly-diimidazo-pyridinylene-dihydroxy-phenylene (Bhatnagar, 2006; Zaera, 2011). These are all organic fibers. Mechanical properties of ballistic-grade fibers are presented in Table 8.1. Such fibers should have low density, high Young's modulus, and high strength.

Mechanical properties of Dyneema (Bhatnagar, 2006; DSM Dyneema Product Specification Sheet, 2014; DSM Dyneema, 2014; Goodfellow Cambridge, 2014;

Table 8.1 Mechanical properties of ballistic grade fibers

Material	Density (Kg/m^3)	Tensile strength (MPa)	Ultimate strain (%)	Tensile modulus (GPa)	References
Dyneema	970	–	3.5	87	Goodfellow Cambridge (2014)
Dyneema	970–980	2200–3900	3.0–4.0	52–132	DSM Dyneema (2014)
Spectra	970	–	2.7–3.5	120–170	Goodfellow Cambridge (2014)
Spectra	970	2180–3680	2.9–3.9	66–35	Honeywell Advanced Fibers and Composites (2014)
Spectra	970	2610–3510	2.9–3.6	79–124	Bhatnagar (2006, p. 212)
Spectra	970	2600	3.5	120	Zaera (2011, p. 327)
Aramid	1440	2794–3429	3.3–4.3	64–96	Bhatnagar (2006, p. 212)
Aramid	1440	2900	2.95	91	Zaera (2011, p. 327)
Nylon	1140	1006	18.2	5	Bhatnagar (2006, p. 212)
Zylon	1540–1560	5800	2.5–3.5	180–270	Bhatnagar (2006, p. 212)
Zylon	1540	5800	3.5	180	Zaera (2011, p. 327)
Carbon	1770	2400	1.1	220	Gibson (1994, p. 80)
S-glass	2500	4200	5.0	90	Zaera (2011, p. 327)

Zaera, 2011), Spectra (Bhatnagar, 2006; Goodfellow Cambridge, 2014; Honeywell Advanced Fibers and Composites, 2014; Prevorsek, 1996; Zaera, 2011), aramid (Bhatnagar, 2006; Zaera, 2011), nylon (Bhatnagar, 2006), Zylon (Bhatnagar, 2006; Zaera, 2011), carbon (Gibson, 1994), and S-glass (Zaera, 2011) are given in Table 8.1. These are the main fibrous materials used for ballistic protection applications. It can be seen that these materials are characterized by high specific strength, high specific tensile modulus, and relatively higher ultimate strain.

Ballistic protection targets are made either by using fabrics of various forms, nonwoven fabrics, or prepreg ballistic composites. The primary requirement of ballistic protection target is energy absorption of the projectile striking on to it. This can be achieved either by using high-performance fibrous materials or by effective design of target structure. As given in Table 8.1, some of these fibrous materials have excellent energy-absorption characteristics. Effective design considerations for enhancing ballistic impact performance of composite targets are presented in Section 8.8. Further details regarding fibrous materials for ballistic protection can be obtained in Bhatnagar (2006) and Zaera (2011). High strain rate response of ultrahigh molecular weight polyethylene is given in Russell et al. (2013).

8.3 Composites for high-performance applications

Marissen et al. (2005) studied mechanical behavior of Dyneema fibers. They observed that these fibers have an extremely high tensile strength, but a low compressive strength. This makes them unsuitable for a number of composite applications. Hence, even though they are characterized by high energy absorption capability, they do not find wide usage in other structural applications.

Polymeric composites are characterized by high specific strength, high specific stiffness, dimensional stability, corrosion resistance, and ease of manufacturing. Primary focus on material selection is on high specific strength and high specific stiffness for aircraft, aerospace, defense, automotive, and naval applications. Generally, carbon composites and glass composites are used for such high-technology applications. Even though carbon composites and glass composites are not the ideal choice for ballistic impact protection applications, they are used in a variety of structural applications. During their service life, such structures would undergo different loading conditions. Ballistic impact is one of the critical loading conditions. The objective of the present study is on ballistic impact behavior of glass/epoxy and carbon/epoxy composites.

8.4 Ballistic impact on composite targets

8.4.1 Penetration and perforation process

Transverse ballistic impact onto a composite target by a projectile generates compressive and shear stress waves along the thickness direction and tensile and shear stress waves along the in-plane direction (Meyers, 1994).

Different stages of penetration/perforation of a rigid cylindrical projectile with a flat end into a 2D woven composite target are presented schematically in Fig. 8.1.

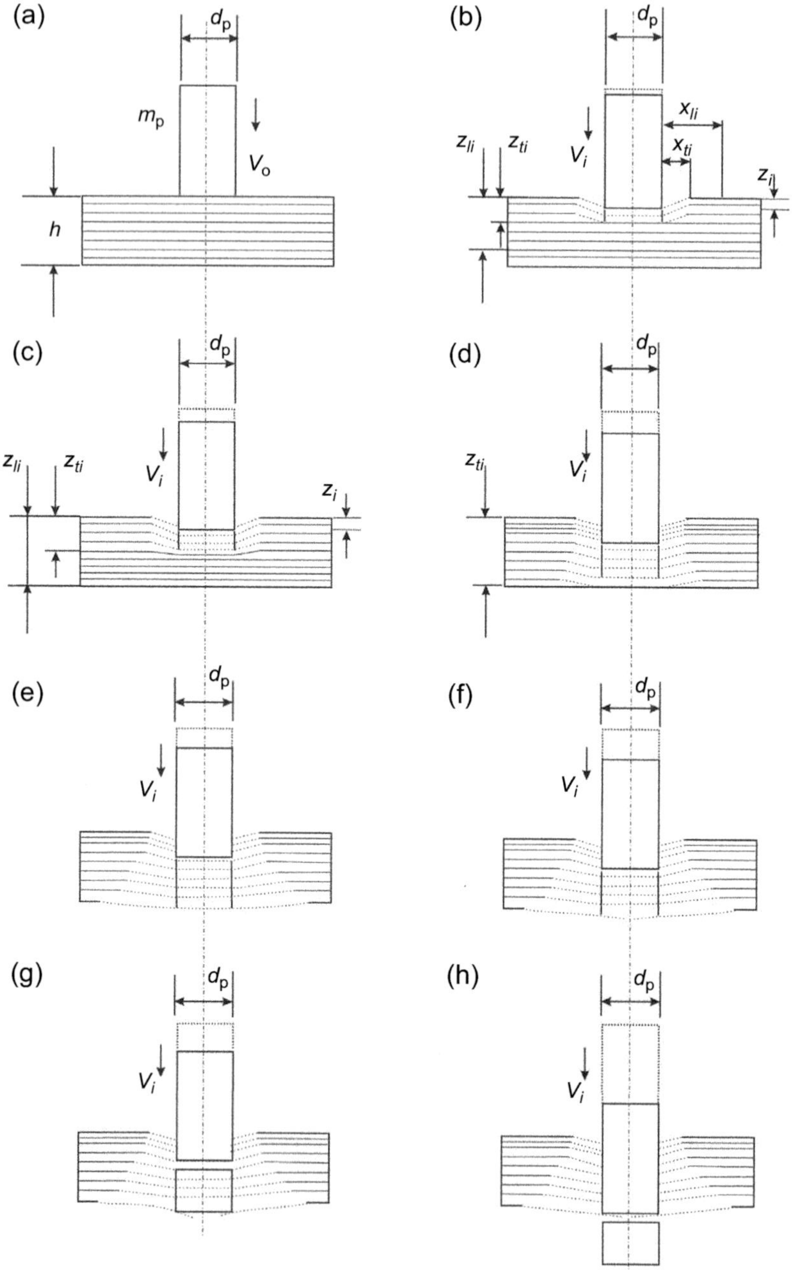

Figure 8.1 Penetration and perforation stages of 2D woven fabric composite target during ballistic impact.

Schematic arrangement of the target during ballistic impact is shown in Fig. 8.2. The front view can be subdivided into two regions as shown in Fig. 8.2(b). The region directly below the projectile is referred to as Region 1. The surrounding region up to which the transverse stress wave travels along the in-plane directions is referred to as Region 2. The yarns that are in contact with the projectile during the ballistic impact event are referred to as primary yarns. The primary yarns are along the warp and fill directions. The remaining yarns within the surrounding region up to which the transverse stress wave travels along the in-plane directions are referred to as secondary yarns. Note that only primary yarns are present in Region 1, whereas both primary and secondary yarns are present in Region 2.

Fig. 8.1(a) indicates the beginning of the ballistic impact event. The impact event can be subdivided into three stages. During Stage 1, the compressive and shear stress waves travel along the thickness direction. The layers of the composite target undergo compression directly below the projectile and also in the surrounding region, as shown in Fig. 8.1(b) and (c). The compression of layers in Region 2 is because of a transverse shear wave propagating in the in-plane directions. Compression of layers also produces tension along the in-plane direction in the surrounding region. The shear

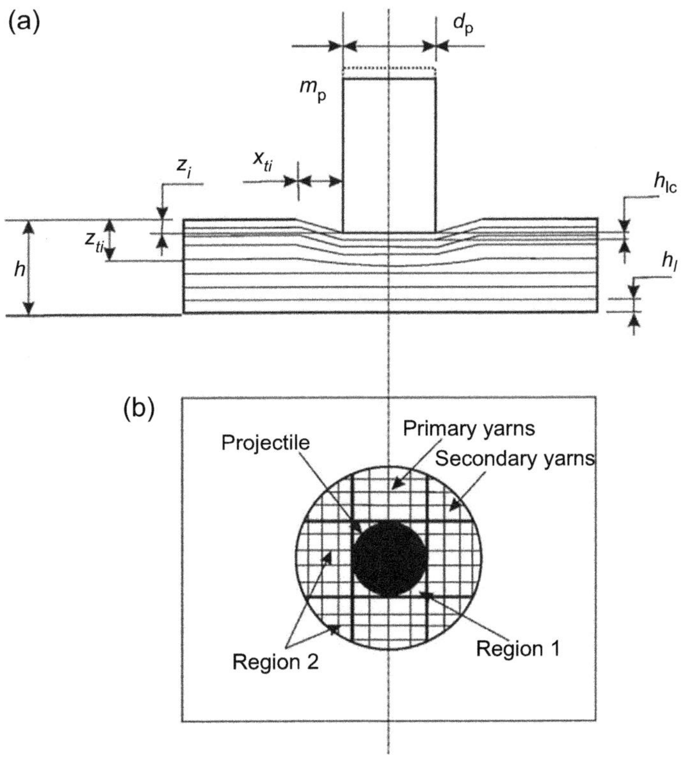

Figure 8.2 Schematic arrangement of a typical 2D woven fabric composite target during ballistic impact: tensile deformation (a) side view and (b) front view.

wave follows the compressive wave. The stress waves generate strains in the target. As the compressive and shear waves travel along the thickness direction, the layers could fail under compression, tension, or shear plugging whenever the induced strains exceed the corresponding failure strains.

Stage 2 starts when the shear wave reaches the back face of the target (Fig. 8.1(d)). Depending on the number of layers failed in Stage 1 and the kinetic energy available with the projectile, conical deformation of the target could take place on the back face, as shown in Fig. 8.1(e). Thickness of the laminate and the incident impact velocity of the projectile influence the number of layers failed and the kinetic energy available with the projectile at the end of Stage 1. The layers that do not fail in Stage 1 undergo tension as a result of conical deformation and could fail when the induced tensile strain exceeds the failure strain (Fig. 8.1(f) and (g)). Stage 2 ends when the material is completely failed either by shear plugging or by tension.

Even after the complete failure of the target takes place either by shear plugging or tension, there can be friction between the target and the moving projectile. This stage is referred to as Stage 3. Some energy can be absorbed because of friction. Stage 3 ends when the projectile tip reaches the back face of the target as shown in Fig. 8.1(f). At this stage, if the projectile is having some residual kinetic energy, it would exit from the target with a certain residual velocity. The incident impact velocity of the projectile that leads to the tip of the projectile reaching the back face of the target is referred to as ballistic limit velocity.

8.4.2 Damage- and energy-absorbing mechanisms

During a ballistic impact event, the target offers resistance to penetration/perforation of the projectile into itself. The incident kinetic energy of the projectile is absorbed by the target through various damage- and energy-absorbing mechanisms. In other words, energy transfer takes place from the projectile to the target. As a result of this, the kinetic energy of the projectile and, in turn, the velocity of the projectile decrease. Compression of the target directly below the projectile, compression in the region surrounding the impacted zone, shear plugging, stretching, and tensile failure of yarns/layers in the region consisting of primary yarns, tensile deformation of yarns/layers in the region consisting of secondary yarns, conical deformation on the back face of the target, delamination, matrix cracking, and friction between the moving projectile and the target are the damage- and energy-absorbing mechanisms.

8.4.3 Analytical formulation

The main objective of the analytical formulation is to predict the energy absorbed by various mechanisms, contact force, projectile velocity, and projectile tip displacement as a function of time. A further objective is to predict ballistic limit velocity and contact duration for a range of laminate thicknesses. The formulation is based on stress wave propagation and energy balance between the rigid cylindrical projectile and the target.

8.4.3.1 Assumptions

The following assumptions are made in the analytical formulation:

1. Projectile impact is normal to the surface of the target.
2. The projectile is considered to be rigid and remains undeformed during the ballistic impact event.
3. The projectile is cylindrical with a flat end.
4. The region of the target up to which the shear stress wave reaches in the in-plane direction around the point of impact offers resistance to penetration.
5. The impact event is subdivided into smaller time intervals for the analysis.
6. During any time interval, the velocity of the projectile remains constant.
7. Longitudinal and transverse stress wave velocities are the same in all the layers.
8. Compressive strain is uniform in all the layers up to which compressive wave has traveled.
9. Shear plugging stress is uniform in all the layers up to which shear wave has traveled.
10. Tensile failure of the layers/yarns takes place at the periphery of the projectile.

8.4.3.2 Projectile velocity through energy balance

At any instant during the ballistic impact event, incident kinetic energy of the projectile can be expressed as the sum of residual kinetic energy of the projectile at that instant and the total energy absorbed by the target till that instant. The energy balance at the end of ith time interval is represented as follows:

$$\begin{aligned}\mathrm{KE}_{\mathrm{p}0} = \mathrm{KE}_{\mathrm{p}i} + E_{\mathrm{CD}i} + \{E_{\mathrm{CR1}i} + E_{\mathrm{CR2}i} + E_{\mathrm{SP}i} + E_{\mathrm{TPY}i} + E_{\mathrm{TSY}i} \\ + E_{\mathrm{DL}i} + E_{\mathrm{MC}i}\} + E_{\mathrm{F}i}\end{aligned} \quad [8.1]$$

Here, $\mathrm{KE}_{\mathrm{p}0}$ refers to incident kinetic energy of the projectile, $\mathrm{KE}_{\mathrm{p}i}$ refers to residual kinetic energy of the projectile at the end of ith time interval, and E refers to energy absorbed.

Suffixes CD, CR1, CR2, SP, TPY, TSY, DL, MC, and F refer to conical deformation, compression in Region 1, compression in Region 2, shear plugging, tension in the region consisting of primary yarns, tensile deformation in the region consisting of secondary yarns, delamination, matrix cracking, and friction, respectively. Suffix i refers up to ith time interval.

Expanding kinetic energy terms and rearranging the other energy terms for simplification gives Eq. [8.2]:

$$\frac{1}{2}m_{\mathrm{p}}V_0^2 - \sum_{i=1}^{i} E_i = \frac{1}{2}\left[m_{\mathrm{p}}V_i^2 + M_{\mathrm{CD}i}(K_1 V_i)^2\right] \quad [8.2]$$

Where the following is defined:

$$K_1 = \frac{1}{x_{\mathrm{tl}_i}} \int_{r=0}^{r=x_{\mathrm{tl}_i}} \frac{(x_{\mathrm{tl}_i} - r)}{x_{\mathrm{tl}_i}} dr \quad [8.3]$$

Here, m_p and V_0 refer to mass and incident impact velocity of the projectile, respectively. M_{CDi} and V_i refer to mass of the moving cone and projectile velocity at the end of ith time interval, respectively. x_{tl_i} refers to the distance traveled by the transverse wave in each layer along the in-plane direction at the end of ith time interval. r refers to distance along in-plane direction.

K_1 is a parameter which takes into account the integrated average velocity of the moving cone. Note that the velocity of the projectile is V_i, whereas the velocity of the moving cone at the root is zero.

Linear variation of velocity from the tip of the projectile to the root of the moving cone is considered in the present case. Note that, considering linear variation, $K_1 = 0.5$.

The total energy absorbed by the target is calculated in Eq. [8.4]:

$$E_i = E_{CDi} + \{E_{CR1i} + E_{CR2i} + E_{SPi} + E_{TPYi} + E_{TSYi} + E_{DLi} + E_{MCi}\} + E_{Fi} \quad [8.4]$$

Hence, the velocity of the projectile at any instant during the ballistic impact event is as follows:

$$V_i = \sqrt{\frac{\left(\frac{1}{2}m_p V_0^2 - E_i\right)}{\frac{1}{2}\left(m_p + \frac{M_{CDi}}{4}\right)}} \quad [8.5]$$

8.4.3.3 Formulation for the first time interval

The evaluation of projectile velocity as given in Eq. [8.5] needs some energy terms, which in turn, require the velocity of the projectile at the end of first time interval. This formulation needs to be supplemented by another approach to derive the velocity of the projectile for the first time interval. An iterative approach is used to determine the velocity of the projectile for the first time interval.

8.4.3.4 Contact force on the target and projectile displacement for the first time interval

The forces acting on the projectile during the first time interval are inertial force and compressive force. The inertial force is calculated by equating the work done by the reaction of the inertial force on the target to the change in the kinetic energy of the displaced material.

The inertial force as given by Awerbuch and Bodner (1974) follows:

$$F_{Ik} = \frac{1}{2}\rho K A_p V_k^2 \quad [8.6]$$

Here, K is a constant which depends on the shape of the projectile. $K = 1$ for a flat-ended cylindrical projectile.

Suffix k indicates the iteration number in the first time interval. $k = 1$ indicates first iteration of the first time interval. A_p, V, and ρ refer to cross-sectional area, velocity of the projectile, and density of the target, respectively.

For the first iteration of the first time interval, $F_{Ik} = F_{I1}$ and $V_k = V_0$.

For the first iteration of the first time interval, the following is given:

$$F_{I1} = \frac{1}{2}\rho K A_p V_0^2 \quad [8.7]$$

The compressive force is shown in Eq. [8.8]:

$$F_{C1} = \sigma_{cz} A_p \quad [8.8]$$

Initially, a trial value of compressive stress σ_{cz} is taken for the first iteration of the first time interval. If the trial value of σ_{cz} considered is less than the actual value of σ_{cz}, the derived total contact force would be less than the actual value of the total contact force. After each iteration, the value of total contact force would increase till it converges. The value of σ_{cz} at the converged value of total contact force is the actual value.

It was observed that any trial value of σ_{cz} considered resulted in the same total contact force at the end of the first time interval.

The total contact force acting on the target is obtained as follows:

$$F_1 = F_{I1} + F_{C1} \quad [8.9]$$

Note that the total contact force given in Eq. [8.9] is not the actual value of total contact force at the end of the first time interval. This is because the energy absorbed by various mechanisms is not considered in this expression. Therefore, an iterative approach is used to determine the total contact force and velocity of the projectile for the first time interval (Udatha et al., 2012; Pandya et al., 2015).

Starting with the total contact force obtained using Eq. [8.9], energy absorbed by various mechanisms is computed, the details of which are explained later. Considering energy absorbed by various mechanisms, a revised total contact force is evaluated. This procedure is continued until a converged value of total contact force is obtained. The converged value of total contact force would be the actual value of total contact force at the end of the first time interval.

Total contact force can also be expressed as shown in Eq. [8.10]:

$$F_1 = m_p \mathrm{dc}_1 \quad [8.10]$$

Here, dc is the deceleration of the projectile, when rearranged giving the following:

$$\mathrm{dc}_1 = \frac{F_1}{m_p} \quad [8.11]$$

Projectile displacement is shown next:

$$z_1 = V_0\Delta t - \frac{1}{2}\mathrm{dc}_1(\Delta t)^2 \quad [8.12]$$

Here, Δt is the time interval.

8.4.3.5 *Energy absorbed by compression of the target directly below the projectile (Region 1)*

As the compressive wave propagates in the thickness direction, compression of the target takes place in the region directly below the projectile, Region 1 (Fig. 8.1(b)). The projectile displacement results in compressive strain in the layers up to the distance traveled by the compressive wave.

The compressive wave velocity and the distance traveled by compressive wave along thickness direction are as given next (Pandya et al., 2015):

$$V_{z1} = \lambda_2\sqrt{\frac{1}{\rho}\frac{d\sigma_{cz}}{d\varepsilon_{cz}}} \quad [8.13]$$

where $\lambda_2 = \left[\dfrac{(1-\upsilon_{wf}\upsilon_{fw})}{(1-\upsilon_{wf}\upsilon_{fw}-\upsilon_{fz}\upsilon_{zf}-\upsilon_{zw}\upsilon_{wz}-2\upsilon_{fw}\upsilon_{zf}\upsilon_{wz})}\right]^{\frac{1}{2}}$

$$z_{11} = V_{z1}\Delta t \quad [8.14]$$

Here, σ_{cz} and ε_{cz} are stress and strain, respectively, along z-direction. V_{z1} and z_1 refer to velocity of compressive wave and distance traveled by compressive wave in z-direction, respectively. υ_{ij} refers to Poisson's ratio. Suffixes w and f refer to warp and fill, respectively.

The methodology used for the prediction of stress wave propagation velocity in an unbound laminated composite medium is based on in-plane stress wave propagation velocity in laminated composites (Daniel and Liber, 1976; Sierakowski and Chaturvedi, 1997). Stress wave propagation velocity is presented in terms of stiffness matrix C_{ij} and density ρ, and it is obtained using appropriate expressions for C_{ij} (Chou, 1992; Jones, 1999). Plastic wave velocity is obtained using the local slope of the stress–strain diagram instead of modulus of elasticity. The details regarding stress wave propagation velocity in an unbound composite medium are given in Pandya et al. (2015).

The projectile displacement results in compressive strain in the layers up to which the compressive wave has traveled.

Compressive strain:

$$\varepsilon_{cz_1} = \frac{z_1}{z_{11}} \quad [8.15]$$

Here, z refers to projectile displacement.

Strain rate:

$$\dot{\varepsilon}_{cz} = \frac{\varepsilon_{czk+1} - \varepsilon_{czk}}{t_{k+1} - t_k} \quad [8.16]$$

Here, t refers to time.

The total energy absorbed due to compression in Region 1 in the first iteration of the first time interval is shown in Eq. [8.17]:

$$E_{CR11} = A_p \left(\int_{\varepsilon_{cz}=0}^{\varepsilon_{cz1}} \sigma_{cz}\, d\varepsilon_{cz} \right) z_{l1} \quad [8.17]$$

8.4.3.6 Energy absorbed by compression in the region surrounding the impacted zone (Region 2)

The region surrounding the impacted zone up to which the transverse stress wave travels along the in-plane directions is referred to as Region 2 (Fig. 8.2(b)). The layers in Region 2 also experience compressive strain along the thickness direction. The compression of layers in Region 2 is because of transverse shear wave propagating in the in-plane directions.

The number of layers through which the compressive wave has traveled is calculated next:

$$n_{lsc1} = \frac{z_{l1}}{h_l} \quad [8.18]$$

Here, h_l is thickness of each layer.

At the end of the first time interval, if the wave front lies within any layer, that layer is not considered in that time interval.

Compression of the yarn/fabric varies from maximum at the periphery of the projectile to zero at the point up to which the transverse stress wave has reached in the in-plane direction.

The total energy absorbed due to compression in Region 2 in the first iteration of the first time interval is as follows:

$$E_{CR21} = 2\pi h_l K_2 \sum_{j=0}^{n_{lsc1}} \int_{(0.5d_p)}^{x_{t1}} \left(\int_{\varepsilon_{cz}=0}^{\varepsilon_{cz1}} \sigma_{cz}\, d\varepsilon_{cz} \right) r dr \quad [8.19]$$

For linear variation of compressive strain, $K_2 = 0.5$.

Here, h_l, x_t, d_p, and r refer to thickness of the layer, distance up to which the transverse wave has traveled in the in-plane direction, projectile diameter, and radial distance, respectively.

8.4.3.7 Energy absorbed due to stretching and tensile failure of yarns/layers in the region consisting of primary yarns

The primary yarns resist the projectile penetration into the target. After the shear wave reaches the back face of the target, conical deformation takes place at the back face. As a result, the layers which have not failed because of shear plugging stress experience tensile strain. If the induced tensile strain exceeds the ultimate tensile strain, the yarns/layers would fail.

The transverse wave velocity and the distance traveled by the transverse wave along the in-plane direction are as given next (Pandya et al., 2015):

$$V_{\mathrm{wt}} = \sqrt{\frac{1}{\rho}\frac{d\tau_{\mathrm{zw}}}{d\gamma_{\mathrm{zw}}}} \qquad [8.20]$$

$$x_{\mathrm{t1}} = V_{\mathrm{wt}}\,\Delta t \qquad [8.21]$$

Here, τ_{zw} and γ_{zw} are shear stress and shear strain of the target, respectively. V_{wt} and x_{t} refer to velocity of the transverse wave and distance traveled by the transverse wave in x-direction, respectively.

Upon impact by the projectile, the layers undergo compression as shown in Fig. 8.2(a). Independent experiments showed that the maximum compressive strain of the 2D plain-weave composite along thickness direction is 5%. Hence, thickness by which a layer has been compressed is as follows:

$$\Delta h_{\mathrm{lc}} = 0.05h_{\mathrm{l}} \qquad [8.22]$$

Thickness of the compressed layer, $h_{\mathrm{lc}} = h_{\mathrm{l}} - \Delta h_{\mathrm{lc}}$.

Distance by which a layer has moved in the z-direction from its original position is calculated based on the layer position from the top surface:

$$z_{\mathrm{pl1}j} = z_1 - j\Delta h_{\mathrm{lc}}; \quad j = 0 \text{ to } n_{\mathrm{lsc1}} \qquad [8.23]$$

Here, j refers to layer number from the top surface.

The distance traveled by the transverse wave in each layer along the in-plane direction is given in Eq. [8.24]:

$$x_{\mathrm{tl1}j} = x_{\mathrm{t1}}\frac{(n_{\mathrm{lsc1}} - j)}{n_{\mathrm{lsc1}}} \qquad [8.24]$$

Note that for $j = 0$, that is, for the top layer, $x_{\mathrm{tl}} = x_{\mathrm{t}}$.

The projectile displacement along thickness results in differential tensile strains in the layers along the in-plane direction. The magnitude of tensile strain varies within the distance traveled by the compressive wave, from maximum at the topmost layer to minimum at the bottommost layer under compression. Length of the layer after tension and the tensile strain along warp direction in each layer is calculated next:

$$x_{\mathrm{llc}1j} = \sqrt{x_{\mathrm{tl}1j}^2 + z_{\mathrm{pl}1j}^2} \quad [8.25]$$

$$\varepsilon_{\mathrm{lTl}1j} = \frac{x_{\mathrm{llc}1j} - x_{\mathrm{tl}1j}}{x_{\mathrm{tl}1j} + 0.5d_{\mathrm{p}}} \quad [8.26]$$

Note that $j = 0$ refers to strain in the top layer.

The strain is assumed to be uniform along the layer in the strain calculation shown previously. But, due to stress wave attenuation, the strain varies from maximum at the point of impact to zero at a distance up to which the tensile stress wave has reached along warp direction. Considering simplified linear strain variation, the maximum strain at the point of impact is expressed as follows (Naik and Doshi, 2005):

$$\varepsilon_{\mathrm{lTl}1}^{\mathrm{max}} = 2\varepsilon_{\mathrm{lTl}1} \quad [8.27]$$

Strain rate:

$$\dot{\varepsilon}_{\mathrm{lTl}} = \frac{\varepsilon_{\mathrm{lTl}k+1} - \varepsilon_{\mathrm{lTl}k}}{t_{k+1} - t_k} \quad [8.28]$$

After the shear wave reaches the back face of the target, conical deformation takes place at the back face. As a result, the layers which have not failed by shear plugging experience tensile strain. The total energy absorbed due to tension in the region consisting of primary yarns is calculated:

$$E_{\mathrm{TPY}1} = A \sum_{j=n_{\mathrm{lf}1}}^{n_{\mathrm{lsc}1}} \int_{(0.5d_{\mathrm{p}})}^{x_{\mathrm{l}1}} \left(\int_{\varepsilon_{\mathrm{lTl}}=0}^{\varepsilon_{\mathrm{lTl}1}} \sigma_{\mathrm{lT}}\, d\varepsilon_{\mathrm{lTl}} \right) dx \quad [8.29]$$

where $A = 4d_{\mathrm{p}}h_{\mathrm{l}}$.

Here, σ_{lT}, $\varepsilon_{\mathrm{lT}}$, x_{l}, n_{lf}, and x refer to tensile stress and tensile strain along warp direction, distance up to which the tensile stress wave has traveled in the in-plane direction, number of layers failed, and a variable along the warp direction, respectively.

8.4.3.8 Energy absorbed due to tensile deformation of yarns/layers in the region consisting of secondary yarns

The yarns/layers in the region consisting of secondary yarns deform along with the primary yarns and undergo tensile strain.

The energy absorbed by the secondary yarns:

$$E_{\mathrm{TSY1}} = \sum_{j=n_{\mathrm{lf1}}}^{n_{\mathrm{lsc1}}} \int_{(0.5d_{\mathrm{p}})}^{x_{\mathrm{t1}}} \left(\int_{0}^{\varepsilon_{r\mathrm{T1}}} \sigma_{r\mathrm{T}}\, d\varepsilon_{r\mathrm{T}} \right) h_{\mathrm{l}}\{2\pi r - 4d_{\mathrm{p}}\} dr \qquad [8.30]$$

Here, $\varepsilon_{r\mathrm{T}}$, $\sigma_{r\mathrm{T}}$, and r refer to tensile strain and tensile stress along the radial direction and a variable in the radial direction, respectively.

8.4.3.9 Energy absorbed by shear plugging

Immediately upon impact, the contact force between the projectile and the target results in through-the-thickness shear plugging stress within the target around the periphery of the projectile. If the induced shear plugging stress exceeds the permissible shear plugging strength, the yarns/layers fail due to shear plugging.

The shear wave velocity and the distance traveled by shear wave along thickness direction are given next (Pandya et al., 2015):

$$V_{z\mathrm{t}} = \sqrt{\frac{1}{\rho} \frac{d\tau_{\mathrm{l}z}}{d\gamma_{\mathrm{l}z}}} \qquad [8.31]$$

$$z_{\mathrm{t1}} = V_{z\mathrm{t}}\, \Delta t \qquad [8.32]$$

Here, $\tau_{\mathrm{l}z}$, $\gamma_{\mathrm{l}z}$, and $V_{z\mathrm{t}}$ refer to shear stress and shear strain of the target and velocity of shear stress wave in z-direction, respectively. z_{t} refers to distance traveled by the shear wave in z-direction.

The number of layers through which the shear wave has traveled in z-direction:

$$n_{\mathrm{lss1}} = \frac{z_{\mathrm{t1}}}{h_{\mathrm{l}}} \qquad [8.33]$$

If the induced shear plugging stress exceeds the permissible shear plugging strength, the yarns/layers fail due to shear plugging.

Shear plugging stress:

$$\tau_{\mathrm{SP1}} = \frac{F_1}{\pi d_{\mathrm{p}} z_{\mathrm{t1}}} \qquad [8.34]$$

Number of layers failed due to shear plugging:

$$n_{\mathrm{lfS1}} = \frac{z_{\mathrm{t1}}}{h_{\mathrm{l}}} \quad \text{for } \tau_{\mathrm{SP1}} \geq S_{\mathrm{sp}} \qquad [8.35]$$

where S_{sp} is shear plugging strength.

The energy absorbed by shear plugging of layers:

$$E_{SP1} = n_{lfS1} h_l S_{sp} \pi d_p h_{01} \quad [8.36]$$

where $h_{01} = n_{lfs1} h_l$.

8.4.3.10 Energy absorbed by delamination and matrix cracking

The tensile strain varies from a maximum value at the point of impact to zero at the location up to which the planar tensile stress wave has reached. The matrix in the region around the point of impact fails when the induced strain exceeds the damage threshold strain for the material under consideration. As a result, matrix cracking takes place which, in turn, leads to delamination in Mode II.

From quasistatic tensile tests, it was observed that damage spread is up to 40% of the gage length of the 2D plain-weave composite specimen at ultimate failure (Eq. [8.37]).

$$x_{d1} = 0.5 d_p + 0.4 x_{tl1} \quad [8.37]$$

Here, x_d denotes size of the delamination zone along planar direction.

Energy absorbed by delamination and matrix cracking:

$$E_{DL1} = \left[P_d \pi x_{d1}^2 A_{ql} G_{IIcd} (N-1) \right] / 100 \quad [8.38]$$

$$E_{MC1} = \left[P_m \, \pi \, x_{d1}^2 A_{ql} E_{mt} \, z_{l1} \, (1 - V_f) \right] / 100 \quad [8.39]$$

Here, P_d, P_m, A_{ql}, G_{IIcd}, E_{mt}, and V_f refer to the percentage of delaminating interfaces, percentage of matrix cracking layers, quasilemniscate area reduction factor, critical strain energy release rate in Mode II, energy absorbed by matrix cracking per unit volume, and fiber volume fraction, respectively.

As the elastic modulus along the warp and fill directions is higher compared to the other directions, the wave velocity and distance reached by the wave are higher along the warp and fill directions. Hence the shape of the planar damage zone would be quasilemniscate and not circular. In the analysis, the damage shape is assumed to be circular for ease of calculations. Quasilemniscate area reduction factor is introduced to take the area of the actual shape into account. Quasilemniscate area reduction factor is defined as the ratio of area of quasilemniscate region to corresponding circular area.

8.4.3.11 Velocity and contact force at the end of first iteration of the first time interval

Incident kinetic energy of the projectile can be expressed as the sum of residual kinetic energy of the projectile and the total energy absorbed by the target until that time interval. Energy balance can be obtained using Eqs. [8.1]–[8.4]. Eq. [8.5] refers to velocity of the projectile at the end of *i*th time interval. Velocity of the projectile at the end of first iteration of the first time interval can be obtained using Eq. [8.5] by replacing i with k and further k with 1. As given in Eq. [8.9], the total contact force acting on the target is evaluated with a trial value of compressive stress, σ_{cz}. Starting with this contact force, energy absorbed by various mechanisms can be calculated for the first iteration of the first time interval. Considering the energy absorbed by various mechanisms, deceleration, tip displacement, and velocity of the projectile can be calculated. Further, contact force at the end of the first iteration of the first time interval can be calculated (Naik et al., 2006).

8.4.3.12 Velocity and contact force during second and subsequent iterations of the first time interval

Based on the contact force obtained at the end of the first iteration of the first time interval, corresponding energy absorbed by various mechanisms is evaluated. In turn, the revised total contact force acting on the target is evaluated. Further, deceleration, projectile tip displacement, and velocity of the projectile are evaluated.

This iterative process is continued until the total contact force converges. The converged value of total contact force would be the actual value of total contact force at the end of the first time interval. Further, deceleration, projectile tip displacement, and velocity of the projectile are evaluated. These values correspond to the end of the first time interval.

As the penetration of the projectile takes place into the target, a plug would be formed directly below the projectile up to the distance up to which the shear wave has reached. The material that is sheared and pushed forward is referred to as a plug. The plug formed would also start moving with the projectile.

Effective mass of the projectile:

$$m_{e1} = m_p + \rho A_p z_{em1} \qquad [8.40]$$

where z_{em1} is the thickness of the moving mass along with the projectile.

If penetration has already taken place, m_p should be replaced by m_{e1} in Eq. [8.10].

Contact force at the end of the first time interval:

$$F_1 = m_{e1} \frac{V_0 - V_1}{\Delta t} \qquad [8.41]$$

8.4.3.13 Formulation from the second time interval up to the end of the ballistic impact event

The velocity of the projectile and the total contact force at the end of the first time interval correspond to the projectile velocity and the total contact force at the beginning of the second time interval, and so on.

8.4.3.14 Projectile tip displacement

The distance traveled by the projectile during *i*th time interval:

$$\Delta z_i = V_{i-1}\Delta t - \frac{1}{2}\left(\frac{F_{i-1}}{m_{e(i-1)}}\right)(\Delta t)^2 \qquad [8.42]$$

The distance traveled by the projectile till the end of *i*th time interval:

$$z_i = \sum_{n=0}^{n=i} \Delta z_n \qquad [8.43]$$

8.4.3.15 Energy absorbed by compression

Compressive strain at the end of *i*th time interval and energy absorbed due to compression in Region 1 are calculated using Eqs. [8.15] and [8.17]. Note that suffix 1 should be replaced by suffix *i*.

8.4.3.16 Total number of layers failed

As the compressive and shear waves travel through the thickness, the layers might fail due to either shear plugging or tension. The total number of layers failed is shown next:

$$n_{lfi} = n_{lfTi} + n_{lfSi} \qquad [8.44]$$

Here, n_{lfT} and n_{lfS} refer to number of layers failed due to tension and shear plugging, respectively.

Note that penetration of the projectile into the target takes place because of either shear plugging or tensile failure of the target. Once shear plugging failure of a layer has taken place, there would not be tensile failure of the same layer. In a similar way, once tensile failure of a layer has taken place, there would not be shear plugging failure of the same layer.

Energy absorbed by the intact layers due to compression in Region 2 till the end of *i*th time interval is shown in Eq. [8.45]:

$$E_{CR2i} = 2\pi h_l K_2 \sum_{j=n_{lfi}}^{n_{lsci}} \int_{0.5d_p}^{x_{ti}} \left(\int_{\varepsilon_{cz}=0}^{\varepsilon_{czi}} \sigma_{cz}\, d\varepsilon_{cz} \right) r dr \qquad [8.45]$$

8.4.3.17 Energy absorbed by tension

Until the shear wave reaches the back face of the target, the tensile strain in each layer is calculated using Eqs. [8.23]–[8.26]. Note that suffix 1 should be replaced by suffix *i*.

After the shear wave reaches the back face of the target, conical deformation takes place on the back face of the target. During conical deformation, it is assumed that the tensile strain in all the intact layers would be the same. The slant height of the cone formed may not be straight. The actual shape of the slant height is considered in the formulation, and the curved length of the slant height formed is taken for strain calculation. Shape of the slant height is calculated using the shape function as described by Naik et al. (2006) for tensile strain variation along the in-plane direction.

Tensile strain in all the layers during conical deformation:

$$\varepsilon_{\mathrm{ITl}i} = \left(\frac{\mathrm{CL}_i - x_{\mathrm{t}i}}{x_{\mathrm{t}i} + 0.5 d_{\mathrm{p}}} \right) \quad [8.46]$$

Here, CL refers to curved length of slant height of the cone formed.

The energy absorbed due to tension in the region consisting of primary yarns till the end of ith time interval is shown next:

$$E_{\mathrm{TPY}i} = A \sum_{j=n_{\mathrm{lf}i}}^{n_{\mathrm{lsc}i}} \int_{0.5d_{\mathrm{p}}}^{x_{\mathrm{l}i}} \left(\int_{\varepsilon_{\mathrm{ITl}}=0}^{\varepsilon_{\mathrm{ITl}i}} \sigma_{\mathrm{IT}} \, d\varepsilon_{\mathrm{IT}} \right) dx \quad [8.47]$$

where $A = 4d_{\mathrm{p}}h_{\mathrm{l}}$.

The energy absorbed due to tensile deformation in the region consisting of secondary yarns till the end of ith time interval is as follows:

$$E_{\mathrm{TSY}i} = \sum_{j=n_{\mathrm{lf}i}}^{n_{\mathrm{lsc}i}} \int_{(0.5d_{\mathrm{p}})}^{x_{\mathrm{t}i}} \left(\int_{0}^{\varepsilon_{r\mathrm{Tl}i}} \sigma_{r\mathrm{T}} \, d\varepsilon_{r\mathrm{T}} \right) h_{\mathrm{l}} \{2\pi r - 4d_{\mathrm{p}}\} dr \quad [8.48]$$

8.4.3.18 Energy absorbed by shear plugging

Shear plugging of the first few layers absorbs some energy, resulting in a sharp decrease in the projectile kinetic energy and total contact force. Shear plugging of the layers would stop when the projectile kinetic energy is less than the energy required to break the yarns by shear plugging. Number of layers that would fail by shear plugging depends upon the target material properties, laminate thickness, and incident impact parameters.

Energy absorbed by shear plugging is calculated using Eqs. [8.34]–[8.36]. Note that suffix 1 should be replaced by suffix i.

8.4.3.19 Energy absorbed by delamination and matrix cracking

Energy absorbed by delamination and matrix cracking is calculated using Eqs. [8.37]–[8.39]. Note that suffix 1 should be replaced by suffix i.

8.4.3.20 Mass of the moving cone and energy absorbed by conical deformation

Upon impact of the projectile onto the target, the upper few layers could fail by shear plugging as the shear wave propagates along the thickness direction.

When the shear wave reaches the back face of the target, conical deformation takes place at the back face depending on the amount of residual kinetic energy available with the projectile. In a typical thin laminated composite, the first few layers fail by shear plugging as the shear wave propagates along the thickness direction. There would be a sudden drop of contact force because of shear plugging failure of a few layers placed toward the incident side. As a result of this, shear plugging failure of further layers may not take place. But the projectile would possess significant kinetic energy even after the shear wave reaches the back face of the target. The layers that are intact absorb the residual kinetic energy of the projectile by conical deformation at the back face. At this stage, the projectile would be moving forward along with the moving cone. The total mass would be the mass of the projectile plus the mass of the moving cone. Mass of the moving cone increases as the transverse wave propagates further along the thickness direction. The velocity of the moving cone can be calculated based on Eqs. [8.2] and [8.3]. Energy absorbed by the moving cone would depend upon the total mass and the velocity of the projectile. The energy absorbed by conical deformation increases initially as the mass of the moving cone increases, but it decreases later as the velocity of the projectile decreases. If the incident impact velocity is lower than the ballistic limit velocity, conical deformation stops as and when the projectile velocity becomes zero within the target. If the incident impact velocity is higher than the ballistic limit velocity, conical deformation continues until all the remaining layers fail under tension.

In a typical thick laminate, many layers would fail by shear plugging before the shear wave reaches the back face of the target. The remaining layers that have not failed by shear plugging undergo conical deformation at the back face. This is also referred to as bulge formation on the back surface.

Mass of the moving cone at the end of ith time interval:

$$M_{\mathrm{CD}i} = \pi x_{\mathrm{t}i}^2 (h - n_{\mathrm{lfS}i} h_1)\rho \qquad [8.49]$$

Energy absorbed by conical deformation till the end of ith time interval:

$$E_{\mathrm{CD}i} = \frac{1}{2} M_{\mathrm{CD}i} (V_i K_1)^2 \qquad [8.50]$$

K_1 is defined in Eq. [8.3].

8.4.3.21 Energy absorbed by friction between the projectile and the target

Even after all the layers fail under shear plugging or tension, the damaged target still provides frictional resistance to the projectile motion. If the projectile does not have enough kinetic energy to overcome the frictional resistance, the projectile would get stuck within the target.

The frictional resistance depends on the projectile dimensions, target material properties, size of the hole formed in the target, and hole surface condition. The energy absorbed due to frictional resistance is obtained experimentally (Naik and Doshi, 2008).

8.4.3.22 Velocity of the projectile, contact force, and projectile tip displacement

Energy balance at the end of ith time interval is calculated using Eq. [8.1]. Note that suffix i in the terms for energy absorbed by various mechanisms except conical deformation should be replaced by $(i - 1)$.

Effective mass of the projectile is calculated using Eq. [8.40]. Note that suffix 1 should be replaced by suffix i.

Projectile velocity at the end of the ith time interval is obtained using Eqs. [8.2]–[8.5]. E_i and m_p should be replaced by E_{i-1} and m_{ei}, respectively, in Eqs. [8.2] and [8.5]. Suffix i should be replaced by suffix $(i - 1)$ except for conical deformation in Eq. [8.4].

Contact force at the end of ith time interval can be calculated using Eq. [8.41]. Here, V_0 and V_1 should be replaced by V_{i-1} and V_i, respectively. Suffix 1 should be replaced by suffix i.

Projectile tip displacement at the end of ith time interval can be calculated using Eq. [8.43].

Contact duration at the end of ith time interval,

$$T_c = n_i \Delta t \qquad [8.51]$$

The formulation given above is used until either the target material is completely penetrated/perforated or the projectile velocity becomes zero within the target.

With certain modifications, the formulation presented can be used for ballistic impact analysis of 3D woven composites (Udatha et al., 2012) and laminated composites made using unidirectional layers. Strip formation is another energy-absorbing mechanism in the case of laminated composites made using unidirectional layers.

8.5 Input parameters

The geometric parameters of the composite target, mechanical properties of 2D plain-weave E-glass/epoxy (Appendix A) and 2D 8H satin-weave T300 carbon/epoxy (Appendix B) at high strain rates and frictional behavior (Appendix C) are the input data required (Table 8.2). Mechanical properties at higher strain rates are obtained as explained in Naik et al. (2011) and Ravikumar et al. (2013).

Table 8.2 Geometry and physical and mechanical properties of the target material

Target	Material	2D plain-weave E-glass/epoxy	2D 8H satin weave T300 carbon/epoxy
Geometric properties	Layer thickness, h_l	0.25 mm	0.34 mm
	Laminate thickness, h	3, 5, 6, and 20 mm	3, 5 mm
	Fiber volume fraction, V_f	0.51	0.56
Physical properties	Density, ρ	1850 kg/m^3	1435 kg/m^3
Mechanical properties	Young's modulus in the in-plane direction, E_l	16 GPa	40 GPa
	Young's modulus in the z-direction, E_z	8 GPa	6.8 GPa
	In-plane shear modulus, G_{lt}	2 GPa	3 GPa
	Shear modulus in thickness direction, G_{lz}	2 GPa	3 GPa
	Tensile failure strain along fiber, ε_{lTf}	0.03	0.0132
	Tensile failure strain in in-plane direction, ε_{45tf}	0.035	0.026
	Compressive failure strain in z-direction, ε_{zCf}	0.05	0.027
	Shear plugging strength, S_{sp}	120 MPa	120 MPa
	Dynamic strain energy release rate, G_{IICd}	1000 J/m^3	980 J/m^3
	Matrix cracking energy, E_{mt}	0.9 MJ/m^3	0.9 MJ/m^3
	Maximum compression in layer, Δh_{lc}	5%	2.7%
	Stress transmission factor, b	0.825	0.83
	Quasilemniscate area reduction factor, A_{ql}	0.9	0.9
	Percent delaminating layers, P_d	90	90
	Percent matrix cracking, P_m	90	90

8.6 Experimental studies

8.6.1 Experimental details

Experimental studies were performed on composite laminates. The projectile and the target details used for experimental studies are given below:

- Projectile: cylindrical, flat-ended, hardened steel projectile with mass, $m_p = 5.0$ g; diameter, $d_p = 6.35$ mm
- Target: 2D plain-weave E-glass/epoxy composite; nominal thickness = 4.25 mm; unsupported area of the target 125 × 125 mm

8.6.2 Experimental observations and validation

There are several ballistic impact test standards used internationally (MIL-STD-662F, 1997; NATO STANAG 2920 PPS, 2003; NIJ Standard–0101.06, 2008; ASTM F1233-08, 2008). In principle, the test facility used in the present study is similar to ASTM F1233-08 test standard. The facility used in the present study is at a laboratory scale and is scaled down compared to the ASTM test standard.

A composite laminate of 4.25-mm nominal thickness was tested using a ballistic impact test apparatus (Pandya et al., 2012a,b, 2013). The projectile velocity was measured using a chronograph.

The specimen impacted with a projectile velocity of 148 m/s is shown in Fig. 8.3. It can be seen that complete penetration has taken place. But, the projectile was stuck within the laminate because of friction between the target and the projectile. The size of damage around the projectile impact zone on the front face and back face of the target can be seen in Fig. 8.3(a) and (b), respectively. It was observed that the damage shape was nearly circular with a diameter of 20 mm on the front face and 40 mm on the back face. The projectile was stuck by friction in the target as shown in Fig. 8.3(c). The plug formed is shown in Fig. 8.3(d). The average plug thickness and diameter were 2.8 and 6.1 mm, respectively.

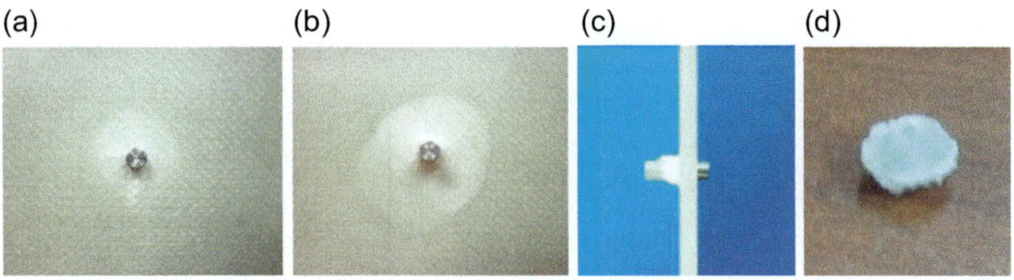

Figure 8.3 Composite specimen impacted at ballistic limit velocity, $h = 4.25$ mm: (a) front view, (b) back view, (c) side view, (d) plug formed.
Shaktivesh, Nair, N.S., Sesha Kumar, Ch.V., Naik, N.K., 2013. Ballistic impact performance of composite targets. Mater. Des. 51, 833–846.

Ballistic limit velocity, V_{BL}, is defined as the incident impact velocity for a specific projectile and target combination which would lead to complete penetration of the target with the projectile tip reaching the back face of the target with zero velocity. It may not be possible to obtain V_{BL} experimentally. Hence, V_{50} is used instead of V_{BL} for experimental studies. V_{50} is defined as the average of equal number of highest partial penetration velocities and lowest complete penetration velocities of a specific projectile and target combination, which occur within a specified velocity range. A minimum of three partial and three complete penetration velocities are used to compute V_{50}. In other words, V_{50} defines incident impact velocity at which there is 50% probability of partial penetration and 50% probability of perforation.

Ballistic impact tests were performed on six specimens. It was observed experimentally that the ballistic limit velocity, V_{50}, is equal to 148 m/s with the velocity range of 145–152 m/s. The ballistic limit velocity was also predicted using the generalized method explained in Section 8.4.3. The analytically predicted ballistic limit velocity, V_{BL}, is 155 m/s. A good match can be seen between the analytical prediction and the experimental result.

For the case of laminate thickness of 4.25 mm, 17 number of plain-weave E-glass layers of thickness 0.25 mm were used. The analytical prediction shows that, at ballistic limit velocity, the top 13 layers failed by shear plugging, whereas the bottom four layers failed by tension. The total thickness of the 13 layers failed by shear plugging, that is, plug thickness is 3.25 mm. The diameter of the plug is 6.35 mm. These values match well with the experimentally obtained plug thickness of 2.8 mm and average diameter of 6.1 mm.

8.6.3 Experimental observations and comparison with analytical predictions

Ballistic impact tests were also performed on 2D plain-weave E-glass/epoxy and 2D 8H satin-weave T300 carbon/epoxy in the thickness range of 3–5 mm (Pandya et al., 2013). In this case, $d_p = 6.35$ mm and $m_p = 6.5$ g. For 2D plain-weave E-glass/epoxy, the analytically predicted ballistic limit velocity (V_{BL}) is 92 and 136 m/s for laminate thickness of 3 and 5 mm, respectively. This is in good agreement with the experimental result (V_{50}) of 98 m/s (within the velocity range of 95–100 m/s) and 130 m/s (within the velocity range of 126–133 m/s) for laminate thickness of 3 and 5 mm, respectively.

For 2D 8H satin-weave T300 carbon/epoxy, the analytically predicted ballistic limit velocity (V_{BL}) is 78 m/s for laminate thickness of 3 mm. This is in good agreement with the experimental result (V_{50}) of 80 m/s (within the velocity range of 78–83 m/s).

8.7 Results and discussion

Ballistic impact studies have been carried out for a typical 2D plain-weave E-glass/epoxy composite with target thickness of 6 and 20 mm using the analytical method presented. Specifically, energy absorbed by different mechanisms, kinetic energy of the projectile, contact force, projectile velocity, and projectile tip displacement are evaluated as a function of time (Table 8.3, Figs. 8.4–8.7).

Analytical studies have also been carried out for two types of composites for comparison: 2D plain-weave E-glass/epoxy and 2D 8H satin-weave T300 carbon/epoxy. The target thicknesses considered are in the range of 3–20 mm. Energy absorbed by different mechanisms is evaluated (Table 8.3).

The geometry and physical and mechanical properties of the target are given in Table 8.2. Details regarding high strain rate properties are presented in Figs. 8.8

Table 8.3 Percentage energy absorbed by different mechanisms for different laminate thicknesses at ballistic limit velocity, $m_p = 6.5$ g, $d_p = 6.35$ mm

Material	2D plain weave E-glass/epoxy				2D 8H satin-weave T300 carbon/epoxy	
Laminate thickness (mm)	**3**	**5**	**6**	**20**	**3**	**5**
Ballistic limit velocity (m/s) energy %	92.0	136.0	171.7	287.3	78.0	119.0
Incident kinetic energy of the projectile (total energy absorbed by the target), KE_{p0}	100.0	100.0	100.0	100.0	100.0	100.0
Energy absorbed due to compression in region 1, E_{CR1}	4.88	4.91	5.02	5.50	5.16	4.31
Energy absorbed due to compression in region 2, E_{CR2}	1.14	1.21	1.32	10.72	0.78	1.16
Energy absorbed due to tension by primary yarns, E_{PY}	9.74	11.22	11.68	16.77	9.50	11.32
Energy absorbed due to tension by secondary yarns, E_{SY}	14.21	14.64	12.99	3.81	13.12	14.15
Energy absorbed due to shear plugging, E_{SP}	63.32	60.53	61.01	40.64	65.92	62.73
Energy absorbed due to delamination, E_{DL}	2.18	1.07	2.19	0.12	1.68	1.21
Energy absorbed due to matrix cracking, E_{MC}	4.53	6.42	5.79	1.13	3.84	5.12
Energy absorbed due to friction, E_F	0	0	0	21.31	0	0
Peak energy carried by moving cone, E_{CF}	9.76	10.43	11.58	0	10.16	8.47

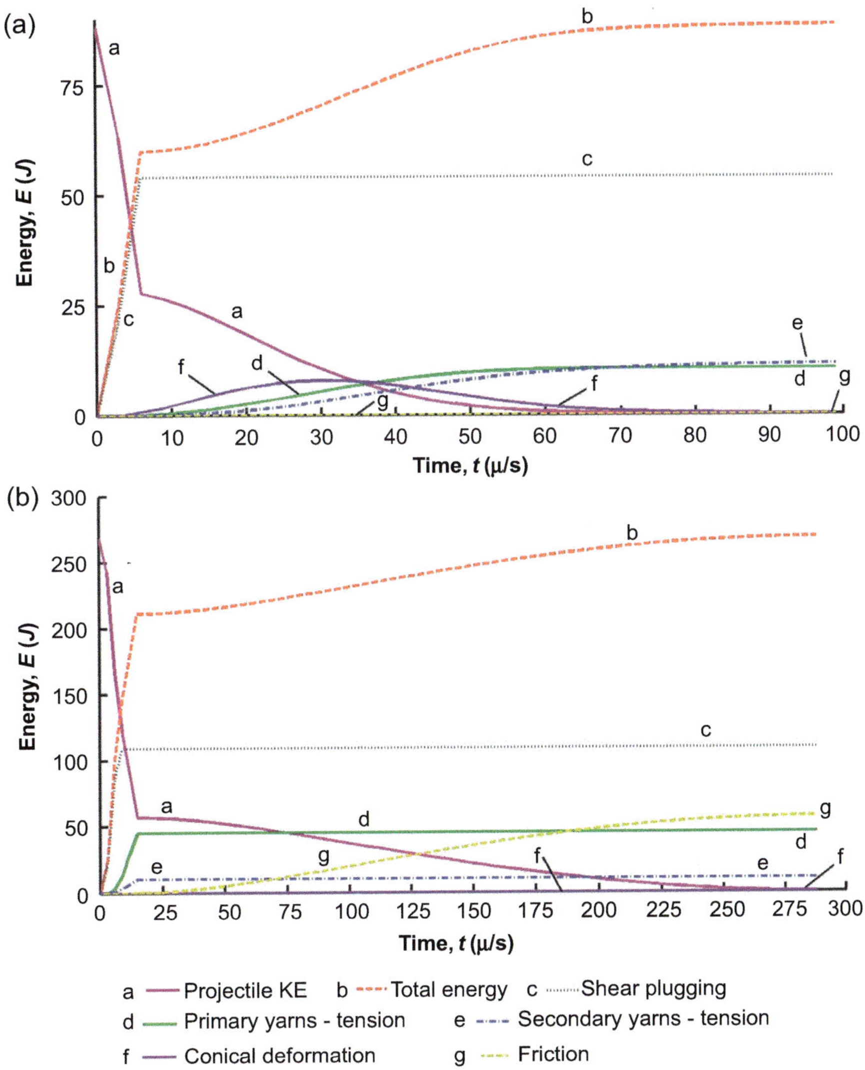

Figure 8.4 Energy absorbed by different mechanisms during ballistic impact event, 2D plain-weave E-glass/epoxy: (a) $h = 6$ mm, $V_i = 171.7$ m/s, (b) $h = 20$ mm, $V_i = 287.3$ m/s.

and 8.9. Mechanical properties at higher strain rates are obtained as explained in Naik et al. (2011) and Ravikumar et al. (2013). A cylindrical, flat-ended rigid projectile of diameter, d_p, of 6.35 mm and mass, m_p, of 6.5 g has been considered for the analytical studies.

The analysis was carried out with different time intervals up to 0.2 μs. The converged contact force obtained at the end of first time interval iteratively was the same for all the cases. Hence, for the present study, the time interval of 0.2 μs is used.

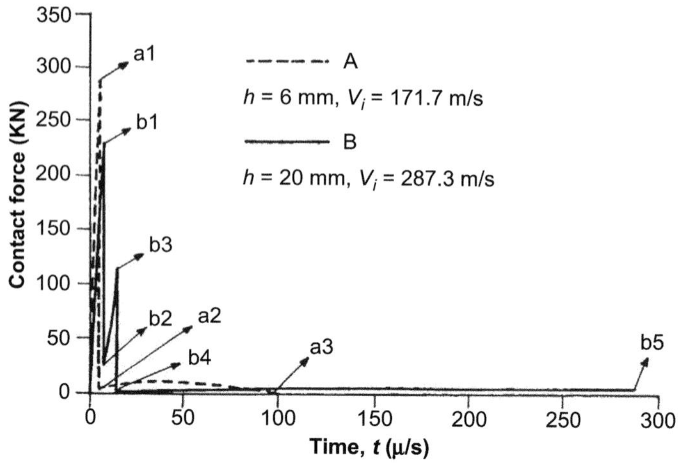

Figure 8.5 Ballistic impact performance of 2D plain-weave E-glass/epoxy: contact force versus time plots.

8.7.1 Energy absorbed by different mechanisms

Fig. 8.4(a) and (b) present energy absorbed by different mechanisms and kinetic energy of the projectile as a function of time for target thickness of 6 and 20 mm, respectively. The plots are for $V_i = 171.7$ m/s for target thickness of 6 mm and $V_i = 287.3$ m/s for target thickness of 20 mm. Percentage energy absorbed by different mechanisms at ballistic limit velocity for different laminate thicknesses for 2D plain-weave E-glass/epoxy is presented in Table 8.3. The energy carried by the moving cone is not included in the total energy absorbed by the target. Note that the incident impact velocities considered are the corresponding ballistic limit velocities.

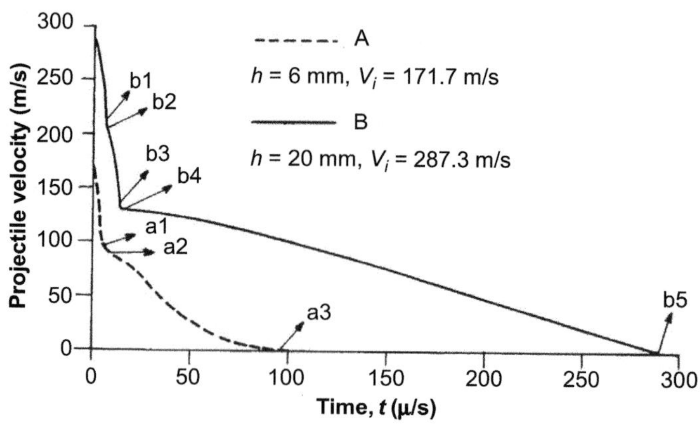

Figure 8.6 Ballistic impact performance of 2D plain-weave E-glass/epoxy: projectile velocity versus time plots.

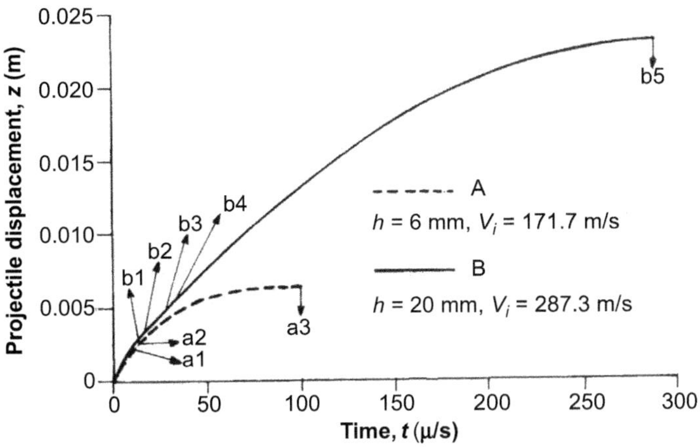

Figure 8.7 Ballistic impact performance of 2D plain-weave E-glass/epoxy: projectile tip displacement versus time plots.

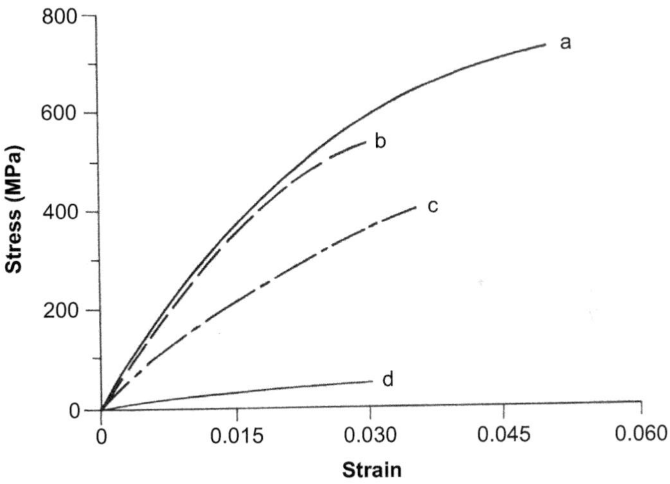

Figure 8.8 High strain rate stress—strain curves for 2D plain-weave E-glass/epoxy: (a) compressive along thickness direction, σ_{cz} versus ε_{cz}, (b) tensile along warp direction, σ_{xt} versus ε_{xt}, (c) tensile along in-plane 45° direction, σ_{xt45} versus ε_{xt45}, (d) interlaminar shear stress versus interlaminar shear strain, τ versus γ.

The contact duration and the incident impact energy are 99 μs and 95 J for target thickness of 6 mm (Fig. 8.4(a)), whereas they are 289 μs and 268 J for target thickness of 20 mm (Fig. 8.4(b)).

The major energy-absorbing mechanisms for target thickness of 6 mm are shear plugging, stretching, and tensile failure in the region consisting of primary yarns, tensile deformation of yarns/layers in the region consisting of secondary yarns, and

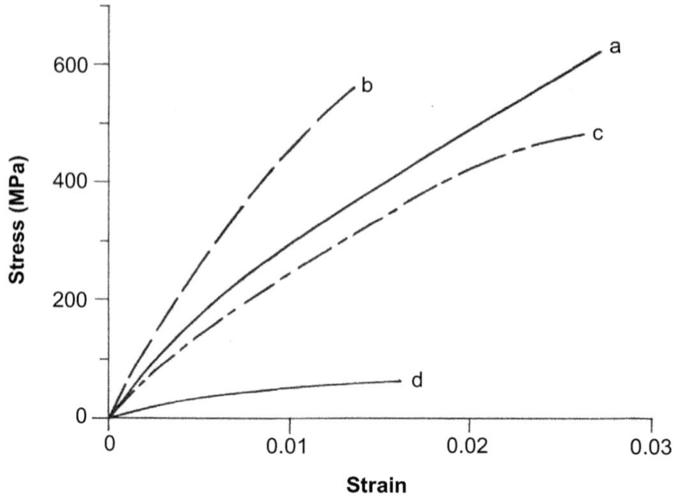

Figure 8.9 High strain rate stress–strain curves for 2D 8H satin-weave T300 carbon/epoxy: (a) compressive along thickness direction, σ_{cz} versus ε_{cz}, (b) tensile along warp direction, σ_{xt} versus ε_{xt}, (c) tensile along in-plane 45° direction, σ_{xt45} versus ε_{xt45}, (d) interlaminar shear stress versus interlaminar shear strain, τ versus γ.

energy carried by the moving cone. Energy absorbed by the other mechanisms is not significant.

Peak energy of 11 J is carried at a time interval of 32 μs by the moving cone for a target thickness of 6 mm. This is 11.58% of the total energy absorbed (95 J) by the target. Energy absorbed by conical deformation is a function of mass and velocity of the moving cone. The mass of the moving cone increases with time, but the velocity decreases. Hence, the energy absorbed by conical deformation increases initially and then decreases. When the velocity of the projectile and, in turn, the velocity of the moving cone is zero, the energy carried by the moving cone would be zero. During conical deformation, the residual kinetic energy of the projectile would be decreasing because of the energy carried by the moving cone. As a result of this, the velocity of the projectile would be decreasing.

The major energy-absorbing mechanisms for target thickness of 20 mm are shear plugging, stretching, and tensile failure in the region consisting of primary yarns, tensile deformation of yarns/layers in the region consisting of secondary yarns, and friction between the projectile and the target. Energy absorbed by the other mechanisms is not significant. Note that, in this case, energy carried by the moving cone is negligible.

The energy carried by the moving cone is not included in the total energy absorbed by the target (Fig. 8.4). At any given instant of time, the sum of residual kinetic energy of the projectile, energy absorbed by the target, and energy carried by the moving cone would be equal to the incident kinetic energy of the projectile.

8.7.2 Contact force, projectile velocity, and tip displacement

Figs. 8.5–8.7 present contact force versus time, projectile velocity versus time, and projectile tip displacement versus time plots, respectively. The plots are for $V_i = 171.7$ m/s for target thickness of 6 mm and $V_i = 287.3$ m/s for target thickness of 20 mm. Note that the incident impact velocities considered are the corresponding ballistic limit velocities.

For target thickness of 6 mm, there is a sudden increase in contact force initially up to a time interval of 4.6 μs because of the resistance offered by the target (Fig. 8.5). The peak contact force reached is 278 kN (point a1). As the shear plugging failure of the upper layers takes place, contact force starts decreasing.

Consequently, shear plugging failure of further layers may not take place. Point a2 indicates the end of shear plugging failure. This corresponds to a time interval of 4.8 μs.

The next phase is conical deformation on the back face and tensile failure of the yarns of the composite target up to the end of the ballistic impact event (99 μs). The end of the ballistic impact event is indicated by point a3 (Fig. 8.5).

Projectile velocity as a function of time is presented in Fig. 8.6. There is a sudden decrease in projectile velocity up to point a2. The corresponding time duration is 4.8 μs. The sudden decrease in velocity is because of a significant amount of energy absorbed by shear plugging. Point a1 corresponds to peak contact force.

The region from point a2 to point a3 represents conical deformation on the back face and tensile failure of yarns of the composite target. During this period, the rate of change of velocity is much less than the rate of change of velocity in the region up to point a2. For this case, the exit velocity is 4 m/s.

Fig. 8.7 presents displacement of projectile tip as a function of time. The maximum displacement is 6.36 mm, which is more than the thickness of the target. This is because of conical deformation on the back face of the target.

For a target thickness of 20 mm also, it can be seen that there is a sudden increase in contact force initially up to a time interval of 6.6 μs because of the resistance offered by the target (Fig. 8.5). The peak contact force reached is 226 kN (point b1). As the shear plugging failure of the upper layers takes place, contact force starts decreasing.

Consequently, shear plugging failure of further layers may not take place. Point b2 indicates the end of shear plugging failure. This corresponds to a time interval of 6.8 μs.

The next phase is tensile failure of the yarns of the composite target up to a time interval of 14.2 μs. The peak contact force reached during this phase is 111 kN (point b3).

Complete failure of the target takes place at a time duration of 14.4 μs as indicated by point b4. As the complete target failure takes place immediately after point b4 is reached, the contact force is nearly zero. After this, only frictional resistance is offered during further movement of the projectile. The end of the ballistic impact event is indicated by point b5 (Fig. 8.5). At point b5, the contact force is 4 kN and time duration is 289 μs.

Projectile velocity as a function of time is presented in Fig. 8.6. There is a sudden decrease in projectile velocity up to point b2. The corresponding time duration is 6.8 μs. The sudden decrease in velocity is because of a significant amount of energy absorbed by shear plugging. Point b1 corresponds to peak contact force.

The region from point b2 to point b3 represents tensile failure of yarns of the composite target. During this period, the rate of change of velocity is less than the rate of change of velocity in the region up to point b2. Point b4 indicates complete failure of the target either by shear plugging or by tensile failure. After this, only frictional resistance is offered during further movement of the projectile up to the end of the ballistic impact event (point b5). For this case, the exit velocity is 1 m/s.

Fig. 8.7 presents displacement of projectile tip as a function of time. The maximum displacement is 23.04 mm, which is more than the thickness of the target. This is because of bulge formation on the back face of the target.

It is interesting to compare the ballistic impact performance of targets with thickness of 6 and 20 mm. With a target thickness of 6 mm, contact force generated is higher and contact duration is lower compared to target thickness of 20 mm (Fig. 8.5). Details regarding the effect of target thickness on ballistic impact performance are presented in Section 8.7.6. Further details on this subject can be obtained from Shaktivesh et al. (2013).

8.7.3 Ballistic impact behavior of different materials

Analytical studies have also been carried out for two types of composites for comparison: 2D plain-weave E-glass/epoxy and 2D 8H satin-weave T300 carbon/epoxy. The target thicknesses considered are in the range of 3–20 mm. Energy absorbed by different mechanisms is presented in Table 8.3.

The ballistic limit velocity is observed to be higher for plain-weave E-glass/epoxy than for 8H satin-weave T300 carbon/epoxy for target thickness of 3 and 5 mm. The analytically obtained ballistic limit velocities (V_{BL}) are in good agreement with the experimental ballistic limit velocities (V_{50}), as explained in Section 8.6.

The major energy-absorbing mechanisms are shear plugging, stretching, and tensile failure in the region consisting of primary yarns and tensile deformation of yarns/layers in the region consisting of secondary yarns for both types of composites considered. In the range of target thicknesses considered, energy carried by the moving cone is also significant for both types of composite specimens.

8.7.4 Strain rate during ballistic impact event

Ballistic impact behavior of 2D plain-weave E-glass/epoxy target of thickness 6 mm along the thickness direction is presented in Fig. 8.10. The incident impact velocity is 171.7 m/s. Fig. 8.10(a) shows the distance traveled by compressive wave, distance traveled by shear wave, and projectile tip displacement as a function of time. At a time duration of 4.32 μs, the through-the-thickness compressive wave reaches the back face of the target (point a), whereas at a time duration of 5.8 μs, the shear wave reaches the back face of the target (point c). Note that variation of projectile tip displacement with

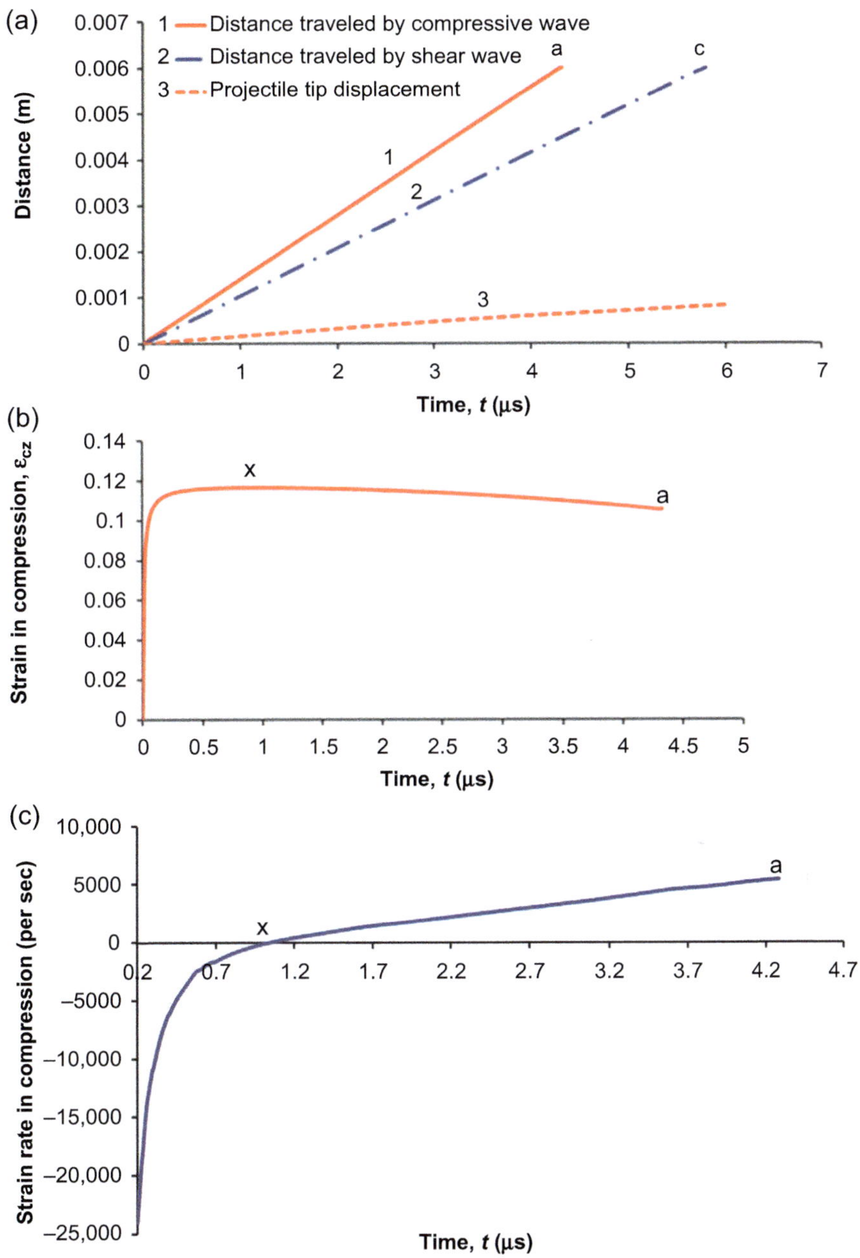

Figure 8.10 Ballistic impact behavior of 2D plain-weave E-glass/epoxy, $V_i = 171.7$ m/s, $h = 6$ mm: (a) distance versus time, (b) compressive strain along z-direction versus time, (c) compressive strain rate along z-direction versus time.
Pandya, K.S., Sesha Kumar, Ch.V., Nair, N.S., Patil, P.S., Naik, N.K., 2015. Analytical and experimental studies on ballistic impact behavior of 2D woven fabric composites. Int. J. Damage Mech. 24, 471–511. doi:10.1177/1056789514531440.

time is not linear (plot 3). Fig. 8.10(b) shows compressive strain along the z-direction as a function of time. Peak strain of 0.117 is reached at a time interval of 0.98 μs (point x). Through-the-thickness compressive strain rate as a function of time is shown in Fig. 8.10(c). It can be seen that strain rate is very high initially. But, at a time interval of 0.48 μs, strain rate is 4211 per sec. Strain rate changes sign at point x.

Ballistic impact behavior of 2D plain-weave E-glass/epoxy target of thickness 6 mm along in-plane direction is presented in Fig. 8.11. Fig. 8.11(a) shows distance traveled by a transverse wave in the in-plane direction and projectile tip displacement as a function of time. At time duration of 4.6 μs, shear plugging of the layers starts (point b). Fig. 8.11(b) shows tensile strain along the in-plane direction as a function of time. Failure of the topmost layer would take place in tension when the induced tensile strain reaches 0.03 (point d). This would take place at a time interval of 11.84 μs. But shear plugging failure of the topmost layer has taken place at a time interval of 4.6 μs (point b). Tensile strain rate along the in-plane direction as a function of time is shown in Fig. 8.11(c). Peak strain rate of 3141 per sec. is reached at a time interval of 3.6 μs.

8.7.5 Effect of incident impact velocity on projectile tip displacement

Fig. 8.12 presents projectile velocity and projectile tip displacement as a function of time. For the analytical studies, incident impact velocities considered are 168 and 150 m/s. These velocities are lower than the ballistic limit velocity of 171.7 m/s. Hence perforation of the target by the projectile would not take place. The studies were performed on a 2D plain-weave E-glass/epoxy target of thickness 6 mm. A projectile mass of 6.5 g and projectile diameter of 6.35 mm were considered. It was observed that the maximum projectile tip displacement, that is, penetration of the projectile into the target was 3.9 and 3.4 mm with incident impact velocity of 168 and 150 m/s, respectively, and the corresponding time duration was 86.2 and 152.4 μs. Projectile velocity was zero when the projectile tip displacement was as indicated above. Point 1 corresponds to peak contact force and point 2 indicates end of shear plugging failure.

Ballistic impact experiments were also performed with incident impact velocities (V_{50}) of 168 and 151 m/s. It was observed experimentally that projectile tip displacement was 4.0 and 3.5 mm with incident impact velocity of 168 and 151 m/s, respectively. Note that there is a good match between the analytical predictions and the experimental observations in terms of projectile tip displacement and projectile velocity.

8.7.6 Effect of target thickness on ballistic impact performance

Ballistic limit velocity as a function of target thickness is presented in Fig. 8.13. It is interesting to note the trend of ballistic limit velocity as a function of target thickness. Target thickness can be subdivided into three zones. In zone 1, that is, up to target thickness of 12 mm, ballistic limit velocity increases with the increase in target

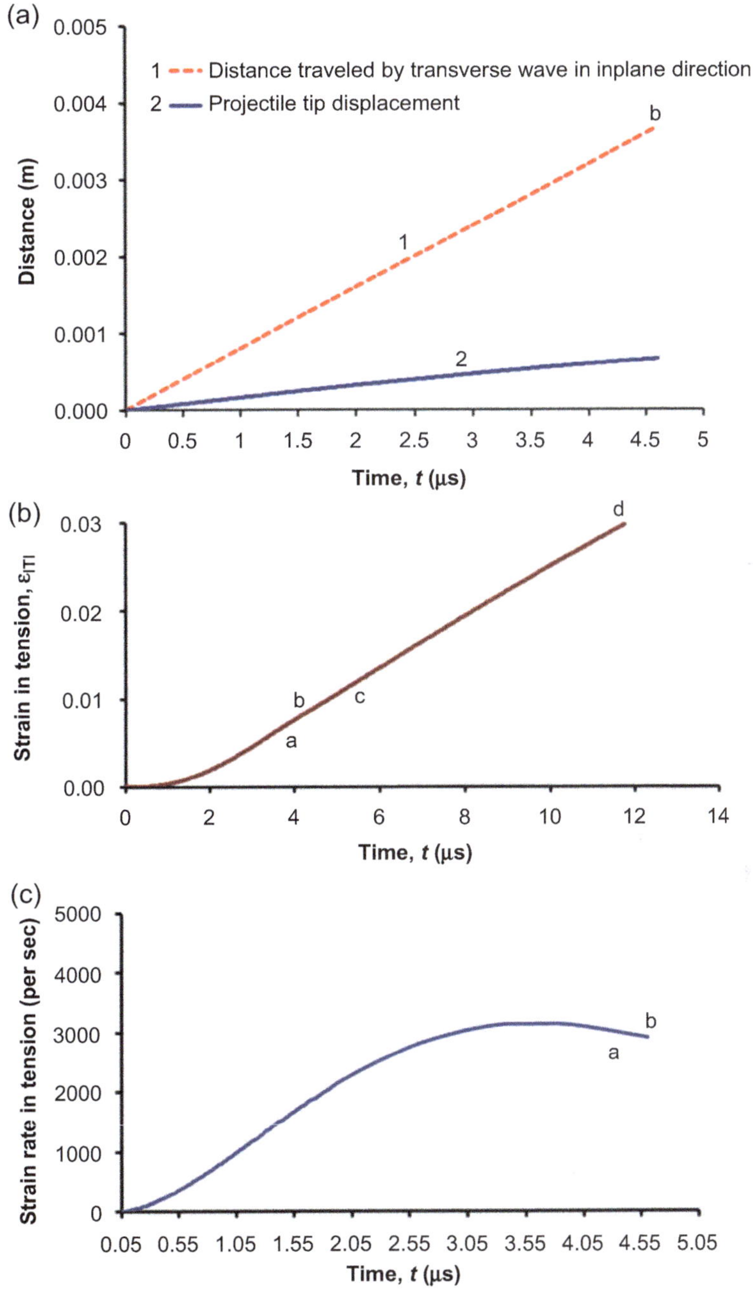

Figure 8.11 Ballistic impact behavior of 2D plain-weave E-glass/epoxy, $V_i = 171.7$ m/s, $h = 6$ mm: (a) distance versus time, (b) tensile strain along in-plane direction versus time, (c) tensile strain rate along in-plane direction versus time.
Pandya, K.S., Sesha Kumar, Ch.V., Nair, N.S., Patil, P.S., Naik, N.K., 2015. Analytical and experimental studies on ballistic impact behavior of 2D woven fabric composites. Int. J. Damage Mech. 24, 471–511. doi:10.1177/1056789514531440.

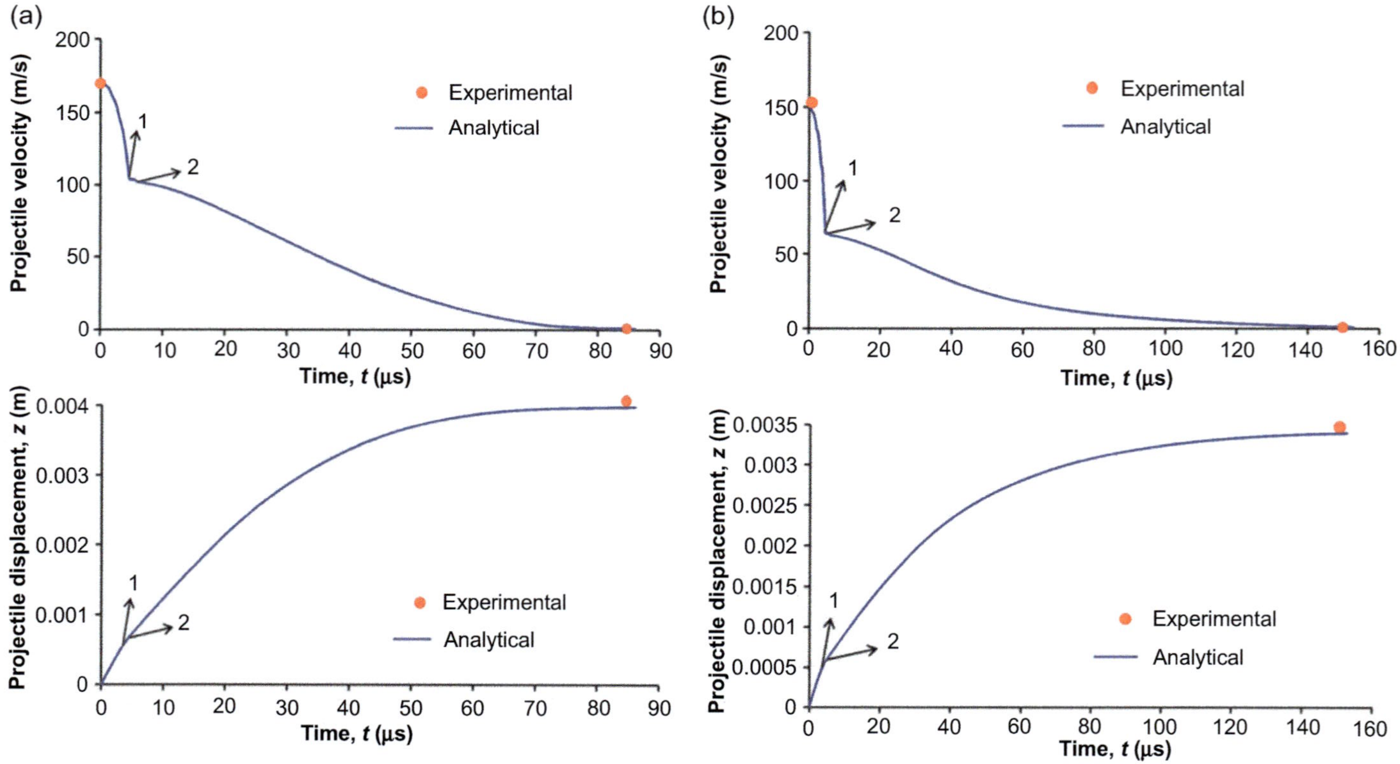

Figure 8.12 Ballistic impact performance of 2D plain-weave E-glass/epoxy, $h = 6$ mm: (a) $V_i = 168$ m/s, (a1) velocity versus time, (a2) projectile tip displacement versus time, (b) $V_i = 150$ m/s, (b1) velocity versus time, (b2) projectile tip displacement versus time.
Pandya, K.S., Sesha Kumar, Ch.V., Nair, N.S., Patil, P.S., Naik, N.K., 2015. Analytical and experimental studies on ballistic impact behavior of 2D woven fabric composites. Int. J. Damage Mech. 24, 471–511. doi:10.1177/1056789514531440.

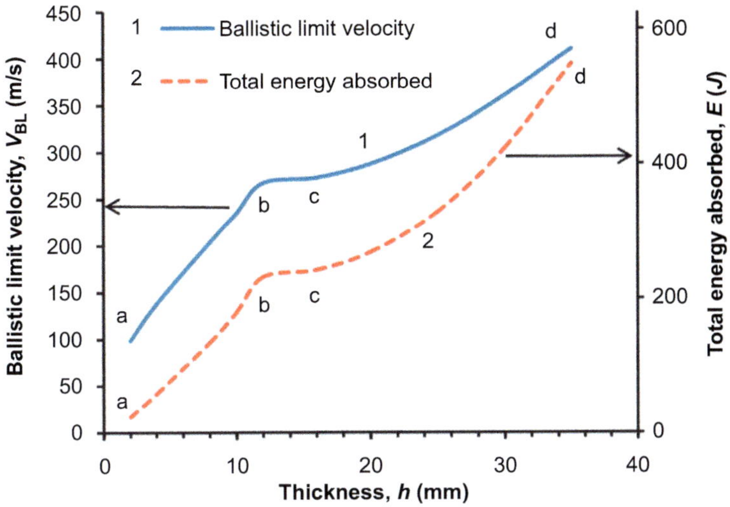

Figure 8.13 Ballistic limit velocity and total energy absorbed by the target at ballistic limit velocity as a function of target thickness.
Shaktivesh, Nair, N.S., Sesha Kumar, Ch.V., Naik, N.K., 2013. Ballistic impact performance of composite targets. Mater. Des. 51, 833–846.

thickness. In zone 2, (from target thickness of 12–16 mm), ballistic limit velocity remains nearly constant. It implies that the target with thickness from 12 to 16 mm would give the same ballistic impact performance. A similar observation that the target with a thickness in a certain range can give the same ballistic impact performance was made for metallic targets as well (Krishna Teja Palleti et al., 2012). In zone 3, ballistic limit velocity increases with the increase in the target thickness above 16 mm. The plot of total energy absorbed by the target at ballistic limit velocity as a function of target thickness also has a similar trend as given for ballistic limit velocity as a function of target thickness (Fig. 8.13). Behavior as presented in Fig. 8.13 for ballistic limit velocity can be explained based on different prominent energy-absorbing mechanisms with different target thicknesses (Shaktivesh et al., 2013). The sum of energy absorbed by different mechanisms leads to an increasing, nearly constant, and then increasing trend for the ballistic limit velocity as the target thickness increases.

8.8 Enhancing ballistic protection capability of composite targets

There are several techniques for enhancing ballistic protection capability of composites. The representative studies include the use of hybrid composites (Ellis et al., 1998; Muhi et al., 2009; Hazell and Appleby-Thomas, 2009; Pandya et al., 2013), dispersion of nanoparticles with high energy-absorption characteristics in polymers/polymer matrix composites (Grujicic et al., 2008; Pandya et al., 2012a; Sun et al., 2009), development of sandwich structures with foam core (Hou et al., 2010) or honeycomb core (Feli and Namdari Pour, 2012), and use of 3D composites (Udatha et al., 2012).

8.8.1 Hybrid composites

Experimental studies are presented on ballistic impact behavior of hybrid composites by Pandya et al. (2013). The materials considered for the studies were four types of symmetric hybrid composites made using plain-weave E-glass fabric and 8H satin-weave T300 carbon fabric with epoxy resin. Studies were also carried out on plain-weave E-glass/epoxy and 8H satin-weave T300 carbon/epoxy for comparison. It was observed that ballistic limit velocity, V_{50}, can be enhanced for the composites by adding E-glass layers to T300 carbon layers, compared with only carbon composites for the same laminate thickness. Placing E-glass layers in the exterior and carbon layers in the interior provided higher ballistic limit velocity than placing carbon layers in the exterior and E-glass layers in the interior. Experimental damage patterns during ballistic impact have been obtained for the materials considered (Pandya et al., 2013).

8.8.2 3D composites

Performance of 3D orthogonal woven E-glass/epoxy composites under high-velocity impact was studied by Udatha et al. (2012). The analytical method used was based on wave propagation in the target and energy balance between the target and projectile. It was observed that ballistic limit velocity for complete penetration for a 3D orthogonal woven composite is higher than that for a 2D plain-weave composite.

8.8.3 Composites dispersed with nanoparticles

Experimental studies were performed on the ballistic impact behavior of nanoparticle-dispersed materials viz symmetric balanced cross-ply laminates made using unidirectional E-glass fabric with epoxy resin and neat epoxy resin by Pandya et al. (2012a). The nanoparticles used were nanosilica and multiwalled carbon nanotube (MWCNT) for polymer matrix composites and nanosilica for epoxy resin. For comparison, studies were carried out on symmetric balanced cross-ply E-glass/epoxy and neat epoxy resin without nanoparticles. Effect of nanoparticle dispersion on ballistic limit velocity, V_{50}, and energy absorbed was investigated. It was observed that V_{50} can be enhanced up to 6.3% for polymer matrix composites and up to 7.3% for neat resin with the addition of nanoparticles. Also, energy absorbed can be increased up to 13.0% for polymer matrix composites and up to 15.2% for neat resin with the addition of nanoparticles. Damage- and energy-absorbing mechanisms for different types of materials studied are also presented. Further, it was observed that the damage size on the target around the point of impact decreases with the addition of nanoparticles. Quantitative data is given for high-velocity impact behavior of the five types of specimens studied (Pandya et al., 2012a).

8.9 Conclusions

A generalized analytical formulation is presented for the prediction of ballistic impact behavior of 2D woven fabric composite laminates impacted with a rigid, cylindrical projectile. The formulation is valid for a wide range of laminate thicknesses. The formulation presented considers both shear plugging and tensile failure during conical deformation at the back face of the target. The specific observations are as follows:

- V_{BL} for 2D plain-weave E-glass/epoxy is 92, 136, 171.7, and 287.3 m/s for target thicknesses of 3, 5, 6, and 20 mm, respectively.
- V_{BL} for 2D 8H satin-weave T300 carbon/epoxy is 78 and 119 m/s for target thicknesses of 3 and 5 mm, respectively.
- V_{BL} is higher for 2D plain-weave E-glass/epoxy compared to 2D 8H satin-weave T300 carbon/epoxy for the same specimen thickness.
- There is a good match between the analytically predicted ballistic limit velocity, V_{BL}, and experimentally obtained ballistic limit velocity, V_{50}, for both types of composites considered for different laminate thicknesses.
- The major energy-absorbing mechanisms are shear plugging, stretching, and tensile failure in the region consisting of primary yarns and tensile deformation of yarns/layers in the region consisting of secondary yarns for both types of composites considered.
- For a 2D plain-weave E-glass/epoxy with a target thickness of 6 mm, energy carried by the moving cone is also significant.
- For a 2D plain-weave E-glass/epoxy with a target thickness of 20 mm, energy absorbed due to friction between the moving projectile and the target is also significant.
- In a certain range of thicknesses, the target would give the same ballistic impact performance.
- There is a good match between the analytically predicted and experimentally observed projectile tip displacements.

Appendices

Appendix A: stress–strain data at high strain rates—2D plain-weave E-glass/epoxy

High strain rate stress–strain curves were obtained using a compressive split Hopkinson pressure bar apparatus (Naik and Kavala, 2008), tensile split Hopkinson pressure bar apparatus (Naik and Perla, 2008; Naik et al., 2010), and torsional split Hopkinson bar apparatus (Naik et al., 2007). Experimental details are given in the respective references. Mechanical properties at higher strain rates are obtained as explained in Naik et al. (2011) and Ravikumar et al. (2013). Typical high strain rate stress–strain curves are presented in Fig. 8.8.

Material: 2D plain-weave E-glass/epoxy composite, fiber volume fraction = 0.5 (Pandya et al., 2015)

- Compressive stress–strain curve along thickness direction is represented by Eq. [A8.1]:

$$y = -5.7 \times 10^6 x^4 + 2.35 \times 10^6 x^3 - 4.1 \times 10^5 x^2 + 2.98 \times 10^4 x + 2 \quad [A8.1]$$

Here, y represents compressive stress along the thickness direction, and x represents compressive strain along the thickness direction.

Compressive strength in z-direction, $Z_c = 725$ MPa

Compressive failure strain in z-direction, $\varepsilon_{zcf} = 0.05$

Strain rate = 2210 per sec.

- Tensile stress–strain curve along the in-plane direction is represented by the following:

$$y = 7.72 \times 10^6 x^3 - 87.29 \times 10^4 x^2 + 368.41 \times 10^2 x - 3 \quad [A8.2]$$

Here, y represents tensile stress along the warp direction, and x represents tensile strain along the warp direction.

Tensile strength in in-plane direction, $X_{lt} = 525$ MPa

Tensile failure strain along fiber, $\varepsilon_{ltf} = 0.03$

Strain rate = 2050 per sec.

- Tensile stress–strain curve along in-plane 45° direction is represented by Eq. [A8.3]:

$$y = -2.91 \times 10^8 x^4 + 2.62 \times 10^7 x^3 - 8.57 \times 10^5 x^2 + 2.17 \times 10^4 x - 1.3 \quad [A8.3]$$

Here, y represents tensile stress along the in-plane 45° direction, and x represents tensile strain along the in-plane 45° direction.

Tensile strength in in-plane 45° direction, $X_{45t} = 395$ MPa

Tensile failure strain along 45° direction, $\varepsilon_{45tf} = 0.035$

Strain rate = 1840 per sec.

- Interlaminar shear stress–shear strain curve is represented by the following:

$$y = -1.9 \times 10^7 x^4 + 2.2 \times 10^6 x^3 - 1.01 \times 10^5 x^2 + 3.07 \times 10^3 x \quad [A8.4]$$

Here, y represents the interlaminar shear stress, and x represents the interlaminar shear strain.

Interlaminar shear strength, $S = 45$ MPa

Interlaminar shear strain, $\gamma = 0.03$

Strain rate = 1005 per sec.

Appendix B: stress–strain data at high strain rates—2D 8H satin-weave T300 carbon/epoxy

High strain rate stress–strain curves were obtained using a compressive split Hopkinson pressure bar apparatus (Naik and Kavala, 2008), tensile split Hopkinson pressure bar apparatus (Naik and Perla, 2008; Naik et al., 2010), and torsional split Hopkinson bar apparatus (Naik et al., 2007). Experimental details are given in the respective references. Mechanical properties at higher strain rates are obtained as explained in Naik et al. (2011) and Ravikumar et al. (2013). Typical high strain rate stress–strain curves are presented in Fig. 8.9.

Material: 2D 8H satin-weave T300 carbon/epoxy, fiber volume fraction = 0.56 (Pandya et al., 2015).

- Compressive stress–strain curve along the thickness direction is represented by Eq. [B8.1]:

$$y = -9 \times 10^8 x^4 + 7 \times 10^7 x^3 - 2 \times 10^6 x^2 + 43615\,x + 1 \quad [B8.1]$$

Here, y represents compressive stress along the thickness direction, and x represents compressive strain along the thickness direction.

Compressive strength in z-direction, $Z_c = 621$ MPa

Compressive failure strain in z-direction, $\varepsilon_{zcf} = 0.027$

Strain rate = 2020 per sec.

- Tensile stress–strain curve along in-plane direction is represented by the following:

$$y = -2 \times 10^9 x^4 + 8 \times 10^7 x^3 - 2 \times 10^6 x^2 + 59254\,x + 1.2 \quad [B8.2]$$

Here, y represents tensile stress along the warp direction, and x represents tensile strain along the warp direction.

Tensile strength in in-plane direction, $X_{lt} = 555$ MPa

Tensile failure strain along fiber, $\varepsilon_{ltf} = 0.0132$

Strain rate = 2110 per sec.

- Tensile stress–strain curve along the in-plane 45° direction is represented by Eq. [B8.3]:

$$y = 4 \times 10^8 x^4 - 2 \times 10^7 x^3 - 44667x^2 + 26018x + 3.25 \quad [B8.3]$$

Here, y represents tensile stress along the in-plane 45° direction, and x represents tensile strain along the in-plane 45° direction.

Tensile strength in in-plane 45° direction, $X_{45t} = 480$ MPa

Tensile failure strain along 45° direction, $\varepsilon_{45tf} = 0.026$

Strain rate = 1890 per sec.

- Interlaminar shear stress—shear strain curve is represented by the following:

$$y = -10^9x^4 + 5 \times 10^7x^3 - 10^6x^2 + 11361\,x + 1 \qquad \text{[B8.4]}$$

Here, y represents the interlaminar shear stress, and x represents the interlaminar shear strain.

Interlaminar shear strength, $S = 66$ MPa

Interlaminar shear strain, $\gamma = 0.016$

Strain rate = 1020 per sec.

Appendix C: frictional behavior of composites—2D plain-weave E-glass/epoxy and 2D 8H satin-weave T300 carbon/epoxy

The frictional behavior of composites is obtained experimentally (Naik and Doshi, 2008). Quasistatic loading at a loading rate of 1 mm/min was applied through a Universal Testing Machine on the two types of composite specimens. The corresponding punch tip displacement was noted until the punch tip displacement reached 25 mm. The thickness, h, of the specimens considered was 19 mm for frictional studies. The punch diameter, d, was 8 mm, whereas the hole diameter, dh, was 7.9 mm. The punch length was 26 mm.

Load displacement plots for frictional behavior of the two materials considered can be represented using polynomial expressions (Pandya et al., 2015).

- For 2D plain-weave E-glass/epoxy:

$$y = -0.0345x^4 + 2.001x^3 - 45.752x^2 + 582.37x \qquad \text{[C8.1]}$$

- For 2D 8H satin-weave T300 carbon/epoxy:

$$y = 3.7763x^3 - 78.973x^2 + 825.33x \qquad \text{[C8.2]}$$

Here, y is load (in N), and x is punch tip displacement (in mm).

Acknowledgments

The author would like to acknowledge the contributions of students of Composites Laboratory, Aerospace Engineering Department, Indian Institute of Technology Bombay toward composites research. Special thanks to Kedar S. Pandya, Ch. V. Sesha Kumar, N. S. Nair, P. S. Patil and Shaktivesh for their contribution toward the current work. Participation of A. Ahilan is highly appreciated toward the preparation of this chapter.

References

ASTM F1233-08, 2008. Standard Test Method for Security Glazing Systems and Materials.

Awerbuch, J., Bodner, S.R., 1974. Analysis of the mechanics of perforation of projectiles in metallic plates. Int. J. Solids Struct. 10, 671–684.

Bhatnagar, A. (Ed.), 2006. Lightweight Ballistic Composites. Woodhead Publishing Limited, Cambridge.

Chocron Benloulo, I.S., Rodriguez, J., Sanchez Galvez, V., 1997. A simple analytical model for ballistic impact in composites. J. Phys. IV 7, 821–826.

Chou, T.W., 1992. Microstructural Design of Fiber Composites. Cambridge University Press, Cambridge, pp. 46–48.

Daniel, I.M., Liber, T., 1976. Wave Propagation in Fiber Composite Laminates, Part II, NASA CR-135086.

DSM Dyneema Product Specification Sheet, 2014. Product Specification Sheet, PSS HB 50. DSM Dyneema Office, Netherlands.

DSM Dyneema, 2014. Dyneema High-Strength, High-Modulus Polyethylene Fiber, Dyneema Fact Sheet. Pelican Rope Works, DSM Dyneema Office, Netherlands.

Ellis, R.L., Lalande, F., Jia, H.Y., 1998. Ballistic impact resistance of SMA and spectra hybrid composites. J. Reinf. Plast. Comp. 17, 147–164.

Feli, S., Namdari Pour, M.H., 2012. An analytical model for composite sandwich panels with honeycomb core subjected to high-velocity impact. Compos. Part B-Eng. 43, 2439–2447.

Gama, B.A., Gillespie Jr., J.W., 2008. Punch shear based penetration model of ballistic impact of thick-section composites. Compos. Struct. 86, 356–369.

Gellert, E.P., Cimpoeru, S.J., Woodward, R.L., 2000. A study of effect of target thickness on the ballistic perforation of glass fiber reinforced plastic composites. Int. J. Impact Eng. 24, 445–456.

Gibson, R.F., 1994. Principles of Composite Material Mechanics. McGraw-Hill, Inc., NewYork.

Goodfellow Cambridge, 2014. Polyethylene–U.H.M.W. Fibre Material Information. Goodfellow, Huntington, England.

Grujicic, M., Bell, W.C., Thompson, L.L., Koudela, K.L., Cheeseman, B.A., 2008. Ballistic-protection performance of carbon-nanotube-doped poly-vinyl-ester-epoxy matrix composite armor reinforced with E-glass fiber mats. Mat. Sci. Eng. A-Struct. 479, 10–22.

Haijun, X., Lulu, L., Guangtao, C., Na, Z., Yiming, F., Weirong, H., 2013. Impact response and damage evolution of triaxial braided carbon/epoxy composites. Part I: ballistic impact testing. Text. Res. J. 83, 1703–1716.

Hazell, P.J., Appleby-Thomas, G., 2009. A study on the energy dissipation of several different CFRP based targets completely penetrated by a high velocity projectile. Compos. Struct. 91, 103–109.

Hazell, P.J., Appleby-Thomas, G.J., 2012. The impact of structural composite materials. Part 1: ballistic impact. J. Strain Anal. Eng. 47, 396–405.

Honeywell Advanced Fibers and Composites, 2014. Spectra–Physical Properties. Honeywell International Inc., Morristown, New Jersey.

Hou, W., Zhu, F., Lu, G., Fang, D., 2010. Ballistic impact experiments of metallic sandwich panels with aluminium foam core. Int. J. Impact Eng. 37, 1045–1055.

Jia, X., Sun, B., Gu, B., 2012. A numerical simulation on ballistic penetration damage of 3D orthogonal woven fabric at microstructure level. Int. J. Damage Mech. 21, 237–266.

Jones, R.M., 1999. Mechanics of Composite Materials. Taylor & Francis, Inc., Philadelphia, pp. 63–67.

Kasano, H., 1999. Recent advances in high-velocity impact perforation of fiber composite laminates. JSME Int. J. A-Mech. Mater. 42, 147–157.

Krishna Teja Palleti, H.N., Gurusamy, S., Kumar, Santosh, Soni, R., John, B., Vaidya, R., Bhoge, A., Naik, N.K., 2012. Ballistic impact performance of metallic targets. Mater. Des. 39, 253–263.

Lin, C., Hoo Fatt, M.S., 2006. Perforation of composite plates and sandwich panels under quasi-static and projectile loading. J. Compos. Mater. 40, 1801–1840.

Marissen, R., Smit, L., Snijder, C., 2005. Dyneema fibers in composites, the addition of special mechanical functionalities. In: The Proceedings of Advancing with Composites 2005, Naples, Italy.

Meyers, M.A., 1994. Dynamic Behavior of Materials. John Wiley and Sons, Inc., New York, pp. 23–65.

MIL-STD-662F, 1997. V50 ballistic test for Armour. US Army Research Lab.

Mines, R.A.W., Roach, A.M., Jones, N., 1999. High velocity perforation behavior of polymer composite laminates. Int. J. Impact Eng. 22, 561–588.

Morye, S.S., Hine, P.J., Duckett, R.A., Carr, D.J., Ward, I.M., 2000. Modelling of the energy absorption by polymer composites upon ballistic impact. Compos. Sci. Technol. 60, 2631–2642.

Muhi, R.J., Najim, F., deMoura, M.F.S.F., 2009. The effect of hybridization on the GFRP behavior under high velocity impact. Compos. Part B-Eng. 40, 798–803.

Naik, N.K., Doshi, A.V., 2005. Ballistic impact behavior of thick composites: analytical formulation. AIAA J. 43, 1525–1537.

Naik, N.K., Doshi, A.V., 2008. Ballistic impact behavior of thick composites: parametric studies. Compos. Struct. 82, 447–464.

Naik, N.K., Kavala, V.R., 2008. High strain rate behavior of woven fabric composites under compressive loading. Mat. Sci. Eng. A-Struct. 474, 301–311.

Naik, N.K., Perla, Y., 2008. Mechanical behaviour of acrylic under high strain rate tensile loading. Polym. Test. 27, 504–512.

Naik, N.K., Shrirao, P., 2004. Composite structures under ballistic impact. Compos. Struct. 66, 579–590.

Naik, N.K., Shrirao, P., Reddy, B.C.K., 2005. Ballistic impact behavior of woven fabric composites: parametric studies. Mat. Sci. Eng. A-Struct. 412, 104–116.

Naik, N.K., Shrirao, P., Reddy, B.C.K., 2006. Ballistic impact behavior of woven fabric composites: formulation. Int. J. Impact Eng. 32, 1521–1552.

Naik, N.K., Asmelash, A., Kavala, V.R., Veerraju, Ch, 2007. Interlaminar shear properties of polymer matrix composites: strain rate effect. Mech. Mater. 39, 1043–1052.

Naik, N.K., Yernamma, P., Thoram, N.M., Gadipatri, R., Kavala, V.R., 2010. High strain rate tensile behavior of woven fabric E-glass/epoxy composite. Polym. Test. 29, 14–22.

Naik, N.K., Parimi, J.S., Kavala, V.R., Ravikumar, G., Pothnis, J.R., Arya, H., 2011. High strain rate mechanical behavior of epoxy under compressive loading: experimental and modeling studies. Mat. Sci. Eng. A-Struct. 528, 846–854.

Naik, N.K., Kumar, S., Ratnaveer, D., Joshi, M., Akella, K., 2013. An energy-based model for ballistic impact analysis of ceramic-composite armors. Int. J. Damage Mech. 22, 145–187.

National Institute of Justice. NIJ Standard–0101.06, 2008. Ballistic Resistance of Body Armor. US Department of Justice, Washington, DC.
NATO STANAG 2920 PPS, 2003. Ballistic Test Method for Personal Armour Materials and Combat Clothing, 2nd ed. NATO Standardization Agency.
Pandya, K.S., Akella, K., Joshi, M., Naik, N.K., 2012a. Ballistic impact behavior of carbon nanotube and nanosilica dispersed resin and composites. J. Appl. Phys. 112 (Article no. 113522).
Pandya, K.S., Dharmane, L., Pothnis, J.R., Ravikumar, G., Naik, N.K., 2012b. Stress wave attenuation in composites during ballistic impact. Polym. Test. 31, 261–266.
Pandya, K.S., Pothnis, J.R., Ravikumar, G., Naik, N.K., 2013. Ballistic impact behavior of hybrid composites. Mater. Des. 44, 128–135.
Pandya, K.S., Sesha Kumar, Ch.V., Nair, N.S., Patil, P.S., Naik, N.K., 2015. Analytical and experimental studies on ballistic impact behavior of 2D woven fabric composites. Int. J. Damage Mech. 24, 471–511. http://dx.doi.org/10.1177/1056789514531440.
Phoenix, S.L., Porwal, P.K., 2003. A new membrane model for the ballistic impact response and V50 performance of multi-ply fibrous systems. Int. J. Solids Struct. 40, 6732–6765.
Potti, S.V., Sun, C.T., 1996. A simple model to predict penetration of thick composite laminates subjected to high velocity impact. Int. J. Impact Eng. 18, 339–353.
Potti, S.V., Sun, C.T., 1997. Prediction of impact induced penetration and delamination in thick composite laminates. Int. J. Impact Eng. 19, 31–48.
Prevorsek, D.C., 1996. Spectra: the latest entry in the field of high performance fibers. In: Menachem, L., Preston, J. (Eds.), Handbook of Fiber Science and Technology, High Technology Fibers, Part D, vol. III. Marcel Dekker Inc., New York, pp. 1–170.
Ravikumar, G., Pothnis, J.R., Joshi, M., Akella, K., Kumar, Santosh, Naik, N.K., 2013. Analytical and experimental studies on mechanical behavior of composites under high strain rate compressive loading. Mater. Des. 44, 246–255.
Rojek, M., et al., 2013. Composite materials with the polymeric matrix applied to ballistic shields. Arch. Mater. Sci. Eng. 63, 26–35.
Russell, B.P., Karthikeyan, K., Deshpande, V.S., Fleck, N.A., 2013. The high strain rate response of ultra high molecular-weight polyethylene: from fibre to laminate. Int. J. Impact Eng. 60, 1–9.
Shaktivesh, Nair, N.S., Sesha Kumar, Ch.V., Naik, N.K., 2013. Ballistic impact performance of composite targets. Mater. Des. 51, 833–846.
Shaktivesh, Nair, N.S., Naik, N.K., 2015. Ballistic impact behavior of 2D plain weave fabric targets with multiple layers: analytical formulation. Int. J. Damage Mech. 24, 116–150.
Sierakowski, R.L., Chaturvedi, S.K., 1997. Dynamic Loading and Characterization of Fiber-reinforced Composites. John Wiley and Sons, Inc., New York, pp. 99–133.
Smith, J.C., McCrackin, F.L., Schiefer, H.F., 1958. Stress-strain relationships in yarns subjected to rapid loading-Part V: wave propagation in long textile yarns impacted transversely. Text. Res. J. 288–302.
Sun, L., Gibson, R.F., Gordaninejad, F., Suhr, J., 2009. Energy absorption capability of nanocomposites: a review. Compos. Sci. Technol. 69, 2392–2409.
Udatha, P., Sesha Kumar, Ch.V., Nair, N.S., Naik, N.K., 2012. High velocity impact performance of three-dimensional woven composites. J. Strain Anal. Eng. 47, 419–431.
Wen, H.M., 2000. Predicting the penetration and perforation of FRP laminates struck normally by projectiles with different nose shapes. Compos. Struct. 49, 321–329.
Wen, H.M., 2001. Penetration and perforation of thick FRP laminates. Compos. Sci. Technol. 61, 1163–1172.

Wu, E., Chang, L.C., 1995. Woven glass/epoxy laminates subjected to projectile impact. Int. J. Impact Eng. 16, 607–619.

Zaera, R., 2011. Ballistic impacts on polymer matrix composites, composite armor, personal armor. In: Abrate, S. (Ed.), Impact Engineering of Composite Structures. Springer Wien, New York, pp. 305–403.

Zee, R.H., Hsieh, C.Y., 1998. Energy absorption process in fibrous composites. Mat. Sci. Eng. A-Struct. 246, 161–168.

Zhang, D., Sun, Y., Chen, L., Zhang, S., Pan, N., 2014. Influence of fabric structure and thickness on the ballistic impact behavior of ultrahigh molecular weight polyethylene composite laminate. Mater. Des. 54, 315–322.

Zhu, G., Goldsmith, W., Dharan, C.H., 1992a. Penetration of laminated Kevlar by projectiles–I. Experimental investigation. Int. J. Solids Struct. 29, 339–420.

Zhu, G., Goldsmith, W., Dharan, C.H., 1992b. Penetration of laminated Kevlar by projectiles–II. Analytical model. Int. J. Solids Struct. 29, 421–436.

Failure mechanisms and engineering of ballistic materials

9

X. Chen
University of Manchester, Manchester, United Kingdom

Y. Chu
University of Manchester, Manchester, United Kingdom; Zhongyuan University of Technology, Zhengzhou, China

9.1 Introduction

Human beings have been confronted with greater threats and disasters since the invention of the firearm in comparison with other weapons. In World War I and World War II, thousands of people lost their lives or were injured due to insufficient body protection. Since the two world wars, although no large-scale wars have occurred, the regional wars and conflicts, for example, the Iraq war and Syrian conflict, still continue to happen, causing numerous deaths. Additionally, due to potential and uncertain threats at any time coming from terrorists or hostile leaders of the country, law enforcement officers need to take precautions against such threats. Consequently, it is imperative to develop and manufacture better ballistic materials to save more lives.

To improve and optimize the protection of the ballistic materials, a comprehensive understanding of the failure mechanism of the ballistic material is essential. The ballistic material is a broad concept. This chapter will discuss the failure mechanism from the view of fiber, yarn, to a single-ply fabric, and then to multi-ply fabric and some 3D fabrics, and finally extend to fabric composites. Based on the failure mechanism analyses, several approaches for engineering light and reliable ballistic fabric are proposed.

9.2 Analysis approaches for ballistic impact

It is understood that the impacting process from the transverse direction with a high-velocity projectile is a dynamic, complicated, and instantaneous physical process. It is far different from the quasistatic mechanical action. Currently, three analysis approaches are frequently adopted in the analysis of this complex physical phenomenon: analytical, experimental, and numerical. These three approaches can be applied individually in research or combined to provide more understanding about the damage and failure mechanism of the fabric subjected to ballistic impact.

Advanced Fibrous Composite Materials for Ballistic Protection. http://dx.doi.org/10.1016/B978-1-78242-461-1.00009-1

9.2.1 Analytical method

Analytical methodology is mainly from a physical perspective and relies on general mechanic laws to set up governing equations using various parameters involved in the ballistic impact process. In the analysis, the whole process is divided into n steps by a small increment of time. At the first increment, the equations are usually derived based on energy conservation, impulse theory, and Newton's second law. Analogously, step by step, until the ultimate time, all the equations at every time increment are derived.

Up till now, numerous researchers devoted themselves to making contributions to this field and applied this method to provide insight into the mechanism of damage and failure of a fabric even in the impact [1–14]. The results predicted by this method are usually validated by comparing with experimental results. Compared with the empirical method, the analytical method consumes fewer materials and relies on less labor. However, it needs complete understanding of the ballistic impact process. Additionally, to simplify the process, some of the parameters tend to be neglected. Interyarn friction is one of those factors. The truth is that the interyarn friction plays a significant role in energy absorption during the impact [15,16].

9.2.2 Experimental method

Experimental method usually acquires the first-hand data based on a certain experimental design and the device provided. Through analyses of data, the response characters of a fabric can be examined and the constitutive relations can be established. The analysis process may consist of curve fitting, nonlinear regression analysis of experimental data, and the use of statistical distributions [11]. It is powerful and useful when a few numbers of variables correlate [17–22]. Majumdar et al. [23,24] and Park et al. [25,26] used it to investigate the ballistic performance of treated fabrics. However, the disadvantage is also obvious that it is a time-consuming and material-costly process. Additionally, the accuracy of the obtained results to a large degree depends on the correctness and completeness of collected data.

9.2.2.1 National Institute of Justice (NIJ) test

The NIJ test is used to understand ballistic resistance of soft body armor in real situations. Normally, in the test, a ballistic panel is positioned on the surface of a backing material, which is made of oil-based clay and used for simulating human skin. After impacting, a hole would be left on the surface of the backing material due to the transverse deflection of the panel. The hole is an indication of blunt trauma brought to a wearer, which is also harmful to the wearer. Although the ballistic panel deflection is a manner to dissipate the projectile energy, it is still anticipated that the panel would transversely deflect as little as possible at the end of impacting.

The blunt trauma of a ballistic panel is usually evaluated by the depth of the hole, depressed on the backing material. The depth of the hole is measured from the plane defined by the front edge of the backing material fixture [27] as shown in Fig. 9.1.

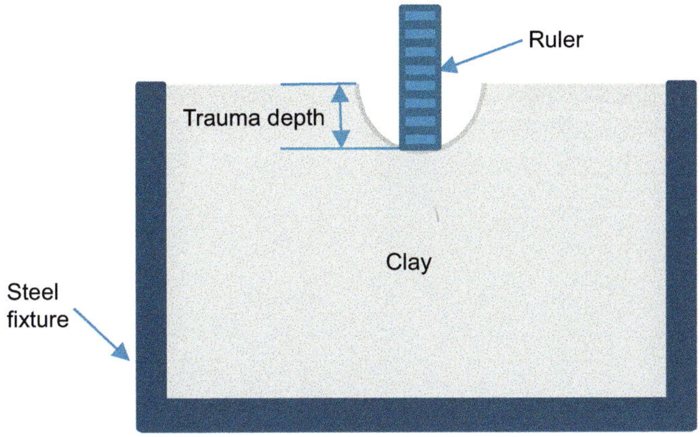

Figure 9.1 Methods of measuring back-face deformation.

According to NIJ standard 0101.04, ballistic armor is classified into seven levels, as shown in Table 9.1. Type I, IIA, II, and IIIA offer increasing levels of protection from handgun threats. Type III and IV armor are levels for measuring protection against high-powered rifle rounds, which are intended for application only in tactical situations [27]. The maximum deflection a soft armor vest can undergo without perforation is 44 mm for all levels of ballistic threats [27]. The apparatus for the test is shown in Fig. 9.2.

HOSDB standard (Home Office Scientific Development Branch standard) is another standard in parallel to the NIJ standard [28]. It is a method used for evaluating ballistic performance of a body armor system for the British Police. Different from the NIJ standard, the number of the ballistic performance levels is five, including HG1/A, HG1, HG2, SG1, and RF1. The maximum indentation depth permitted from any test shot is 25 mm measured from the top edges of a steel tray containing the backing material, except that the maximum indentation depth for HG1/A is 44 mm.

9.2.2.2 Energy absorption test

Energy absorption in perforation test

The perforation test is described as a projectile or a bullet fully getting through a testing target during the impact process. In the perforation test, the impact velocity and the residual velocity can be measured. Thus, variation in projectile energy loss can be calculated from these two velocities [29–33], as shown in Eq. [9.1]. To compare the projectile energy losses of panels with different weights, an indicator of total energy loss divided by the areal density of the fabric panel is also used:

$$\Delta E = \frac{1}{2}m\left(v_{\mathrm{i}}^2 - v_{\mathrm{r}}^2\right) \qquad [9.1]$$

where ΔE is projectile energy loss; m is the mass of a projectile; v_{i} is impact velocity, and v_{r} is residual velocity.

Table 9.1 **The NIJ Standard 0101.04 P-BFS performance test summary [27]**

Armor type	Test round	Test bullet	Bullet weight	Reference velocity (±30 ft/s)	Hits per armor part at 0 degree angle of incidence	BFS depth maximum	Hits per armor part at 30 degree angle of incidence	Shots per panel	Shots per sample	Shots per threat	Total shots req'd
I	1	0.22 caliber LR LRN	2.6 g 40 gr	329 m/s (1080 ft/s)	4	44 mm (1.73 in)	2	6	12	24	48
	2	0.380 ACP FMJ RN	6.2 g 95 gr	322 m/s (1055 ft/s)	4	44 mm (1.73 in)	2	6	12	24	
IIA	1	9 mm FMJ RN	8.0 g 124 gr.	341 m/s (1120 ft/s)	4	44 mm (1.73 in)	2	6	12	24	48
	2	40 S&W FMJ	11.7 g 180 gr.	322 m/s (1055 ft/s)	4	44 mm (1.73 in)	2	6	12	24	
II	1	9 mm FMJ RN	8.0 g 124 gr.	367 m/s (1205 ft/s)	4	44 mm (1.73 in)	2	6	12	24	48
	2	357 Mag JSP	10.2 g 158 gr.	436 m/s (1430 ft/s)	4	44 mm (1.73 in)	2	6	12	24	
IIIA	1	9 mm FMJ RN	8.2 g 124 gr	436 m/s (1430 ft/s)	4	44 mm (1.73 in)	2	6	12	24	48
	2	44 Mag JHP	15.6 g 240 gr.	436 m/s (1430 ft/s)	4	44 mm (1.73 in)	2	6	12	24	
III	1	7.62 mm NATO FMJ	9.6 g 148 gr.	838 m/s (2780 ft/s)	6	44 mm (1.73 in)	0	6	12	12	12
IV	1	0.30 caliber M2 AP	10.8 g 166 gr.	869 m/s (2880 ft/s)	1	44 mm (1.73 in)	0	1	2	2	2
Special	*	*	*	*	*	44 mm (1.73 m)	*	*	*	*	*

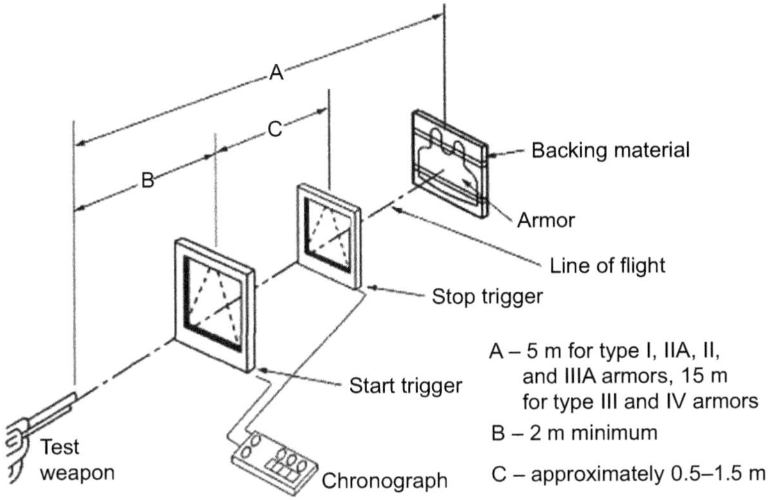

Figure 9.2 Test range configuration.
Adapted from National Institute of Justice USDoJ. Ballistic resistance of body armor NIJ Standard-0101.06. 2000.

Without any other external force action and without considering energy loss by heat, intermolecular friction, air resistance, acoustic energy, etc., the projectile energy loss can be assumed to be fully absorbed by the fabric based on the law of energy conservation. The energy dissipation in the fabric takes place mainly in three ways [12–15,34–36]: (1) kinetic energy due to the movement of the fabric caused by a longitudinal wave and transverse wave, (2) strain energy due to the deformation of yarns stressed, and (3) frictional dissipation energy because of interyarn friction. The energy transfer between the projectile and fabric can be simply described in the following equation:

$$E_p = E_f = E_{KE} + E_{SE} + E_{FDE} \qquad [9.2]$$

where E_p means projectile energy loss; E_f means energy absorption of a fabric; E_{KE} means kinetic energy (KE); E_{SE} indicates strain energy (SE); and E_{FDE} represents frictional dissipation energy.

Energy absorption in nonperforation test

In the nonperforation test, the projectile energy loss is partly transferred into the fabric and partly to the backing material. The energy transmitted to the fabric E_f is equal to value obtained through total projectile energy loss E_{pl} minus the energy absorbed by the backing clay E_c, as given in Eq. [9.3]. The energy absorbed by the backing material is associated with the volume of the hole formed on the baking clay. Generally speaking, the shape of the hole is irregular, and it is difficult to acquire this value. Karahan et al. [37] tried to regenerate the trauma geometry of the signature by method of spline curve fitting technique and compute out the volume of the hole. With the known of the volume of the hole, the energy absorption of the clay can be referred to Karahan et al. [37]. For the given same impact energy, it is hoped that the projectile energy is transmitted to the back clay as little as possible.

$$E_f = E_{pl} - E_c \qquad [9.3]$$

9.2.2.3 V_{50} test

To evaluate the fail or pass property of body armor in a ballistic impact event, a ballistic limit test is applied. Ballistic limit designates a velocity at which a projectile completely perforates specific armor when hitting at a specified angle of obliquity. This velocity is called ballistic limit velocity. It is the incident impact velocity for a specific projectile and target combination, which would lead to complete perforation of the target with the projectile tip reaching the back face of the target with zero residual velocity [38,39]. Villanueva and Cantwell [40] derived an equation for estimating the ballistic limit of laminates, expressed in Eq. [9.4]:

$$V_b = \frac{\pi\tau\sqrt{\rho_t\sigma_e}D^2T}{4m\left(1+\sqrt{1+\frac{8m}{\pi\tau^2\rho_t D^2 T}}\right)} \qquad [9.4]$$

where V_b is the ballistic limit, τ is a projectile constant determined experimentally, ρ_t is the density of the laminate, σ_e is the static linear elastic compression limit, D is the diameter of the projectile, T is the thickness of the laminate, and m is the mass of the projectile.

Due to the cost of ballistic impact tests, the impossibility of controlling impact velocity exactly and the difficulty to distinguish complete perforation from partial perforation under apparently identical conditions, statistical approaches are necessary, based upon limited firings [41]. Hence, V_{50} is commonly used instead of V_b. V_{50} is defined as the average of equal number of highest partial perforation velocities and lowest complete perforation velocities for a specific projectile and target system, which occurs within a specified velocity range. In other words, V_{50} defines incident impact velocity at which there is 50% probability of partial and 50% probability of nonperforation [38,39]. The normal up-and-down firing procedure is applied. A 0.51-mm-thick 2024 T3 sheet of aluminum is positioned 152 $\pm$ 12.7 mm behind and parallel to the target to witness complete penetrations. Normally, a minimum of two partial and two complete perforation velocities are used to compute V_{50}. Four, six, and ten-round ballistic limits are frequently used. The maximum allowable velocity span depends on the armor material and test conditions and spans of 18, 27, 30, and 38 m/s are commonly used [41].

9.2.3 Numerical method

Continued efforts have been made to reduce the need for experimental test programmes. Investigation about ballistic impact is developing toward the direction of simulating. Owing to the advent of the computer, study of ballistic impact with computer-based analyses and simulations is possible. The numerical investigation is an investigation approach based on the finite element (FE) theory and commercial computer software, such as Abaqus, Ansys, and LS-DYNA, to establish a projectile-fabric simulation model for elucidating the mechanism behind a fabric subjected to impact. FE theory is better suited to analyze dynamic mechanic problems, and a bullet striking a piece of fabric is rightly a dynamic mechanics problem. The conception of FE theory is that the integrity is divided into limited and small units, which is called mesh, and the displacement, strain, and stress of individual mesh are computed step by step when loaded. Eventually, varieties of stress and strain experienced by a yarn can be acquired. It is an efficient and effective approach because the complex and instantaneous process would be converted into a controllable and visual process through the simulation. Consequently, a large number of researchers started to investigate ballistic behaviors of fabrics using FE simulation.

Up till now, three mature methods are commonly used in modeling ballistic material structure, and they are the pin-jointed model, full 3D model, and mesoscale unit cell–based model. The pin-jointed model is usually used to model the pure fabric structure. It models a fabric as a 2D assembly of pin-jointed yarns, resulting in a net-like structure, as shown in Fig. 9.3. Discrete nodal masses are located at the crossings of yarns, and the masses are interconnected by cable (string) elements in orthogonal in-plane directions representing warp and weft yarns of the fabric [42].

The full 3D model is a more detailed modeling technique, which is able to capture the precise geometry of each yarn. The full 3D model for modeling a fabric is shown in Fig. 9.4. Shockey et al. [44,45] first attempted to use this model to explicitly simulate individual yarns and combined them to form a fabric. This modeling method has also been used by Duan et al. [16,36], Rao et al. [50], Talebi et al. [46], and Nilakantan et al. [47–49] in fabric modeling. For the UD material and fabric composite, the full 3D model is shown in Figs. 9.5 and 9.6, respectively.

To both account for yarn geometry and computation efficiency, a mesoscale unit cell–based model is developed. The term mesoscale denotes yarn-level details about millimeter-length scale of an architecture. For the fabric, the unit cell means the basic structural unit in a woven single-ply fabric. For example, the unit cell representing a

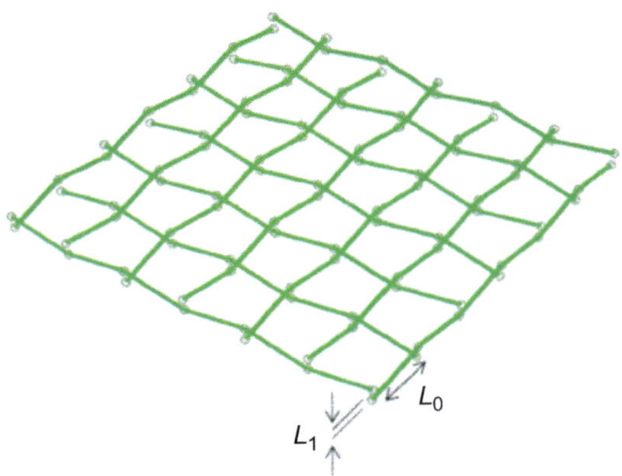

Figure 9.3 Pin-jointed model for the fabric [88].
Adapted from Chen X, Chu Y, Wang N. Study on inter-yarn friction and engineering of ballistic fabrics. In: Paper presented at the 14th AUTEX world textile conference, Bursa, Turkey; 2014.

Figure 9.4 3D continuum solid model for a fabric.
Adapted from Duan Y, Keefe M, Bogetti T, Cheeseman B, Powers B. A numerical investigation of the influence of friction on energy absorption by a high-strength fabric subjected to ballistic impact. Int J Impact Eng 2006;32(8):1299–312.

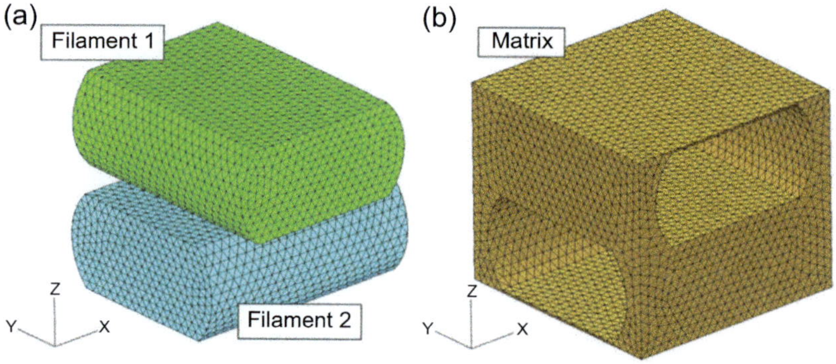

Figure 9.5 3D model for UD fabric.
Adapted from Grujicic M, Arakere G, He T, Bell WC, Cheeseman BA, Yen CF, et al. A ballistic material model for cross-plied unidirectional ultra-high molecular-weight polyethylene fiber-reinforced armor-grade composites. Mater Sci Eng A 2008;498:231–41.

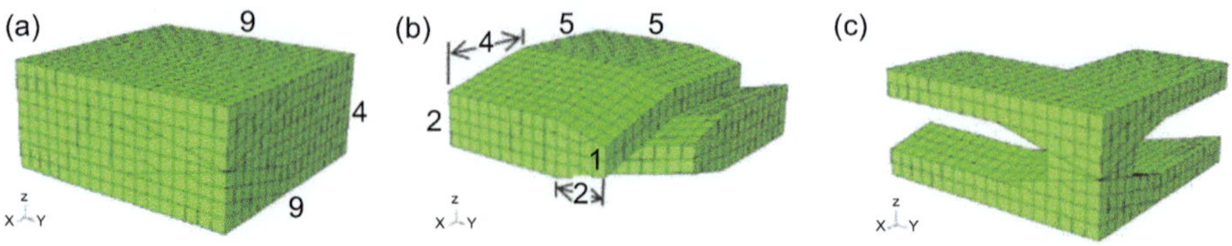

Figure 9.6 3D model for fabric composite. (a) Unit cell, (b) yarns, and (c) matrix.
Adapted from Li S, Zhou C, Yu H, Li L. Formulation of a unit cell of a reduced size for plain weave textile composites. Comput Mater Sci 2011;50:1770–80.

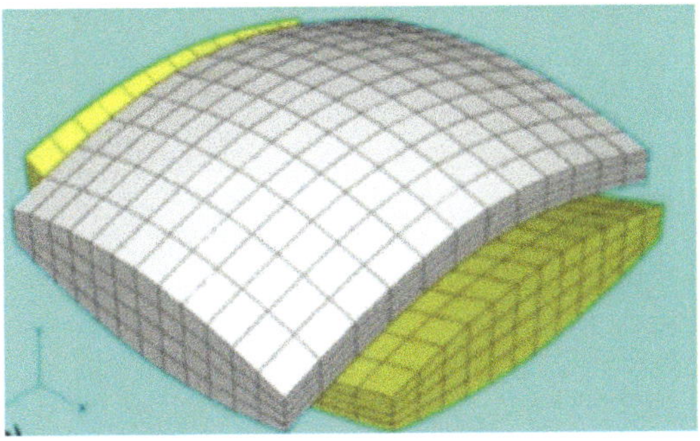

Figure 9.7 The unit cell representative of crossover.
Adapted from Grujicic M, Bell W, Arakere G, He T, Xie X, Cheeseman B. Development of a meso-scale material model for ballistic fabric and its use in flexible-armor protection systems. J Mater Eng Perform 2010;19(1):22–39

single-ply woven plain fabric structure is allotted to a single-yarn crossover in its initial undeformed configuration. In the mesoscale unit cell, the yarn geometry is taken into consideration as shown in Fig. 9.7. The properties of the unit cell are computed based on mechanical analysis from Kawabata et al. [53,54], and the properties include (1) biaxial in-plane response, (2) in-plane shear response, and (3) out-of-plane (transverse) shear response. For the fiber-reinforced composite, a unit cell model for it with opened-packing plain weave as an example is shown in Fig. 9.8(a) using a bold square, where a small square covered with light lines indicates that the warp yarn that crosses

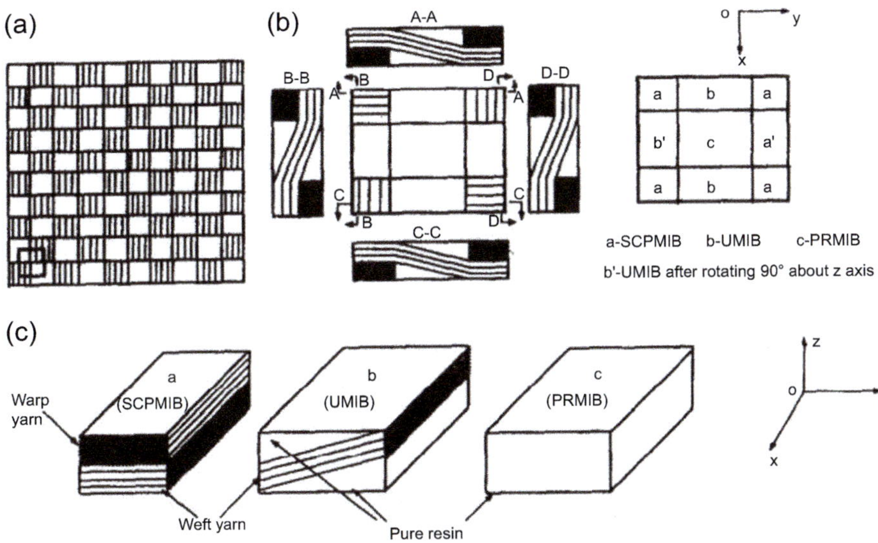

Figure 9.8 A schematic of plain weave–based composite. (a) Plain weave ($n_g = 2$), (b) plain weave unit cell architecture, and (c) architectures of the micro blocks SCPMIB, UMIB and PRMIB.
Adapted from Tan P, Tong L, Steven GP. A three-dimensional modelling technique for predicting the linear elastic property of opened-packing woven fabric unit cells. Compos Struct 1997; 38(1–4):261–71.

over the weft yarn at that intersection can be considered to be an assemblage of three types of micro-blocks, as shown in Fig. 9.8. They are the straight cross-ply micro-block (SCPMIB), the undulated yarn micro-block (UMIB), and the pure resin micro-block (PRMIB) (see Fig. 9.8(c)). The micro-block SCPMIB is composed of two straight fiber yarn segments which are cross-banded, while the micro-block UMIB consists of an undulated fiber yarn segment and pure resin. The constituent of the micro-block PRMIB is pure resin. The mechanical property of the unit cell can be calculated according to the proportion of the volume of resin and fiber, respectively [55].

The mesoscale unit cell—based material model described is written into a material user subroutine, VUMAT, the commercial FE program in Abaqus [56]. This subroutine is compiled and linked with the FE solver. During the analysis, this unit cell—based material will be implicitly implanted into a large-scale model on behalf of the fabric [57,58].

9.3 Failure mechanisms of ballistic materials

A bullet or projectile striking a ballistic material is a complex physical process because of the instantaneity of impact process and the complexity of ballistic material structure. The better cognition of wave propagation manifested in ballistic impact would start from a single yarn subjected to ballistic impact.

9.3.1 Stress propagation in fibers and yarns

Many researchers have covered the field of investigating a single yarn transversely impacted by a high-speed projectile. Smith et al. [1] developed the theory for describing the transverse impact on a single yarn. Morrison [96] tried to verify the accuracy of Smith's equation through using computer simulation methods, and they found that the two methods agreed in all given cases. The mechanics of a yarn impacted by a bullet have been described in the work done by many researchers [59—62]. When a single yarn is subjected to ballistic impact, a tent-shaped transverse deflection develops in the yarn because the yarn is forced to have a forward movement with the projectile. Simultaneously, a longitudinal wave gradually builds up in the yarn and rapidly propagates away from the impact center at the velocity of sound in the material, traveling in the direction of the axis of the yarn. In the longitudinal wave front, the yarn material is set in motion inwardly toward the impact center due to the stretching. The inwardly flowing material continues to feed the advancing transverse deflection until the strain in the yarn reaches its breaking strain. The responses of a yarn subjected to ballistic impact is illustrated in Fig. 9.9. The existence of longitudinal wave propagation in the yarn has been corroborated by Stewart et al. [63,64], and the velocity of it has been measured by Freeston and Claus [65].

According to the theoretical work done by Smith [34], the behaviors of a single yarn subjected to ballistic impact are governed by the following equations. If the yarn is elastic, Eq. [9.5] can be simplified as Eq. [9.6].

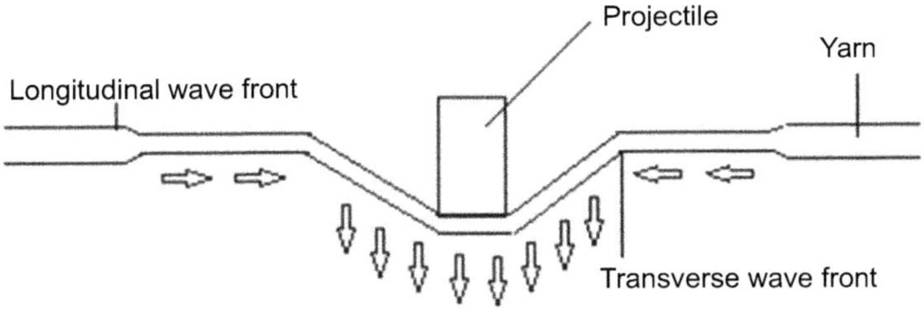

Figure 9.9 A yarn responds to ballistic impact.
Adapted from Cheeseman BA, Bogetti TA. Ballistic impact into fabric and compliant composite laminates. Compos Struct 2003;61(1):161–73.

$$C(\varepsilon) = \sqrt{\frac{1}{\rho}\frac{d\sigma}{d\varepsilon}} \quad [9.5]$$

$$C = \sqrt{\frac{E}{\rho}} \quad [9.6]$$

$$u_{lag} = C\sqrt{\frac{\varepsilon}{(1+\varepsilon)}} \quad [9.7]$$

$$u_{lab} = C\left(\sqrt{\varepsilon(1+\varepsilon)} - \varepsilon\right) \quad [9.8]$$

$$V = C\sqrt{2\varepsilon\sqrt{\varepsilon(1+\varepsilon)} - \varepsilon^2} \quad [9.9]$$

where C is longitudinal wave velocity; E is fiber tensile modulus, and ρ is yarn density; u_{lag} is the transverse wave velocity in Lagrangian coordinates, which is the coordinate system used in fluid dynamics in which the coordinate is fixed to a given parcel of fluid but moves in space. u_{lab} is transverse the wave velocity in laboratory coordinates; ε is the strain level between the strain wave front and point of impact; σ is stress; V is impact velocity.

9.3.2 Failure mechanisms of single-ply materials

9.3.2.1 Plain-woven fabrics

Normally, yarns in a fabric are divided into primary yarns (or principal yarns) and secondary yarns according to their relative positions to the projectile, as depicted in Fig. 9.10. The yarns directly contacting with the projectile are termed as primary yarns, and the others are defined as secondary yarns. Cheeseman and Bogetti [62] and Cunniff [61] noted that the responses of a fabric to the ballistic impact have some similarity to that of the single yarn. When a projectile strikes a single-layer fabric, analogously, the impact produces the transverse deflection in the primary yarns and the longitudinal wave, propagating away from the impact center along the axes of the

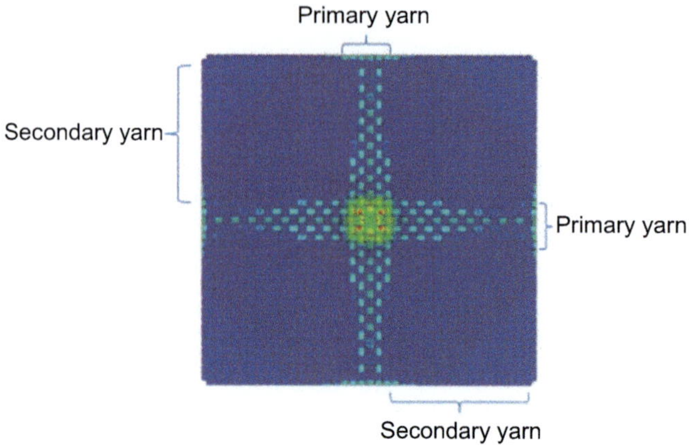

Figure 9.10 Primary yarn and secondary yarn in the fabric during ballistic impact.
Adapted from Duan Y, Keefe M, Bogetti T, Cheeseman B. Modeling friction effects on the ballistic impact behavior of a single-ply high-strength fabric. Int J Impact Eng 2005;31(8):996–1012.

primary yarns [61]. At the same instant, the secondary yarns, intersecting with the primary yarns away from the impact center, are then driven out of the original fabric plane by the primary yarns to different extents depending on the distances between the secondary yarns and the impact center. These secondary yarns experience a deformation and develop a strain wave like those in the primary yarns, and these secondary yarns then pull yarns with which they interlace. These yarn–yarn interactions, which by nature are a function of interyarn friction and introduced by the woven structure, force the secondary yarns to engage in the transverse deflection. The transverse deflection proceeds until the strain of the yarn at the impact center rises up to the breaking strain [39]. Fig. 9.11 describes the responses of a fabric subjected to a high-velocity impact by a spherical projectile. Due to the increased linear density at the crossovers, the longitudinal wave velocity in the yarns in a woven fabric structure was found slower than the expected value in a single yarn subjected to ballistic impact by a factor of $\sqrt{2}$ [66].

9.3.2.2 UD laminates

The UD fabric is a lightweight composite for ballistic protection, which utilizes a resin matrix to adhere outstanding impact resistance of high-modulus, high-strength polymeric fibers such as aramid or ultra-high molecular weight polyethylene (UHMWPE) fibers together by means of lamination with the direction of fibers in each layer at a certain angle to the direction of the fibers in the adjacent layers [67]. The percentage of the resin taken is approximately less than 20%. However, for the woven fabric counterpart, the motion between yarn and yarn is restricted by the interlacement of woven structure instead of the adherence of resin, it is more flexible compared to the UD fabric, which would reduce the entire rigidity of the body armor and provide less constrain for the wearers. Because of the different fixation

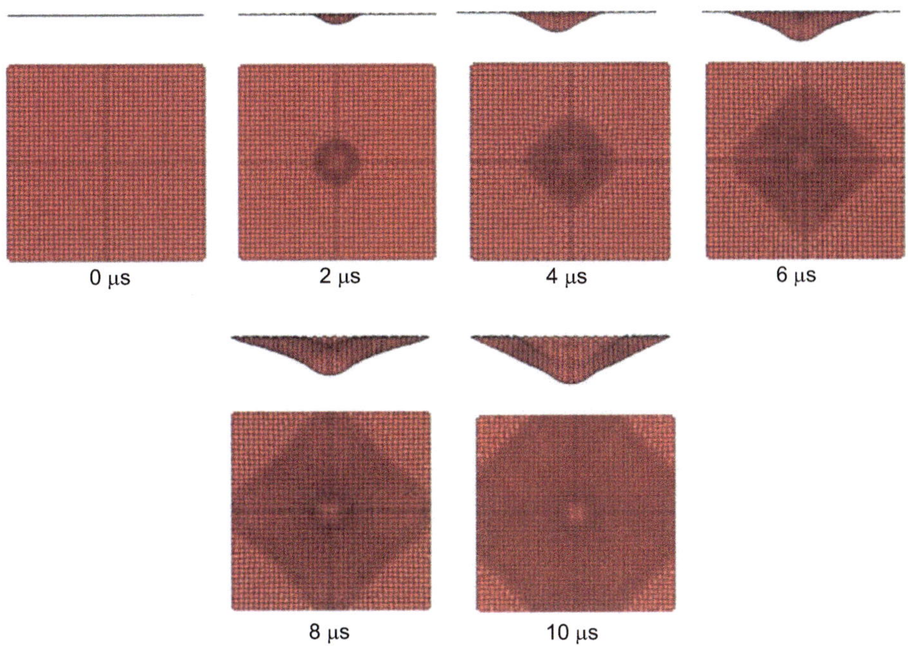

Figure 9.11 Sphere impacting a single ply of fabric. (a) Side view and (b) top view of z displacement.
Adapted from Duan Y, Keefe M, Bogetti T, Cheeseman B. Modeling friction effects on the ballistic impact behavior of a single-ply high-strength fabric. Int J Impact Eng 2005;31(8):996–1012.

of the yarn in the two constructions, the dominant fabric damages for the two constructions are different. It is reported that for the UD fabric, fabric damage is dominant by shear failure of the fiber because of the constraint of resin matrix [68], whereas for the woven fabric, fiber tensile failure coupled with yarn pull-out feature are the reasons [69].

9.3.3 Failure mechanisms of multi-ply panels

9.3.3.1 Nonperforation

In the nonperforation case, the investigation of the failure of a fabric panel is concentrated on the depth of trauma. Based on the NIJ standard, the maximum depth allowed for a commercial body armor is 44 mm. The depth of the trauma is associated with the transverse stiffness of a fabric panel. The higher the stiffness of a fabric panel is, the less trauma depth is. A reduction of 6.7% in the trauma depth was found with more stitching in a fabric panel because stitching can increase the panel stiffness [36–70]. In addition, Joo and Kang [71] reported that the energy absorption is the highest for the first layer followed by the rest of the layers while the sequence is reversed in the mature-perforation case.

9.3.3.2 Perforation

In a fabric panel, the mechanisms for the failure of different layers in a perforation case vary according to their positions in the system. Cunniff [72] reported that the response of the impact face portion is dominated by inelastic impact, and the rear layers are elastic. In a 15-layer Kevlar® plain-woven fabric panel system, it is stated that the front layers mainly fail because of shear action, and the rear layers are broken by tensile load [13], as given in Fig. 9.12. In addition, the back layer absorbs more SE than the front layers because of the longer engagement time of the rear layers with the front layers. However, the relationship between the number of fabric layers and the resulting energy absorption capability is not linear [73]. An addition of a new layer would make the increased percentage of the fabric SE decrease [74]. Each layer plays different roles in the multilayer system. Adding a fabric layer is beneficial only to a certain degree, exceeding which the add-on layer is redundant because more layers make the panel bulky but contribute less to energy absorption.

9.3.4 Failure of some 3D woven panels

The traditional weaving principle is also capable of producing structures that have substantial dimension in the thickness direction, generally termed as 3D fabrics. In other words, the 3D fabric is a type of fabric constructed by interlacing the yarns in the network-forming fashion, introducing the third dimension other than the planar dimension. It is suggested that the 3D fabric could be grouped into four types: solid, hollow, shell, and nodal [75]. The most common 3D fabric under research in the ballistic impact field is the solid one. According to the manner of incorporating

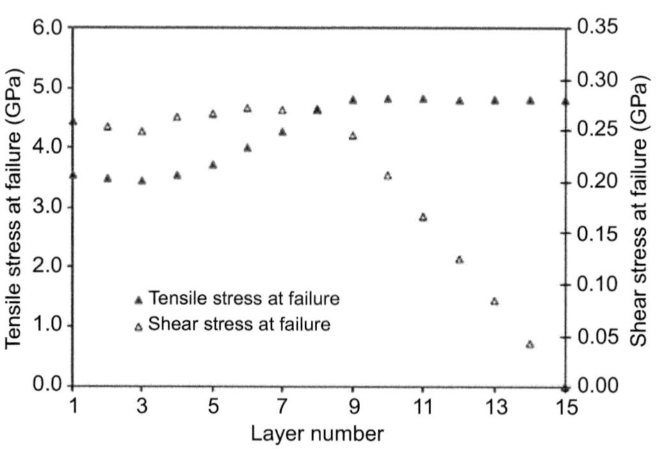

Figure 9.12 Tensile and shear stresses at the moment of failure in the primary yarns of different layers in a 15-layer Kevlar® plain-woven fabric subjected to a 1-g cylindrical projectile at a speed of 514 m/s.
Adapted from Chen X, Zhu F, Wells G. An analytical model for ballistic impact on textile based body armour. Compos Part B Eng 2013;45(1):1508–14.

through-the-thickness direction yarn, the 3D solid fabric can be categorized into three groups: multilayer woven architecture, orthogonal woven architecture, and angle-interlock woven architecture.

9.3.4.1 Orthogonal

The orthogonal woven architecture is characterized by the straight yarns. Warp yarns and weft yarns are basically straight and the through-the-thickness yarns travel vertically between any two weft yarns. The most common structure is that they travel between the top and bottom layer, as shown in Fig. 9.13. The orthogonal structures that apply one set of through-the-thickness warp to bind the structures together are called the ordinary orthogonal structures, as shown in Fig. 9.14(a), and those that use two sets of opposite-traveling binding warp yarns are called the enhanced orthogonal structures, as shown in Fig. 9.14(b). More binding yarns in the thickness direction are desired for more resistance to composite delamination. It is also possible that the orthogonal structures are made using some weft yarns as the binder.

9.3.4.2 Angle-interlock

An angle-interlock structure contains a set of straight yarns and another set of yarns that weave with the straight yarns in a diagonal direction through the thickness direction. Mostly, the weft yarns are straight, and the warp yarns bind diagonally from the top to the bottom. In some cases, the warp yarns can bind diagonally not for the full thickness. The cross-sections of the angle-interlock fabric structures bond in different ways and are displayed in Fig. 9.15. Fig. 9.16 is a model of an angle-interlock fabric through the thickness. Straight warp yarns can also be added to the angle-interlock

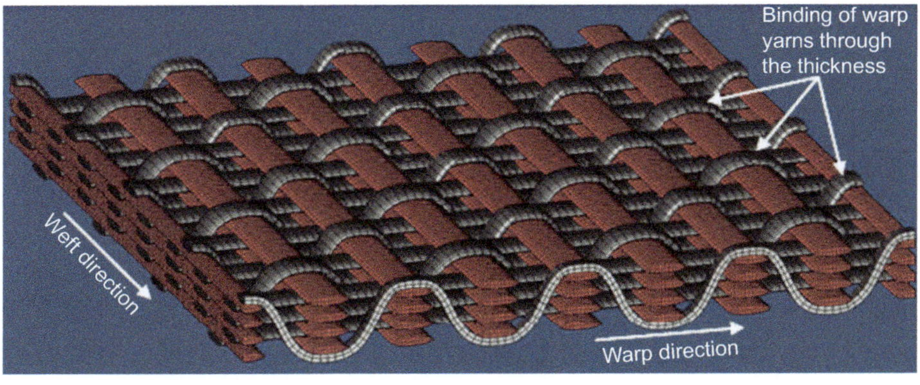

Figure 9.13 A four-layer 3D orthogonal fabric with binding of warp yarns through the thickness.
Adapted from Ha-Minh C, Boussu T, Kanit T, Crépin D, Imad A. Effect of frictions on the ballistic performance of a 3D warp interlock fabric: numerical analysis. Appl Compos Mater 2012; 19(3–4):333–47.

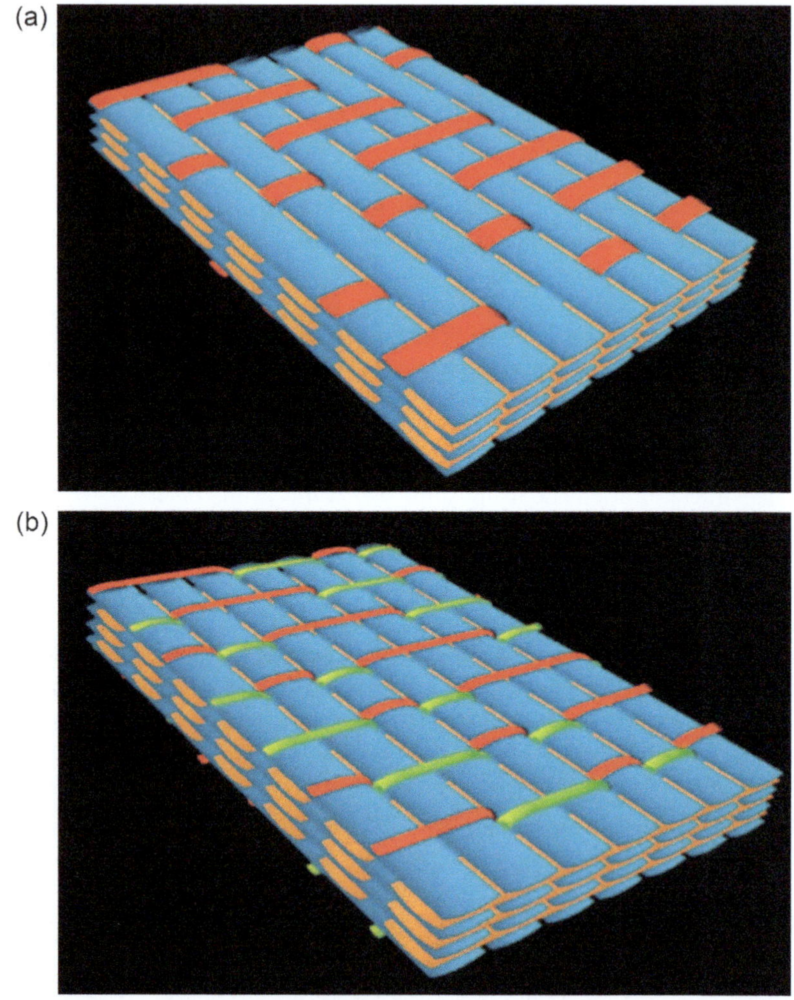

Figure 9.14 Ordinary and enhanced orthogonal structures with three warp yarn layers.
(a) Image of a three warp–layer ordinary orthogonal fabric. (b) Image of a three warp–layer enhanced orthogonal fabric.
Adapted from Chen X, Taylor LW, Tsai LJ. An overview on fabrication of three-dimensional woven textile preforms for composites. Text Res J 2011;81(9):932–44.

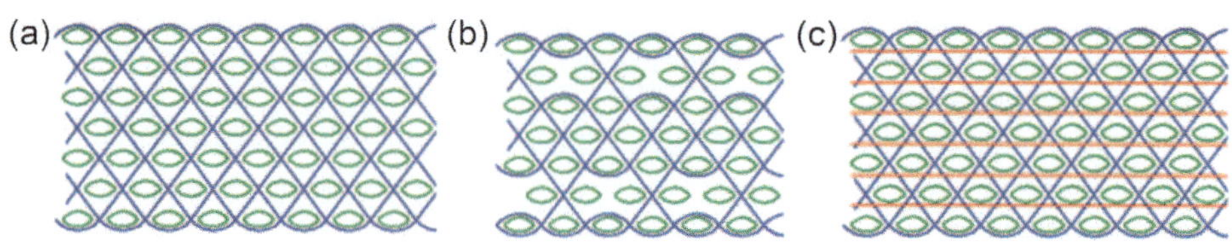

Figure 9.15 Cross-sectional views of angle-interlock fabrics (along the warp direction) with seven layers of weft yarns: (a) top-to-bottom binding, (b) binding to the fifth layer, and (c) warp-wadded fabric.
Adapted from Chen X, Taylor LW, Tsai LJ. An overview on fabrication of three-dimensional woven textile preforms for composites. Text Res J 2011;81(9):932–44.

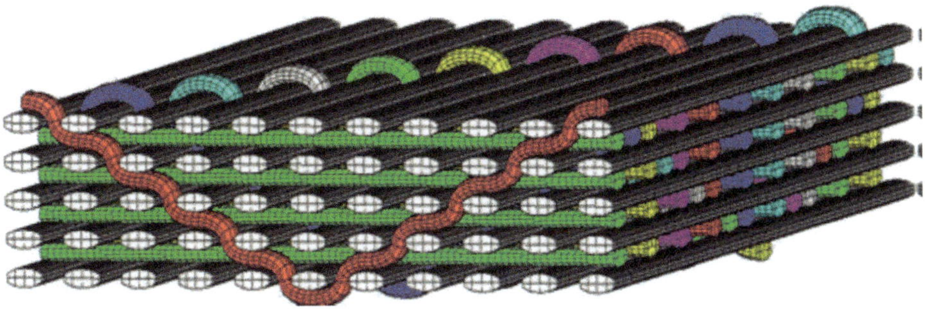

Figure 9.16 Angle-interlock fabric through the thickness.
Adapted from Ha-Minh C, Boussu T, Kanit T, Crépin D, Imad A. Effect of frictions on the ballistic performance of a 3D warp interlock fabric: numerical analysis. Appl Compos Mater 2012; 19(3–4):333–47.

structure, and this fabric structure is reported beneficial to increasing the tensile modulus and tensile strength in the warp direction [77]. Compared with equivalent multilayer woven fabrics, the angle-interlock woven fabrics have lower shear modulus. The moldability of a fabric depends greatly on its shear modulus, that is, the lower is the shear modulus, the superior the moldability is. The moldability of 3D angle-interlock woven fabric is quite attractive for researchers in ballistic fabric investigation. It has been reported that the angle-interlock fabric can be developed to manufacture women's ballistic body armor to better fit the bust curves [78]. Apart from making use of the moldability of the 3D interlock fabric for designing fitter female body armor, this brilliant characteristic, moldability, can also be taken advantage of in manufacturing a helmet for riot police officers or for military application. Owing to the unique moldability, the 3D angle-interlock fabric can be draped to the mold surface by applying forces to enable smooth conformation and avoid the trimming which may produce the weakness. At the same time the surface of the fabric shows no wrinkles or unsmooth areas. The developed single-piece riot helmet is shown in Fig. 9.17 [79,80].

9.3.5 Failure of ballistic composite panels

Ballistic composite panels are usually used under high level of threats. The failure mechanisms are more complicated than ballistic fabric panels for those composite panels since they contain two constituents: fiber and resin. Normally, the fail of the composite is local because of the higher stiffness of it. The failure behavior is influenced by the type of weave, fabric geometry, fiber column fraction, laminate configuration, and the material system [82]. The failure of a fiber- or fabric-enforced composite may fail due to delamination, matrix cracking, fiber micro-buckling, fiber breakage, etc., wherein the delamination is dominated by through-thickness stress waves for high-velocity impacts [83].

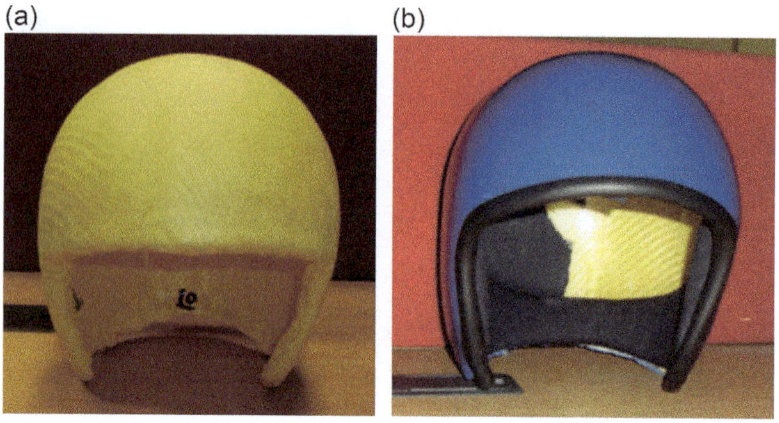

Figure 9.17 (a) Developed single-piece riot helmet shell and (b) finished look of developed single-piece riot helmet shell.
Adapted from Zahid B, Chen X. Manufacturing of single-piece textile reinforced riot helmet shell from vacuum bagging. J Compos Mater 2012;47(19):2343–51.

9.4 Engineering design of ballistic materials

9.4.1 Fibers and interyarn friction

According to the capstan equation, the interyarn friction F_f between the two interlaced yarns can be expressed in Eq. [9.10]. The expression of interyarn friction in the fabric indicates two manners to change the interyarn friction in the fabric, one being changing the coefficients of friction and the other one changing the wrapping angle. Normally, changing the coefficients of friction is based on the chemical surface coating methods since the coefficient of friction to a great degree relies on the surface morphology. For changing the wrapping angle, weaving-based technology can be used.

$$F_f = T \times e^{\mu\theta} \qquad [9.10]$$

where T is the tension of the yarn within the fabric; μ is the frictional coefficient between the warp and weft yarns; θ is the wrap angle (in radians) of weft yarn over the warp yarn (or otherwise).

9.4.2 Engineering of 2D ballistic fabrics

9.4.2.1 Yarn and fabric surface modifications

Plasma-enhanced vapor deposition (PCVD)

Two types of aramid fabric and yarn samples were treated by PCVD method for surface modification to enhance the yarn gripping by Sun and Chen [84]. A N_2 plasma with and without precursor dichlorodimethylsilane were used to modify the surface of the Kevlar® fabric. Fig. 9.18 shows scanning electron microscopy (SEM) images of the fiber surfaces with (a) the untreated fiber surface; (b) treated with N_2; and (c) treated with $(CH_3)_2Cl_2Si$. It is clear that the N_2 treatment caused only small surface changes to the aramid fiber, and the plasma treatment with $(CH_3)_2Cl_2Si$ modified the fiber surface much more. A quasistatic yarn pull-out test is a well-accepted method used to evaluate the interyarn friction in woven fabrics and has been used by many researchers. In the present study, a single yarn was pulled out from a 35-mm length of a fabric at a constant rate of 100 mm/min. Fig. 9.19 shows the load−displacement relationship for the three fabric samples. The load indicates the resistance to the yarn pulling from fabrics, while the N_2-treated fabric sample showed little difference from the untreated fabric; the load in the $(CH_3)_2Cl_2Si$ treated one is almost three times that in the original one.

The treatment of PCVD method using plasma with $(CH_3)_2Cl_2Si$ as precursor was also extended to the Twaron® yarn, another aramid yarn commonly used in ballistic impact, by Chu et al. [85]. The treatment time on the surface modification has been explored. Fig. 9.20 shows the Twaron® yarns treated by different times. Clearly, longer treatment time results in a much rougher fiber surface because of the plasma itching and the substance deposition formed from plasma. The coefficients of friction

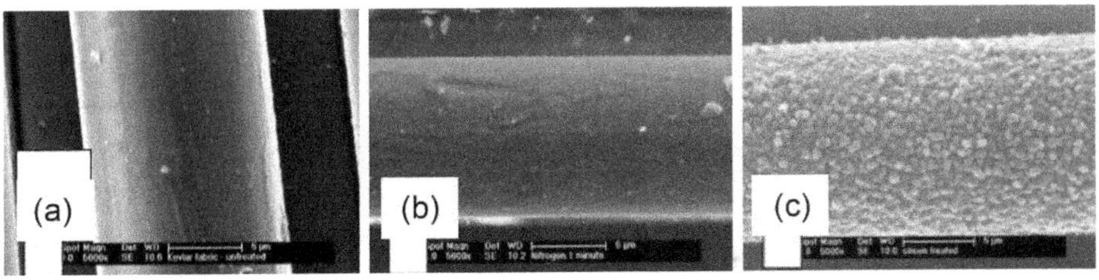

Figure 9.18 SEM images of Kevlar® fiber. (a) Untreated, (b) N_2 plasma treated, and (c) $(CH_3)_2Cl_2Si$ treated.
Adapted from Sun D, Chen X. Plasma modification of Kevlar fabrics for ballistic applications. Text Res J 2012;82(18):1928−34. http://dx.doi.org/10.1177/0040517512450765.

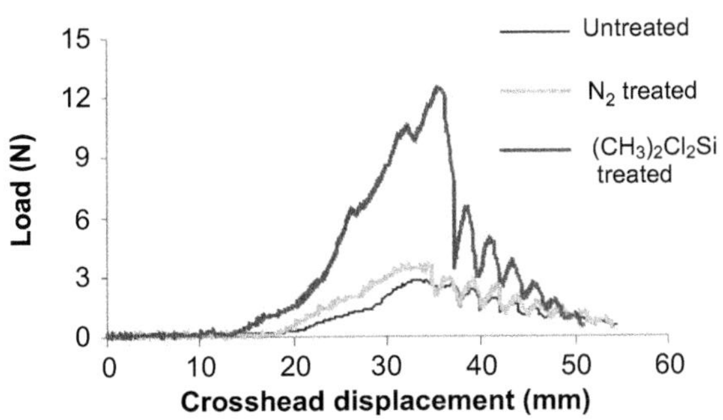

Figure 9.19 Resistance of pulling out a yarn from fabric.
Adapted from Sun D, Chen X. Plasma modification of Kevlar fabrics for ballistic applications. Text Res J 2012;82(18):1928−34. http://dx.doi.org/10.1177/0040517512450765.

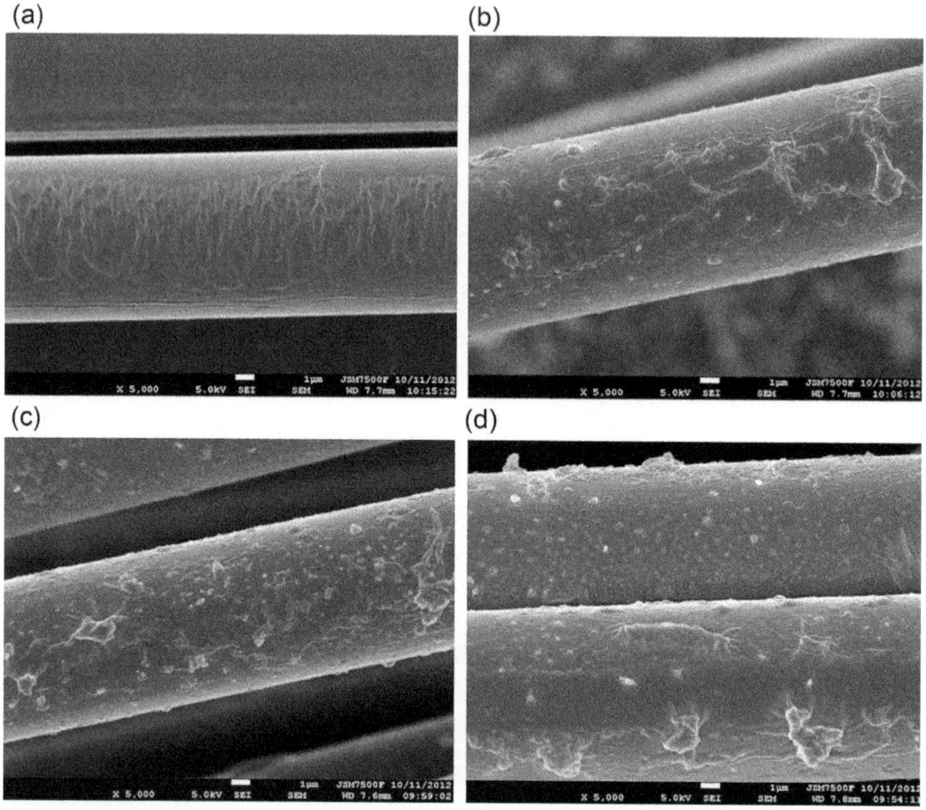

Figure 9.20 The surface morphology of the Twaron® Fibers: (a) original, (b) 21 s, (c) 2 min, and (d) 4 min.

Adapted from Chu Y, Chen X, Sheel DW, Hodgkinson JL. Surface modification of aramid fibers by atmospheric pressure plasma-enhanced vapor deposition. Text Res J 2014;84(12):1288–97.

test for the treated fabric is shown in Fig. 9.21. Similarly, as treatment time goes longer, the coefficients of static and dynamic friction become larger.

Sol–gel technology

Chu et al. reported on modifying surface frictions using sol–gel technology [86]. Submicro-sized and nano-sized TiO_2/ZnO hydrosols prepared from the precursors zinc acetate and titanium butoxide were investigated for the yarn treatment in the sol stage with and without curing process. Fig. 9.22 compares the surface of the yarn treated by different treatment processes. Clearly, the surfaces of the yarn are covered with some irregular substances. The submicro-sized TiO_2/ZnO hydrosol tends to form a lump-like substance attached to the surface of the yarn, while in the case of nano-sized TiO_2/ZnO hydrosol, surfaces of yarns are covered by a layer of laminar film-like substance. The surface of the yarn treated by TiO_2/ZnO submicro-sized hydrosol is much rougher and more nonuniform than these treated by nano-sized

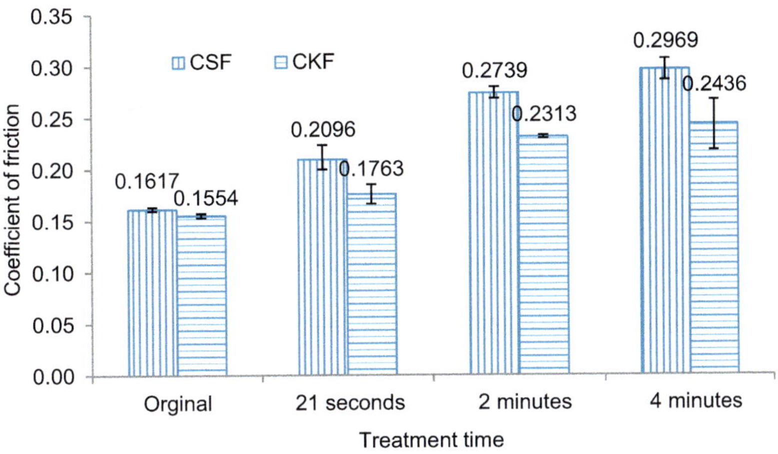

Figure 9.21 The coefficient of friction with different treatment time.
Adapted from Chu Y, Chen X, Sheel DW, Hodgkinson JL. Surface modification of aramid fibers by atmospheric pressure plasma-enhanced vapor deposition. Text Res J 2014;84(12):1288–97.

hydrosol. The treatment without curing is more likely to shape a bigger piece, while the treatment with curing is inclined to make the piece smaller. The effects of particle size in the sol and the curing process on the coefficient of friction are examined and are given in Fig. 9.23. The coefficients of friction for the yarns treated by submicro-sized hydrosol present an approximately 50% increase, while these for the yarns with nano-sized hydrosol show an around 10% increase. That is because the coating formed from submicro-sized hydrosol is more irregular, much harder, and produces a much rougher surface, while the coating on the fiber with nano-sized hydrosol is relatively flat. The process with curing or without curing gives rise to the same level of increase of interyarn friction.

9.4.2.2 Fabrics with enhanced interyarn gripping

Gripping method in a single-layer fabric

Three weave methods are attempted to increase the interyarn friction, involving introducing leno structure, weft cramming, and double picks to the normal plain weave, as described in Figs. 9.24 and 9.25. Leno is special type of weave structure where the warp yarns, instead of going parallel with one another in the fabric, are arranged to cross each other, leading to a firmer gripping of the weft yarns. Weft cramming is performed by periodically stopping the take-up process but keeping all other actions as usual during weaving. Another approach to increase the wrapping angle is to insert two weft yarns and to do the cramming at the same time. As the two weft yarns are combined into one, the warp yarn wrapping angle is believed to increase. These three methods have been applied to Kevlar® fabric [87], Twaron® fabric [88], and Dyneema® fabric [89], and the yarn pull-out test shows that the interyarn friction for the fabrics with these gripping methods has been improved.

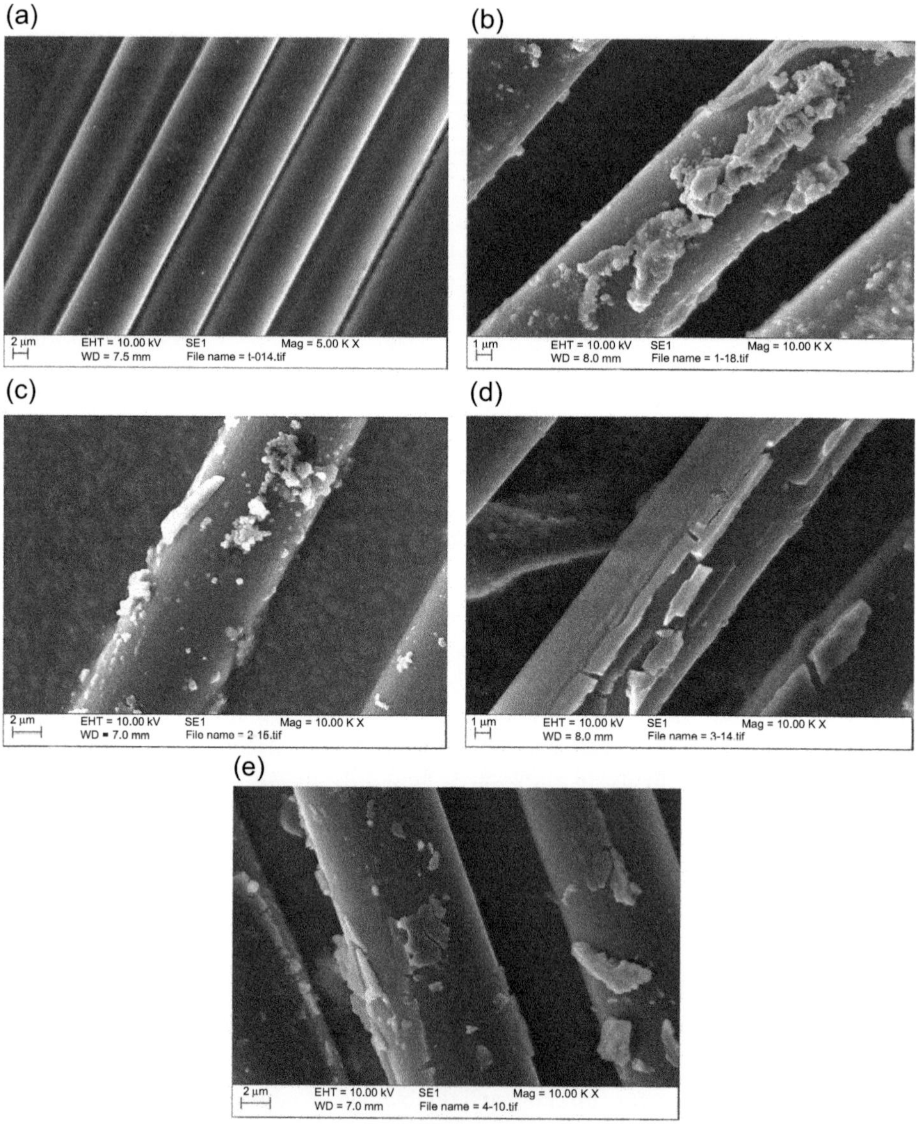

Figure 9.22 The surface coating by the sol–gel method on Twaron® yarns. (a) Original, (b) WDPD, (c) WDPDC, (d) TDPD, and (e) TDPDC.

Adapted from Chu Y, Chen X, Wang Q, Cui S. An investigation on sol–gel treatment to aramid yarn to increase inter-yarn friction. Appl Surf Sci 2014;320:710–7.

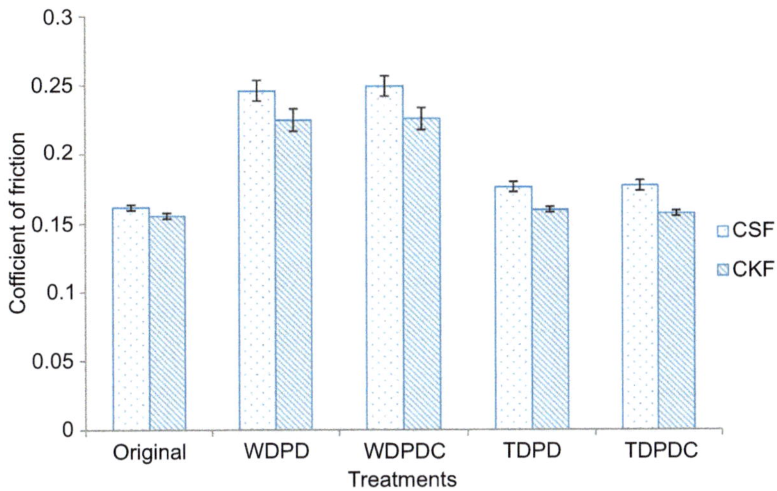

Figure 9.23 The coefficients of interyarn friction for different sol–gel treatments.
Adapted from Chu Y, Chen X, Wang Q, Cui S. An investigation on sol–gel treatment to aramid yarn to increase inter-yarn friction. Appl Surf Sci 2014;320:710–7.

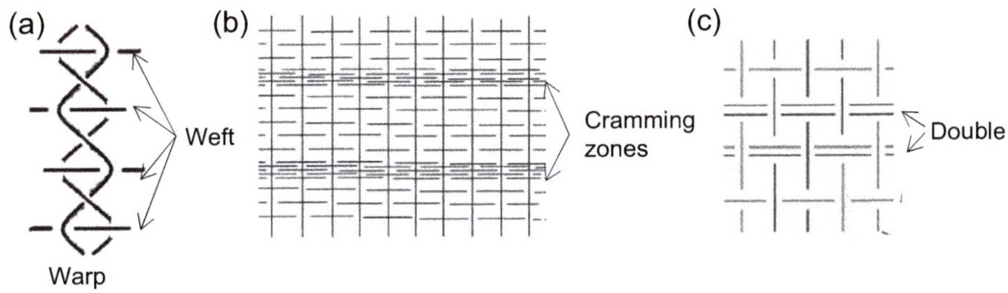

Figure 9.24 The gripping methods in a single-layer fabric. (a) Leno, (b) weft cramming, and (c) double pick.
Adapted from Zhou Y, Chen X, Wells G. Influence of yarn gripping on the ballistic performance of woven fabrics from ultra-high molecular weight polyethylene fibre. Compos Part B Eng 2014; 62:198–204.

Gripping methods in multilayer fabric

Fig. 9.26 illustrates a couple two-layered fabrics, with Fig. 9.26(a) showing the regular construction where the two layers join into one and then separate to their original positions, and Fig. 9.26(b) displays a layer interchange construction, where the two layers swap position after joining together. These woven fabrics were designed with the aim to gain an understanding on yarn gripping in the fabric and strain distribution from one layer to other layers.

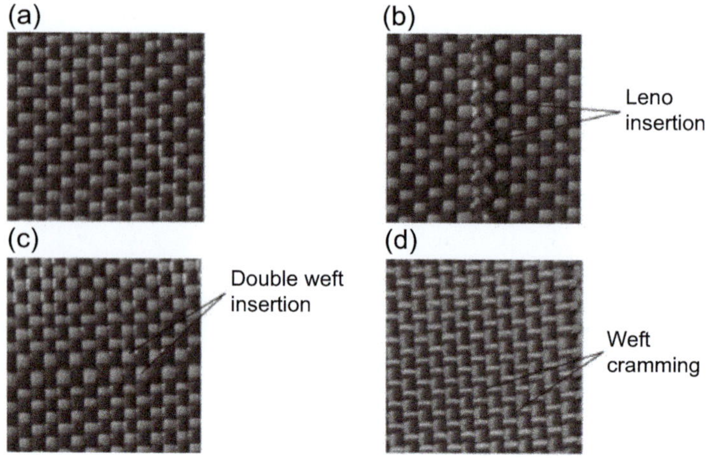

Figure 9.25 The appearance of the fabric with gripping insertion.
Adapted from Zhou Y, Chen X, Wells G. Influence of yarn gripping on the ballistic performance of woven fabrics from ultra-high molecular weight polyethylene fibre. Compos Part B Eng 2014; 62:198–204.

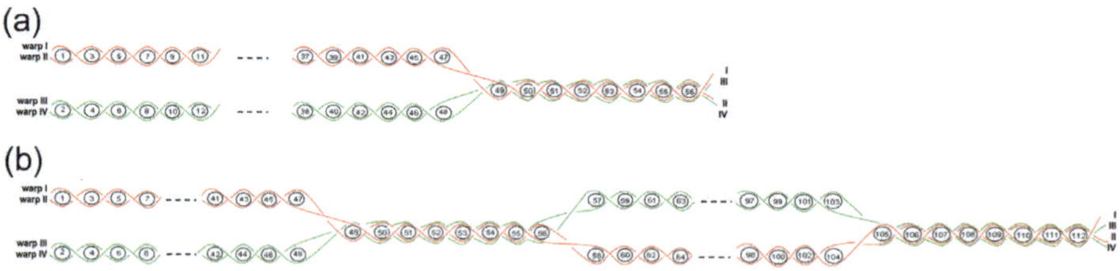

Figure 9.26 Two-layer fabric designed for ballistic protection. (a) A 2-layered fabric with regular construction and (b) a 2-layered fabric with interchange construction.
Adapted from Sun D, Chen X, Wells G. Engineering and analysis of gripping fabrics for improved ballistic performance. J Compos Mater 2014;48(11):1355–64.

9.4.2.3 Ballistic performance of fabrics with enhanced yarn gripping

V_{50} test

The V_{50} test was carried out by an independent institution using a 1.1-g chisel-nosed fragment simulating projectile. The panel size is 40 × 40 cm. The plain fabric panel is composed of 19 plies leading to an overall areal density of 3.0 kg/m^2, and the panel made from fabric with leno insertions involves 18 layers with the overall areal density being 2.9 kg/m^2. The gap between adjacent leno lines is 2 cm. The fabric with leno insertions was randomly layered, and there was no specific attempt to correlate the leno to leno, or leno to plain weaves. Twenty shots were made to each panel.

Fig. 9.27(a) shows the V_{50} test result, and Fig. 9.27(b) the normalized V_{50}. It is clear that the ballistic panel made from the plain-woven fabric with leno insertions

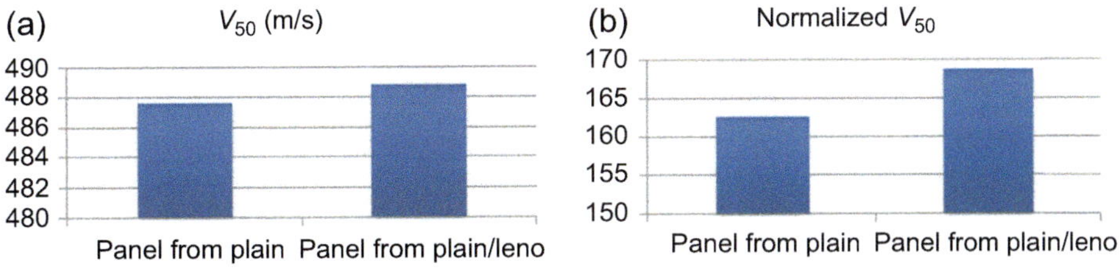

Figure 9.27 Comparison of V_{50} results for panels from fabric with and without leno insertions. (a) Direct results and (b) normalized results.
Adapted from Chen X, Chu Y, Wang N. Study on inter-yarn friction and engineering of ballistic fabrics. In: Paper presented at the 14th AUTEX world textile conference, Bursa, Turkey; 2014.

outperforms its plain-woven fabric counterpart. For the same amount of ballistic protective material used, the panel from plain fabric with leno insertions will be more protective, in this case about 4%. To satisfy the same level of ballistic protection, the employment of leno insertion in the plain-woven fabric will lead to savings of materials, resulting in lighter weight body armor.

Energy absorption

Single-layer aramid fabrics The energy absorptions for the fabrics with different gripping methods are shown in Figs. 9.28 and 9.29. Considering the deviation of

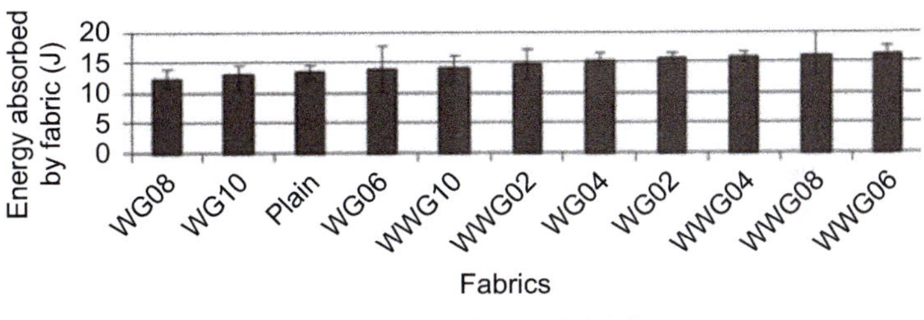

Figure 9.28 Energy absorbed by various manufactured fabrics.
Adapted from Sun D, Chen X, Mrango M. Investigating ballistic impact on fabric targets with gripping yarns. Fibers Polym 2013;14(7):1184–9.

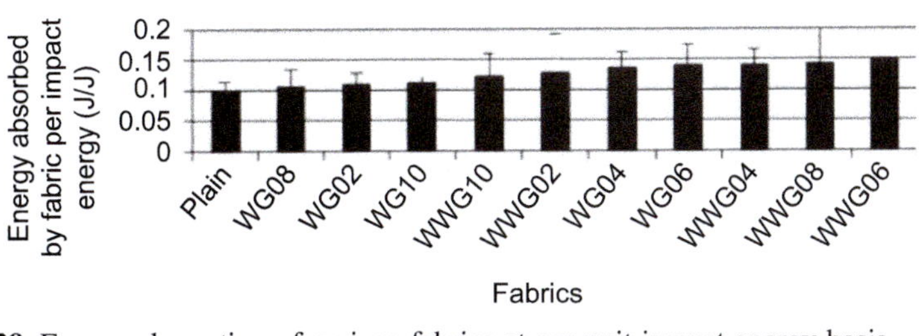

Figure 9.29 Energy absorption of various fabrics at per unit impact energy basis.
Adapted from Sun D, Chen X, Mrango M. Investigating ballistic impact on fabric targets with gripping yarns. Fibers Polym 2013;14(7):1184–9.

the impact energy, the vertical axis indicates the energy absorption of fabric normalized to the impact energy. All the fabrics were made from Kevlar® 49 yarns with linear density of 158 tex. The warp density and weft density are 7.5/cm. In Figs. 9.28 and 9.29, the letters WG mean warp gripping and letters WWG represent both warp and weft gripping. The number means the distance between each interval. Warp gripping yarns are designed in certain centimeter intervals along fabric width. Both warp and weft gripping yarns are designed in every centimeter interval along the warp and weft directions. As anticipated, the ballistic performance of all the gripping fabrics has been improved. Also, the fabrics with both warp and weft gripping outstand the warp gripping counterparts. Moreover, the ballistic distance between each insertion affects the energy absorption of the fabric. In both cases, the interval of 6 cm is the optimum. Fig. 9.30 compares the energy absorption of fabric with leno structure and the fabric with combined two insertions, and it is clearly showing that the fabrics with combined insertion are better in energy absorption than the corresponding fabrics with only leno structure.

Fig. 9.31 shows energy absorption of the FE modeled fabric corresponding to the fabric used in experimental work. It shows that fabric absorbs more energy with the increase of impact time, and more energy is absorbed by the fabrics with higher coefficients of friction compared with that of the lower values. This indicates that ballistic performance can be improved by enhanced yarn gripping.

Single-layer UHMWPE fabrics A UHMWPE fiber was used to make woven fabrics with interyarn gripping features. Energy absorption at a designated impact velocity (500 m/s) for each fabric was used for comparison [89]. The energy absorption was normalized by fabric areal density, and the results are shown in Fig. 9.32. PW refers to the plain fabric. PWL stands for the leno insertion in the warp direction of plain fabric. The number following the PWL symbol means the distance between each

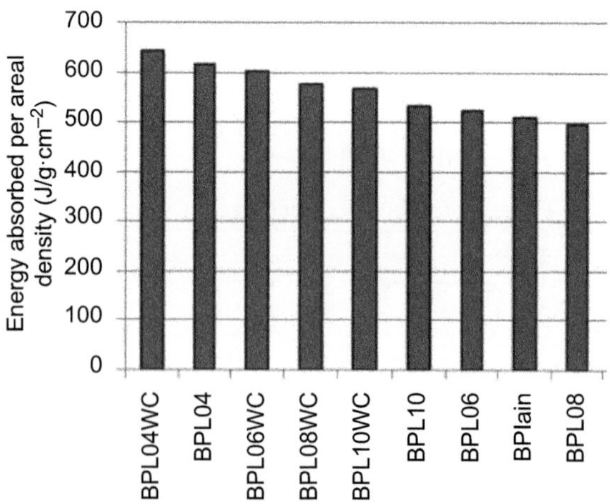

Figure 9.30 Energy absorption of fabric with leno structure and fabric with combined leno and weft cramming.

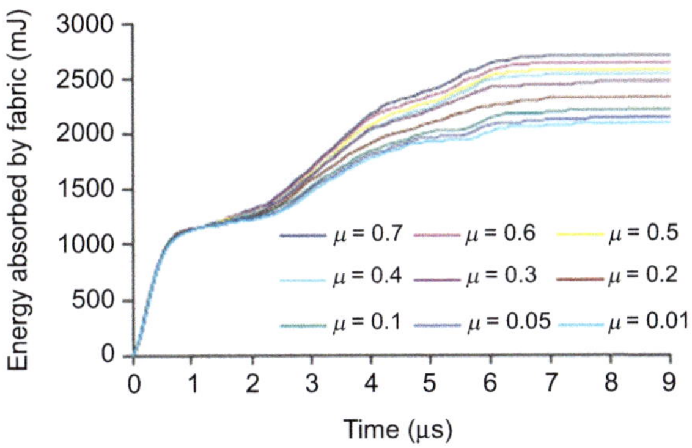

Figure 9.31 The FE simulation of the energy absorption by fabric as a function of coefficient of interyarn friction.
Adapted from Sun D, Chen X. Plasma modification of Kevlar fabrics for ballistic applications. Text Res J 2012;82(18):1928–34. http://dx.doi.org/10.1177/0040517512450765.

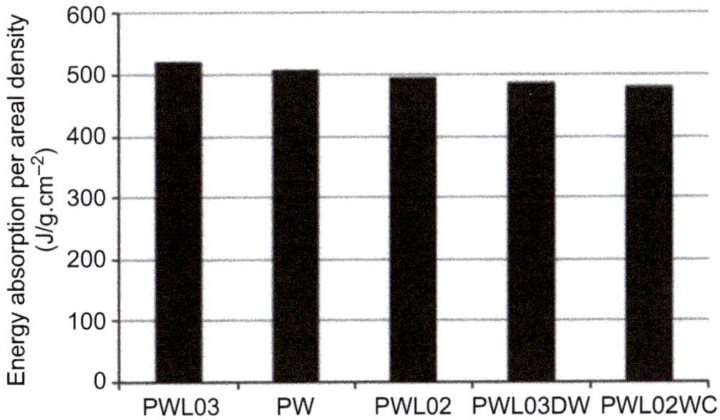

Figure 9.32 Energy absorption per areal density for Dyneema® woven fabrics at the impact velocity of 500 m/s.
Adapted from Zhou Y, Chen X, Wells G. Influence of yarn gripping on the ballistic performance of woven fabrics from ultra-high molecular weight polyethylene fibre. Compos Part B Eng 2014; 62:198–204.

leno line in the fabric. PWL03DW and PWL02WC refer to plain fabric with combined leno and double pick and plain fabric with combined leno insertion and weft cramming, respectively. The results show that enhanced interyarn gripping works to improve the energy absorption of the UHMWPE fabrics. It was also found that when the gripping is too high, the energy absorption capacity of the fabric would decrease. The FE simulation results also shows such a trend, as given in Fig. 9.33. The peak yarn pull-out load is shown in Table 9.2.

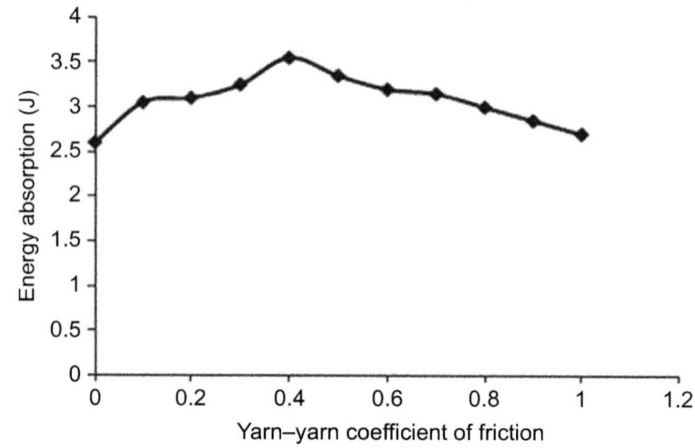

Figure 9.33 Fabric energy absorption as a function of coefficient of interyarn friction. Adapted from Zhou Y, Chen X, Wells G. Influence of yarn gripping on the ballistic performance of woven fabrics from ultra-high molecular weight polyethylene fibre. Compos Part B Eng 2014; 62:198–204.

Table 9.2 The peak yarn pull-out load in Dyneema® fabric with different gripping methods [35]

Fabric labels	Peak load in yarn pull-out test (*N*)
PW	2.4
PWL02	3.0
PWL03	2.5
PWL03DW	3.1
PWL02WC	3.2

Two-layer networked fabrics

Fig. 9.34 presents the results of energy absorbed by two-layered networked fabrics (2LRe and 2LIn) compared to a two-layer panel made from a plain fabric. 2LRe refers to the two-layer fabric with regular structure, and 2Lin refers to the two-layer fabric with interchange structure. Due to the difference in yarn density of the plain-weave fabric and the two-layered fabrics, the energy absorbed by the fabric is normalized by the areal density. Apparently, both joint two-layered fabrics show better ballistic performance than the two-layer plain-weave fabrics in terms of energy absorbed by fabric at per areal density and per unit impact energy basis. In both these cases, the regular joint two-layered fabric absorbs more energy than the interchanging two-layered fabric [95].

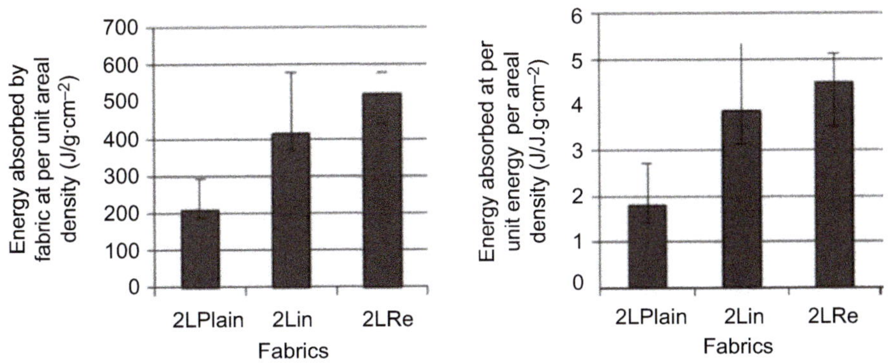

Figure 9.34 Energy absorption of the two-layer fabric with different gripping methods. Adapted from Sun D, Chen X, Mrango M. Investigating ballistic impact on fabric targets with gripping yarns. Fibers Polym 2013;14(7):1184–9.

9.4.3 Ballistic panel engineering

9.4.3.1 Quasiisotropic panels

For the commonly used plain fabric, it always shows an orthotropic behavior due to the structure nature, leading to a square-based pyramid developing at the local deformation zone during the striking event by a high-velocity projectile, as shown in Fig. 9.35(a). The yarns that are in direct contact with the projectile are known as the principal yarns with the other yarns known as secondary yarns, which intersect the principal yarns proceeding until the material strain at the impact point reaches a breaking strain. The energy absorption of the fabric is to some degree in relation to the deformation zone. In plain-woven fabric, the central deformation zone always takes a pyramid shape on the fabric in an orientation shown in Fig. 9.35(b). However, if the fabric panel is an isotropic one, for example, with the structure of Fig. 9.35(c), the

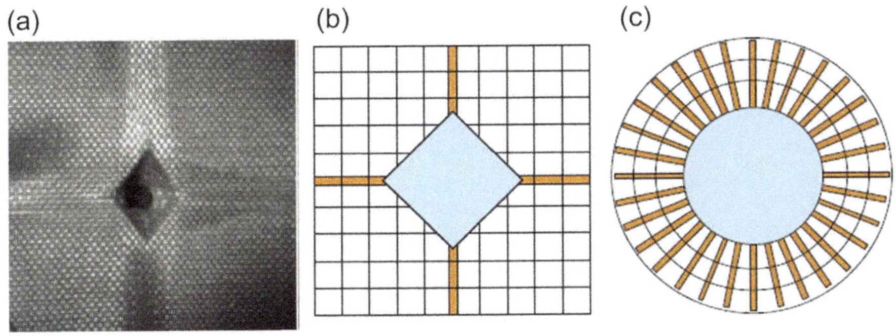

Figure 9.35 Images of a projectile impacting a plain-woven fabric: (a) image from high-speed camera, (b) deformation of plain-woven fabric (biaxial), and (c) deformation of multiaxial. Adapted from Wang Y, Chen X, Young R, Kinloch I, Wells G. A numerical study of ply orientation on ballistic impact resistance of multi-ply fabric panels. Compos Part B Eng 2015;68:259–65.

central deformation zone would be approximately more circle-like, making the impact energy distributed on a larger area. Obviously, the amount of SE absorbed in Fig. 9.35(c) would be larger than that in Fig. 9.35(b) since more material is involved in the impact deformation.

Hearle et al. [91] proposed the use of triaxial fabrics, where three sets of yarns intersect at 60 degree intervals, instead of the traditional biaxial plain fabrics where the warps and wefts intersect perpendicularly. Although the triaxial ply structure managed to spread the impact loading more evenly along three directions, the ballistic performance actually is reduced due to the inherent openness of the structure. It is hard to make the fabric isotropic in one layer of the fabric without other parameters affected. However, since the fabric panel is generally constructed by several layers of fabric, it would be acceptable to make the fabric panel more isotropic, for example, through orientating each ply in different directions through the thickness direction. In other words, it is to angle-lay a ply toward the neighboring ply, and the yarn in a different ply would have different orientations. Finally, the orientation yarns are distributed much more uniformly in different directions in the entire panel.

Angle-laid panels

The x, y, z orthogonal coordinate system is used in describing the structure of the multi-ply fabric panels, as shown in Fig. 9.36. The orientations of the plies are specified by the angle θ with respect to the x-axis. The angle θ is between the warp yarn and x-axis and is positive in the counterclockwise direction. For example, the fabric panel consisting of four plies shown in Fig. 9.37 is designated as [45/45/0/0]. This fabric panel contains two-ply groups, the first containing two plies in the 45 degree

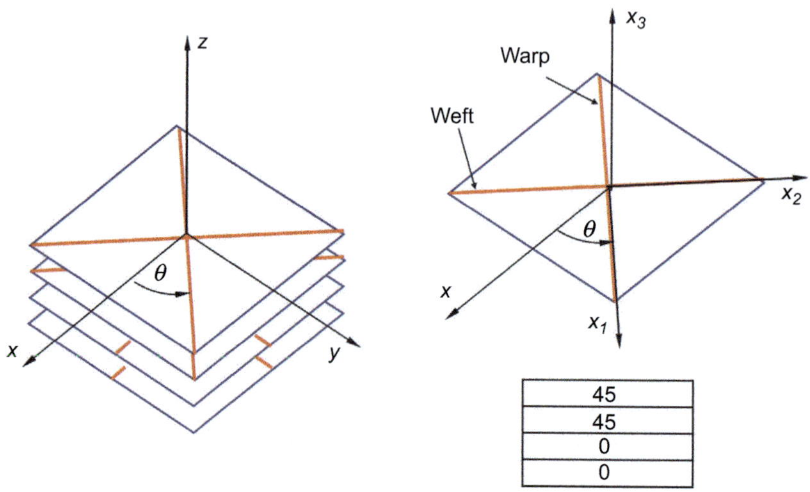

Figure 9.36 A schematic illustration of the coordinate system of each ply in a panel. Adapted from Wang Y, Chen X, Young R, Kinloch I, Wells G. A numerical study of ply orientation on ballistic impact resistance of multi-ply fabric panels. Compos Part B Eng 2015;68:259–65.

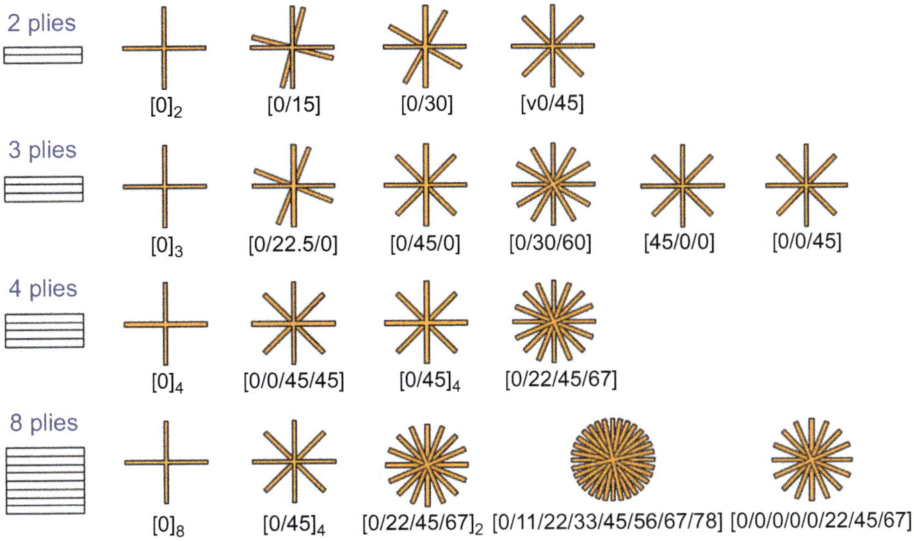

Figure 9.37 Angle-laid fabric panel design.
Adapted from Wang Y, Chen X, Young R, Kinloch I, Wells G. A numerical study of ply orientation on ballistic impact resistance of multi-ply fabric panels. Compos Part B Eng 2015;68:259–65.

direction, and the second containing two plies in the 0 degree direction. When all the plies (all warp yarns in the fabric panel) are in the same direction, the panel is defined as an aligned fabric panel, for example, [0/0/0/0]. When the warp yarns are in different directions, the panel is defined as an angled fabric panel, for example, [0/45] or [0/22/45/67]. Based on this design code, four groups of fabric panels with different angle orientations have been presented below.

Ballistic performance of angle-laid panels

Numerical An FE model is established to analyze the energy-absorption capacity of the designed fabric panels with different angle orientations. The fabric in the simulation is modeled at yarn level. The Young's modulus of the yarns and the elongation are set as 93.5 GPa and 3%, following the aramid fiber. The density of the yarn is 1440 kg/m^3, and the Poisson's ratio is 0.35. To investigate the effects of angle-laying on the energy absorption of the four groups of fabric panels, the impact velocity in the following is set at the same level of 500 m/s. Fig. 9.38 shows the energy absorption of the designed fabric panels in the four-layer groups. Unquestionably, all the angle-laid fabric panels demonstrated a superior energy absorption capacity compared with the aligned one in each group. In the two-ply group, compared with aligned ply panel, the energy absorption in the angled panel [0/45] increased by 11.4%. In the three-ply panel group, the most energy absorption panel is [0/30/60], absorbing about 16% more than the aligned one. In the four-ply panel group, the angle-laid design [0/22.5/45/67.5] absorbs the highest energy, almost 11.6% more than the corresponding aligned one. In the eight-ply fabric panel group, the most

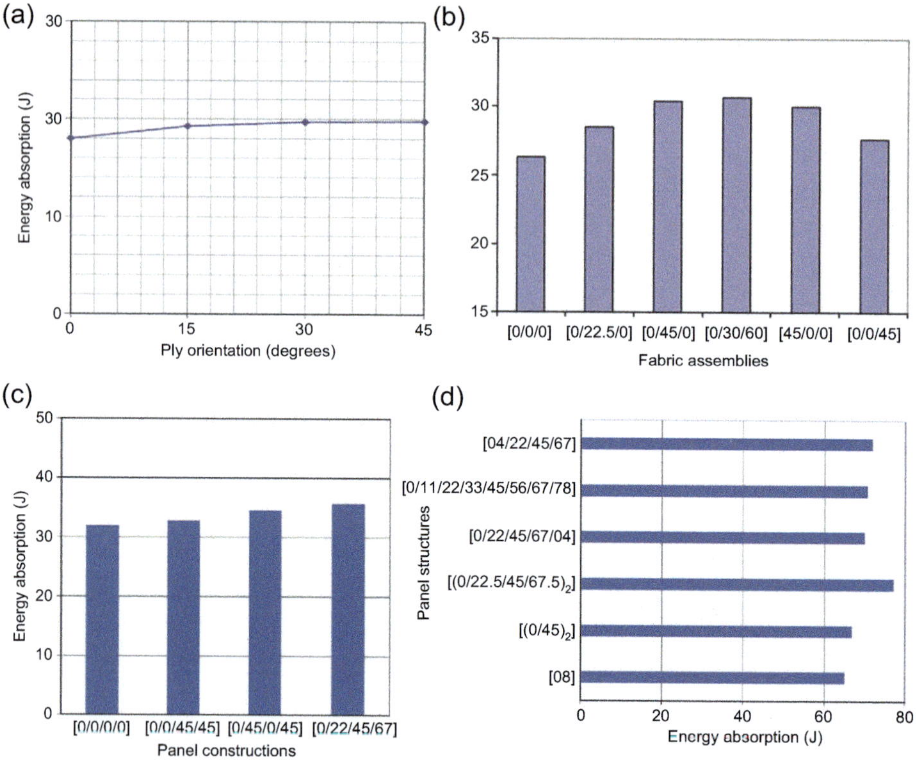

Figure 9.38 The energy absorption for designed angle-laid panels. (a) 2-ply, (b)3-ply, (c) 4-ply, and (d) 8-ply.
Adapted from Chen X, Sun D, Wang Y, Zhou Y. 2D/3D woven fabrics for ballistic protection. In: Proceedings to 4th world conference on 3D fabrics and their applications, Aachen; 2012.

energy absorption lies in the angle-laid case [(0/22.5/45/67.5)$_2$], 18.5%. It is also found that in each fabric panel group, the most energy absorption rests on the relatively evener distribution of angle case. This is ascribed to the even distribution of angle leading to the panel being much more isotropic. However, it should avoid the over-even distribution. As it is shown in Fig. 9.38(d), the eight-ply case, the construction design [0/11/22/33/45/56/67/78] does not present the highest energy absorption.

Experimental Fig. 9.39 shows the energy absorption values of the designed angle-laid panels calculated from the experimental test results for the eight-ply group, respectively. Clearly, the experimental results show a similar trend with the results from the FE analysis. Moreover, the experimental results also show that all the energy absorptions for the angle-laid fabric panels are larger than that of the aligned one, and the fabric panel with the construction of [(0/22.5/45/67.5)$_2$] from the experimental test displays the most energy absorption.

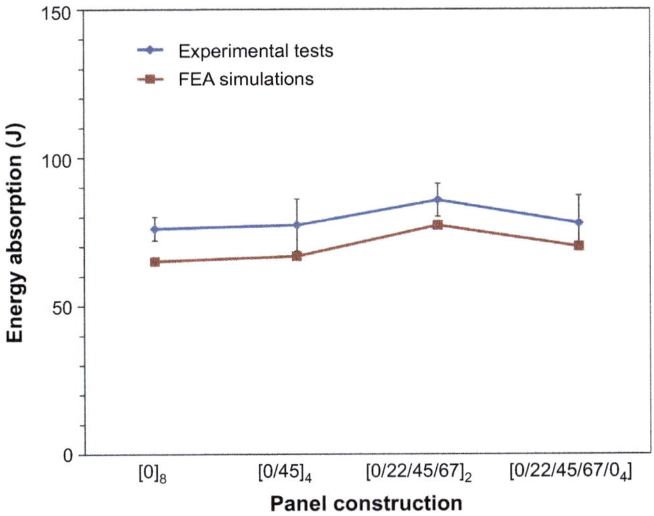

Figure 9.39 Comparison of the energy absorption for designed angle-laid panels between FE simulation and experimental tests.
Adapted from Wang Y, Chen X, Young R, Kinloch I, Wells G. A numerical study of ply orientation on ballistic impact resistance of multi-ply fabric panels. Compos Part B Eng 2015;68:259–65.

9.4.3.2 Hybrid panels

Previous study also shows that hybrid panels formed by UD fabrics and woven fabrics perform better than single-material panels. Nonpenetration ballistic impact tests carried out by Thomas [93] showed that single-material aramid filament panels gave deeper back-face signature than hybrid panels. Karahan et al. [94] showed that around 13.9% less energy was transmitted to the backing materials through a hybrid panel than through a single-material UD fabric panel, and an improvement of 8.5% in the energy absorbed per unit panel weight was obtained with hybrid panels compared to 100% woven fabric panels. As the trauma concerned, it was shown that in hybrid panels, trauma depth was decreased around 4.5%, and trauma diameter was increased around 3%, compared to 100% unidirectional fabric panels.

According to the fact that the type of material from one layer in a fabric panel would result in different responses, an advantage in combining more than one type of material in a panel is suggested. A hybrid fabric panel from two types of components is proposed, one being a unidirectional shield (UD) or UD fabric, also known as fiber-reinforced laminates, and the other a woven fabric.

Hybrid fabric panel

Two types of fabric panel (panel A and panel B) were designed from a plain-woven fabric made of Dyneema® SK75 and a UD fabric made of Dyneema® SB21. In type A panels, woven fabrics were placed in front of the UD fabrics, and the sequence of the two types of fabrics was reverse in B panels. To better compare the performance of the panels, the areal densities of different panels were made as similar as possible. The panel details are given in Tables 9.3 and 9.4. The size of fabric panels is 23 × 23 cm.

Ballistic performance of hybrid panels

Fig. 9.40 compares the ballistic performance of hybrid type A panels and hybrid type B panels. Clearly, all the hybrid type B panels give a deeper hole than the corresponding type A panel. This result suggests the woven fabric group positioned at the face of the panel and the UD fabric group positioned at the back of the panel are much more beneficial to reduce the depth of the trauma. Damage morphology for different fabrics in a panel is very similar to those for single-layer tests. For the UD fabric, the damage is dominant by shear failure of the fiber because of the constraint of resin matrix, whereas for the woven fabric, fiber tensile failure coupled with yarn pull-out feature are the reasons. In a panel, the front layers of fabric are more likely to be sheared out by a projectile, while the rear layers of fabric tend to be stretched to failure and exhibited larger transverse deflection. This indicated that using shear resistant materials for the front layer and tensile resistant materials for the rear layer might improve the ballistic performance of the fabric panel. This is the reason for the improvement in ballistic performance of the hybrid panel with woven fabric placed in the front and the UD fabric placed behind.

Table 9.3 Designed type A panel [73]

Type A panel	Panel model	Areal density (g/m^2)
40 layers of UD fabric		5800
6 layers of woven fabric +30 layers of UD fabric		5790
12 layers of woven fabric +20 layers of UD fabric		5780
18 layers of woven fabric +10 layers of UD fabric		5770
24 layers of woven fabric		5760

Table 9.4 Designed type B panel [73]

Type B panel	Panel model	Areal density (g/m^2)
24 layers of woven fabric		5760
10 layers of UD fabric +18 layers of woven fabric		5770
20 layers of UD fabric +12 layers of woven fabric	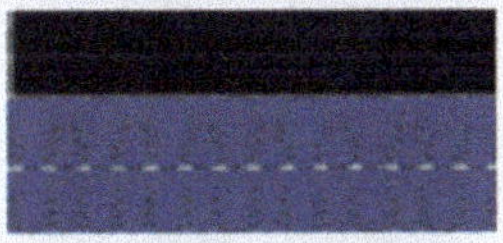	5780
30 layers of UD fabric +6 layers of woven fabric		5790
40 layers of UD fabric		5800

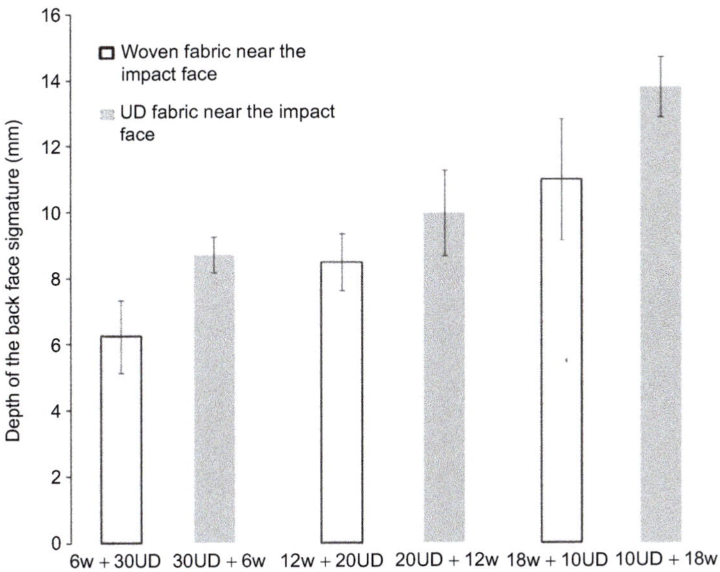

Figure 9.40 Comparison between type A panels and type B panels.
Adapted from Chen X, Zhou Y, Wells G. Numerical and experimental investigations into ballistic performance of hybrid fabric panels. Compos Part B Eng 2014;58:35–42.

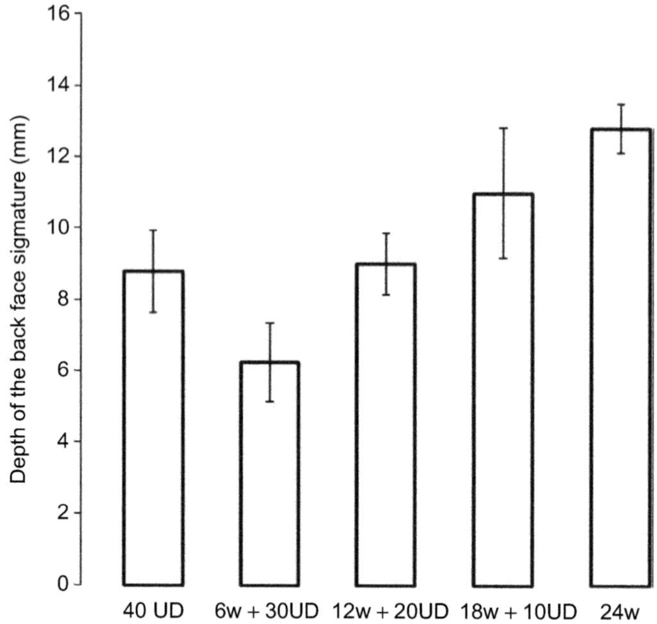

Figure 9.41 Depth of the back-face signature for type A panels.
Adapted from Chen X, Zhou Y, Wells G. Numerical and experimental investigations into ballistic performance of hybrid fabric panels. Compos Part B Eng 2014;58:35–42.

Fig. 9.41 compares the depths of the back-face signature between the designed hybrid panels and the net woven fabric and the net UD fabric. As anticipated, the net woven fabric panel leads to the largest depth. All the hybrid panel designs show smaller depths of the back-face signature than the net woven fabric panel. The hybrid panel with six layers of woven fabric and 30 layers of UD fabric results in a much smaller back-face signature than the net UD fabric panel. The results from the FE simulation also show the similar trend, as plotted in Fig. 9.42. This indicates that the proportion of woven fabric in the hybrid panel should be less than or around 25% to achieve the best hybrid effect.

9.5 Future trends

In ballistic material investigations, lightweight and reliable ballistic performance has always been the target for research. Manufacturing better ballistic materials depends on a good understanding of the failure mechanism of ballistic materials. However, the failure mechanism for ballistic material is rather complicated because of the diversity of materials and structures used for making ballistic fabrics and panels. In addition, the failure mechanisms of ballistic materials are not fully answered due to the limitation in the research techniques analytically, experimentally, and numerically. Much effort will still be seen in the future to reveal how the ballistic materials fail in different ballistic events.

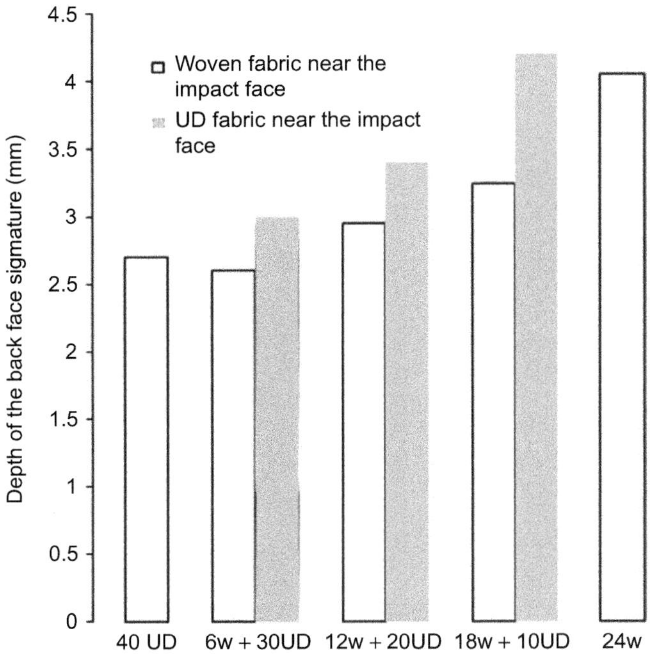

Figure 9.42 FE results for type A panels and type B panels.
Adapted from Chen X, Zhou Y, Wells G. Numerical and experimental investigations into ballistic performance of hybrid fabric panels. Compos Part B Eng 2014;58:35–42.

For ballistic materials made from UD laminates, woven fabrics, or in the form of composites, light weight and reliability are the two primary requirements for the modern military protection. Owing to the progress made in high-performance fiber, the weight of the ballistic protective equipment has been reduced greatly. Despite that, the weight of the ballistic materials is still far from satisfactory. Reducing the weight of ballistic materials is imperative. Engineering design of ballistic materials and development of ballistic panels using the hybrid method and quasiisotropic approach will be further experimented on as effective measures to achieve both light weight and reliability for ballistic protection. High-performance fibers with better ballistic properties will continue to be researched. Novel materials other than fibrous materials for ballistic protection will continue to be explored.

References

[1] Smith JC, McCrackin FL, Schiefer HF. Stress-strain relationships in yarns subjected to rapid impact loading Part V: wave propagation in long textile yarns impacted transversely. Text Res J 1958;28(4):288–302.

[2] Vinson JR, Taylor WJ. Modeling ballistic impact into flexible materials. AIAA J 1990; 28(12):2098–103.

[3] Vinson J, Zukas J. On the ballistic impact of textile body armor. J Appl Mech 1975;42(2): 263–8.
[4] Parga-Landa B, Hernandez-Olivares F. An analytical model to predict impact behaviour of soft armours. Int J Impact Eng 1995;16(3):455–66.
[5] Hetherington J. Energy and momentum changes during ballistic perforation. Int J Impact Eng 1996;18(3):319–37.
[6] Chocron-Benloulo IS, Rodriguez J, Sanchez-Galvez V. A simple analytical model to simulate textile fabric ballistic impact behavior. Text Res J 1997;67(7):520–8.
[7] Billon H, Robinson D. Models for the ballistic impact of fabric armour. Int J Impact Eng 2001;25(4):411–22.
[8] Gu B. Analytical modeling for the ballistic perforation of planar plain-woven fabric target by projectile. Compos Part B Eng 2003;34(4):361–71.
[9] Porwal PK, Phoenix SL. Modeling system effects in ballistic impact into multi-layered fibrous materials for soft body armor. Int J Fract 2005;135(1–4):217–49.
[10] Porwal P, Phoenix S. Effects of layer stacking order on the V_{50} velocity of a two-layered hybrid armor system. J Mech Mater Struct 2008;3(4):627–39.
[11] Phoenix S, Porwal PK. A new membrane model for the ballistic impact response and V_{50} performance of multi-ply fibrous systems. Int J Solids Struct 2003;40(24):6723–65.
[12] Mamivand M, Liaghat G. A model for ballistic impact on multi-layer fabric targets. Int J Impact Eng 2010;37(7):806–12.
[13] Chen X, Zhu F, Wells G. An analytical model for ballistic impact on textile based body armour. Compos Part B Eng 2013;45(1):1508–14.
[14] Ha-Minh C, Imad A, Boussu F, Kanit T. On analytical modelling to predict of the ballistic impact behaviour of textile multi-layer woven fabric. Compos Struct 2013;99:462–76.
[15] Duan Y, Keefe M, Bogetti T, Cheeseman B. Modeling friction effects on the ballistic impact behavior of a single-ply high-strength fabric. Int J Impact Eng 2005;31(8):996–1012.
[16] Duan Y, Keefe M, Bogetti T, Cheeseman B, Powers B. A numerical investigation of the influence of friction on energy absorption by a high-strength fabric subjected to ballistic impact. Int J Impact Eng 2006;32(8):1299–312.
[17] Backman ME, Goldsmith W. The mechanics of penetration of projectiles into targets. Int J Eng Sci 1978;16(1):1–99.
[18] Cunniff PM. A semiempirical model for the ballistic impact performance of textile-based personnel armor. Text Res J 1996;66(1):45–58.
[19] Tan V, Khoo K. Perforation of flexible laminates by projectiles of different geometry. Int J Impact Eng 2005;31(7):793–810.
[20] Cork C, Foster P. The ballistic performance of narrow fabrics. Int J Impact Eng 2007;34(3): 495–508.
[21] Drodge DR, Mortimer B, Holland C, Siviour CR. Ballistic impact to access the high-rate behaviour of individual silk fibres. J Mech Phys Solids 2012;60(10):1710–21.
[22] Sakaguchi S, Carr D, Horsfall I, Girvan L. Protecting the extremities of military personnel: fragment protective performance of one-and two-layer ensembles. Text Res J 2012;82(12): 1295–303.
[23] Majumdar A, Butola BS, Srivastava A. An analysis of deformation and energy absorption modes of shear thickening fluid treated Kevlar fabrics as soft body armour materials. Mater Des 2013;51:148–53.
[24] Majumdar A, Butola BS, Srivastava A. Development of soft composite materials with improved impact resistance using Kevlar fabric and nano-silica based shear thickening fluid. Mater Des 2014;54:295–300.

[25] Park JL, Yoon BI, Paik JG, Kang TJ. Ballistic performance of p-aramid fabrics impregnated with shear thickening fluid; part I—effect of laminating sequence. Text Res J 2011; 82(6):527—41.
[26] Park JL, Yoon BI, Paik JG, Kang TJ. Ballistic performance of p-aramid fabrics impregnated with shear thickening fluid; Part II—Effect of fabric count and shot location. Text Res J 2012;82(6):542—57.
[27] National Institue of Justice USDoJ. Ballistic resistance of body armor NIJ Standard-0101.06. 2000.
[28] Croft J, Longhurst D. HOSDB body armour standards for UK police (2007) Part 2: ballistic resistance. St Albans (Herfordshire): Sandridge; 2007.
[29] Zee RH, Hsieh CY. Energy loss partitioning during ballistic impact of polymer composites. Polym Compos 1993;14(3):265—71.
[30] Lee B, Song J, Ward J. Failure of spectra® polyethylene fiber-reinforced composites under ballistic impact loading. J Compos Mater 1994;28(13):1202—26.
[31] Shim V, Tan V, Tay T. Modelling deformation and damage characteristics of woven fabric under small projectile impact. Int J Impact Eng 1995;16(4):585—605.
[32] Sun C, Potti S. A simple model to predict residual velocities of thick composite laminates subjected to high velocity impact. Int J Impact Eng 1996;18(3):339—53.
[33] Starratt D, Sanders T, Cepuš E, Poursartip A, Vaziri R. An efficient method for continuous measurement of projectile motion in ballistic impact experiments. Int J Impact Eng 2000; 24(2):155—70.
[34] Grujicic M, Bell W, Arakere G, He T, Xie X, Cheeseman B. Development of a meso-scale material model for ballistic fabric and its use in flexible-armor protection systems. J Mater Eng Perform 2010;19(1):22—39.
[35] Zhou Y. Development of lightweight soft body armour for ballistic protection. Manchester: The University of Manchester; 2013.
[36] Duan Y, Keefe M, Bogetti T, Powers B. Finite element modeling of transverse impact on a ballistic fabric. Int J Mech Sci 2006;48(1):33—43.
[37] Karahan M, Kuş A, Eren R. An investigation into ballistic performance and energy absorption capabilities of woven aramid fabrics. Int J Impact Eng 2008;35(6):499—510.
[38] Tam T, Bhatnagar A. Chapter 7 High-performance ballistic fibers. In: Bhatnagar A, editor. Lightweight ballistic composites. Abington Hall, Abington, Cambridge CBI 6AH (England): Woodhead Publishing Limited; 2006. p. 189—209.
[39] Pandya KS, Pothnis JR, Ravikumar G, Naik N. Ballistic impact behavior of hybrid composites. Mater Des 2013;44:128—35.
[40] Reyes Villanueva G, Cantwell W. The high velocity impact response of composite and FML-reinforced sandwich structures. Compos Sci Technol 2004;64(1):35—54.
[41] Department of Defense USoA V_{50} Ballistic Test for Armour.
[42] Novotny W, Cepuš E, Shahkarami A, Vaziri R, Poursartip A. Numerical investigation of the ballistic efficiency of multi-ply fabric armours during the early stages of impact. Int J Impact Eng 2007;34(1):71—88.
[43] Zeng X, Shim V, Tan V. Influence of boundary conditions on the ballistic performance of high-strength fabric targets. Int J Impact Eng 2005;32(1):631—42.
[44] Shockey DA, Erlich DC, Simons JW. Lightweight fragment barriers for commercial aircraft. In: Proceedings of the 18th international symposium on ballistics, San Antonio, Texas; 1999.
[45] Shockey DA, Erlich DC, Simons JW. Improved barriers to turbine engine fragments: interim report III. Menlo Park (CA): SRI International; 2001.
[46] Talebi H, Wong S, Hamouda A. Finite element evaluation of projectile nose angle effects in ballistic perforation of high strength fabric. Compos Struct 2009;87(4):314—20.

[47] Nilakantan G, Keefe M, Wetzel ED, Bogetti TA, Gillespie Jr JW. Computational modeling of the probabilistic impact response of flexible fabrics. Compos Struct 2011;93(12):3163–74.
[48] Nilakantan G, Keefe M, Wetzel ED, Bogetti TA, Gillespie Jr JW. Effect of statistical yarn tensile strength on the probabilistic impact response of woven fabrics. Compos Sci Technol 2012;72(2):320–9.
[49] Nilakantan G, Wetzel ED, Bogetti TA, Gillespie Jr JW. Finite element analysis of projectile size and shape effects on the probabilistic penetration response of high strength fabrics. Compos Struct 2012;94(5):1846–54.
[50] Rao M, Duan Y, Keefe M, Powers B, Bogetti T. Modeling the effects of yarn material properties and friction on the ballistic impact of a plain-weave fabric. Compos Struct 2009; 89(4):556–66.
[51] Grujicic M, Arakere G, He T, Bell WC, Cheeseman BA, Yen CF, et al. A ballistic material model for cross-plied unidirectional ultra-high molecular-weight polyethylene fiber-reinforced armor-grade composites. Mater Sci Eng A 2008;498:231–41.
[52] Li S, Zhou C, Yu H, Li L. Formulation of a unit cell of a reduced size for plain weave textile composites. Comput Mater Sci 2011;50:1770–80.
[53] Kawabata S, Niwa M, Kawai H. 4-The finite-deformation theory of plain-weave fabrics. Part II: the uniaxial-deformation theory. J Text Inst 1973;64(2):47–61.
[54] Kawabata S, Niwa M, Kawai H. 5-The finite-defrmation theory of plain-weave fabrics. part III shear-deformation theory. J Text Inst 1973;64(2):62–85.
[55] Tan P, Tong L, Steven GP. A three-dimensional modelling technique for predicting the linear elastic property of opened-packing woven fabric unit cells. Compos Struct 1997; 38(1–4):261–71.
[56] User documentation. Abaqus version 6.7. Dassault Systems; 2007.
[57] Grujicic M, Bell W, Arakere G, He T, Cheeseman B. A meso-scale unit-cell based material model for the single-ply flexible-fabric armor. Mater Des 2009;30(9):3690–704.
[58] Shahkarami A, Vaziri R. A continuum shell finite element model for impact simulation of woven fabrics. Int J Impact Eng 2007;34(1):104–19.
[59] Fd L. Dynamic response of a constrained fibrous system subjected to transverse imapct. Part 2. Watertown (MA): a mechanical model Army materials and mechanicals research centre; 1970.
[60] Roylance D. Wave propagation in a viscoelastic fiber subjected to transverse impact. J Appl Mech 1973;40(1):143–8.
[61] Cunniff PM. An analysis of the system effects in woven fabrics under ballistic impact. Text Res J 1992;62(9):495–509.
[62] Cheeseman BA, Bogetti TA. Ballistic impact into fabric and compliant composite laminates. Compos Struct 2003;61(1):161–73.
[63] Stewart G, Patterson D, Hamburger T. Dynamics of body-armor materials under high speed impact, Part II. single and triple microflash instrumentation for single yarn studies. 1957. Army Chemical Center Report CWLR 2142.
[64] Petterson D, Stewart G, Odell F, Maheux R. Dynamic distribution of strain in textile materials under high-speed impact part I: experimental methods and preliminary results on single yarns. Text Res J 1960;30(6):411–21.
[65] Freeston WD, Claus WD. Strain-wave reflections during ballistic impact of fabric panels. Text Res J 1973;43(6):348–51.
[66] Roylance D, Wilde A, Tocci G. Ballistic impact of textile structures. Text Res J 1973; 43(1):34–41.

[67] Chen X. Ballistic protection. In: Ccott RA, editor. Textiles for Protection. Abington Hall, Abington: Cambridge CBI 6AH: Woodhead Pbulishing limited in association with The Textile Institute Woodhead Publishing Limited; 2005. p. 529–56.

[68] Lee BL, Walsh TF, Won ST, Patts HM, Song JW, Mayer AH. Penetration failure mechanisms of armor-grade fiber composites under impact. J Compos Mater 2001;35(18):1605–33.

[69] Tabiei A, Nilakantan G. Ballistic impact of dry woven fabric composites: a review. Appl Mech Rev 2008;61(1):1–12.

[70] Bilisik K, Korkmaz M. Multilayered and multidirectionally-stitched aramid woven fabric structures: experimental characterization of ballistic performance by considering the yarn pull-out test. Text Res J 2010;80(16):1697–720.

[71] Joo K, Kang TJ. Numerical analysis of energy absorption mechanism in multi-ply fabric impacts. Text Res J 2008;78(7):561–76.

[72] Cunniff PM. Decoupled response of textile body armor. In: Proceedings of the 18th international symposium of ballistics. Boca Raton (FL): CRC; 1999.

[73] Chen X, Zhou Y, Wells G. Numerical and experimental investigations into ballistic performance of hybrid fabric panels. Compos Part B Eng 2014;58:35–42.

[74] Gogineni S, Gao X-L, David N, Zheng J. Ballistic impact of Twaron CT709® plain weave fabrics. Mech Adv Mater Struct 2012;19(6):441–52.

[75] Chen X, Taylor LW, Tsai LJ. An overview on fabrication of three-dimensional woven textile preforms for composites. Text Res J 2011;81(9):932–44.

[76] Ha-Minh C, Boussu T, Kanit T, Crépin D, Imad A. Effect of frictions on the ballistic performance of a 3D warp interlock fabric: numerical analysis. Appl Compos Mater 2012; 19(3–4):333–47.

[77] Han Z. Production and tensile properties of 3D woven structures as composite reinforcements. In: Institute of science and technology. Manchester: Department of Textiles; 1999.

[78] Chen X, Yang D. Use of 3D angle-interlock woven fabric for seamless female body armor: Part 1: ballistic evaluation. Text Res J 2010;80(15):1581–8.

[79] Zahid B, Chen X. Manufacturing of single-piece textile reinforced riot helmet shell from vacuum bagging. J Compos Mater 2012;47(19):2343–51.

[80] Zahid B, Chen X. Impact performance of single-piece continuously textile reinforced riot helmet shells. J Compos Mater 2013;48(6):761–6.

[81] Rocdel C, Chen X. Innovation and analysis of police riot helmets with continuous textile reinforcement for improved protection. In: Computational engineering in systems application; 2006. Beijing (China).

[82] Ganesh VK, Naik NK. Failure behavior of plain weave fabric laminates under in-plane shear loading: effect of fabric geometry. Compos Struct 1995;30:179–92.

[83] Bland PW, Dear JP. Observations on the impact behaviour of carbon–fibre reinforced polymers for the qualitative validation of models. Compos Part A Appl Sci Manuf 2001; 32:1217–27.

[84] Sun D, Chen X. Plasma modification of Kevlar fabrics for ballistic applications. Text Res J 2012;82(18):1928–34. http://dx.doi.org/10.1177/0040517512450765.

[85] Chu Y, Chen X, Sheel DW, Hodgkinson JL. Surface modification of aramid fibers by atmospheric pressure plasma-enhanced vapor deposition. Text Res J 2014;84(12):1288–97.

[86] Chu Y, Chen X, Wang Q, Cui S. An investigation on sol–gel treatment to aramid yarn to increase inter-yarn friction. Appl Surf Sci 2014;320:710–7.

[87] Sun D, Chen X, Wells G. Engineering and analysis of gripping fabrics for improved ballistic performance. J Compos Mater 2014;48(11):1355–64.

[88] Chen X, Chu Y, Wang N. Study on inter-yarn friction and engineering of ballistic fabrics. In: Paper presented at the 14th AUTEX world textile conference, Bursa, Turkey; 2014.

[89] Zhou Y, Chen X, Wells G. Influence of yarn gripping on the ballistic performance of woven fabrics from ultra-high molecular weight polyethylene fibre. Compos Part B Eng 2014;62:198–204.

[90] Wang Y, Chen X, Young R, Kinloch I, Wells G. A numerical study of ply orientation on ballistic impact resistance of multi-ply fabric panels. Compos Part B Eng 2015;68:259–65.

[91] Hearle JWS, Leech CM, Adeyefa A, Cork CR. Ballistic impact resistance of multi-layer textile fabrics. Manchester: University of Manchester Institute of Science and Technology; 1981.

[92] Chen X, Sun D, Wang Y, Zhou Y. 2D/3D woven fabrics for ballistic protection. In: Proceedings to 4th world conference on 3D fabrics and their applications, Aachen; 2012.

[93] Thomas GA. Non-woven fabrics for military applications. In: Wilusz E, editor. Military textiles. Cambridge CB21 6AH, England: Woodhead Publishing Limited in Associated with The Textile Institute Great Abington; 2008. p. 17–49.

[94] Karahan M, Eren R, Karahan N. Investigation of ballistic properties of woven-unidirectional hybrid panels according to energy absorption capabilities. In: Paper presented at the SAMPE'07, Baltimore; 2007.

[95] Sun D, Chen X, Mrango M. Investigating ballistic impact on fabric targets with gripping yarns. Fibers Polym 2013;14(7):1184–9.

[96] Morrison CE. The mechanical response of an aramid textile yarn to ballistic impact. Doctoral thesis. University of Surrey; 1984.

Narrow fabrics for enhanced ballistic performance

10

C.R. Cork
Nottingham Trent University, Nottingham, United Kingdom

10.1 Introduction

There is increasing evidence that, in many cases, the boundary conditions are of importance when fabric assemblies undergo ballistic impact. It has been known for over half a century that boundary conditions are of importance for strain reflection when yarns were impacted. However, the effects of the boundary when fabrics are impacted have often been ignored by researchers. Impact velocity, fabric dimensions, and other factors determine the importance of boundary conditions in each case. Typically, boundary conditions have been shown to have an effect at impact velocities of up to 500 m/s where fabric widths are of the order of 30 cm or below. The boundary conditions can be defined by the size of fabric under impact, the nature of the fabric edge and the type of clamping, if present. There is evidence to suggest that woven narrow fabrics clamped on two sides rather than four provide enhanced ballistic performance and that, when clamped along the short edge, the fabric width and weave structure are important factors that also affect performance.

Narrow fabrics can be defined as ribbon-like fabrics where the fabric edge plays a part in their function. Examples of these in the field of technical textiles are pulleys, safety belts, and parachute webbing. Usually, but not always, woven narrow fabrics include a selvedge with a continuous weft. The selvedge is the edge of a fabric parallel to the warp direction as it comes off the loom. If the fabric is woven from a shuttle loom then the weft yarns are continuous and reverse at the selvedge. This is the arrangement used for most, but not necessarily all, narrow fabrics. In contrast, most modern fabrics are woven using looms where the weft yarns are not continuous and are trimmed off at the selvedge. The use of a continuous-weft selvedge prevents fraying and has been shown to improve ballistic performance.

10.2 Ballistic armor

Ballistic armor can be made from any material as it is always possible to include sufficient mass to stop a projectile, whether it be a bullet or a piece of shrapnel. For example, a vest could theoretically be made from lead plate or bird feathers, but in both cases the resultant vest would be too heavy. In contrast, multiple-layer textile fabrics can provide good ballistic resistance while retaining the flexibility and lightness which is characteristic of many textile structures.

Advanced Fibrous Composite Materials for Ballistic Protection. http://dx.doi.org/10.1016/B978-1-78242-461-1.00010-8

Textile body armor is used extensively by the military and police forces in addition to private security personnel, politicians, and hunters. Widespread use of ballistic armor based on textiles did not begin until the Korean War in the 1950s when it was realized that shrapnel was one of the major causes of wounds and fatalities and that significant protection could be offered by flak jackets composed of multilayered, high-tenacity nylon fabrics. In the 1970s, DuPont introduced the first commercial para-aramid, Kevlar, which offered far greater performance-to-weight ratios than nylon. Since then, advanced para-aramids have been tailored to produce still greater increases in performance. In addition, ultra-high molecular weight polyethylene (UHMWPE) is also extremely strong and stiff and competes with the para-aramids in ballistic applications.

Despite the progress that has been made over the years, armor can have serious consequences for the comfort and mobility of the wearer. Ballistic armor has become lighter, but compared to standard clothing it is still relatively heavy, bulky, and sometimes uncomfortable. Therefore, in most applications a compromise must be reached between the level of protection and the comfort and mobility of the wearer. In practice, reduced weight is of particular importance in military applications where impaired mobility can itself lead to greater casualties. Consideration must also be given to the possibility that the garment will not be worn on a particular occasion due to discomfort. The choice will depend on the level of the threat, the climate, and the operational requirements. Compare for instance the armor worn by an infantry soldier and a bomb disposal engineer. For the soldier, mobility over an extended period is important, whereas this is not the case when a bomb is to be made safe.

The shape of the garment is an important factor in minimizing risk. Lightweight garments are usually designed to protect the vital organs contained in the upper torso. Analyses of the anatomical location of wounds during conflict have shown that 45% of all hits were to this part of the body (Gofrit et al., 1996). To minimize the damage, garments can also include a soft armor collar to shield the neck and protect the spinal column and the vital blood supply to the head. As more protection is required, garments can be enlarged to enclose more of the body. Hard armor, such as ceramic plates which protect against high-velocity impacts, can be inserted into pockets to provide enhanced protection. Many manufacturers offer a range of concealable garments than can be worn under normal clothing. These are generally lighter and less bulky that conventional armor but offer less protection. As there are increasing numbers of women entering armies and police forces, there is a need to consider the mass and gender of the wearer (Bir and Wilhelm, 2004). It is seen as increasingly important to contour the garment to the body shape. It is possible that narrow fabrics can play a part in creating shaped garments, perhaps personalized for each individual.

As no armor can protect against all threats, a question arises as to whether the degree of injury can be exacerbated when a projectile perforates textile armor. There are indications that this is not the case and that even in these circumstances the level of injuries is reduced compared to the non-armor case (Lanthier et al., 2004). Where the armor is not penetrated, it is very important that excessive deformations to the back face of the fabric do not lead to serious injury. Indeed, it is known that severe tissue crushing behind armor can be caused by nonpenetrating bullets (Liden et al., 1988). Thus when considering narrow fabrics and narrow fabric assemblies, it is important that fabric deformations are minimized.

10.3 Importance of fiber type

The mechanical properties of the fiber are one of the most important factors as a fabric undergoes ballistic impact. High-strength, high-modulus materials offer the best performance-to-weight ratios. Weight-for-weight, high-tenacity textiles are stronger under tensile loading than many other strong materials, including steel. This is especially true of high-tenacity nylon, the para-aramids, polybenzobisoxazole (PBO), and UHMWPE. The strength of these fibers is due to their composition where long chain carbon polymers are aligned along the fiber length, with a high degree of molecular orientation and with high crystallinity. Although these textiles are strong and stiff under tensile loads, they are relatively weak under compression. Therefore, textiles can offer good ballistic properties when impacted at velocities where the material undergoes tensile loading but can perform relatively poorly at higher velocities where they can fail under transverse forces.

The performance of fibers can be significantly different for the quasistatic and ballistic cases. For example, although aramid fibers are stronger than nylon, the work of rupture of nylon under quasistatic loading is greater than that of aramid fibers. However, in the ballistic case the aramids are able to dissipate more energy due to their higher modulus.

Overall, the mechanics of high-speed impact are different from those at lower impact velocities. For example, at high speeds a fabric might fail instantaneously, and stress will not have time to propagate outward from the impact zone. At slower speeds, strain might travel outward but might not reach the fabric boundaries. At still slower speeds, strain might be reflected back and forth between the projectile and the fabric boundary many times. At quasistatic speeds, this repeated reflection leads to a more uniform distribution of strain across the fabric. For speeds below ballistic velocities, the quasistatic work of rupture can be an important factor in absorbing energy. At higher speeds the effects of fiber modulus take precedence.

10.4 Importance of fabric construction

Despite the importance of fiber type, the effects of fabric construction should not be underestimated. For example, Fig. 10.1 shows the projectile energy loss upon penetration as a function of the areal density of a multilayer sample for three fabrics, namely a woven aramid, a woven nylon, and a knitted aramid. The woven nylon outperforms the knitted aramid, although, of course, the woven aramid outperforms both.

Generally, woven nylon and para-aramid fabrics are constructed from flat, untwisted yarns with low crimp. Plain weaves, simple satins, and twills predominate. In contrast, UHMWPE is used in composite laminate form where unidirectional sheets of yarns are cross-plied and bonded with a thermoplastic matrix.

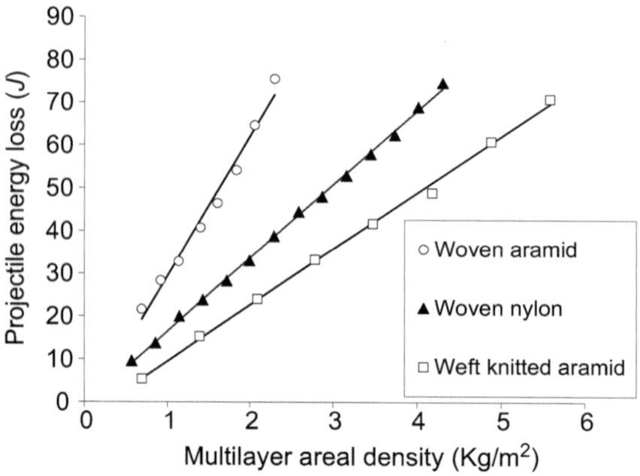

Figure 10.1 Effect of fabric structure.
Reproduced from Cork, C.R., Foster, P.W., 2007. The ballistic performance of narrow fabrics. International Journal of Impact Engineering 34 (3), 495–508.

A typical woven para-aramid fabric might consist of 1670 dtex yarn woven with a thread density of 6.7 threads per centimeter in both warp and weft. These weaves are surprisingly open in structure compared to everyday textile fabrics. For example, fabrics that can stop a bullet can be easily penetrated by the point of a pen.

It might be thought that weaving a tighter fabric would increase ballistic performance, but this is not the case; perhaps because crimp is increased, tightness at yarn intersects leads to greater strain reflections, or yarn mobility is constrained. An unfortunate effect of the openness of the weave is that fabrics can rarely offer both stab resistance and resistance to ballistic impact.

Finer weaves, such as 240 dtex yarns woven with 23 threads per centimeter exhibit better ballistic performances but carry a price premium, as finer yarns are more expensive to produce, and finer fabrics are more expensive to weave. More picks have to be inserted for any given length of fabric, and more layers of fabric have to be produced to bring the multilayer weight of the ballistic panel up to the required level.

Normally, ballistic fabrics perform better if symmetric in warp and weft to form square fabrics. However, more research is required to determine whether narrow fabric assemblies should be constructed from square weaves.

10.5 Ballistic testing

Ballistic testing is slow, laborious, and expensive, and the possible fabric and yarn parameters are infinite. In addition, parameters are rarely independent. For example, increasing the weft thread density will affect areal density, fabric squareness, and warp crimp. Therefore, large-scale analyses of a wide range of fibers, fabrics, and fabric boundary conditions are rarely possible. Thus, the results from the literature can appear to be contradictory mainly due to the almost infinite choice of parameters available for both simulations and empirical studies. Since 2005, however, there has been a rising consensus in the literature that boundary conditions are of importance and worthy of study.

There are a number of ballistic test methods for textile armor. The first test is the V_{50} and its variants such as the V_0 and V_{100} tests (US Department of Defense, 1997). Here, a multilayer assembly is impacted at increasing velocities with a number of firings at each velocity. As the velocity increases the number of cases where the projectile penetrates the assembly will also increase. The V_{50} is the velocity at which 50% of firings penetrate the assembly. In practice, the V_{50} is found from interpolation of the data. The V_0 is the highest velocity at which no projectiles penetrate the assembly, and the V_{100} is the lowest velocity at which the assembly is penetrated by all firings. The V_{50} is a useful research tool as the performance of an assembly can be quantified on a continuum scale. However, fabrics are normally clamped, and this must have some bearing on the ballistic performance. Although performance-to-weight ratios are of prime importance, the weight of the clamp is rarely considered when reporting results. Sometimes, this is legitimate where the ballistic panel is to be attached to a preexisting structure, but in the case of body armor the weight of the clamp must be considered.

The second, widely adopted test is the NIJ method for testing personal body armor (US National Institute of Justice, 2008). The test is designed as a pass/fail test for garments. The NIJ is practically useful in that it defines which threats a garment will protect against, but it is less useful in comparative research where it is not possible to distinguish between the performances of fabric assemblies that, for example, pass the test.

In ballistic testing, it is sometimes difficult to control the impact velocity. For the V_{50} test, cartridges have to be made up individually. An alternative approach is to use a fixed impact velocity, vary the number of layers, and observe when penetration occurs. These tests can be conducted against clay and the depth of crater noted for cases where penetration does not occur. Alternatively, for fixed-impact velocities the energy loss when penetrated can be correlated with the multilayer areal density.

10.6 High-speed photography

In research, it is possible to gain a better understanding of the impact process using high-speed photography. A range of high-speed video cameras are available but tend to be relatively expensive. A simple method is to use a single-shot technique to capture the impact. The procedure is to fire at the target in a darkened room. A single flash of light from a source with a suitable short duration in the microsecond range is triggered by the passage of the projectile through a laser beam. A slight delay is introduced to allow the projectile to progress to the target, and the image is captured using a digital camera with the shutter open. Most modern digital cameras with an open-shutter setting are suitable for this work. This technique was used for the pictures of narrow fabric impact in this chapter.

10.7 Effect of boundary conditions on transverse yarn impact

The key role that boundary conditions can play was demonstrated by the earliest work on the transverse impact of yarns (Smith et al., 1956; Roylance et al., 1973). Under impact a fiber or yarn develops a transverse deformation that progresses outward at a fixed velocity, U. In addition, a longitudinal strain wave propagates outward at the velocity of sound in the material (c_0) which is a function of the tensile modulus at zero strain (E_0) and the material density (ρ):

$$c_0 = \sqrt{\frac{E_0}{\rho}} \qquad [10.1]$$

The tensile modulus is therefore important as it determines how quickly strain is dissipated away from the point of impact. If the modulus is too low, strain is localized at the point of impact, and failure occurs more quickly. Most textiles are rate dependent so that E must be specified as the dynamic modulus, that is, the modulus at high rates of strain.

As the impact process proceeds, additional increments of strain generated at the point of impact propagate outward through an already strained yarn at a velocity (c_ε) determined by the modulus of the material at the preexisting strain level. c_ε is given by the generalized form of Eq. [10.1]:

$$c_\varepsilon = \sqrt{\frac{1}{\rho}\frac{d\sigma}{d\varepsilon}}$$

where $\frac{d\sigma}{d\varepsilon}$ is the slope of the dynamic stress—strain curve at a given strain level. Thus the moduli at all strain levels have an effect on the ballistic performance.

The propagation velocity of the transverse deformation (U) is also a function of the material moduli, but the relation is more complex than for the longitudinal wave.

At high velocities the boundary conditions play no part, as failure occurs before the strain wave reaches the boundary, and strain development at the point of impact is the same as for a fiber of infinite length. However, at lower velocities the boundary conditions are of importance as both the longitudinal and transverse deformations are reflected at the boundary. These return waves are also reflected at the point of impact (Smith et al., 1956), so as the impact process continues, strain levels increase. Furthermore, if the strain wave is reflected at a boundary, it will travel back and interact with the transverse wave-front, causing it to increase in velocity (Smith et al., 1960).

10.8 Effect of boundary conditions on fabric impact

On impact, a fabric undergoes a localized transverse displacement in the direction of flight of the projectile and conforms to the shape of the projectile. As it propagates outward the transverse deformation takes on a shape that is dependent on the fabric construction, number of layers, and boundary conditions. For example, it is pyramidal for a single-layer, biaxially woven fabric fitted to a circular clamp. This is illustrated in Fig. 10.2 where the reverse of a single-layer aramid fabric is shown undergoing ballistic impact and where the distinct pyramidal deformation can be seen with the in-plane distortions proceeding further afield.

As in the case of the impact of a single fiber or yarn, the modulus of the constituent fibers is important as it affects stress dissipation. In the case of fabrics, both the transverse and longitudinal propagation velocities are influenced by the macroscopic structure and boundary conditions. For example, strain can be reflected at yarn-to-yarn intersections, and rigid boundaries can lead to strain reflection.

The modulus of a fabric is also important as stiffer fabrics deform less and deformations to the back of the armor must be minimized to reduce tissue injury on impact, so-called blunt trauma. It would be of no use to stop a projectile if it were to take a meter to do so. However, some flexibility is required so that the fibers bend and undergo tensile loading and are not subject to brittle failure on impact.

Due to the high complexity of the projectile–fabric interaction, precise prediction of fabric performance from fiber properties is difficult to achieve. The problem is made more complicated as, in addition to the dynamic moduli, other properties, such as the coefficient of friction at high strain rates, are often unknown. In addition, the projectile, whether it be a bullet or shrapnel, can be relatively small compared to the thread

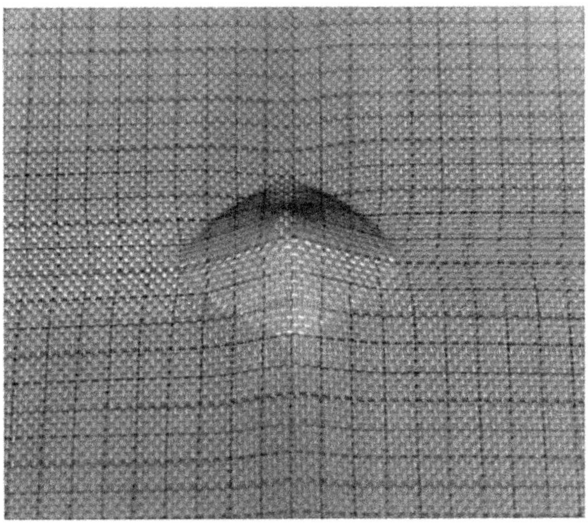

Figure 10.2 Reverse of a single-layer woven fabric undergoing impact.

spacing of a typical ballistic fabric, and projectiles do not generally penetrate fabrics by breaking all the yarns in their path. During the impact process, yarns might be pushed aside or only partially severed. For example, Fig. 10.3 shows the reverse of a penetrated fabric where the conditions were such that no yarns were broken. The number of yarns broken when a fabric is penetrated will depend on the impact velocity, the projectile size and shape, the fabric structure, and the boundary conditions. It has been noted that friction can be helpful in maintaining the integrity of the weave at the point of impact, with the result that more yarns are involved with the impact process (Duan et al., 2005a). Conversely, sharper projectiles result in less energy absorption (Tan et al., 2003), and ammunition manufacturers produce armor-piercing bullets that depend on this vulnerability.

In some cases the boundary conditions need not be considered as fabrics can be penetrated, or the projectile arrested, before the strain wave reaches the fabric edges. In other cases a strain wave will be dampened on encountering a fiber free end or reflected at a fixed boundary with the result that strain levels are doubled.

Boundary conditions can influence the degree of yarn pull and affect these processes. Yarn movement relative to the fabric as a whole can play an important role in reducing strain levels at the point of impact. It is reported that modest changes in friction have a major effect on ballistic performance (Briscoe and Motamedi, 1992), with fabrics with the highest yarn-to-yarn friction leading to higher energy dissipation. Conversely, loosely woven fabrics or fabrics with low yarn-to-yarn friction are reported to exhibit reduced ballistic performances. It is also suggested that low friction leads to poor weave integrity, whereas high friction reduces relative sliding between yarns, which results in high stress concentrations at the point of impact (Zeng et al., 2006). The dimensions of the pull-out zone, that is, the locus of fabric where yarn slippage occurs, have been correlated with ballistic performance (Bazhenov, 1997).

Figure 10.3 Fabric penetration without yarn failure.

Cepuš et al. (1999) confirmed that boundary conditions were often overlooked and concluded that they have significant effects on the energy-absorption mechanisms and ballistic performances. Zeng et al. (2005) reported the results of a numerical model of impact onto fabrics clamped on all edges and two edges with the yarns aligned with the edges. Empirical tests were also undertaken to validate the model. It was found that ballistic resistance was sensitive to both boundary conditions. Fabrics clamped on two edges absorbed more energy than fabrics clamped on four edges. The reason was ascribed to reflections of stresses at the clamped boundaries.

Duan et al. (2005a) used finite element analysis to investigate the effects of friction when fabrics with different boundary conditions are impacted. The simulations show that the boundary conditions significantly affect fabric deformation, stress distribution, and time history of energy absorption. Under the conditions investigated, the fabrics with four edges clamped performed better than those with two edges clamped. Duan et al. (2005b) used finite element code to simulate impact onto fabrics with four and two edges clamped. It was found that the fabric with two edges clamped was more effective at reducing the projectile residual velocity, even though the fabric with four edges clamped slowed down the projectile more quickly. This was because the time needed to perforate the fabric was much less for the case with four edges clamped. Duan et al. (2006) used a 3D finite element analysis model to simulate impact onto woven Kevlar fabrics with different boundary conditions. These were four edges free, two edges clamped, and four edges clamped. The authors conclude that at low-impact velocities, the boundary conditions play an important part in the impact process and affect the fabric deformation, stress distribution, energy absorption, and failure modes. However, when the impact velocity is sufficiently high, the yarns break instantaneously, and the boundary conditions play no part. In the model, the boundary conditions significantly affect the stress distributions and transverse deformations in an impacted fabric. The simulations show that the unclamped fabrics perform best, and the clamped fabric performs worst.

Zhang et al. (2008) used a numerical simulation of the ballistic impact of fabric armor to show that clamping on two sides, rather than four, produced superior ballistic performances. Nilakantan and Nutt (2014a) investigated the V_{50} performance of aramid fabrics of various sizes under a range of clamping conditions. The sample with two sides clamped showed the lowest V_{50}. Nilakantan et al. (2015) tested fabrics that were mounted in a novel way that left the principal yarns unclamped. The results indicated that yarn sliding and pull-out were the primary energy dissipating mechanisms for samples clamped in this manner. The results raise the possibility that these mechanisms might be relevant to impact onto narrow fabrics where essentially the principal yarns in the narrow direction are unclamped. Nilakantan and Nutt (2014b) concluded that the clamping specifications are just as important as the material and threat specifications. In addition, the authors note that the various mechanisms of fabric deformation and energy dissipation can be promoted or suppressed. Note that for yarn pull-out to be a significant factor for energy absorption, two edges must be clamped.

Dong and Sun (2009) studied yarn pull-out in plain Kevlar weaves and concluded that it has a positive effect on ballistic performance. Nilakantan and Gillespie (2012) looked at impact onto samples clamped on two and four sides. They indicate that yarn pull-out is the primary mechanism for energy dissipation for the case where two edges are clamped.

Overall boundary conditions and yarn pull are shown to influence ballistic performance with indications that asymmetric clamping is an important factor.

10.9 Impact of narrow fabrics

Different fabric structures perform differently at different impact velocities, and this is likely to be particularly true for narrow fabrics. Intuitively, it would be expected that fabric width would have an effect on ballistic performance. For a fabric of infinite width the selvedges would play no part in arresting the projectile. Conversely, if the width were very small, then the principal yarns could be easily pulled from the structure. In addition, the tightness of the weave is bound to have an effect as, for the more open weave fabrics, yarns can be more easily removed.

The shape of the out-of-plane deformations of fabrics can be influenced by the boundary conditions. For example, Fig. 10.4 shows the face of a woven narrow fabric under ballistic impact. Initially the deformation takes the circular shape of the projectile. However, as impact progresses the transverse deformation takes on a peanut shape (Fig. 10.5) rather than a pyramid. In this case the boundaries of the fabric, which has a continuous weft, are pulled inward.

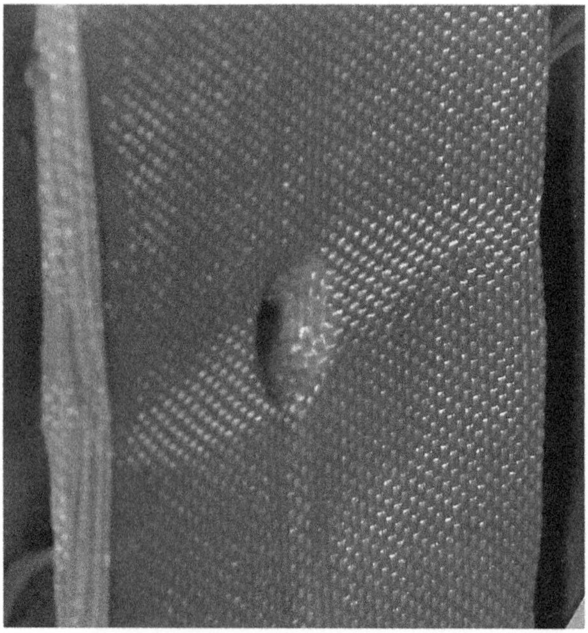

Figure 10.4 Face of a narrow fabric at the point of initial impact.

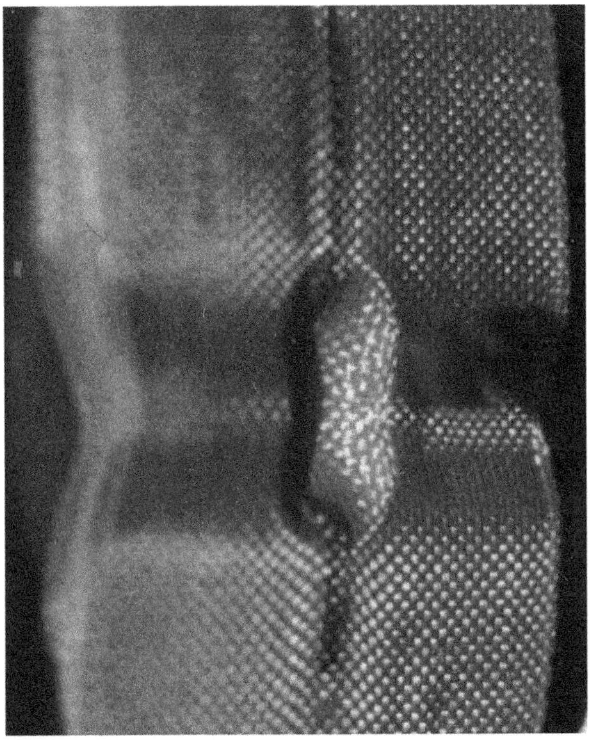

Figure 10.5 Peanut-shaped deformation at a later stage during impact.
Reproduced from Cork, C.R., Foster, P.W., 2007. The ballistic performance of narrow fabrics. International Journal of Impact Engineering 34 (3), 495–508.

The effects of a continuous-weft selvedge can also be seen in the residual damage shown in Fig. 10.6 where the principal yarns that intersect the impact point have nipped in the selvedge. It seems likely that selvedges can play a part in preventing removal of yarns when narrow fabrics are impacted.

Shockey et al. (1999) looked at the design of fragment barriers for commercial aircraft. Fabrics gripped on two sides were found to perform better than fabrics gripped on four sides by margins of between 25% and 60%. Shockey et al. (2002) also demonstrated the effectiveness of holding fabrics at the corners. Tan et al. (2005) compared fabric deformations of fabric strips under impact from both empirical tests and computer modeling, and they demonstrated good correspondence that showed more extensive transverse deformations toward the clamped edges, as in Fig. 10.5.

A patent (Dischler, 1995, 1996) suggests improving ballistic performance by using narrow fabric strips. The mechanism suggested is the control of peak stresses during impact. The authors propose implementing the strips in practicable structures by setting them side by side or by winding or weaving strips to form an assembly. Another patent (Hartert and Bottger, 2012) suggests the use of fabric with different constructions in different areas. Perhaps a similar approach could be used to produce

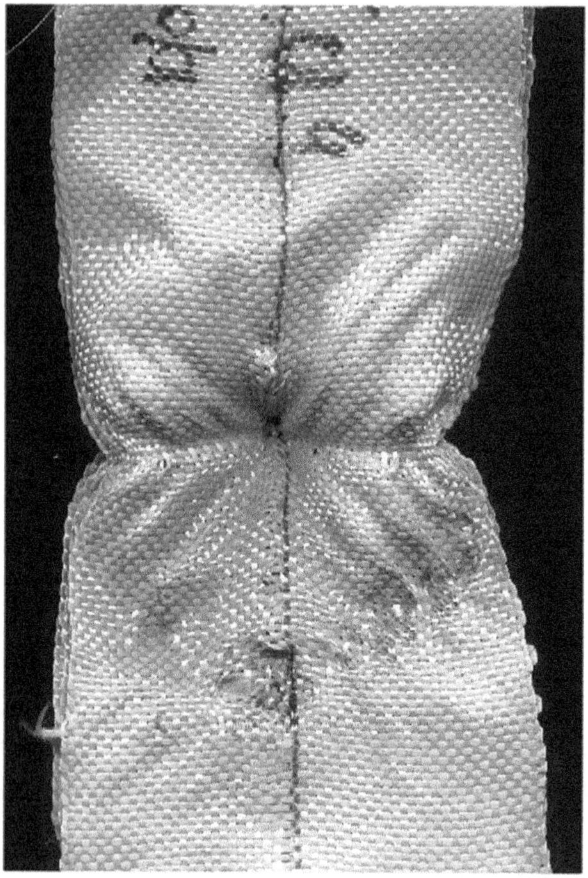

Figure 10.6 Residual damage to a narrow fabric.

pseudo-narrow fabrics in a continuous fabric sheet with high crimp in strips forming the boundaries between strips and allowing the necessary yarn pull.

10.10 Effect of clamping on the ballistic performance of narrow fabrics

It has been suggested that ballistic performance can be enhanced by the application of yarn tension (Chitrangad, 1994). Rabb (2007) has tested Kevlar fabrics cut into strips of dimensions 5.08 cm × 25.4 cm. The short edge was clamped but the long edge was unconstrained. It was observed that yarn pull-out is most prominent in the direction where the yarn length is short and the edges are unrestricted. At low velocities the fabric deformations reached the fabric boundary before failure. Rabb's results confirm that the boundary conditions affect ballistic performance. Clamping on two edges resulted in better ballistic performances than clamping on all edges. It was established that boundary conditions have more effect at lower velocities. Rabb also concluded that boundary conditions have more influence at low-impact velocities because more of the fabric is deformed before the projectile penetrates.

Sun et al. (2013) looked at the effect of sample grip on ballistic performance of narrow fabrics. They tested narrow fabric assemblies with widths of 6, 8, 12, and 24 cm. The narrow fabrics with smaller widths had better ballistic performances than wider fabrics. Narrow fabrics were made by cutting and folding fabrics. Empirical data showed that the folded fabrics performed best. However, the results from wide panels made from the narrow fabrics did not improve on standard ballistic panels.

A case where the effect of clamping of narrow fabrics has a positive effect on ballistic performance can be seen in Fig. 10.7 where low values indicate superior performance-to-weight ratios. The clamped samples shown in Fig. 10.7 were mounted on a substantial aluminum clamp. If the weight of the clamp had been taken into account, then the performance-to-weight ratio of the clamped samples would have been very poor. Therefore, Cork and Foster (2007) investigated the effects of simply gluing the narrow fabrics to a lightweight polystyrene frame and found that the level of clamping was sufficient to produce a good performance-to-weight ratio. Shockey et al. (2001) also note that "gluing exploits the high fiber strength of the fabric about as much as tight gripping" when referring to attaching Zylon (PBO) to an existing aircraft structure for fragment protection.

The performance of narrow fabrics is highly dependent on fabric structure. Fig. 10.8 shows the multilayered areal density required to prevent penetration, and again low values indicate better performance-to-weight ratios. The results are similar at a fabric width of 24 cm but diverge significantly as the narrow fabric widths are

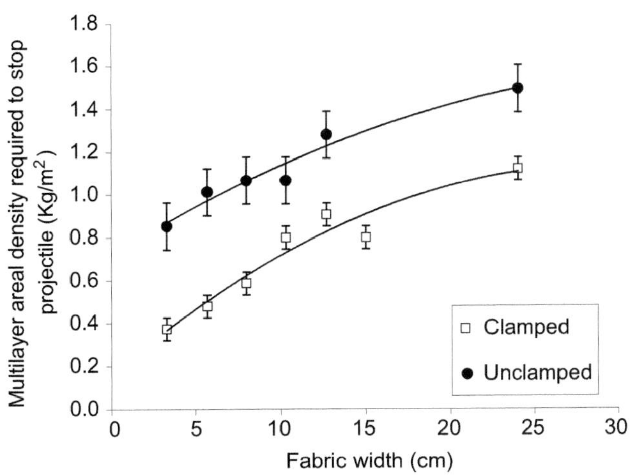

Figure 10.7 Effect of clamping on narrow fabric performance.
Reproduced from Cork, C.R., Foster, P.W., 2007. The ballistic performance of narrow fabrics. International Journal of Impact Engineering 34 (3), 495–508.

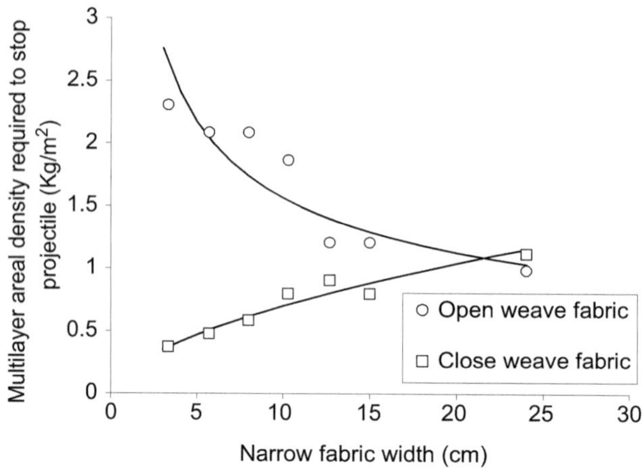

Figure 10.8 Effect of fabric structure on narrow fabric performance.
Reproduced from Cork, C.R., Foster, P.W., 2007. The ballistic performance of narrow fabrics. International Journal of Impact Engineering 34 (3), 495–508.

reduced. The open weave fabric requires increasingly more layers, while the close weave fabric requires significantly fewer layers at lower widths. This is probably because, for the open weave fabric, the principal yarns were more easily removed by impact.

10.11 The design of practicable armor using narrow fabrics

No matter what the performance of a narrow fabric is, it cannot be used alone to offer protection over a large area. Therefore, narrow fabric assemblies must be arranged in an assembly to offer protection. One option would be to overlap narrow fabrics in each layer, but a better solution would be to butt joint strips and offset the joints in subsequent layers. However, a compromise has to be made on the performance of the narrow fabric and that of the ballistic panel. For example, the narrowest fabrics might offer the best ballistic performances when impacted directly, but they might be not as good when made into a panel with numerous butt joints. Conversely, slightly wider narrow fabrics might offer marginally lower ballistic performances but might perform better in a panel due to fewer joints.

Cork and Foster (2007) found that suitable narrow fabrics arranged in this way and adhered to a polystyrene holder could confer weight savings of between 6% and 31%. In some cases, where narrow fabric assemblies are attached to a preexisting structure, the corresponding weight savings would be between 24% and 63%. This demonstrated the potential for the technology, but there is a need to test a far greater range of narrow fabric and panel parameters.

10.12 Conclusions

Although the study of the transverse impact of yarn has identified the crucial effects of boundary conditions for moderate impact velocities, the effects of boundary conditions on the ballistic impact of fabrics has only been studied in recent years. The evidence is contradictory in places, and this is to be expected given the infinite number of variables. However, the effects of boundary conditions for moderate impact velocities above the critical velocity are undeniable. Fiber type, fabric dimensions, the type of clamping, the fabric selvedge, yarn pull, and fabric structure have all been shown to contribute to ballistic performance.

Improvements to ballistic performance using woven narrow fabrics have been demonstrated, and practicable narrow fabric panels have shown their potential for significant improvements in performance-to-weight ratios. Where existing structures exist, such as in aircraft, narrow fabric assemblies have the potential to provide still greater improvements in ballistic performances.

10.13 Future trends

The literature suggests a whole new area for ballistic research where perceived wisdoms might not apply. For example, are square fabrics still best when narrow fabrics are clamped on the short sides? Is it best to maximize thread densities in the long direction to resist yarn pull, or is it best to minimize thread densities to minimize crimp in the long direction? What is the ideal length of the narrow fabrics? How firmly do the edges have to be gripped, and what is the minimum weight and stiffness of the gripping mechanism? These factors and many more have yet to be established.

Sources of further information and advice

Recommended books on the ballistic impact of textiles are by Lyons (1963), Laible (1980), and Wilusz (2008). Further information can also be found in review papers by Bajaj and Sriram (1997), Cheeseman and Bogetti (2003), and Tabiei and Nilakantan (2008). In addition, an NIJ guide to personal body armor is available online (US National Institute of Justice, 2001) as are the documentation for the NIJ (US National Institute of Justice, 2008) and the V_{50} (US Department of Defense, 1997) tests.

References

Bajaj, P., Sriram, 1997. Ballistic protective clothing-an overview. Indian Journal of Fibre and Textile Research 22, 274–291.

Bazhenov, S., 1997. Dissipation of energy by bulletproof aramid fabric. Journal of Materials Science 32 (15), 4167–4173.

Bir, A.B., Wilhelm, M., 2004. Female armor assessment: current methods and future techniques. In: Proceedings of the Personal Armour System Symposium 2004, The Hague, The Netherlands.

Briscoe, B.J., Motamedi, F., 1992. The ballistic impact characteristics of aramid fabrics: the influence of interface friction. Wear 158 (1), 229–247.

Cepuš, E., Shahkarami, A., Vaziri, R., Poursartip, A., 1999. Effect of boundary conditions on the ballistic response of textile structures. In: 12th International Conference on Composite Materials, pp. 5–9.

Cheeseman, B.A., Bogetti, T.A., 2003. Ballistic impact into fabric and compliant composite laminates. Composite Structures 61 (1–2), 161–173.

Chitrangad, 1994. Ballistic Structure. U.S. Patent 5,275,873.

Cork, C.R., Foster, P.W., 2007. The ballistic performance of narrow fabrics. International Journal of Impact Engineering 34 (3), 495–508.

Dischler, L., 1995. U.S. Patent No. 5,466,503. U.S. Patent and Trademark Office, Washington, DC.

Dischler, L., 1996. U.S. Patent No. 5,589,254. U.S. Patent and Trademark Office, Washington, DC.

Dong, Z., Sun, C.T., 2009. Testing and modeling of yarn pull-out in plain woven Kevlar fabrics. Composites Part A: Applied Science and Manufacturing 40 (12), 1863–1869.

Duan, T., Keefe, M., Bogetti, T.A., Cheeseman, B.A., 2005a. Modeling friction effects on the ballistic impact behavior of a single-ply high-strength fabric. International Journal of Impact Engineering 31, 996–1012.

Duan, Y., Keefe, M., Bogetti, T.A., Cheeseman, B.A., 2005b. Modeling the role of friction during ballistic impact of a high-strength plain-weave fabric. Composite Structures 68 (3), 331–337.

Duan, Y., Keefe, M., Bogetti, T.A., Powers, B., 2006. Finite element modeling of transverse impact on a ballistic fabric. International Journal of Mechanical Sciences 48 (1), 33–43.

Gofrit, O.N., Kovalski, N., Leibovici, D., Shemer, J., O'Hana, A., Shapira, S.C., 1996. Accurate anatomical location of war injuries: analysis of the Lebanon war fatal casualties and the proposition of new principles for the design of military personal armor system. Injury-International Journal of the Care of the Injured 27 (8), 577–581.

Hartert, R., Bottger, C., 2012. U.S. Patent No. 8,293,665. U.S. Patent and Trademark Office, Washington, DC.

Laible, R.C. (Ed.), 1980. Ballistic Materials and Penetration Mechanics. Elsevier, Oxford.

Lanthier, J.M., Iremonger, M.J., Lewis, E.A., Horsfall, I., Gotts, P.L., 2004. Is the wounding potential of high velocity military bullets increased after perforation of textile body amour. In: Proceedings of the Personal Armour System Symposium 2004, The Hague, The Netherlands.

Liden, E., Berlin, R., Janzon, B., Schantz, B., Seeman, T., January 1988. Some observations relating to behind-body armor blunt trauma effects caused by ballistic impact. Journal of Trauma-Injury Infection and Critical Care 28 (Suppl. 1), S145–S148.

Lyons, W.J., 1963. Impact Phenomena in Textiles. MIT Press, Cambridge, Massachusetts.

Nilakantan, G., Gillespie Jr., J.W., 2012. Ballistic impact modeling of woven fabrics considering yarn strength, friction, projectile impact location, and fabric boundary condition effects. Composite Structures 94 (12), 3624–3634.

Nilakantan, G., Nutt, S., 2014a. Effects of fabric target size, shape, and clamping on the V_{50} ballistic impact performance of aramid soft body armor. In: SAMPE Conference Proceedings. Seattle, WA, June 2–5, 2014. Society for the Advancement of Material and Process Engineering.

Nilakantan, G., Nutt, S., 2014b. Effects of clamping design on the ballistic impact response of soft body armor. Composite Structures 108, 137–150.
Nilakantan, G., Merrill, R.L., Keefe, M., Gillespie, J.W., Wetzel, E.D., 2015. Experimental investigation of the role of frictional yarn pull-out and windowing on the probabilistic impact response of Kevlar fabrics. Composites Part B: Engineering 68, 215–229.
Rabb, R.J., 2007. A Mesomechanical Particle-Element Model of Impact Dynamics in Neat and Shear Thickening Fluid Kevlar. The University of Texas at Austin (Ph.D. thesis).
Roylance, D., Wilde, A., Tocci, G., 1973. Ballistic impact of textile structures. Textile Research Journal 43 (1), 34–41.
Shockey, D.A., Erlich, D.C., Simons, J.W., 1999. Lightweight fragment barriers for commercial aircraft. In: Proceedings of the 18th International Symposium on Ballistics, San Antonio, Texas.
Shockey, D.A., Erlich, D.C., Simons, J.W., 2001. Improved Barriers to Turbine Engine Fragments: Interim Report III. U.S. Department of Transportation.
Shockey, D.A., Erlich, D.C., Simons, J.W., 2002. Improved Barriers to Turbine Engine Fragments: Interim Report IV. U.S. Department of Transportation.
Smith, J.C., McCrackin, F.L., Schiefer, H.F., Stone, W.K., Towne, K.M., 1956. Stress-strain relationships in yarns subjected to rapid impact loading part IV: transverse impact tests. Textile Research Journal 26 (11), 821–828.
Smith, J.C., Blandford, J.M., Schiefer, H.F., 1960. Stress-strain relationships in yarns subjected to rapid impact loading part VI: velocities of strain waves resulting from impact. Textile Research Journal 30 (10), 752–760.
Sun, D., Chen, X., Wells, G., 2013. Engineering and analysis of gripping fabrics for improved ballistic performance. Journal of Composite Materials 48 (11), 1355–1364.
Tabiei, A., Nilakantan, G., 2008. Ballistic impact of dry woven fabric composites: a review. Applied Mechanics Reviews 61 (1), 010801.
Tan, V.B.C., Lim, C.T., Cheong, C.H., 2003. Perforation of high-strength fabric by projectiles of different geometry. International Journal of Impact Engineering 28 (2), 207–222.
Tan, V.B.C., Shim, V.P.W., Zeng, X., 2005. Modelling crimp in woven fabrics subjected to ballistic impact. International Journal of Impact Engineering 32 (1), 561–574.
US Department of Defense, 1997. V_{50} Ballistic Test for Armor, MIL-STD-662F.
US National Institute of Justice, 2008. Ballistic Resistance of Personal Body Armor, NIJ Standard-0101.06.
US National Institute of Justice, 2001. Selection and application guide to personal body armor. National Institute of Justice Guide 100–101.
Wilusz, E. (Ed.), 2008. Military Textiles. Elsevier.
Zhang, G.M., Batra, R.C., Zheng, J., 2008. Effect of frame size, frame type, and clamping pressure on the ballistic performance of soft body armor. Composites Part B: Engineering 39 (3), 476–489.
Zeng, X.S., Shim, V.P.W., Tan, V.B.C., 2005. Influence of boundary conditions on the ballistic performance of high-strength fabric targets. International Journal of Impact Engineering 32 (1), 631–642.
Zeng, X.S., Tan, V.B.C., Shim, V.P.W., 2006. Modelling inter-yarn friction in woven fabric armour. International Journal for Numerical Methods in Engineering 66 (8), 1309–1330.

Multiscale modeling of polymeric composite materials for ballistic protection

11

M. Grujicic
Clemson University, Clemson, SC, United States

11.1 Introduction and synopsis

The present chapter deals with multilength-scale modeling of polymer-matrix composites reinforced with high specific strength, high specific stiffness polymeric fibers/filaments (eg, Kevlar® or Twaron®). These materials are commonly used in various protective systems whose main requirement is a high level of penetration resistance against high kinetic energy projectiles (eg, bullets, exploded mine–propelled soil ejecta, fragments of improvised explosive devices, or turbine blades). The high mass efficiency (ie, mass-normalized performance) of these composite materials makes them particularly suitable for use in applications such as protective garments for personnel extremity protection, interior spall liners in infantry vehicles, and a lining/shroud for turbine fragment containment. In the remainder of this section, brief descriptions are provided of the three main topics: (1) definition of the armor-grade composite materials; (2) the need for and utility of composite-material modeling; and (3) the basics of multilength-scale composite-material modeling.

11.1.1 The definition of armor-grade composite materials

Continuous fiber–reinforced polymer-matrix composites are among the most advanced commercially available materials and are widely used in aerospace, construction, automotive, sporting goods, and defense-related industries. The main reason for the aforementioned widespread use of the composite materials is their ability to simultaneously meet a variety of functional and manufacturing requirements. For example, the new Boeing 787 Dreamliner is constructed with a large number of carbon fiber–reinforced epoxy-matrix composite components that, in addition to having outstanding mechanical properties, result in airframes that weigh less and are stronger than conventional airframes, resulting in improved operating efficiency and performance. Additionally, carbon fiber–reinforced epoxy-matrix composites tend to resist corrosion and fatigue, two phenomena which are well established to cause gradual degradation and ultimate failure of metallic airframes. The composite materials just described are commonly referred to as structural composite materials. The main figures of merit for this class of composite materials are their density-normalized stiffness and strength (ie, specific stiffness and specific strength, respectively).

Advanced Fibrous Composite Materials for Ballistic Protection. http://dx.doi.org/10.1016/B978-1-78242-461-1.00011-X

In many blast and ballistic protection systems in military and civilian applications, another class of composite materials (commonly referred to as armor-grade composites) are used (eg, Grujicic et al., 2008c, 2009a,b). The latter class of composites generally is optimized with respect to its ballistic impact/penetration resistance, that is, with respect to its energy absorbing capability, and its most commonly cited figure of merit is its ballistic limit, which is defined as the critical velocity below which no full perforation takes place (eg, Grujicic et al., 2008c, 2009a,b). To maximize this figure of merit, armor-grade composite materials are generally resin starved. That is, they are not fully impregnated with resin. Consequently, the reinforcing fibers are not highly constrained by the surrounding polymeric matrix and, hence, do not suffer extensive cutting/shearing during the transverse impact of a projectile onto an armor-grade composite-material panel. As a result, during such an impact, the following three stages of projectile/panel interactions are commonly observed: (1) *initial penetration phase*—dominated by filament shearing/cutting by the projectile and, to a significant extent, by plastic deformation of the projectile; (2) *intermediate stage*—characterized by extensive delamination of the composite laminate accompanied by destabilization and fragmentation of the projectile; and (3) *final stage*—characterized by highly pronounced bulging of the laminate back face accompanied by extensive stretching of the filaments (which have been freed/debonded from the surrounding matrix), resulting in high absorption of the projectile's kinetic energy.

11.1.2 The need for and utility of composite-material modeling

Development of the aforementioned protection systems is traditionally carried out using legacy knowledge and extensive fabricate and test procedures. Since this approach is not only economically unattractive, but is often associated with significantly longer lead times, it has gradually become complemented by the appropriate cost- and time-efficient computer-aided engineering (CAE) analyses. This trend has been accelerated by the recent developments in the numerical modeling of transient nonlinear dynamics phenomena such as those accompanying blast and ballistic loading conditions. However, the tools used in these analyses themselves suffer from a number of deficiencies/limitations which prevent these analyses from being more widely utilized. In the context of the use of CAE analyses for development of the aforementioned protection systems, it is well established that one of the main deficiencies stems from the inability of the most currently available material models to realistically represent the response of these materials under high-deformation rate, large-strain, high-pressure loading conditions, the conditions typically encountered during projectile impact events. One of the reasons for the indicated shortcomings of the present material models is a lack of inclusion of the contribution of various phenomena and processes occurring at different length scales to the overall behavior/performance of the material. In the next subsection, a brief overview will be provided of the main microstructural scales encountered in the case of armor-grade polymer-matrix fiber-reinforced composite materials.

11.1.3 The basics of multilength-scale composite-material modeling

Continuous fiber–reinforced polymer-matrix composites under investigation are quite complex materials. This complexity can be linked to the following main sources: (1) they contain a hierarchical/multilength-scale microstructure/architecture; (2) their mechanical response is often quite nonlinear and rate-/time-/temperature-/pressure-dependent; and (3) they are associated with a multiplicity of complex phenomena/processes (eg, filament twisting/buckling, interfilament friction, and sliding). A detailed examination of the intrinsic makeup of this class of materials typically reveals the existence of eight well-defined microstructural length scales (eg, Grujicic et al., 2011b,c, 2013a,b,c,d, 2014d). Schematics and explanations of these length scales, starting from the finest (ie, molecular chain–level) length scale, is provided in Fig. 11.1. The first column in this figure shows a set of simple schematics of the material microstructure/architecture at the length scale in question along with the labels denoting the main microstructural constituents/features. In the second column, a brief description is provided of the key features of the material models used to capture the material behavior at the same length-scale. An additional brief description of the material microstructure/architecture and the corresponding material models at each of the length scales listed in Fig. 11.1 is provided next.

11.1.3.1 Molecular chain–level length scale

At this length scale, chemical structure and conformation of the individual molecules constituting the chain are analyzed using atomic-/molecular-modeling tools/procedures (eg, Grujicic et al., 2011b,c). Within these procedures, the material is modeled as a collection of constituent particles (atoms or ions) that interact via valence bond (chemical) and nonbond (physical) forces. The main goal of the material modeling at this length scale is to identify the most likely molecular conformations present in the fibrils (defined next). This greatly reduces the computational cost expended at the fibril length scale.

11.1.3.2 Fibril-level length scale

In the case of Kevlar® type fibers (the subject of the present overview), there is a substantial experimental support (eg, Grujicic et al., 2011c) for the existence of fibrils within the fibers. Fibrils are smaller bundles of molecular chains within which chain molecules are tightly bonded into a perfect or nearly perfect crystalline phase. In addition, the material at this length scale (as well as the fiber length scale) may contain a variety of microstructural and topological defects and chemical impurities which may significantly alter its properties (Grujicic et al., 2011b,c). As in the molecular chain length scale case, the material at this length scale is treated as a collection of discrete interacting/bonded particles and analyzed using classical atomic-/molecular-modeling tools/procedures. The information gained at this length scale is then passed to the fiber length scale, where it is used as an input to a homogenization procedure for determination of the fiber-level material mechanical properties.

11.1.3.3 Fiber length scale

At this length scale, fibers are considered as assemblies of nearly coaxial, fairly straight fibrils which are held together by nonbond (van der Waals or Coulomb) forces (eg, Grujicic et al., 2013b,c,d). Due to the prohibitively large computational cost, the material at this length scale is modeled either using the so-called coarse-grained

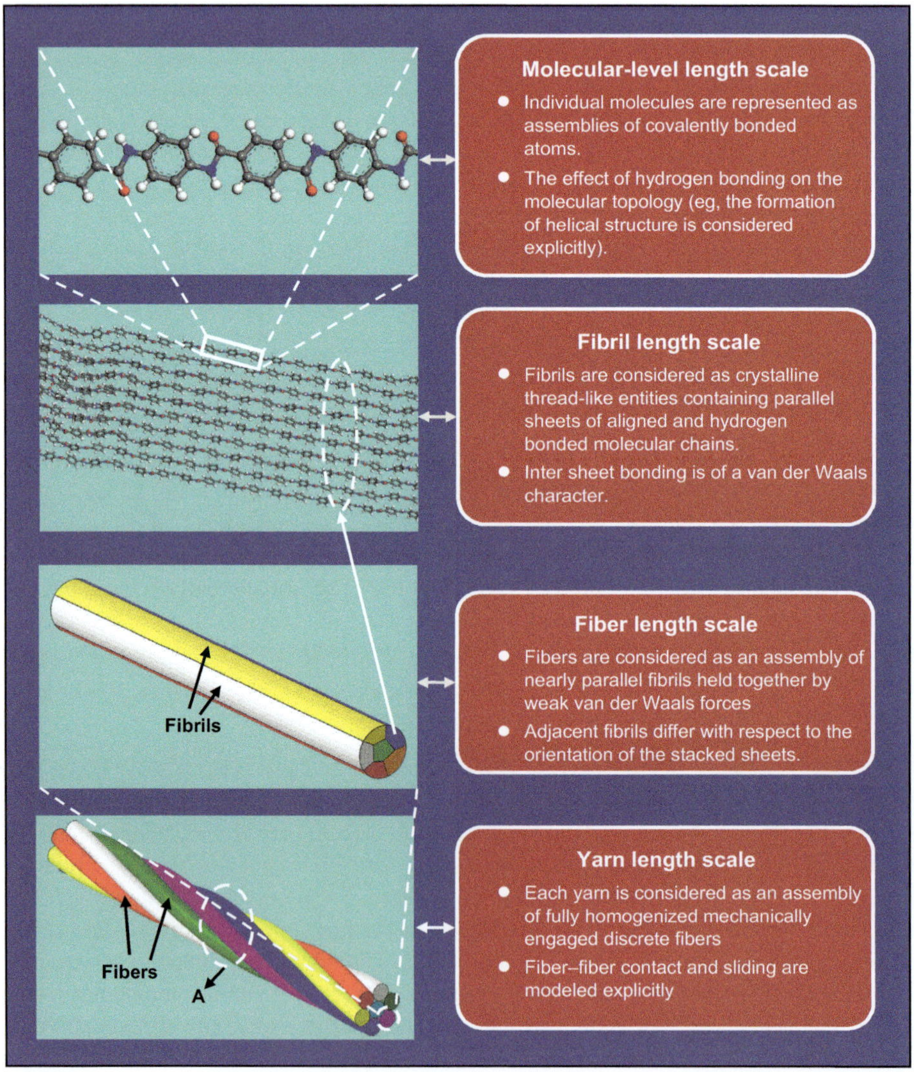

Figure 11.1 Various length scales and the associated material model assumptions/simplifications used in the study of polymer-matrix composite materials with high-performance fiber-based structures.

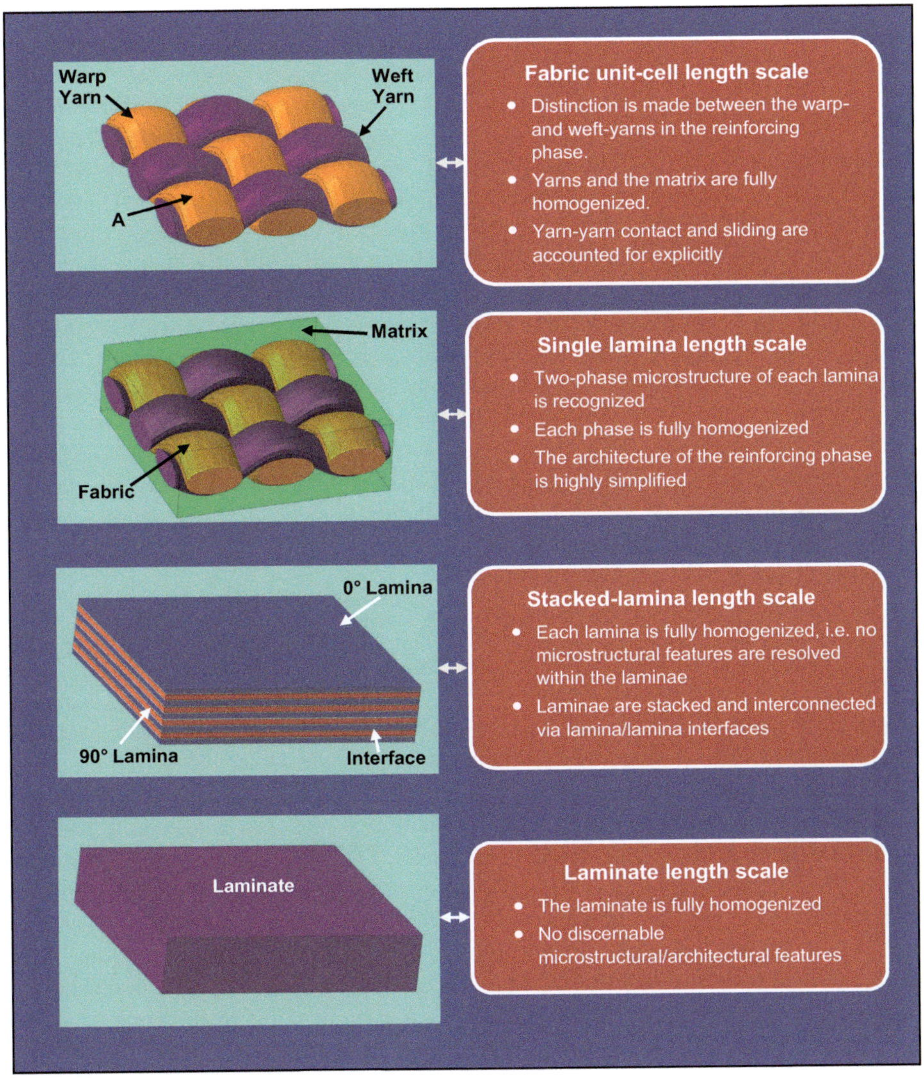

Figure 11.1 cont'd.

computational methods and tools (within which a collection of neighboring atoms/ions is replaced with a hypothetical particle commonly referred to as a bead) (Grujicic et al., 2014a,b) or a continuum-mechanics formulation (the focus of the present work). As in the previous two length scales, (1) the knowledge gained at the fiber length scale is used to enrich the material description at the next level length scale (the yarn length scale, described next); and (2) modeling and simulation are not used in large-scale simulations of projectile/protective structure interactions (due to the prohibitive computational cost).

11.1.3.4 Yarn length scale

At this length scale, the macrostructure/architecture of each yarn is accounted for explicitly (eg, Grujicic et al., 2011c). In other words, yarns are considered as assemblies of nearly parallel fibers/filaments which are mechanically engaged by either the application of a light twist to the yarn or by wrapping a fiber around the fiber/filament assembly. The constitutive response of the yarns is typically obtained computationally by carrying out different virtual mechanical tests, within which individual fibers and their arrangement within the yarn is modeled explicitly, and the material properties of the fibers are inherited from the fiber length scale computational analysis. In addition, the effects of fiber—fiber friction and its dependence on the fiber-surface topology, contact temperature, contact pressure, etc., are taken into account. The results obtained at this length scale are passed to the fabric length scale to help derive the effective/homogenized material properties and the constitutive response of a single reinforcing ply produced using different textile-fabrication processes (eg, weaving or knitting). Detailed computational analyses of the projectile/protection system impact interactions in which yarns are modeled explicitly are not generally employed due to the prohibitive computational cost.

11.1.3.5 Fabric/ply length scale

At this length scale, a closer look is given to the architecture of the reinforcing fabric and to the interaction of its constituents (eg, weft and warp filaments) (eg, Grujicic et al., 2008b,d, 2009d, 2010a, 2012b, 2014d). Specifically, details of yarn weaving and crimping, yarn cross-section change, and yarn sliding at the warp yarn/weft yarn crossover points are taken into account. Each yarn is fully homogenized in this case, and the yarn mechanical properties/constitutive response is inherited from the yarn scale computational analysis. The outcome of this procedure is a set of homogenized mechanical properties and the constitutive responses (ie, the material model) of a single reinforcing ply. This model is sometimes used in direct numerical simulations of the impact of various projectiles onto the single- or multi-ply ballistic fabric (Grujicic et al., 2008b,d, 2009d, 2010a, 2012b). More importantly for the present review, this model is used as an input to the derivation of the corresponding homogenized material model for a single lamina. Note that the term *ply* is used here to denote the fabric structure consisting only of the reinforcing filaments, while the term *lamina* is used to denote a two-phase (fabric/matrix) material unit. Note also that the five length scales covered so far deal exclusively with the reinforcing material structures/architectures at different length scales, and they do not consider the polymeric-matrix phase present in the subject composite materials.

11.1.3.6 Single-lamina length scale

At this length scale, the macrostructure/architecture of the reinforcing fabric and matrix phases are explicitly taken into account, while the two associated materials are treated as featureless/homogenized (Grujicic et al., 2004, 2007a,b, 2008a). To obtain the two-phase material model for a single composite lamina, either a homogenization technique is employed or a series of virtual mechanical tests is carried out. In either case, the two constituent phases (ie, the fabric and the matrix) are represented explicitly. The resulting material model is then passed to the next scale (the stacked-lamina length-scale) and used, in conjunction with the lamina/lamina interface model, to determine the effective mechanical response of a composite laminate (a plate-like structure consisting of a relatively large number of stacked laminae).

11.1.3.7 Stacked-lamina length scale

At this length scale, the presence of discrete stacked laminae is recognized, while the material within each lamina as well as interlaminar boundaries are kept featureless/homogenized (Grujicic et al., 2010c). The material model for each lamina is inherited from the single-lamina length scale computational analysis, while the constitutive model for the lamina/lamina interfaces has to be derived independently. The two types of material models are then used within a large-scale computational analysis to investigate impact of a projectile onto the *stacked-lamina* composite laminate.

11.1.3.8 Laminate length scale

At this length scale, the composite laminate is treated as a featureless plate-like entity (Grujicic et al., 2013a). In other words, no recognition is given to the stacked-lamina architecture of the laminate. Furthermore, the composite material is assumed not to possess any discernable microstructural features and to be homogeneous (but typically not isotropic). The associated material model is often obtained by homogenizing single-lamina and lamina/lamina-interface models. Large-scale computational analyses of the projectile impact onto a composite-laminate protective structure are most frequently investigated using the laminate length scale (eg, Grujicic et al., 2011f, 2012a, 2014c). The main reason for this is the low computational cost. While the laminate architecture and the material microstructure are not accounted for explicitly, enrichment of the material models through incorporation of the details gathered at the smaller length scales is generally believed to be sufficient to overcome apparent limitations of such computational analyses.

Careful examination of the eight length scales described suggests that they could be grouped in the following way:

1. molecular and fibril length scales: At these length scales, the material is represented as an ensemble of interacting constituent particles (atoms/ions);
2. fiber, yarn, single-ply length scales: At these length scales, the materials are treated as being of a continuum character, while the structural constituents (eg, fibers/yarns, fabric unit-cells) are treated as being discrete; and
3. single-lamina, stacked-lamina, and laminate length scales: At these length scales, both the material and the structural elements (eg, laminae, lamina/lamina interfaces, etc.) are treated as being of a continuum character.

This grouping is adopted in the remainder of this chapter and the subsequent sections are formulated accordingly.

11.2 Molecule- and fibril-scale modeling

Material models falling into this class will be discussed within the context of *p*-phenylene terephthalamide (PPTA) polymeric filaments such as Kevlar® and Twaron®. As mentioned earlier, these filaments are commonly used in various ballistic/blast protection systems with the main requirement being a high level of penetration resistance against large kinetic energy projectiles.

11.2.1 Microstructure of PPTA molecules and fibrils

PPTA filaments are made from the family of polymeric materials known as polyamides. Polyamides are typically classified as aromatic polyamides or aramids (eg, Kevlar® or Twaron®) and nonaromatic polyamides (eg, nylon 6,6). A pictorial representation of a single PPTA repeat unit, consisting of two phenylene rings/moieties joined by two amide linkages, is shown in Fig. 11.2(a). Fig. 11.2(b) uses the same type of ball-and-stick representation to schematically display the basic PPTA condensation–polymerization reaction. For improved clarity, the atomic species are labeled in this figure. While, in principle, PPTA can appear both in the (opposite) *trans*- (Fig. 11.2(a)) and (on the same side) *cis*- (not shown for brevity) stereo-isomeric conformations, the latter conformation is rarely observed. This finding is explained by steric hindrance inhibiting the attainment of the *cis*-conformation, while the *trans*-conformation promotes formation of lower energy stretched-out/extended molecules/chains. The presence of large numbers of nearly parallel molecules with a nearly linear backbone structure promotes, in turn, formation of (crystalline or paracrystalline) PPTA fibrils.

Due to a large difference in electronegativity between oxygen and hydrogen, amide linkages possess large dipole moments and, hence, are prone to forming hydrogen bonds. When such bonds are formed laterally between parallel PPTA molecules/chains, *sheet-like* structures are created. PPTA fibrils are commonly found to have a crystalline structure consisting of stacked sheets. However, while hydrogen bonding plays a key role in the formation of the sheets, its contribution to the intersheet bonding is generally considered to be minor. Examination of the crystalline PPTA fibrils reveals that the sheets are not entirely planar but contain small-amplitude (ca. 40 nm), large wavelength (ca. 250–500 nm) accordion-style pleats.

As far as the PPTA fibers are concerned, as mentioned earlier, they generally consist of assemblies of nearly coaxial, fairly straight fibrils which are held together by nonbond (van der Waals or Coulomb) forces.

Due to their molecular structure, PPTA chains possess high bending stiffness and, hence, do not easily flex. High rigidity of the PPTA molecules is believed to be one of the major factors affecting the microstructure of the PPTA fibrils (and, in turn, fibers). That is, in contrast to the flexible polymeric molecules which can undergo extensive folding and give rise to the formation of the commonly observed (crystalline + amorphous) two-phase polymeric microstructure, the PPTA fibrils typically acquire either a paracrystalline or a fully crystalline microstructure. In the case of the paracrystalline structure, PPTA molecules are all aligned in the same direction, but no order

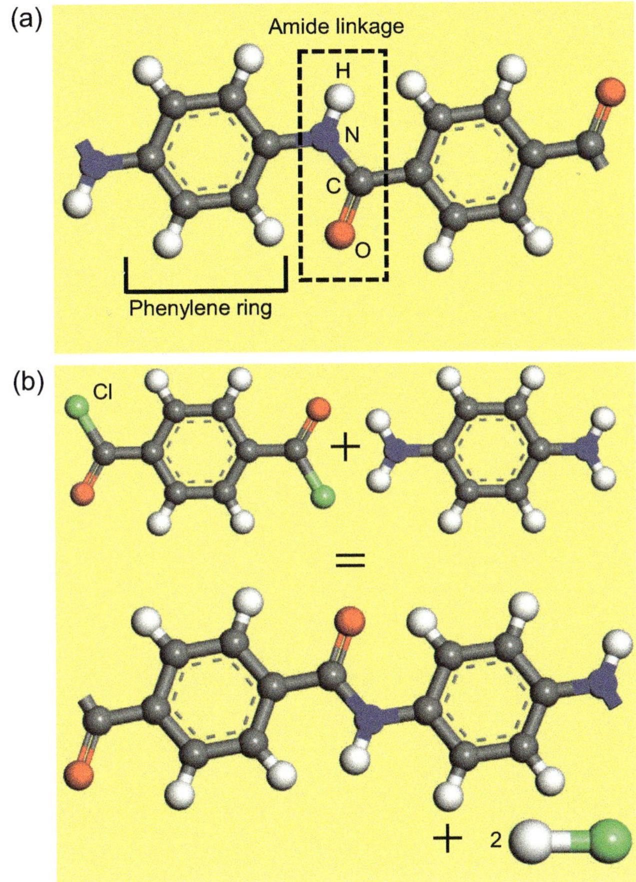

Figure 11.2 (a) *Trans*-molecular conformations in typical *p*-phenylene terephthalamide-based polymeric-material chains/molecules and (b) *p*-phenylene terephthalamide condensation–polymerization reaction.

exists in a plane orthogonal to this direction. In sharp contrast, in the case of the fully crystalline PPTA fibrils, molecules are aligned in all three mutually orthogonal directions. It should also be noted that the formation of paracrystalline or crystalline structures is promoted by the presence of the planar phenylene and amide groups and by the ability of the adjacent chains to form hydrogen bonds. Close examination of the PPTA fibril layered crystal structure reveals that it mainly possesses ABABAB... stacking sequence. As mentioned earlier, the sheets are formed due to hydrogen bonding between the adjacent parallel PPTA molecules while the intersheet bonding is mainly of the van der Waals (and *p*-electron weak chemical bond) type. Due to the low strength of the intersheet bonding, PPTA crystal structure is prone to the formation of stacking faults and kink bands, and consequently, PPTA fibrils/fibers possess inferior longitudinal compressive strength and buckling resistance.

11.2.1.1 Synthesis/processing-induced defects/flaws

As in most engineering materials, properties of PPTA fibers are greatly affected by the presence of various crystallographic and morphological defects/flaws. The character, size, and the number density of these flaws is closely related to the PPTA synthesis and fiber fabrication processes. Details regarding PPTA synthesis and fiber fabrication can be found in the relevant patent literature (Kwolek, 1972; Blades, 1973; du Pont, 1983), and a brief overview of these processes was given in a recent work (Grujicic et al., 2011b,c).

The summary of the PPTA synthesis and fiber fabrication processes presented by Grujicic et al. (2011b,c) clearly revealed that different types of defects/flaws may be and do get generated within PPTA fibrils/fibers. Since these defects have a profound effect on the fibril/fiber properties, as well as on the properties of coarser scale fiber-based structures (eg, yarns, fabrics, laminae, and laminates), they were also reviewed by Grujicic et al. (2011b,c). A summary of the PPTA fiber most common defects, their dimensionality, their cause, ways of reducing their number density, and their typical concentrations is provided in Table 11.1.

11.2.2 Modeling and simulations of PPTA molecules and fibrils

Generally, discrete-material models and simulations (eg, molecular statics and molecular dynamics) are used to infer the constitutive behavior of the reinforcing filaments at the molecular and fibril length scales, and to obtain insight into the effect of various microstructural features (including defects) on this behavior. As far as the fibers are concerned, as will be discussed in the next section, they are more often treated using continuum-level approaches.

In general, the formulation of a molecular-/fibril-level simulation problem requires, at a minimum, specification of the following five items: (1) a molecular-level computational model consisting of atoms, ions, moieties, and functional groups; (2) a set of force-field functions, which describe various bonding and nonbonding interactions between the constituents of the molecular-scale model; (3) types and sequence of the computational method(s) to be used in the simulation; (4) formulation of the physical problem being investigated/simulated; and (5) specification of the methods and procedures to be used in the postprocessing data-reduction analysis. A brief overview of these items is provided in the remainder of this section.

11.2.2.1 Computational model(s)

To gain insight into the basic microstructure-property relations at the molecule and fibril length scales, a relatively large number of molecular-level computational models containing various perfect-crystal PPTA conformations and different microstructural/topological defects have to be used. Fig. 11.3 depicts several examples of such PPTA computational models containing defects.

Table 11.1 Classification of the most common defects found in PPTA fibers

Defect class	Defect type	Cause(s)	Defect formation prevention	Number density range
Isolated chain ends (point defect)	$-COOH$	H_2SO_4 catalyzed hydrolysis causing PPTA chain scission. Na^+ deficiency with respect to complete neutralization of side/end acidic groups.	Use concentrated H_2SO_4 for dope preparation. Shorten the fiber wash time	0.35 per PPTA chain for each defect[a] (~350 ppm-mass-based)
	$-NH_2$	H_2SO_4 catalyzed hydrolysis causing PPTA chain scission. Na^+ deficiency with respect to complete neutralization of side/end acidic groups.	Use higher concentration NaOH solution	0.35 per PPTA chain for each defect[a] (~350 ppm-mass-based)
	$-COO^-Na^+$	COOH neutralization with Na^+	No remedy required since this is one of the preferred chain ends	1.1 per PPTA chain[a] (~1100 ppm-mass-based)
	$-NH_3^+HSO_4^-$	Sulfonation of the NH_2 chain ends	Increase the H_2SO_4 removal and neutralization rate	0.2 per PPTA chain[a] (~200 ppm-mass-based)
Side groups (point defect)	$-SO_3H$	Exposure of PPTA in the dope to concentrated H_2SO_4 (sulfonation)	Reduce the H_2SO_4 concentration in the dope	~1300 ppm (mass-based)

Continued

Table 11.1 Continued

Defect class	Defect type	Cause(s)	Defect formation prevention	Number density range
	$-SO_3^-Na^+$	Neutralization of sulfonic acid side groups by NaOH	Remedy may not be required since this side group improves fiber longevity. However mechanical performance may be compromised	~2500 ppm (mass-based)
Voids and interstitials (point defects)	Microvoids	Swelling induced by hydration of intrafibrillar Na_2SO_4	Increase the extent of sodium salt dissolution by prolonged exposure of fibers to boiling water	~150 ppm (mass-based)
	Mobile Trapped H_2SO_4	Nonneutralized or unwashed intrafibrillar H_2SO_4	Thorough washing in hot solvent aqueous bath	~70 ppm (mass-based)
Defect bands (planar defects)	$NH_3^+HSO_4^-$ agglomerated chain ends	Coulombic attraction–induced clustering of ion-terminated chain ends	The phenomenon is not well understood so no remedy is obvious	One band every 40–60 nm of fibril (ca. 3000 ppm-mass-based)

[a]Extruded fibers.

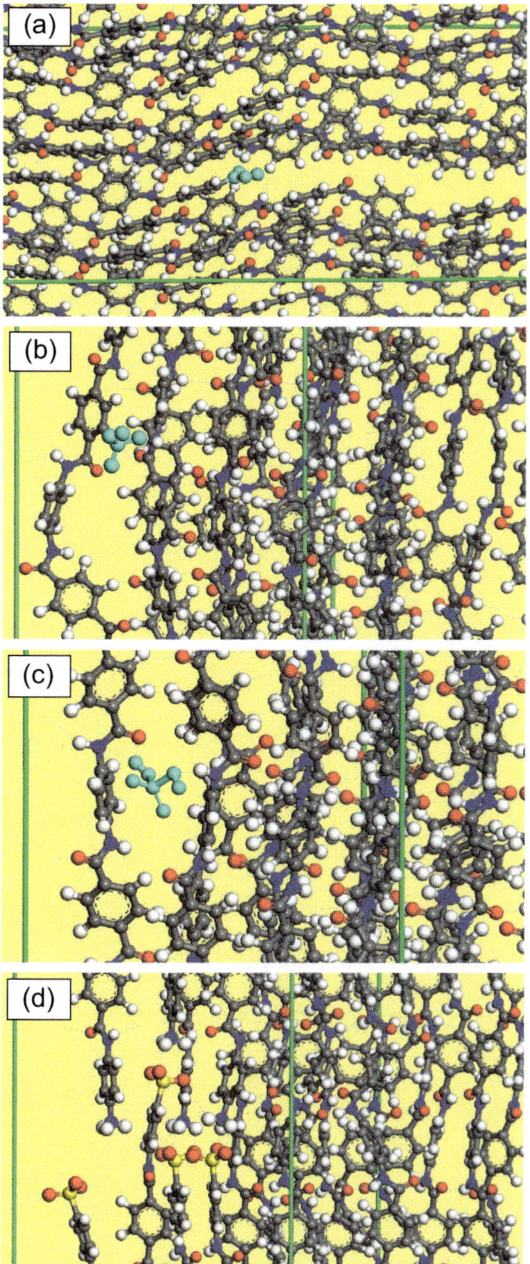

Figure 11.3 Examples of the computational models containing defective *p*-phenylene terephthalamide crystalline structures: (a) isolated −COOH group (a chain-end defect); (b) isolated $-SO_3H$ group (a side-group defect); (c) isolated mobile H_2SO_4 molecule (an interstitial defect); and (d) a defect band (a planar defect). For clarity, atoms within defects are enlarged and highlighted in yellow in (a)−(c), and enlarged in (d).

Construction of the defect-free PPTA computational material models is typically carried out under the following conditions:

1. The fibers (not modeled explicitly) are treated as bundles of parallel, fully crystalline fibrils.
2. As far as the crystalline structure of the fibrils is concerned, it is modeled as a stack of parallel PPTA sheets (formed by hydrogen bonding between nearly parallel PPTA molecular chains, all aligned in the fiber/fibril axis direction).
3. To account for the fact that the vast majority of the molecules inside PPTA fibrils are fully surrounded by other chains: (1) a bulk-like environment is created using the concept of the computational cell and the periodic boundary conditions. The computational cells used were all of the rectangular parallelepiped shape, with one of the computational cell edges aligned with the fiber axis. The other two computational cell edges lie in the PPTA sheet and orthogonal to it, respectively; and (2) the periodic boundary conditions are applied not only in the fiber/fibril axial direction but also in the two transverse directions.
4. Sheet pleating is typically not accounted for (at least, explicitly) since its wavelength typically exceeds the longest dimension of the PPTA computational cells used.

In the case of the computational cells containing imperfect crystalline PPTA, the following classes of crystallographic/topological defects are typically considered: (1) chain ends (treated as point-type defects) terminated with one of four possible end groups ($-COOH$, $-NH_2$, $-COO^-Na^+$, $-NH_3^+HSO_4^-$), Table 11.1; (2) side groups involving two possible moieties ($-SO_3H$, $-SO_3^-Na^+$), Table 11.1; (3) intrafibril voids and interstitial H_2SO_4; and (4) defect bands. Defects are commonly created within the initially perfect crystalline material by using the following procedures: (1) chain scission followed by the attachment of the appropriate chain-end groups, in the case of chain-end defects; (2) chain-side functionalization, in the case of side-group defects; (3) insertion of nitrogen molecules, in the case of (air-filled) voids; (4) insertion of H_2SO_4 molecules, in the case of interstitials; and (5) scission of a number of adjacent chains in a plane orthogonal to the fibril axis, in the case of defect bands.

11.2.2.2 Force fields

The behavior of a material system at the molecular level is governed by the appropriate force fields which define functional relationships between different components of the constituent particle (atom/ion) interaction energies and the particle position, nature, charge, and state of bonding. Due to its critical role in modeling the material behavior at the fiber and subfiber length scales with high fidelity, highly accurate ab-initio quantum mechanics—derived force fields are generally preferred. Within these force fields, the potential energy of a system of interacting atoms/ions is expressed as a sum of a valence (covalent bonding) term, a cross (atomic environment—dependent bonding) term, and a nonbond van der Waals and Coulombic (including hydrogen bonding) pair-interaction term.

11.2.2.3 Computational method

To gain insight into the mechanical response of a perfect and defective crystalline PPTA, computational cells described earlier are subjected to molecular statics and molecular dynamics simulations. The molecular statics approach is effectively a constrained-optimization procedure within which the potential energy of the computational cell (the objective function) is minimized with respect to the positions of the constituent atoms and ions (the design variables), while subjected to the constraints arising from the imposed periodic boundary conditions and loading.

Within the molecular dynamics approach, the following two-step procedure is repeated at each (0.1–1.0 fs-long) time step: (1) the negative gradient of the potential energy with respect to the atom/ion positions is first used to compute forces acting on each atom/ion; (2) then, the associated Newton's equations of motion are integrated (over a fs-long time period) numerically to compute the acceleration associated with each atom/ion and to, in turn, update the velocity and the position of each atom/ion.

Molecular dynamics methods are generally classified into the equilibrium and nonequilibrium methods (Grujicic et al., 2011e, 2012c).

Within the equilibrium molecular dynamics methods, the subject material system is maintained in the state of thermal–mechanical equilibrium by coupling it to the surroundings, such as a constant-pressure bath or a constant-temperature reservoir. This ensures the absence of net flux of the mass, momentum, and energy in any of the three principal coordinate directions.

Within nonequilibrium molecular dynamics, the system is subjected to large mechanical and/or thermal perturbations (eg, finite changes in the axial parameter of the computational cell, accompanying the simulation of a high-rate mechanical test). As a consequence, the system experiences large fluxes of its thermodynamic quantities (mass, momentum, and energy).

11.2.2.4 Virtual mechanical testing

Molecular statics and (equilibrium/nonequilibrium) molecular dynamics simulations are often used to simulate different (molecular-/fibril-scale) mechanical tests, and to determine the associated material mechanical properties. The mechanical tests typically simulated involve axial tension/compression, transverse compression, transverse shear, axial torsion, etc. It should be noted that, typically, PPTA fiber diameter is on the order of 10 μm, a length scale which is 10^3–10^4 times larger than the size of a prototypical molecular-level computational cell. As established earlier, fibers are composed of parallel crystalline fibrils with lateral dimensions being one to two orders of magnitude smaller than those of the fibers. Taking these findings into consideration, it is clear that a PPTA material domain of the size equal to that of the computational cell has the highest probability of residing within a single PPTA fibril. For this reason, the computational cells typically used in the simulations of the mechanical tests are assumed to contain perfect or imperfect single-crystalline material, and the associated virtual mechanical tests to yield fibril (not fiber) properties.

When the low-rate mechanical tests are simulated, the following three-step procedure is commonly employed: (1) loading is applied in an incremental fashion; (2) each loading step is followed first by a molecular statics potential energy minimization step; and (3) the statically equilibrated material is then dynamically relaxed by subjecting it to a prolonged (ca. 50 ns) equilibrium molecular dynamics simulation. When the high-rate mechanical tests are simulated, nonequilibrium molecular dynamics is employed, and no potential energy minimization or dynamic relaxation techniques are used.

The results obtained, either under low-rate or high-rate loading conditions, contained the history of the interacting particles' positions, velocities, and forces. These results are subsequently processed to extract the essential features of the material response when it is subjected to different types of loading at different imposed loading rates. In addition, deformed microstructure is examined in great detail and correlated with the material mechanical response to help establish microstructure-property relations in the subject material.

11.2.2.5 Postprocessing data-reduction analysis

Postprocessing of the results generated within different virtual mechanical tests can be used to infer the associated material stiffness and strength properties.

Both molecular statics and molecular dynamics results can be used to determine the components of the fourth-order elastic stiffness tensor. When molecular statics approaches are used, the contributions of the finite-temperature thermal expansion and the vibrational/configurational entropic effects to the system free energy are ignored. Consequently, the resulting elastic stiffness tensor defined as a second-order derivative of the system's free energy (ie, potential energy, in the zero temperature case) with respect to the unit cell strain components reflects the behavior of the material in question under small deformations at zero absolute temperature. Since these finite-temperature effects can make a significant contribution to the elastic stiffness tensor, they could not be generally neglected. Consequently, a molecular dynamics-based elastic stiffness computational procedure which accounts for the aforementioned thermal and entropic effects is generally employed. Within the molecular dynamics-based procedure, for determination of the material stiffness (a fourth-order tensor), the (equilibrium) history/trajectory results are used to compute the ensemble average for the material stress and strain (both treated as second-order tensors). Then, using the generalized Hooke's law (or more precisely, the time-based correlation functions between different stress and strain components), the corresponding material stiffness tensor is determined and, through the use of the Voigt reduced notation, converted into a (6×6) material stiffness matrix. The results obtained for the PPTA-based materials typically reveal that the C matrix contains nine independent elements, suggesting that the PPTA crystal structure possesses orthotropic symmetry. Since it is customary practice to work with the engineering moduli rather than with the elastic stiffness constants, the corresponding three Young's moduli (E_{11}, E_{22}, E_{33}), three shear moduli (G_{12}, G_{13}, G_{23}), and three Poisson's ratios ν_{12} $(=\nu_{21}{*}E_{11}/E_{22})$, ν_{13} $(=\nu_{31}{*}E_{11}/E_{33})$, and ν_{23} $(=\nu_{32}{*}E_{22}/E_{33})$ are next determined using standard functional relations (eg, Grujicic et al., 2010c). Note that typically the three principal (ie, computational cell edge) directions are respectively aligned with (1) the fibril/fiber longitudinal direction, direction x_1; (2) with the sheet-normal direction, direction x_2; and (3) with the corresponding in-sheet orthogonal direction, direction x_3.

As far as the material strength is concerned, it is generally determined by conducting repeated loading/unloading mechanical tests to progressively higher imposed strains and determining the minimum stress level (for a given type of loading) which produces an irreversible/inelastic strain (Grujicic et al., 2013b,c,d).

11.2.3 Typical results

11.2.3.1 PPTA fibril stiffness

The procedure described in the previous section generally yields the results which reveal that the fibril stiffness is affected by the type, size, concentration/number density, and clustering of microstructural/topological defects. This is exemplified by the results displayed in Figs. 11.4(a)–(d), 11.5(a)–(b), and 11.6(a)–(c). Fig. 11.4(a)–(d) shows, respectively, the effect of the concentration of the $-COOH$, $-NH_2$, $-COO^-Na^+$, and $NH_3^+HSO_4^-$ chain-end groups on the relative change in the nine orthotropic elastic moduli with respect to their values in the defect-free fibrils ($E_{11} = 105.3$ GPa, $E_{22} = 2.0$ GPa, $E_{33} = 31.3$ GPa, $G_{12} = 0.9$ GPa, $G_{13} = 5.6$ GPa, $G_{23} = 2.6$ GPa, $\nu_{12} = 0.46$, $\nu_{13} = 0.147$, and $\nu_{23} = 0.15$). The effect of the concentration of the $-SO_3H$ and the $-SO_3^-Na^+$ side groups on the orthotropic elastic moduli is

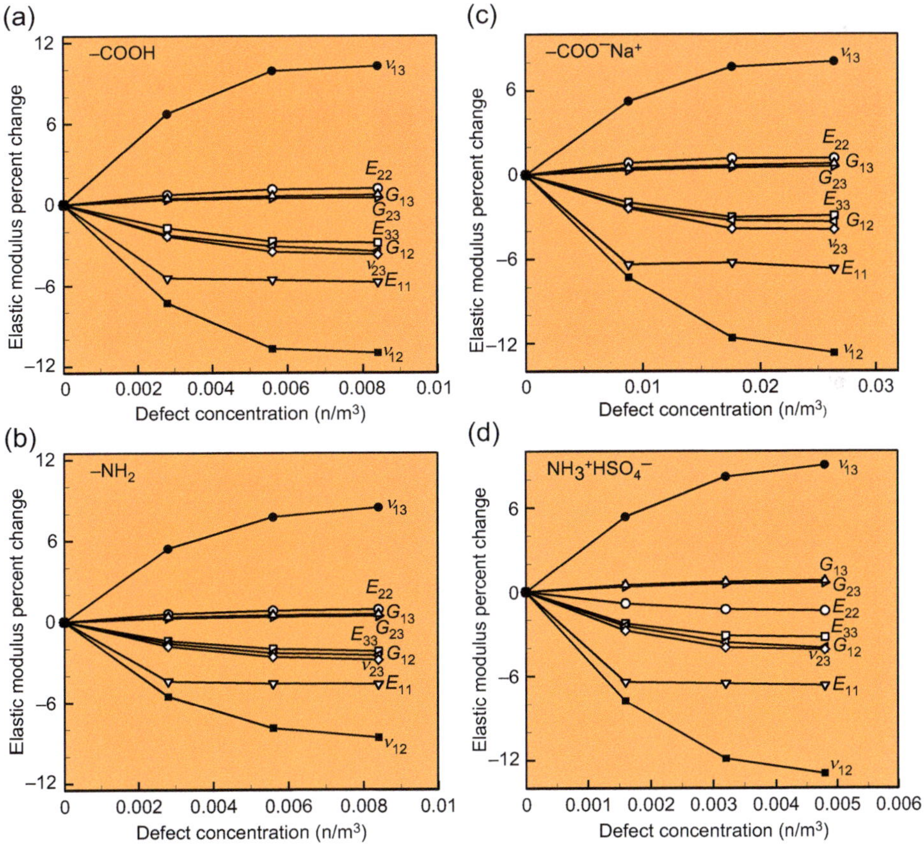

Figure 11.4 The effect of the concentration of the: (a) $-COOH$; (b) $-NH_2$; (c) $-COO^-Na^+$; and (d) $NH_3^+HSO_4^-$ chain-end groups on the orthotropic elastic stiffness moduli of the *p*-phenylene terephthalamide fibrils.

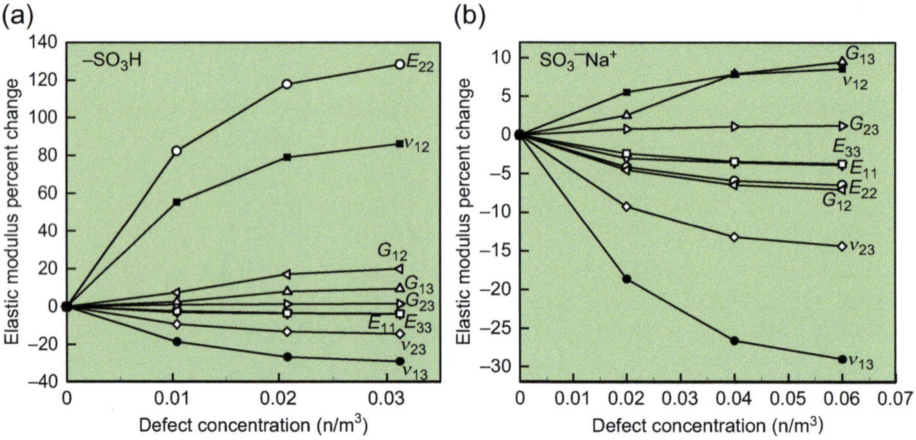

Figure 11.5 The effect of the concentration of the: (a) $-SO_3H$; and (b) $SO_3{}^-Na^+$ side groups on the orthotropic elastic moduli of the *p*-phenylene terephthalamide fibrils.

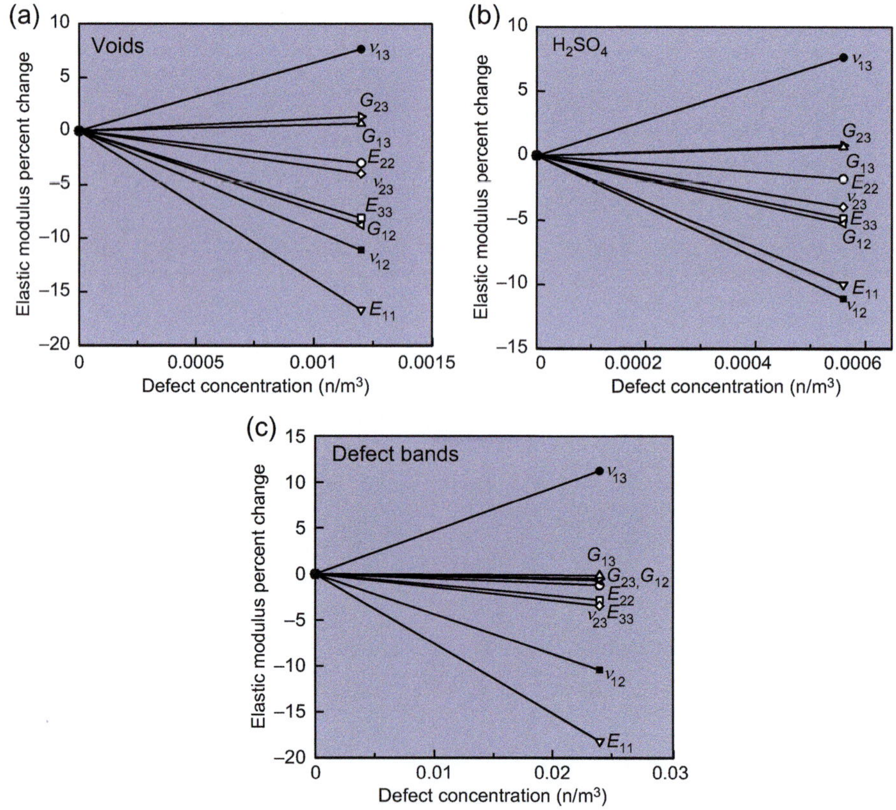

Figure 11.6 The effect of the concentration of the (a) voids; (b) mobile H_2SO_4 molecules; and (c) defect bands on the orthotropic elastic constants of the *p*-phenylene terephthalamide fibrils.

displayed in Fig. 11.5(a)−(b), respectively. The effect of the concentration of the voids and the mobile H_2SO_4 molecules on the same set of elastic moduli is displayed in Fig. 11.6(a)−(c), respectively.

Examination of Figs. 11.4(a)−(d), 11.5(a)−(b), and 11.6(a)−(c) reveals the following: (1) The orthotropic elastic moduli change monotonically with an increase in the defect concentrations; (2) Among the elastic moduli, longitudinal normal stiffness (E_{11}) and the intersheet shear stiffness (G_{12}) are most strongly affected/compromised by the presence of the chain-end defects, Fig. 11.4(a)−(d); (3) In the case of side-group defects, Fig. 11.5(a)−(b), *trans*-sheet normal stiffness increases either strongly due to establishment of intersheet hydrogen bonds (the case of SO_3H, Fig. 11.5(a)), or weakly if such hydrogen bonding is absent (the case of SO_3^- Na^+, Fig. 11.5(b)). Fig. 11.5(a)−(b) further reveals that the shear moduli G_{12} and G_{13} are also increased due to the presence of hydrogen bonding; (4) The presence of voids and interstitial defects has a highly detrimental longitudinal stiffness−reducing effect, Fig. 11.6(a)−(b). This effect is even more pronounced in the case of defect bands, Fig. 11.6(c).

To predict the mean values of the orthotropic elastic constants in fibrils in the commercially produced PPTA fibers, the following deterministic procedure is generally employed: (1) For each of the microstructural/topological defects analyzed in the present work, a mean value of the defect concentration (and its typical range) under prototypical PPTA fiber fabrication conditions is assessed using the available open-literature reports; (2) Due to the very low values of the defect concentrations, it is assumed that the combined effects of different defects can be obtained using a simple linear-superposition procedure; (3) The effect of individual defects on the fibril orthotropic elastic constant moduli is assumed to be represented by the results displayed in Figs. 11.4(a)−(d), 11.5(a)−(b), and 11.6(a)−(c).

11.2.3.2 PPTA fibril strength

When the procedure described in the previous section is applied to perfect PPTA fibrils, the resulting longitudinal/axial-tensile strength is generally found to be of a highly reproducible value (ca. 6.3 GPa). On the other hand, the longitudinal compressive strength of perfect PPTA fibrils is found to be fairly scattered and lower, by at least one order of magnitude, than its tensile counterpart. In the case of PPTA fibrils containing defects/flaws, it is generally found the fibril longitudinal tensile (as well as compressive) strength is effectively controlled by the size and composition of the largest (ie, the most potent) defect or defect cluster. Due to the stochastic character of the defects present within the fibrils, fibril/fiber longitudinal strength is generally found to also be a stochastic quantity, which is in agreement with several experimental reports (eg, Newell and Sagendorf, 1999; Steenbakkers and Wagner, 1988; Knoff, 1987). To determine the associated probability distribution function, the following five-step procedure is generally employed: (1) It is first assumed that defect distribution within the fibrils follows the principles of a Poisson point process; (2) The concept of *volume of influence* is next introduced to establish a criterion for declaring an ensemble of closely-spaced individual defects as a single (high-potency) defect cluster;

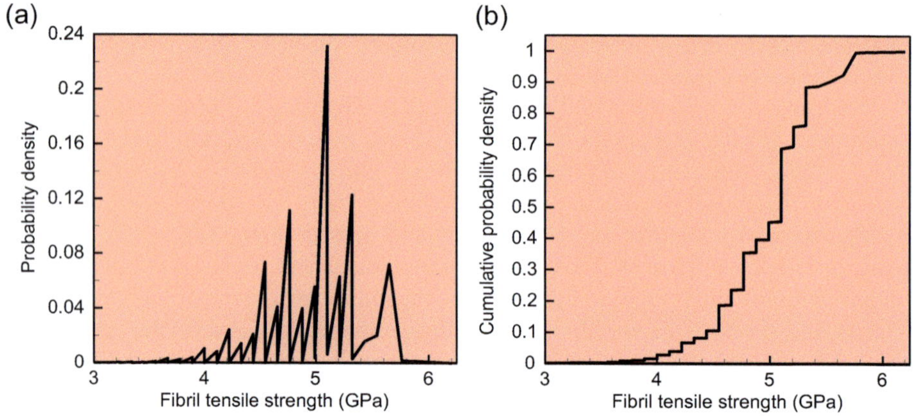

Figure 11.7 (a) *p*-phenylene terephthalamide fibril longitudinal-strength probability density function; and (b) the corresponding cumulative distribution function.

(3) The defect concentrations associated with the prototypical PPTA fiber fabrication conditions are then assumed; (4) A large number of (defective PPTA fibril) models are created stochastically, and for each case, the corresponding longitudinal tensile strength is determined using the procedure described in the previous section; (5) The results obtained are finally processed using a statistical probability analysis.

Application of this procedure to the results obtained in the present work yielded the PPTA fibril/fiber longitudinal tensile strength probability density and cumulative probability density functions depicted, respectively, in Fig. 11.7(a)–(b). Examination of the results displayed in these figures and the fact that the measured PPTA fibril longitudinal tensile strength is typically found to be in a 3–5 GPa range (eg, Newell and Sagendorf, 1999; Cheng et al., 2004) suggests that the computational procedure employed is fairly reliable. The cumulative distribution function displayed in Fig. 11.7(b) defines the probability that the PPTA fibril/fiber longitudinal tensile strength is lower than a given value, and this type of function is often used in probabilistic/reliability-based design analyses (Grujicic et al., 2010b).

As far as the transverse (compressive) strength of PPTA fibrils is concerned, it is generally found to also be a stochastic quantity and to be greatly dependent on (1) the orientation of the PPTA sheet normal within the fibril, and (2) the type, local number density, clustering, and potency of the defects (Grujicic et al., 2013b).

11.3 Fiber-, yarn-, and fabric-level modeling

11.3.1 Fiber-level modeling

Most often, fibers are treated as (continuum) entities consisting of parallel stretched-out fibrils, laterally bonded by weak van der Waals forces, Fig. 11.8(a). Consequently, the fiber properties are derived by proper averaging of the

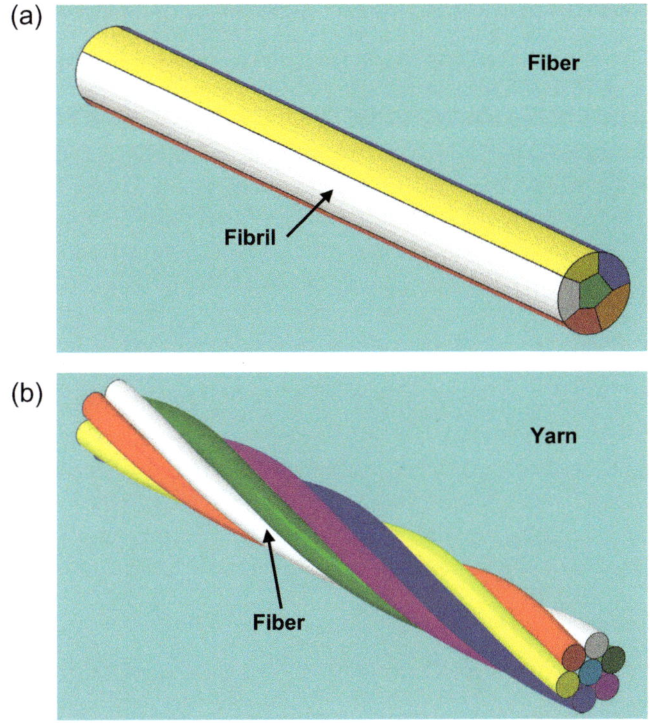

Figure 11.8 Topological relationships between: (a) fibrils within a single fiber; and (b) fibers within a lightly twisted yarn.

corresponding fibril properties. Typically, one or both of the following averaging schemes is applied: (1) a constant-strain upper-bound Voigt averaging scheme or (2) a constant-stress lower-bound Reuss averaging scheme.

11.3.1.1 Stiffness of PPTA fibers

Due to random orientation of fibril-sheet normals (within a plane normal to the fiber axis), the PPTA fibers are generally found to be transversely isotropic (with the fiber axis being the unique material direction). On the other hand, as established earlier, PPTA fibrils are of an orthotropic character.

Application of the aforementioned averaging procedures to the PPTA fiber stiffness properties typically reveals the following (Grujicic et al., 2013b): (1) In the axial direction, the fiber normal stiffness, E_{11}, is quite comparable with its fibril counterpart; (2) In the transverse direction, the normal stiffness properties, $E_{22} = E_{33}$, are generally found to be close to the more compliant E_{33} Young's modulus of the fibrils; (3) Fiber G_{23} shear modulus generally takes on a value close to its fibril counterpart; (4) The two remaining fiber shear moduli G_{12} and G_{13} are generally found to have nearly identical values, the values which are somewhat higher than the fibril G_{12} shear modulus (the modulus which controls the compliant PPTA intersheet sliding).

11.3.1.2 Strength of PPTA fibers

To obtain PPTA fiber (normal and shear) strengths in different loading conditions from the corresponding PPTA fibril strengths, an averaging procedure analogous to the one described earlier for the case of the PPTA fiber stiffness is generally utilized. This procedure typically reveals the following (Grujicic et al., 2013b): (1) The longitudinal strength of the fibers and its distribution are quite comparable to their fibril counterparts; (2) The fiber transverse strength is generally plane-isotropic with a mean value slightly higher than the lowest fibril transverse strength; (3) Fiber shear strength is fairly isotropic with an average value comparable to the mean shear strength of the fibril.

11.3.2 Yarn-level modeling

Yarns are generally modeled as assemblies of nearly parallel fibers which are mechanically engaged by either light twisting of or wrapping a fiber around the fiber assembly, Fig. 11.8(b). In most cases, mechanical properties of the yarns are determined using finite element simulation—based virtual mechanical tests in which (1) yarn internal fiber-based architecture is accounted for explicitly; (2) fiber material properties are inherited from the fiber length—scale analysis (described earlier); and (3) the phenomena such as fiber/fiber friction and fiber juxtaposition are included explicitly.

11.3.2.1 Stiffness of PPTA yarns

The finite element analyses mentioned previously typically reveal that, at the yarn-level, the material is transversely isotropic, with respect to both its stiffness and strength properties. In addition, the yarn material is generally found to be nearly linear-elastic up to the point of fracture. The five transversely isotropic elastic moduli obtained through the use of the single-yarn virtual mechanical tests are generally found to be as follows:

1. The axial elastic modulus, E_{11}, is increased relative to its fiber counterpart by an amount proportional to the extent of yarn twisting. This increase reflects the contribution associated with interfiber friction arising from the twist applied to the yarn to engage the fibers, and it reveals the fact that the yarn-level material properties are dependent on the fibril/fiber (crystallographic/morphological defects-controlled) properties as well as on the details of yarn architecture (eg, the extent of fiber twist).
2. The transverse elastic moduli, $E_{22} = E_{33}$, generally acquire values which are only a small fraction of the axial Young's modulus, which can be linked to the relatively weak transverse interfiber bonding.
3. All shear moduli acquire quite small values, as a result of the interplay of fiber juxtaposition.
4. All Poisson's ratios acquire values near 0.0 because the yarn is an assembly of discrete fibers. Consequently, fibers can experience Poisson's effects without the yarn experiencing these effects.

11.3.2.2 Strength of PPTA yarns

The finite element single-yarn virtual-testing procedures described previously generally yield the following information regarding the PPTA-yarn strength (Grujicic et al., 2011c): (1) the mean axial strength is slightly (by ca. 0.2 GPa) increased relative to its fiber counterpart, while the standard deviation is noticeably smaller than that observed for the PPTA fiber normal strength; (2) the transverse (compressive) strength of the yarn is quite comparable to its fiber counterpart; and (3) yarn transverse shear strength is somewhat higher than its fiber counterpart.

11.3.3 Fabric-level modeling

Modeling of the reinforcing material at the (woven) fabric level is normally done in one of the two following ways:

1. Each ply of the fabric is modeled using conventional shell- or solid-type finite elements, and weaving is done in an explicit manner by snaking the yarns through the fabric. In this case, yarn material properties resulting from the aforementioned yarn-level analysis are used.
2. A mesoscale material model for a single-ply woven fabric is first developed and the fabric is treated as a planar shell-type surface with a constitutive response governed by the mesoscale material model. During the derivation of this material model, yarn-level material properties are used along with the fabric architecture (eg, weaving, crimping, etc.).

The main intent of the model is to attain computational efficiency when calculating the mechanical response of the multi-ply fabric-based flexible armor material during its impact with various projectiles without significantly sacrificing the key physical aspects of the fabric microstructure, architecture, and behavior. In the remainder of this section, a brief description is provided of the basic structure of a prototypical mesoscale fabric model. Note that to include the effect of weaving on the yarn material properties, both approaches to the fabric-level modeling typically make a distinction between the weft and the warp filaments. The material properties in the two types of yarns are degraded (by different amounts) from their reference values obtained in the yarn-level analysis.

A simple schematic of the unit cell which is used to represent the plain-woven single-ply fabric structure/architecture allotted to a single-yarn crossover in its initial (undeformed) configuration is depicted in Fig. 11.9(a). Note that the schematic displayed in Fig. 11.9(a) is a simplification of the corresponding unit cell model based on the 3D fabric structure/architecture whose solid 8-node finite element discretization is depicted in Fig. 11.10(a). A comparison of the models depicted in Figs. 11.9(a) and 11.10(a) shows that the two yarns are simplified in Fig. 11.9(a) as two 2-member truss elements, each with a cross section that is initially uniform and elliptical. A deformed configuration of the simplified unit cell model is depicted in Fig. 11.9(b). The corresponding deformed unit cell of the fabric described using the 3D finite element framework is depicted in Fig. 11.10(b).

It should be noted that the unit cell depicted in Fig. 11.9(a) is generally used to derive a shell-type fabric-level material model. Examination of this unit cell reveals the following:

1. Due to the truss-character of the yarn elements, the contact between the yarns at the crossover is reduced to a point.

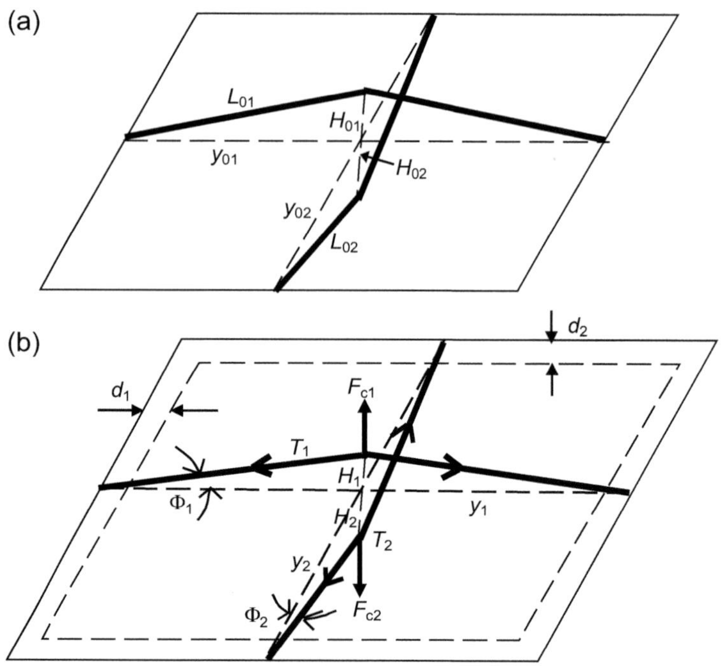

Figure 11.9 Geometrical representation of a single unit cell for a plain-woven single-ply fabric: (a) before and (b) after application of a normal biaxial in-plane deformation.

2. When determining the potential contact between the crossing yarns, their instantaneous half-thicknesses should be taken into account.
3. When the unit cell is stretched in a particular in-plane direction, the corresponding yarn is either decrimped/straightened or stretched. In the former case, no tension is built within the yarn, while in the latter case, tension is created within the yarn and, if sufficiently high, can cause yarn failure.
4. Furthermore, the extent of tension in a given yarn is affected by the failure status of the other crossing yarn. For example, if the crossing yarn is unbroken and in contact with the stretched yarn, tension will develop in the two yarns, the extent of which depends on the extent of contact overclosure (ie, the reduction in distance of the two summit points relative to the initial distance) and the functional relationship between the contact overclosure and the magnitude of the contact force.
5. Due to their low buckling resistance, compression cannot be built within the yarns when the cell is subjected to in-plane normal compressive strains.
6. Many other phenomena/scenarios are encountered in the fabric unit cell when the cell is subjected to different types of in-plane and out-of-plane loading, the loading which can cause breakage of one of the yarns, yarn crimping, yarn/yarn sliding, etc.

To derive the mesoscale fabric-type material model for shell elements, the unit cell is subjected generally to three types of loading: (1) in-plane biaxial tension/compression; (2) in-plane shear; and (3) out-of-plane (transverse) shear. For each of these loading conditions, the kinematic response of the unit cell is analyzed

(a)

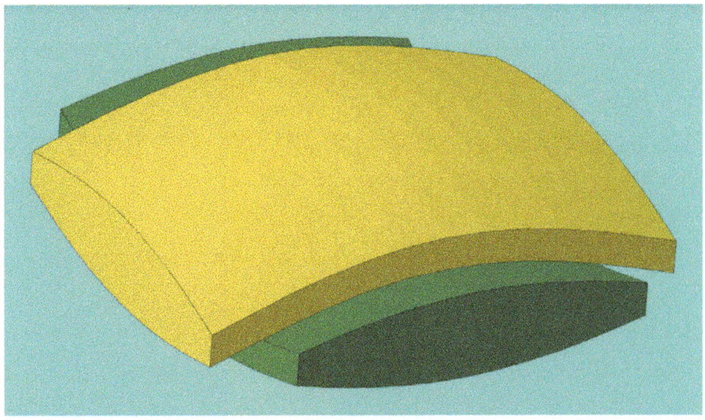

(b)

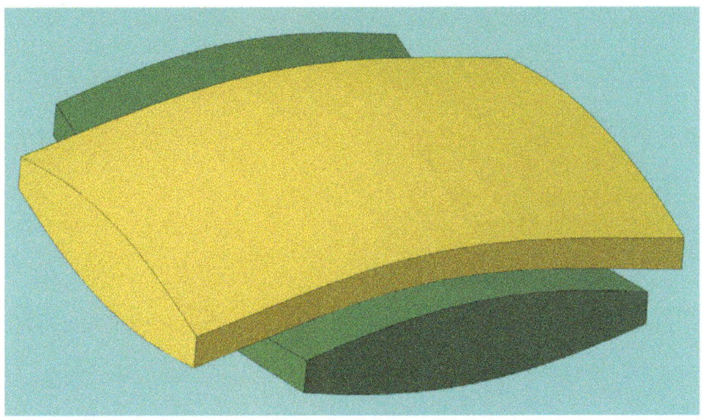

Figure 11.10 Three-dimensional finite element representation of the unit cell in a plain-woven single-ply fabric: (a) initial configuration; and (b) after in-plane biaxial stretching.

and the accompanying phenomena and processes, for example, yarn tension/crimping, yarn/yarn contact, and yarn/yarn sliding are identified. To infer the basic constitutive relations governing these phenomena/processes and to parameterize these relations, a 3D solid finite element method (FEM) model of a single unit cell is constructed, Fig. 11.10(a), and subjected to the same type of loading. For example, balanced biaxial in-plane tension is used to monitor the evolution of the yarn/yarn contact force as a function of the contact overclosure to infer the basic functional form and to parameterize the truss/truss contact force versus summit−point distance constitutive relationship. An example of such a functional relationship is depicted in Fig. 11.11(a). Examples of additional functional relationships obtained through the application of the in-plane shear and out-of-plane/transverse shear are shown in Fig. 11.11(b)−(c), respectively.

The final outcome of the procedure described in this section is a complete material constitutive model for a single ply of the reinforcing fabric. A constitutive model typically includes a generalized Hooke's law (which governs the stress evolution), an inelastic-deformation law (which governs the strain evolution), and a set of material-state evolution equations (equations which describe permanent changes in the state of the material during loading). This material model is subsequently combined, within a homogenization scheme, with the polymeric-matrix

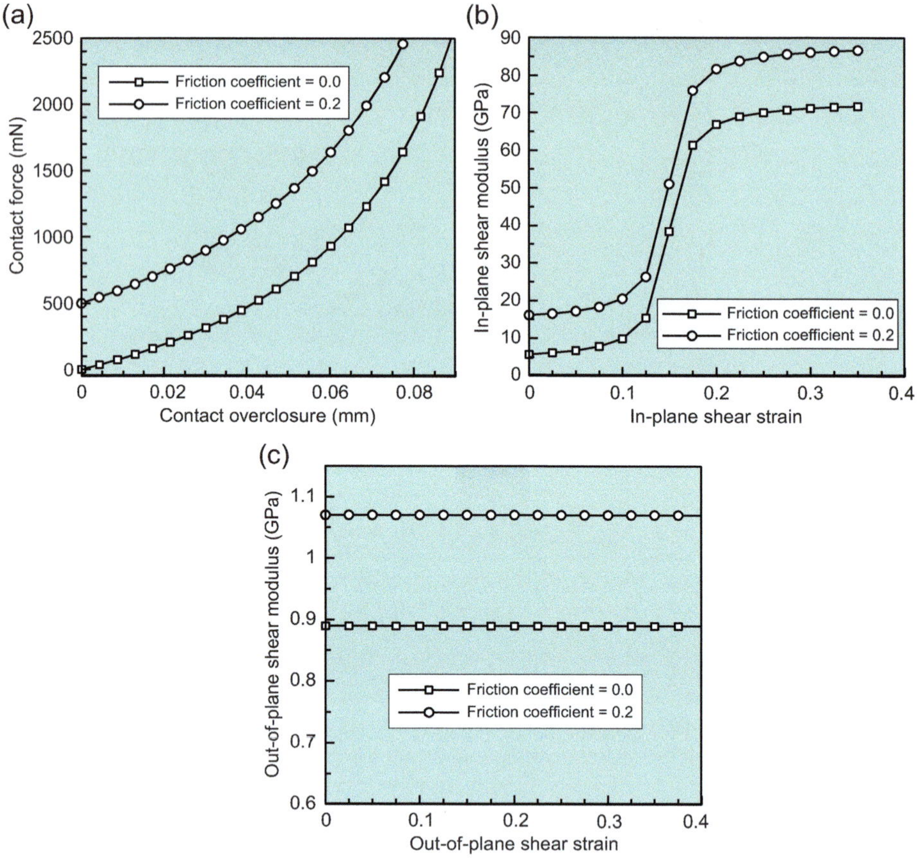

Figure 11.11 Typical results obtained in 3D FEM analyses of the deformation of the single-ply fabric unit cell under (a) in-plane biaxial tension; (b) in-plane shear; and (c) out-of-plane/transverse shear.

material model to determine the corresponding composite single-lamina material constitutive model.

11.4 Single-/stacked-lamina level modeling

11.4.1 Single-lamina material modeling

Within this length scale, the single-ply fabric material model is combined with the polymeric-matrix material model, within a homogenization procedure, to arrive at the effective material model for a single composite-material lamina. Typically, homogenization procedures such as the Generalized Method of Cells (Bednarcyk and Arnold, 2002) or the Voronoi cell–based homogenization procedure (Grujicic and Zhang, 1998) are used. The resulting single-lamina material model is subsequently passed to the stacked-lamina material model.

11.4.2 Stacked-lamina material modeling

Within this type of material model, separate constitutive relations are defined for the (homogenized) single-ply laminae and for the lamina/lamina interfaces, as well as an algorithm for their integration into a laminate-level material constitutive response (Ladeveze and Le Dantec, 1992; Ladeveze et al., 2000; Ladeveze, 2005).

11.4.2.1 Composite lamina material modeling

The single-lamina material constitutive model used in this case is an adaptation, simplification, and refinement of the corresponding material model described earlier in this section. Specifically, the single-lamina material model used is assumed (for the sake of computational efficiency) to be of a linear-elastic character with degradable elastic moduli. The number of the damage variables used, their evolution equations, and the effect of these variables on different elastic moduli are chosen in such a way that the resulting single-lamina material model yields the results which are consistent with the ones yielded by the single-lamina material model, described earlier in this section. The present single-lamina material constitutive model is enriched and further refined under the following basic assumptions and conditions:

1. Due to a relatively small lamina thickness, variations of the stress, strain, and damage fields through the lamina thickness can be ignored.
2. Material failure is controlled by progressive damage (resulting from high local in-plane tensile and shear stresses) and by quasibrittle fracture (controlled by high local compressive and through-the-thickness shear stresses).
3. Since damage nucleation and evolution are controlled by the growth of the associated material flaws/defects (the velocity of which is limited by a relativistic effect—based terminal velocity), the rate of damage progression is also limited.
4. The extent of material damage can be represented by three scalar internal damage variables: (1) one that affects materials mechanical response in the fiber direction; (2) one that affects material response in the transverse and in the through-the-thickness directions; and (3) one that affects shear response of the laminae. While the first variable is directly linked to fiber failure within the yarns, the other two damage variables are assumed to result from all the remaining intralaminar damage processes.

The main single-lamina material model components within the stacked-lamina-scale material model are (1) a damage evolution law and the effect of damage on lamina stiffness/compliance and on the failure; (2) contribution of the material damage to, and its interaction with, inelastic-deformation processes; and (3) integration of the material and stress/strain states over an arbitrary loading path/trajectory. Details of these three aspects of the meso-/lamina-scale material model can be found in Grujicic et al. (2010c).

11.4.2.2 Lamina/lamina-interface material modeling

Lamina/lamina-interfaces are generally modeled as 3D surface entities. Consequently, the interfacial stress (or more precisely, the surface traction) state is represented by one normal and two shear stresses. These stresses are taken to depend, in a nonlinear fashion, on the corresponding interfacial displacement discontinuities (ie, displacement differences between the two sides of the interface). The associated stress versus displacement-discontinuity functional relations are obtained by differentiating the corresponding interfacial potential functions (Grujicic et al., 2013e). Under quasistatic loading conditions, within the meso-/lamina-scale material model, evolution of the (mode I, II, and III) interfacial damage is assumed to be controlled by the corresponding thermodynamic forces (ie, energy release-rate-like quantities). Often, it is assumed that the three basic modes of interfacial damage are mutually related/coupled. This is accomplished within the model by assuming that the extents of all three basic damage modes depend on a single generalized energy release-rate-like quantity. Furthermore, through proper selection of the damage extent versus release rate functional relations, the interfacial response within the model can be made either brittle (instantaneous failure) or ductile (progressive damage and gradual failure). Under dynamic/impact loading conditions, the rate of interfacial damage accumulation may become limited by the terminal velocity at which the damage front can advance along the interface. It should be noted that, in contrast to the behavior of the individual laminae, lamina-/lamina-interfaces are assumed to remain elastic (until failure) except for the progressive loss of their stiffness (due to damage accumulation).

11.4.2.3 Fracture mechanics—enriched stacked-lamina model

Within the (classical) stacked-lamina-scale material model discussed earlier, Fig. 11.12(a), all damage modes are treated as being continuously distributed throughout the lamina interior or over lamina/lamina interfaces. While this assumption can be justified in the case of filament damage, filament/matrix debonding, and lamina-matrix degradation, the same cannot be said for transverse microcracking. That is, this type

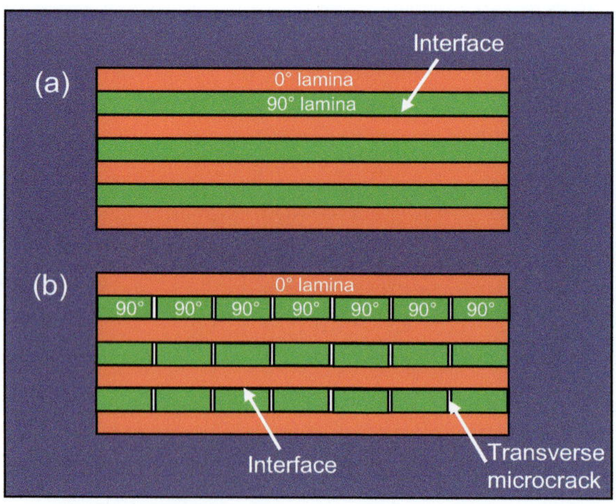

Figure 11.12 Schematic representation of the composite laminate as represented by (a) the classical stacked-lamina model; and (b) the fracture mechanics—enriched stacked-lamina model.

of damage is manifested as discrete cracks running parallel to the yarns and being nearly orthogonal to the lamina/lamina interfaces, Fig. 11.12(b). In transversely loaded laminae, spacing of these cracks decreases with loading by nucleation of new cracks at locations halfway between the existing microcracks. Since this process results in the formation of new (fracture) surfaces, it should be modeled using a fracture mechanics approach. However, in conventional fracture mechanics, one is generally concerned with crack growth under increased loading, while in the case of transverse microcracking, crack formation and their full extension over the width and thickness of the lamina is essentially instantaneous. To handle this challenge, Hashin (1985, 1996) initially proposed the so-called *finite fracture mechanics* approach. In this approach, the following basic assumptions and simplifications are made:

1. The density of microcracks is a function of the so-called *rupture-envelope* potential.
2. The functional relationship for the rupture-envelope potential is then constructed in such a way as to reflect that microcracking is the result of mixed-mode loading.
3. To represent the effect of microcracking on material degradation, three additional scalar damage variables are introduced.
4. Functional relationships between microcrack density and the three additional damage variables are typically established by carrying out the appropriate numerical simulations of microcracking in laminates in which microcracking is treated explicitly through the use of so-called *cohesive elements* (Grujicic et al., 2009c). An example of these functional relationships is provided in Fig. 11.13.

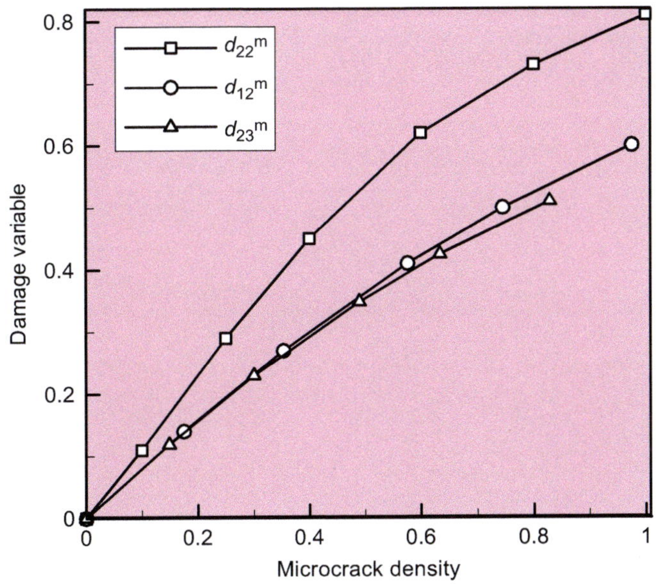

Figure 11.13 Functional relationships between the three damage variables (associated with transverse microcracking) and the microcrack density parameter, ρ. The three damage parameters respectively degrade: (a) transverse Young's modulus, (b) transverse shear modulus associated with shear along the fibers; and (c) transverse shear modulus associated with shear in the through-the-thickness direction.

5. Construction of the rupture-envelope potential requires knowledge of the three basic-mode fracture toughness parameters. These parameters are generally determined through simple mechanical testing of 0 degree/90 degree/0 degree three-ply laminates under pure tension or shear loading conditions, or using double-cantilever beam 0 degree/90 degree composite-laminate specimens and subjecting them to through-the-thickness tension at their free tip.

11.4.2.4 Typical results

Examples of the finite element results obtained using the stacked-lamina material model for a PPTA fabric—reinforced polymer-matrix composite laminate consisting

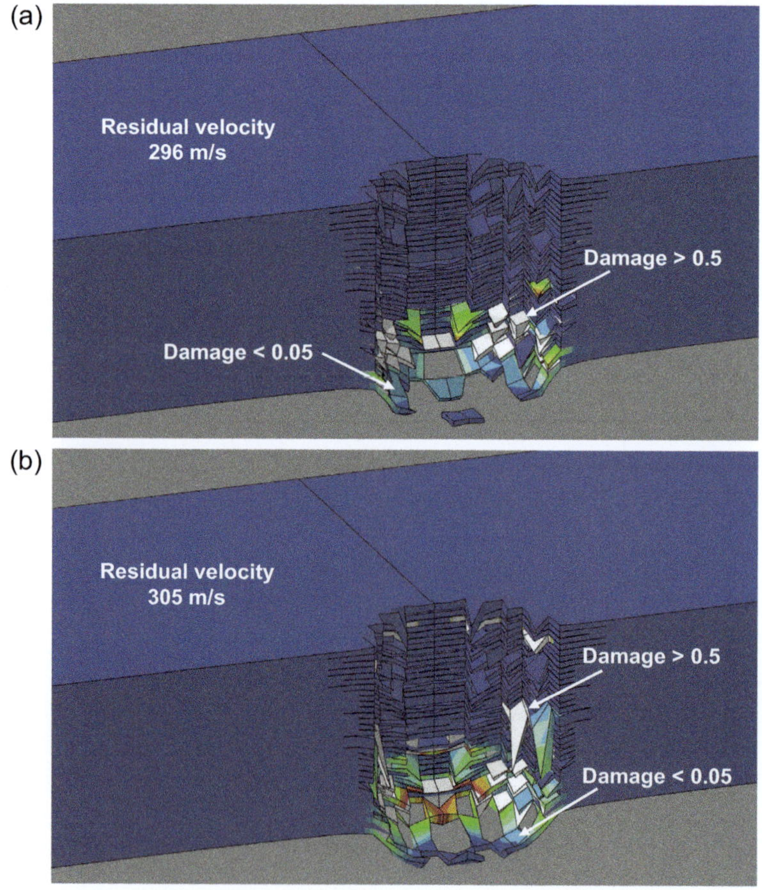

Figure 11.14 Finite element analysis results pertaining to the postmortem spatial distribution of the damage variable characterizing the extent of yarn fracture within the fabric plies, in a 32-laminae *p*-phenylene terephthalamide fabric—reinforced polymer-matrix composite laminate impacted by a 0.30-caliber steel FSP at a 0 degree obliquity angle and an incident velocity of 500 m/s: (a) the classical; and (b) the fracture mechanics—enriched stacked-lamina material models.

of 32 laminae and impacted (at a 0 degree obliquity angle and an incident velocity of 500 m/s) by a 0.30-caliber steel fragment-simulating projectile (FSP) are shown in Fig. 11.14(a)–(b). The results displayed in Fig. 11.14(a)–(b) reveal the postmortem distribution of the damage variable characterizing the extent of yarn fracture within the fabric plies. Examination of the results displayed in these figures reveals that (1) fracture mechanics enrichment of the stacked-lamina material model can significantly affect the response of the composite-laminate plate; and (2) the presence of lamina–lamina interfaces affects the local distribution of the material mechanical response, including damage.

11.5 Laminate-/continuum-level modeling

As in the case of the stacked-lamina-scale material models, laminate-type material models are also of a continuum character. However, in the case of the former material models, a material point is located either within the lamina interior or at the lamina/lamina interface, and the constitutive relations for the material point will differ depending on (lamina vs. interface) the location. In the case of the latter material models, the composite laminate is fully homogenized. Consequently, each material point resides in the same homogenized-composite material, and its behavior is governed by the same constitutive laws. Note, however, that despite the fact that the composite material is homogenized, during the derivation of the laminate-/continuum-scale material models, full recognition is given to the underlying material microstructure as well as to the associated deformation, damage, and failure processes. Furthermore, recognize that most laminate-/continuum-scale material models are constructed using the Continuum Damage Mechanics concepts and principles. That is, damage is treated as a continuum field variable which evolves during loading and degrades locally the material stiffness. The starting point in the derivation of the laminate-level material model is homogenization of the stacked-lamina material model, the model presented in the previous section. The resulting continuum-level material model is subsequently enriched, as presented in the remainder of this section, through incorporation of various phenomena and processes occurring within the composite materials subjected to high-rate, large-strain, high-pressure loading conditions.

11.5.1 Foundation of the laminate-/continuum-level model

The key features of this class of material models can be defined as follows (Grujicic et al., 2013a; Yen, 2012):

1. The material constitutive response is rate-dependent and accounts for the 3D aspects of material microstructure/architecture as well as for the high pressures accompanying impact loading.
2. The material models are strain based and utilize the concept of a ply/lamina-averaged local strain.
3. The models incorporate Mohr-Coulomb (Puck and Schurmann, 1998) shear strength effects; that is, they account for the fact that the (matrix) in-plane shear strength is generally a

function of the (tensile/compressive) character and magnitude of the through-the-thickness lamina strain.

4. Within the models, no distinction is made between fibers and yarns; that is, the terms fiber and yarn are either used interchangeably or replaced by the term filament.
5. The following three fiber and two matrix failure modes are considered: (1) fiber failure due to a combination of axial-tensile and transverse shear strains; (2) fiber buckling failure due to axial-compressive strains; (3) fiber crushing failure due to lamina through-the-thickness compressive strains; (4) matrix in-plane shear failure; and (5) matrix delamination failure due to combined effects of the through-the-thickness tensile strains and transverse shear strains.
6. The models are essentially of a linear-elastic orthotropic type with progressively degradable elastic stiffness constants. In other words, a generalized Hooke's law relation is assumed to hold between stress and strain quantities, while damage is treated as a process which takes place gradually during ballistic impact. Consequently, the models rely on the definition of the damage initiation criteria and the corresponding damage evolution laws.

11.5.2 Damage initiation criteria

Within the laminate-/continuum-scale material models, separate damage initiation criteria, the criteria which control the onset and continuation of material damage, are defined for each of the five aforementioned fiber/matrix failure modes. In addition, where appropriate (ie, fiber damage under combined axial tension and transverse shear and fiber damage under axial compression), distinction is made between weft and warp yarns. Consequently, damage initiation is described by a set of seven equations with the following general form: a strain-based loading function has to be greater than or equal to the particular damage-type material resistance for the damage of this type to initiate or continue to evolve.

11.5.2.1 Fiber damage under combined axial tension and transverse shear

The onset and progression of this damage/failure mode is assumed to be defined by a superposition of the quadratic contributions of the axial-tensile and transverse shear strains. Furthermore, separate damage initiation functions are defined for the weft and the warp fiber directions. Also, it is generally assumed within the models that the laminate transverse shear strength is controlled by the corresponding shear strength of the fiber itself.

11.5.2.2 Fiber damage under axial compression

The onset and progression of this damage/failure mode is assumed to be controlled by the compressive axial strains whose effect is modulated by the lamina through-the-thickness compressive strain. In other words, the tendency of fibers to buckle under axial compression is resisted by the stabilizing effect of the through-thickness compressive strain. For this damage/failure mode, distinction is also made between the weft and the warp fiber directions.

11.5.2.3 Fiber damage under transverse compressive loading

This damage/failure mode involves fiber crushing under sufficiently high through-the-thickness compressive strain. No distinction is made here between the weft and the warp fiber directions.

11.5.2.4 Matrix in-plane shear failure

This mode of damage/failure is associated with the fact that the matrix shear strength is lower than the fiber shear strength and that the composite material in question possesses a laminated architecture. Consequently, sufficiently high in-plane shear strains may cause matrix shear failure without giving rise to any significant fiber damage.

11.5.2.5 Matrix delamination due to through-the-thickness tensile and shear stresses

This mode of damage/failure also gives rise to the matrix in-plane failure. The loading function in this case contains contributions from tensile through-the thickness strain and transverse shear strains. Furthermore, within the models the effect of the (compressive) through-the-thickness strains on the matrix shear strength is taken into account through the application of the Mohr-Coulomb law. Also, nonlocal stress/strain concentration effects are introduced in the non-delaminated portion of the material adjacent to the delamination front. The inclusion of such stress/strain concentration effects have been often found necessary to obtain a satisfactory agreement between the computational and experimental results pertaining to the spatial distribution and temporal evolution of the delamination zone (Yen, 2012).

11.5.3 Damage evolution and material degradation

Each of the seven damage initiation relations mentioned previously defines, within the strain space, a damage initiation/progression hypersurface. The size of each of the seven hypersurfaces scales inversely with the magnitude of six (three Young's and three shear) material stiffness moduli. Consequently, as the damage progresses and the material stiffness degrades, the damage initiation/progression hypersurfaces expand (ie, additional straining is required to produce additional damage). Within most of the laminate-/continuum-scale material models, damage progression is assumed to cause degradation in material stiffness (without an accompanying loss in material strength) properties.

To model (irreversible) damage evolution and the associated (irreversible) degradation in material stiffness, the following procedure is typically applied: (1) First, six stiffness-damage variables, one for each of the six elastic moduli, is introduced. (2) Then, the evolution equations for each of the six damage variables are assembled, and the contribution of the five (or more precisely, seven) damage/failure modes to these equations are identified and quantified. (3) Each of the six damage variables is updated, at a given level of applied strains and material stiffness and strength parameters, by integrating their respective evolution equations. (4) Next, the six updated damage variables are assembled into a damage matrix, and this matrix is then used, through a proper multiplication operation, to update the material compliance matrix; (5) The updated compliance matrix is inverted to obtain the corresponding stiffness matrix. (6) The updated stiffness matrix is used within the accompanying boundary-value problem (eg, transverse impact of a projectile onto a composite-laminate panel) to update the applied strains, and the procedure is resumed starting with step (3).

During identification and quantification of different damage/failure modes to each of the six elastic moduli, the following approach is generally taken: (1) failure of the weft and warp fibers both in tension and transverse shear and in compression degrades the associated Young's modulus and shear moduli; (2) fiber crushing mode degrades all six elastic moduli; (3) in the case of the in-plane shear failure, only the associated shear modulus is affected; and (4) delamination degrades the through-the-thickness Young's modulus as well as the two transverse shear moduli.

In general, the damage variable evolution equation can, in principle, give rise to a value of 1.0 for the corresponding damage variable. In this case, the material completely loses the associated load carrying capacity. While this is physically realistic for some of the failure modes, for example, the fiber failure under axial tension/transverse shear, it may not be fully realistic under other failure modes, such as fiber axial compression. Consequently, selected damage variable evolution equations have to be defined in such a way that they allow a nonzero value of the corresponding material stiffness modulus.

11.5.4 Rate dependency of strength and stiffness

For the laminate-/continuum-scale material models to be applicable in the ballistic/blast computational analyses, they must include (and they generally do) the effects of loading rate on the material stiffness and strength. These rate-sensitivity effects are typically included through the use of appropriate thermal activation-based or scaling laws (Grujicic et al., 2013a; Puck and Schurmann, 1998). The accompanying thermal effects are generally accounted for through the use of temperature-dependent material parameters and by assuming adiabatic conditions. That is, due to the short duration of the blast/ballistic impact events, the effects of heat conduction (and other heat-transfer modes) is neglected.

11.5.5 Typical results

In Fig. 11.15(a)–(d), a set of finite element results is presented pertaining to the spatial distribution of the composite material (modeled using the continuum-level material model) being impacted transversely by a right-circular, cylindrical, rigid projectile at a 0 degree obliquity angle. It should be noted that to enable monitoring of the penetration process, the target plate is made transparent. The results displayed in Fig. 11.15(a)–(d) clearly reveal the development of the back-face bulge. It is generally found that details regarding the spatial distribution of the material during impact as well as the overall penetration resistance of the composite laminate can be greatly

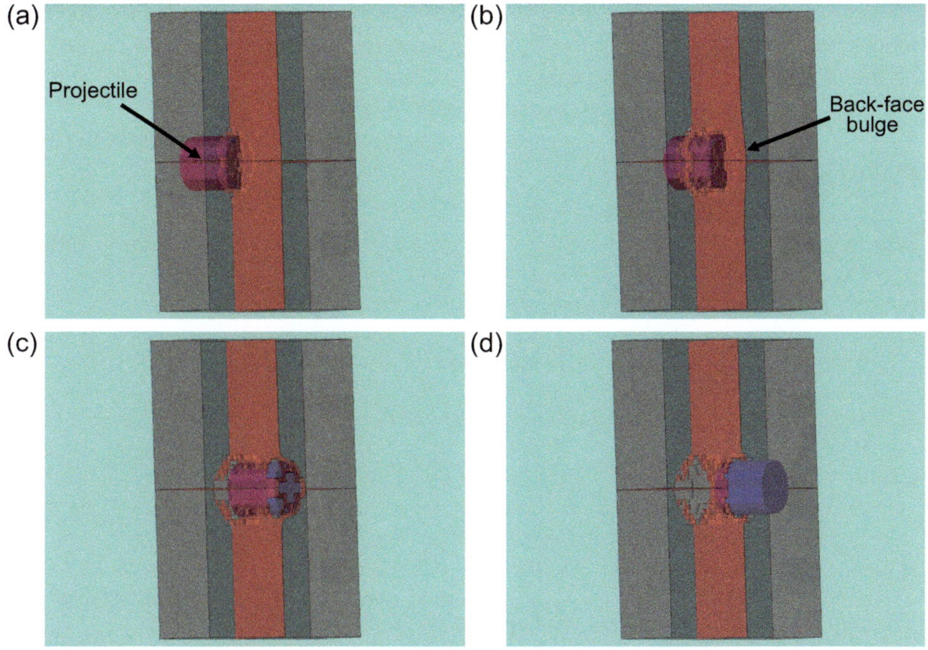

Figure 11.15 Typical results pertaining to the temporal evolution of the spatial distribution of the composite-laminate material during impact by a right-circular, cylindrical, rigid projectile.

affected by the choice of the functional forms and the parameterization of the constitutive relations governing the kinematic and mechanical response of the composite-laminate material. This finding justifies the use of a comprehensive multilength-scale computational approach, such as the one presented in this overview, for inferring and calibrating such relationships.

11.6 Conclusions

It should be noted that the present overview was focused on PPTA-based filaments and the associated polymer-matrix composites. This was done for two main reasons: (1) to display the full richness of the hierarchical nature of the material microstructure and properties; and (2) to be able, within the limited space available for this review, to highlight some specific aspects of the material microstructure and behavior at different length scales. Other types of advanced filaments and associated polymer-matrix composites are also used in various armor applications. Among these filaments, the ones most frequently used include filaments made of nonpolymeric materials (eg, S2-glass composed of silica (SiO_2), alumina (Al_2O_3), and magnesia (MgO)) and other classes of polymeric materials (eg, high molecular weight polyethylene, polybenzobisoxazole, polypyridobisimidazole, Zylon, etc.). For each of these filaments and the associated composite materials, one can identify a hierarchy of microstructural scales and develop the appropriate material models to capture the constituent behavior of the material at the specific length scale. For example, a multilength-scale approach to the material model development for the case of glass-reinforced polymer-matrix composites can be found in Grujicic et al. (2011a,d).

References

Bednarcyk, B.A., Arnold, S.M., 2002. MAC/GMC 4.0 User's Manual. NASA/TM−2002−212077.

Blades, H., 1973. Dry jet wet spinning process. U.S. Patent 3767756.

Cheng, M., Chen, W., Weerasooriya, T., 2004. Experimental investigation of the transverse mechanical properties of a single Kevlar KM2 fiber. International Journal of Solids and Structures 41, 6215−6232.

du Pont, E.I., 1983. Technical Brochure, "Weight Savings for Aircraft Using Kevlar Aramid Fiber".

Grujicic, M., Zhang, Y., 1998. Determination of effective elastic properties of functionally graded materials using Voronoi cell finite element method. Materials Science and Engineering A 251, 64−76.

Grujicic, M., Cao, G., Roy, W.N., 2004. Atomistic modeling of solubilization of carbon nanotubes by non-covalent functionalization with poly (*p*-Phenylenevinylene-*co*-2, 5-dioctoxy-*m*-phenylenevinylene). Applied Surface Science 227, 349−363.

Grujicic, M., Angstadt, D.C., Sun, Y.-P., Koudela, K.L., 2007a. Micro-mechanics based derivation of the materials constitutive relations for carbon nanotube reinforced poly-vinyl-ester-epoxy based composites. Journal of Materials Science 42, 4609−4623.

Grujicic, M., Pandurangan, B., Angstadt, D.C., Koudela, K.L., Cheeseman, B.A., 2007b. Ballistic-performance optimization of a hybrid carbon-nanotube/E-glass reinforced poly-vinyl-ester-epoxy-matrix composite armor. Journal of Materials Science 42, 5347−5359.

Grujicic, M., Bell, W.C., Biggers, S.B., Koudela, K.L., Cheeseman, B.A., 2008a. Enhancement of the ballistic-protection performance of e-glass reinforced poly-vinyl-ester-epoxy composite armor via the use of a carbon-nanotube forest-mat strike face. Journal of Materials: Design and Applications 222, 15−28.

Grujicic, M., Arakere, G., He, T., Gogulapati, M., Cheeseman, B.A., 2008b. A numerical investigation of the influence of yarn-level finite-element model on energy absorption by a flexible-fabric armor during ballistic impact. Journal of Materials: Design and Applications 222, 259−276.

Grujicic, M., Arakere, G., He, T., Bell, W.C., Cheeseman, B.A., Yen, C.-F., Scott, B., 2008c. A ballistic material model for cross-plied unidirectional ultra-high molecular weight polyethylene fiber-reinforced armor-grade composites. Materials Science and Engineering A 498, 231−241.

Grujicic, M., Bell, W.C., He, T., Cheeseman, B.A., 2008d. Development and verification of a meso-scale based dynamic material model for plain-woven single-ply ballistic fabric. Journal of Material Science 43, 6301−6323.

Grujicic, M., Arakere, G., He, T., Bell, W.C., Glomski, P.S., Cheeseman, B.A., 2009a. Multi-scale ballistic material modeling of cross-plied compliant composites. Composites Part B: Engineering 40, 468−482.

Grujicic, M., Glomski, P.S., He, T., Arakere, G., Bell, W.C., Cheeseman, B.A., 2009b. Material modeling and ballistic-resistance analysis of armor-grade composites reinforced with high-performance fibers. Journal of Materials Engineering and Performance 18, 1169−1182.

Grujicic, M., Sellappan, V., Kotrika, S., Arakere, G., Obieglo, A., Erdmann, M., Holzleitner, J., 2009c. Suitability analysis of a polymer metal hybrid technology based on high-strength steels and direct polymer-to-metal adhesion for use in load-bearing automotive body-in-white applications. Journal of Materials Processing Technology 209, 1877–1890.

Grujicic, M., Bell, W.C., Arakere, G., He, T., Cheeseman, B.A., 2009d. A meso-scale unit-cell based material model for the single-ply flexible-fabric armor. Materials and Design 30, 3690–3704.

Grujicic, M., Bell, W.C., Arakere, G., He, T., Xie, X., Cheeseman, B.A., 2010a. Development of a meso-scale material model for ballistic fabric and its use in flexible-armor protection systems. Journal of Materials Engineering and Performance 19, 22–39.

Grujicic, M., Arakere, G., Bell, W.C., Marvi, H., Yalavarthy, H.V., Pandurangan, B., Haque, I., Fadel, G.M., 2010b. Reliability-based design optimization for durability of ground-vehicle suspension-system components. Journal of Materials Engineering and Performance 19, 301–313.

Grujicic, M., He, T., Marvi, H., Cheeseman, B.A., Yen, C.-F., 2010c. A comparative investigation of the use of laminate-level meso-scale and fracture-mechanics enriched meso-scale composite-material models in ballistic resistance analyses. Journal of Materials Science 45, 3136–3150.

Grujicic, M., Bell, W.C., Glomski, P.S., Pandurangan, B., Cheeseman, B.A., Fountzoulas, C., Patel, P., Templeton, D.W., Bishnoi, K.D., 2011a. Multi-length scale modeling of high-pressure induced phase transformations in soda-lime glass. Journal of Materials Engineering and Performance 20, 1144–1156.

Grujicic, M., Bell, W.C., Glomski, P.S., Pandurangan, B., Yen, C.-F., Cheeseman, B.A., 2011b. Filament-level modeling of aramid-based high performance structural materials. Journal of Materials Engineering and Performance 20, 1401–1413.

Grujicic, M., Bell, W.C., Glomski, P.S., Pandurangan, B., Yen, C.-F., Cheeseman, B.A., 2011c. Multi-length scale computational derivation of Kevlar® yarn-level material model. Journal of Materials Science 46, 4787–4802.

Grujicic, M., Pandurangan, B., Bell, W.C., Cheeseman, B.A., Patel, P., Gazonas, G.A., 2011d. Molecular-level analysis of shock-wave physics and derivation of the Hugoniot relations for soda-lime glass. Journal of Materials Science 46, 7298–7312.

Grujicic, M., Pandurangan, B., Bell, W.C., Cheeseman, B.A., Yen, C.-F., Randow, C.L., 2011e. Molecular-level simulations of shock generation and propagation in polyurea. Materials Science and Engineering A 528, 3799–3808.

Grujicic, M., Pandurangan, B., Bell, W.C., Yen, C.-F., Cheeseman, B.A., 2011f. Application of a dynamic-mixture shock-wave model to the metal-matrix composite materials. Materials Science and Engineering A 528, 8187–8197.

Grujicic, M., Pandurangan, B., d'Entremont, B.P., Yen, C.-F., Cheeseman, B.A., 2012a. The role of adhesive in the ballistic/structural performance of ceramic/polymer-matrix composite hybrid armor. Materials and Design 41, 380–393.

Grujicic, M., Hariharan, A., Pandurangan, B., Yen, C.-F., Cheeseman, B.A., Wang, Y., Miao, Y., Zheng, J.Q., 2012b. Fiber-level modeling of dynamic strength of Kevlar® KM2 ballistic fabric. Journal of Materials Engineering and Performance 21, 1107–1119.

Grujicic, M., Yavari, R., Snipes, J.S., Ramaswami, S., Runt, J., Tarter, J., Dillon, G., 2012c. Molecular-level computational investigation of shock-wave mitigation capability of polyurea. Journal of Materials Science 47, 8197–8215.

Grujicic, M., Pandurangan, B., Snipes, J.S., Yen, C.-F., Cheeseman, B.A., 2013a. Multi-length scale enriched continuum-level material model for Kevlar®-fiber reinforced polymer-matrix composites. Journal of Materials Engineering and Performance 22, 681–695.

Grujicic, M., Ramaswami, S., Snipes, J.S., Yavari, R., Lickfield, G.C., Yen, C.-F., Cheeseman, B.A., 2013b. Molecular-level computational investigation of mechanical transverse behavior of *p*-phenylene terephthalamide (PPTA) fibers. Multidiscipline Modeling in Materials and Structures 9, 462–498.

Grujicic, M., Yavari, R., Snipes, J.S., Ramaswami, S., Yen, C.-F., Cheeseman, B.A., 2013c. Molecular-level study of the effect of prior axial compression/torsion on the axial-tensile strength of PPTA fibers. Journal of Materials Engineering and Performance 22, 3269–3287.

Grujicic, M., Ramaswami, S., Snipes, J.S., Yavari, R., Yen, C.-F., Cheeseman, B.A., 2013d. Axial-compressive behavior, including kink-band formation and propagation, of single *p*-phenylene terephthalamide (PPTA) fibers. Advances in Materials Science and Engineering 2013. Article ID 329549. http://dx.doi.org/10.1155/2013/329549.

Grujicic, M., Ramaswami, S., Snipes, J.S., Galgalikar, R., Chenna, V., Yavari, R., 2013e. Computational investigation of roller-bearing premature-failure in horizontal-axis wind-turbine gearboxes. Solids and Structures 2, 46–55.

Grujicic, M., Snipes, J.S., Ramaswami, S., Yavari, R., Ramasubramanian, M.K., 2014a. Meso-scale computational investigation of shock-wave attenuation by trailing release-wave in different grades of polyurea. Journal of Materials Engineering and Performance 23, 49–64.

Grujicic, M., Snipes, J.S., Ramaswami, S., Yavari, R., Runt, J., Tarter, J., Dillon, G., 2014b. Coarse-grained molecular-level analysis of polyurea properties and shock-mitigation potential. Journal of Materials Engineering and Performance 22, 1964–1981.

Grujicic, M., Snipes, J.S., Ramaswami, S., Yavari, R., Yen, C.-F., Cheeseman, B.A., 2014c. Analysis of steel-with-composite material substitution in military-vehicle hull-floors subjected to shallow-buried landmine-detonation loads. Multidiscipline Modeling in Materials and Structures 10 (3), 416–448.

Grujicic, M., Yavari, R., Snipes, J.S., Ramaswami, S., 2014d. The effect of plain-weaving on the mechanical properties of warp and weft *p*-phenylene terephthalamide (PPTA) fibers/yarns. Journal of Materials Science 49 (24), 8272–8293. http://dx.doi.org/10.1007/s10853-014-8536-1.

Hashin, Z., 1996. Finite thermoelastic fracture criterion with application to laminate cracking analysis. Journal of the Mechanics and Physics of Solids 44, 1129–1145.

Hashin, Z., 1985. Analysis of cracked laminates: a variational approach. Mechanics of Materials 4, 121–136.

Knoff, W.F., 1987. A modified weakest-link model for describing strength variability of Kevlar aramid fibers. Journal of Materials Science 22, 1024–1030.

Kwolek, S.L., 1972. Optically Anisotropic Aromatic Polyamide Dopes, U.S. Patent 3671542.

Ladeveze, P., 2005. Multiscale computational damage modelling of laminate composites. In: Sadowski, T. (Ed.), Multiscale Modelling of Damage and Fracture Processes in Composite Materials. Springer-Verlag, Berlin, pp. 171–212.

Ladeveze, P., Le Dantec, E., 1992. Damage modelling of the elementary ply for laminated composites. Composites Science and Technology 43, 257–267.

Ladeveze, P., Allix, O., Deu, J.-F., Leveque, D., 2000. A mesomodel for localisation and damage computation in laminates. Computer Methods in Applied Mechanics Engineering 183, 105–122.

Newell, J.A., Sagendorf, M.T., 1999. Experimental verification of the end-effect Weibull model as a predictor of the tensile strength of Kevlar 29 (poly *p*-phenylene terephthalamide) at different gage lengths. High Performance Polymers 11, 297–305.

Puck, A., Schurmann, H., 1998. Failure analysis of FRP laminates by means of physical based phenomenological models. Composites Science and Technology 58, 1001–1010.

Steenbakkers, L.W., Wagner, H.D., 1988. Elasticity and mechanical breakdown of Kevlar 149 aramid fibers by probabilistic approach. Journal of Materials Science Letters 7, 1209–1212.

Yen, C.-F., 2012. A ballistic material model for continuous-fiber reinforced composites. International Journal of Impact Engineering 46, 11–22.

Stab characterization of STF and thermoplastic-impregnated ballistic fabric composites

12

H.M. Rao, M.V. Hosur, S. Jeelani
Tuskegee University, Tuskegee, AL, United States

12.1 Introduction

Flexible stab- and puncture-resistant materials are required for a range of applications, including industrial safety wear, needle-resistant gloves and garments for medical workers [1], inflatable structures [2,3], tamper-proof packaging, and ruggedized soft carriers. One particularly challenging application is stab-resistant body armor. Flexible, comfortable systems are required to protect law enforcement and security personnel against stab attacks. These concerns are of particular concern in European countries, where stab assaults are much more likely than firearm attacks [4].

Stab attacks refer to impacts in which the direction of thrust of the stabbing weapon is perpendicular to the target. In contrast, *slash* attacks refer to cases where the weapon is swept in a direction parallel to the target during engagement [5]. In general, preventing penetration from stab attacks is significantly more difficult than providing protection from slash attacks [6].

Stab weapons can be classified as edged or pointed [7]. Edged weapons, or knives, include a long, continuous cutting edge. In contrast, pointed weapons consist of a slender rod with a pointed tip. Pointed weapons, also referred to as spikes, include ice picks and awls. General-purpose stab vests, such as for police officers, require both knife and spike protection. Specialty applications, such as for corrections officers, require only spike protection.

Stab-resistant body armors can be characterized according to the National Institute of Justice (NIJ) Standard 0115.00 [8]. This standard defines weapons of two distinct nature; two-edged weapons like a commando knife and one-edged weapons like a spike. The weapons are individually mounted in a damped two-part drop mass, designed to replicate the dynamics of a stab attack [9]. The drop mass is dropped onto targets to achieve six discrete impact velocities, generating stab energies that represent likely stab attacks of increasing severity. Armors are mounted to a prescribed stack of foams and elastomers that provide a damped target response similar to armor mounted on a human torso [9]. Penetration through the armor, and into the target backing, is measured for each stab event. To meet NIJ requirements, the penetration through the armor must

Advanced Fibrous Composite Materials for Ballistic Protection. http://dx.doi.org/10.1016/B978-1-78242-461-1.00012-1

fall below prescribed maximum values, over multiple impacts, and under both normal and oblique impact directions.

The inherent tensile strength and cut resistance of high-performance fibers, such as aramids and ultra-high molecular weight polyethylene, make them good candidates for the construction of flexible, stab-resistant materials [10–13]. One approach is to weave yarns of these fibers into plain-woven fabrics. These fabrics can provide outstanding ballistic resistance, but in general provide poor stab resistance. Special woven geometries, including high yarn count fabrics, such as DuPont™ Correctional Kevlar®, are efficient at stopping spike threats but do not protect against knife threats. Ring mesh is commonly used to add knife resistance to body armors. However, ring mesh is very heavy and does not provide spike resistance. Rigid inserts such as metal or ceramic plates provide excellent stab resistance, but their bulk, weight, and stiffness significantly degrade the comfort and dexterity of the wearer.

An alternative approach is to incorporate polymer coatings or laminated films into woven fabrics. A number of commercial stab-resistant fabrics utilize this approach, under trade names including Argus™ (Barrday Inc.), Kevlar® MTP™ (DuPont), and Twaron® SRM™ (Teijin-Twaron). The polymer matrix provides additional toughness to the woven fabrics, requiring more energy to cut and separate fibers and yarns. In some cases, abrasive fillers are added to the coatings to further enhance cut resistance. An additional advantage of these materials is, because they are based on woven high-performance fibers, they can be used to construct body armors that are resistant to both stab and ballistic attacks. Note that other thermoplastic laminated fabric products, with tradenames such as SpectraShield™ (Honeywell Inc.) and Dyneema® UniDirectional (DSM Dyneema B.V.), utilize unidirectional fibers rather than woven fabrics. These unidirectional laminates have been shown to provide outstanding ballistic resistance but poor stab resistance.

In spite of the presence of commercial stab-resistant fabrics, there are few published studies on stab penetration of body armor materials, and none focus on thermoplastic-coated fabrics. Flambard and Polo [14] characterized the stab resistance of knitted fabrics using a nonstandard, undamped drop mass system. Gadow and Niesen [15] found that a thermally sprayed ceramic coating increased the quasistatic knife penetration resistance of a woven aramid fabric. Anctil et al. [16] compare various armor designs subject to both stab and ballistic impacts. Unfortunately, however, they were not permitted to disclose the composition of the armor designs that were being compared. Termonia [17] presents a numerical model for the quasistatic spike puncture of a woven fabric. Shedden et al. [18] and Wang et al. [19] perform finite element simulations of knife stab into woven fabric armors. Decker et al. [20] compare the stab performance of fabrics with and without shear thickening fluid (STF) treatments using an undamped drop mass. Decker et al. also provide a comprehensive review of previous studies on the puncture and stab testing, under dynamic and quasistatic conditions, for a range of fabrics and engineering materials.

A number of studies on the low-velocity impact performance of thermoplastic-matrix composites have been conducted, but in most cases the composites are fully laminated into relatively rigid plates, and most are based on glass fiber reinforcement

[21–25]. Furthermore, these studies utilize hemispherically tipped impactors, rather than sharp-edged or sharp-tipped weapons. Alcock et al. [26,27] performed low-velocity, penetrating impact studies on woven tapes of polypropylene fibers in a polypropylene matrix. Xu and Farris [28] studied the low-velocity impact of single-layer SpectraShield and mildly compacted woven UHWMPE (Spectra) fabrics. A number of publications on the ballistic behavior of thermoplastic-matrix unidirectional laminates also exist [29–32], but few consider thermoplastics reinforced by woven fabrics [33,34].

Application of STF enhances the impact resistance performance of the Kevlar® fabric [35,36]. STFs are non-Newtonian fluids which show drastic increase in viscosity beyond a critical shear rate [37]. Decker et al. [20] investigated the stab resistance of STF-treated Kevlar® and nylon fabrics and found to exhibit significant improvements over neat fabric targets of equivalent areal density. Specifically, dramatic improvements in puncture resistance (spike threat) are observed under high- and low-speed loading conditions, while slight increases in cut protection (knife threat) are also observed. Studies on the effect of fabric architecture indicate that STF addition provides benefits analogous to the effect of increasing fabric yarn count, with STF addition primarily reducing the mobility of filaments and yarns in the impact zone. Lee et al. [38] studied the ballistic penetration performance of a composite material composed of woven Kevlar® fabric impregnated with a colloidal STF (silica particles (450 nm) dispersed in ethylene glycol). The impregnated Kevlar® fabric yielded a flexible, yet penetration-resistant composite material. Fragment simulation projectile ballistic penetration measurements at ~244 m/s were performed to demonstrate the efficacy of the novel composite material. The results demonstrated a significant enhancement in ballistic penetration resistance due to the addition of STF to the fabric, without any loss in material flexibility. Majumdar et al. [39] investigated optimal design of soft body armor materials by treating Kevlar® fabrics with silica nanoparticle–based STF. They studied the effect of silica concentration, padding pressure, and diluent: STF ratio (solvent ratio) on STF add-on% and impact energy absorption. Silica concentration, solvent ratio, and square of solvent ratio were found to be statistically significant terms influencing the STF add-on% on Kevlar® fabrics. On the other hand, silica concentration, padding pressure, and the square of solvent ratio were the statistically significant terms influencing the impact energy absorption.

For the work reported in this chapter, STF was prepared by evaporation method using polyethylene glycol (PEG) and spherical silica nanoparticles. Ethanol was added to the silica/PEG solution at a 2:1 ratio of ethanol to silica/PEG. Sonication was carried out for 5 h, and ethanol was evaporated using a magnetic stirrer at elevated temperature of about 100°C for 3 h. STF was impregnated in Kevlar® and correctional Kevlar® fabrics. Stab tests were carried out on neat and STF-impregnated fabrics. It is possible that STF is washed away when exposed directly to water or moisture. To overcome this, a hybrid system was developed and studied. This hybrid system consisted of two layers of thermoplastic-impregnated fabrics used as top and bottom layers of the target, thereby protecting the STF-impregnated fabrics in between. Various combinations like neat, STF-impregnated (nonhybrid), STF and thermoplastic integrated (hybrid) were tested to see the change in stab resistance offered. For thermoplastic impregnation, Surlyn (3mil), a high-performance, tough, and expensive film that is manufactured by DuPont, was used based on the study carried out by Mayo et al. [40].

12.2 Experimental procedure

12.2.1 Materials and processing

The thermoplastic film used in this study was Surlyn® poly(ethylene-co-methacrylic acid) (SU). SU, which is manufactured by the DuPont Company, is a high-performance, tough and comparatively expensive thermoplastic film used for higher value packaging applications. The SU films (Cincinnati Printers, Cincinnati, Ohio) used in this study had thicknesses of three mil (125 μm). The composite fabrics used in this study were JPS Composites Style 706 Kevlar® (Anderson, South Carolina), a 13 × 13 yarns-per-cm plain-woven fabric composed of 600 denier KM-2 yarns; JPS Composites Kevlar® K-15 style 779 (Anderson, South Carolina), a 12 × 12 yarns-per-cm plain-woven fabric composed of 200 denier k-15 yarns; nylon manufactured by JPS composites (Anderson, South Carolina); and Snowtex®, a colloidal silica with particle size ranging from 90 to 100 nm, manufactured by Nissan Chemical, USA (Houston, Texas).

Colloidal silica-MP1040 (a mixture of 60% water and 40% silica) and polyethylene glycol (PEG) were used in this study to prepare the STF. Colloidal silica was mixed with PEG and stirred manually for 30 min. The mixture was then kept in an oven for 24 h at a constant temperature of 105°C in a glassware. This was done to evaporate the water in STF. Weight of the sample, pre-/post-heating, was noted to get the exact weight of STF required. The ratio of MP1040 and PEG in mixture was 43% and 57%.

For impregnating with STF, fabrics were cut according to NIJ standards. Each fabric measured 15 inches in width and 15 inches in breadth. Each of these fabric layers were then immersed in the prepared STF which was diluted in ethanol. Ethanol was used to make the process of impregnating the fabric easy. A nip roller was used to remove the excess amount of STF from the fabric. A nip roller basically consists of two steel rollers adjacent to each other, one of which has a rubber coating. Fabrics were dipped in STF for about 60 s, then passed between the rollers to remove excess STF, which also aided in uniform distribution of STF on the fabric. These fabrics were then dried in an oven for 8 h at a temperature of 75°C and later at room temperature for 24 h.

The thermoplastic Kevlar® fabric layer was prepared by melt impregnating TP onto the fabric using a hot press. An investigation of the processing effects was conducted to observe how the TP-film would adhere or wet through the fabric. It was observed that at pressures from 138 to 276 kPa and temperatures from about 93 to 260°C, many different effects would take place due to various processing conditions. For instance, for all pressures with temperatures less than 110°C, there was poor adhesion of TP films to the fabric. For pressures of about 193 kPa with temperatures between 110°C and 150°C, there was moderate adhesion. Pressures of about 193 kPa with temperatures above 150°C yielded a completely saturated TP-Kevlar® fabric. Table 12.1 presents the description of each fabric type used in the study.

Table 12.1 Properties of the single-layer fabric composite sheets

Fabric type	Thickness (cm)	Measured areal density per sheet (g/cm^2)
Kevlar	0.0459	0.0189
STF Kevlar	0.0463	0.0226
Correctional Kevlar	0.0432	0.0131
STF correctional Kevlar	0.0437	0.0167
TP-Kevlar	0.0535	0.0277
3-mil Surlyn	0.0076	0.0085

12.2.2 Target preparation

For the characterization of stab response, fabric layers were selected such that they had similar areal density. Fifteen layers of neat Kevlar® having an areal density of 0.2835 g/cm^2 were chosen to be target for comparison of data. Targets were set according to NIJ standards. In a nonhybrid test, combinations having a single type of fabrics were used as either neat or STF fabrics. In a hybrid test a combination of TP fabric with STF fabric and neat fabric was studied. The number of layers in each target fabric was determined by matching the areal density of each target to the areal density of the 15-layer neat Kevlar® target. Targets were strapped on to backing material using nylon fabric belts. This is to make sure target fabrics are in place during stab testing. Backing material consisted of two layers of rubber at the bottom followed by a layer of polyethylene foam and alternate layers of neoprene sponge and witness paper, as shown in Fig. 12.1. Backing materials were used for both dynamic and quasistatic tests. For dynamic test, a in-house made NIJ stab tower was used. This stab tower was made in strict accordance with NIJ standard 0115.00.

Nonhybrid targets consisted of several layers of a single type of fabric, either neat or STF-impregnated fabrics, stacked one above the other (Table 12.2). Hybrid targets

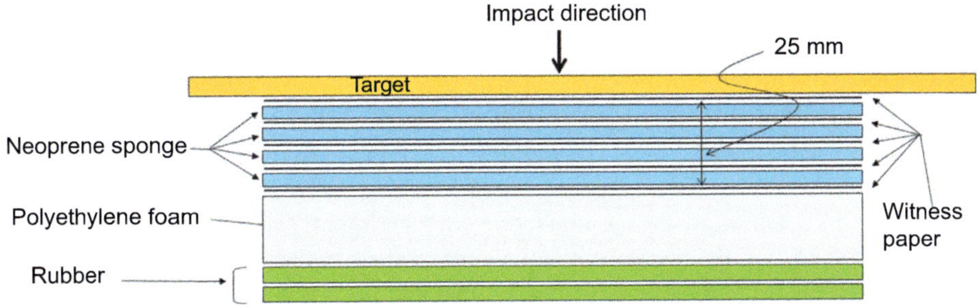

Figure 12.1 Schematic of NIJ target backing.

Table 12.2 Properties of the nonhybrid targets

Label	No. of layers	Layer combination	Thickness (cm)	Areal density (g/cm^2)
15 Kev	15	Neat 706 Kevlar fabric	0.680	0.2835
STF Kev	12	STF-impregnated Kevlar fabric	0.556	0.2712
Neat CK	21	Neat correctional Kevlar	0.907	0.2751
STF CK	17	STF-impregnated correctional Kevlar	0.742	0.2839

Table 12.3 Properties of the thermoplastic Kevlar hybrid target

Label	No. of layers	Layer combination	Thickness (cm)	Areal density (g/cm^2)
15-Kev	15	Neat 706 Kevlar fabric	0.680	0.2835
Thermoplastic Kevlar hybrid				
NK/TPK	12 + 2	Neat Kevlar/thermoplastic Kevlar	0.685	0.2718
STFK/TPK	10 + 2	STF Kevlar/thermoplastic Kevlar	0.680	0.2793
NCK/TPK	16 + 2	Neat correctional Kevlar/ thermoplastic Kevlar	0.685	0.2640
STFCK/TPK	13 + 2	STF correctional Kevlar/ thermoplastic Kevlar	0.680	0.2716

consisted of multiple layers of fabric with the top and bottom layers either TP nylon or TP-Kevlar® with neat or STF-impregnated fabrics stacked in between (Table 12.3).

12.2.3 NIJ stab tower and drop mass

Dynamic stab testing of composites was done with a stab/drop tower that was made in accordance with NIJ Standard 0115.00 [8]. The NIJ Standard 0115.00 outlines protection levels 1, 2, and 3. The three level of protection presented in the standard was derived from a 1997–99 Police Scientific Development Branch program where the frequency distributions of the energy that can be delivered by a male population using several stabbing techniques were characterized. The lowest energy level corresponds to the 85th percentile, the next energy level corresponds to the 90th percentile, and the highest energy level corresponds to the 96th percentile. Table 12.4 is a listing of the stab-resistant level strike energies. The first energy level, labeled $E1$, corresponds to

Table 12.4 **Table of stab-resistant protection level strike energies**

Protection level	*E*1 strike energy, J	*E*2 overtest strike energy, J
1	24 ± 0.50	36 ± 0.60
2	33 ± 0.60	50 ± 0.70
3	43 ± 0.60	65 ± 0.80

the percentiles described earlier in this section. At this condition, a maximum blade or spike penetration of 7 mm (0.28) is allowable. The 7-mm designation is chosen because research indicates that injuries to internal organs are highly unlikely at and below 7-mm depths. The completed tower design is a 4.3-m high, 0.1016-m mean diameter, clear PVC pipe that is attached to a sturdy concrete/cinder block wall with galvanized steel channels. Fig. 12.2 shows photographs of the Tuskegee University NIJ stab tower design. In Fig. 12.2(a), there is a pulley at the top of the tower that allows for lifting and lowering the drop mass that holds the threat weapon (knife or spike) used for testing. The NIJ Standard 0115.00 explains that the drop tube apparatus should have a guide rail in its design. The guide rail in Fig. 12.2(b) is situated inside the

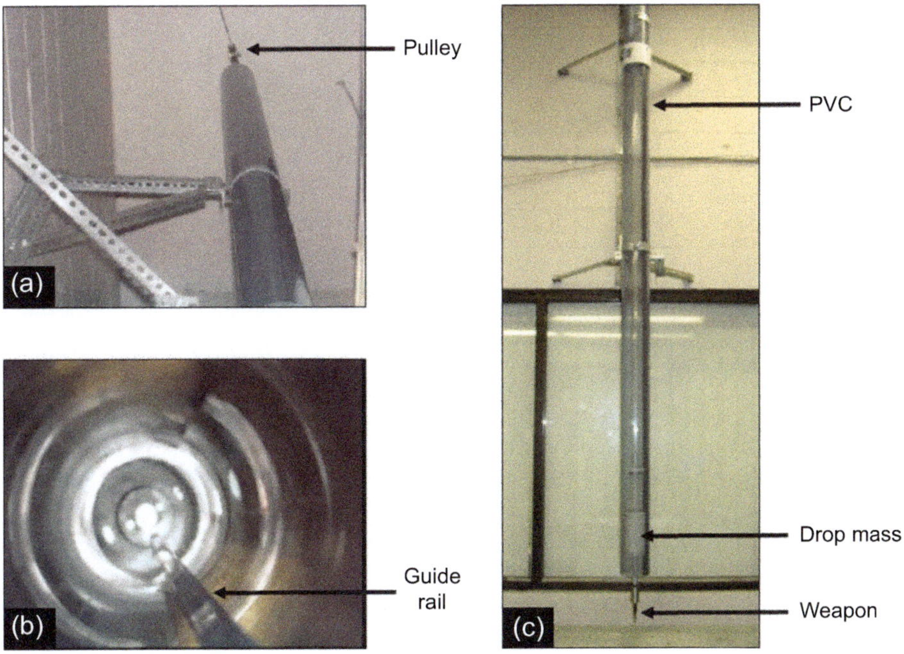

Figure 12.2 NIJ stab tower: (a) top portion of the stab tower indicating the pulley, (b) the guide rail system inside the stab tower, and (c) overview of the stab tower test indicating the outer transparent PVC pipe, drop mass, and weapon.

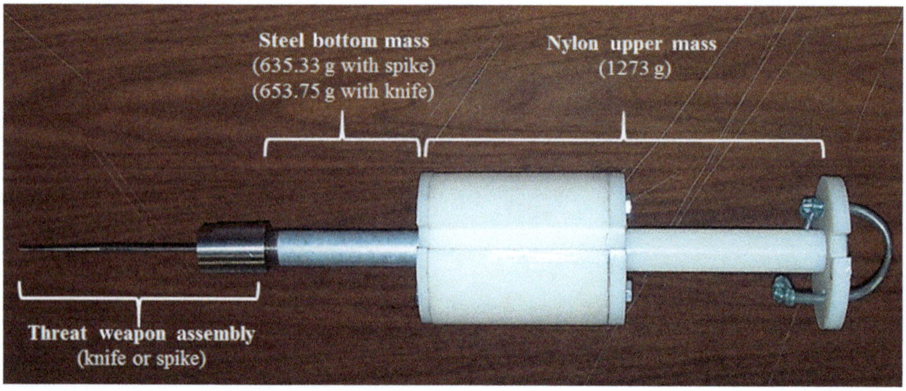

Figure 12.3 Drop mass used for dynamic stab testing.

PVC tube to prevent the drop mass from rotating about its vertical axis during descent to ensure that the test knife blade or spike strikes at the correct orientation and continues in that orientation until it comes to a complete stop. A picture of the PVC guide rail in the NIJ stab tower is shown in Fig. 12.2(c). The drop mass is the most essential part of the design. A picture of the drop mass used for dynamic stab testing is illustrated in Fig. 12.3. The upper part of the drop mass is constructed of nylon 6. Nylon 6 was chosen because it is a low-friction material. The drop mass has a groove so that it fits securely in the tube with the guide rail to resist any rotation. The upper part of the drop mass has a mass of 1273 g.

The bottom part of the drop mass was designed to hold the threat weapon. It is composed of a galvanized steel pipe attached to a stainless-steel weapon holder. The assembly screws together for ease of weapon attachment and disassembly. The bottom part of the drop mass was designed to fit securely into the upper part with discs inside the upper part so that the bottom part does not fall out. The bottom part of the drop mass is 635.33 g with the spike installation and 653.75 g with the knife installation. The NIJ standard states that this lower mass should be 650 ± 7 g. The bottom part of the drop mass with knife installation is within specifications of the NIJ standard. However, the bottom part of the drop mass with the spike installation is not within specifications. Completely assembled, the drop mass is 1908.33 and 1926.75 g with the spike and knife, respectively. The drop mass used for testing explained in this section also has a damping mechanism. The damping mechanism is in place because there is a damping motion incurred during an actual stabbing attack. The wrist actually causes the clenched hand that holds the threat weapon to jerk backward during a stabbing attack/event. The two polyethylene foam rubber discs simulate this damping motion. The impact speed of the drop mass was measured using an Arbor Scientific Speed CheckR 7500. The velocity measurement system uses two lasers that are interrupted by the leading edge of the drop mass to obtain the velocity. Pictures of the velocity measurement system during testing are in Fig. 12.4(a) and (b).

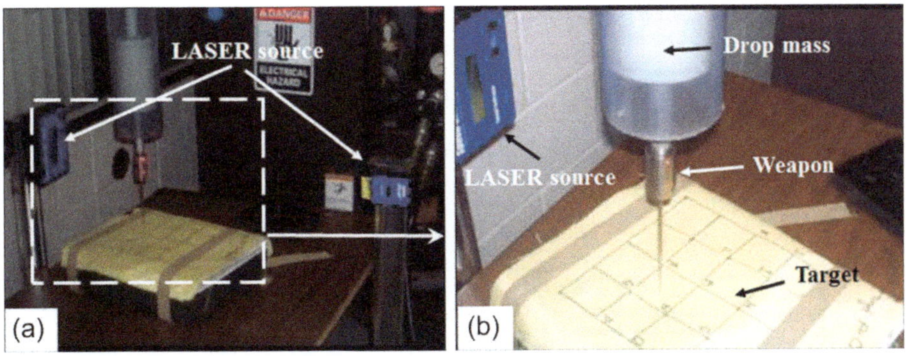

Figure 12.4 Views of the velocity measurement system and thermoplastic Kevlar composite target used for dynamic stab testing: (a) overview of the LASER system and target, and (b) close-up view of the system showing the target, weapon system, and LASER source.

12.3 Stab characterization of nonhybrid target fabric composites

A grid on the top layer of the target was drawn to assure that each stab is at least 50 mm apart. During the stab test, the drop mass was lifted to the required height inside of the drop tube and then was released from that height to achieve the specified velocity so that the energy needed could be calculated. The grid on the top layer of the target included letters from A to R. The letters A to I corresponded to the energies used for dynamic spike testing and the letters J to R correspond to the energies used for dynamic knife testing. Levels A and J correspond to the strike energy 2.7 J for the knife and spike, respectively. Similarly, letters I and R corresponded to the strike energy 24.3 J for the knife and spike, respectively.

The testing grid in Table 12.5 is the complete energy distribution used for testing. The target manufacturing technique was explained in chapter High-performance ballistic fibers: ultra-high molecular weight polyethylene (UHMWPE). The target was secured with nylon straps to the NIJ target backing. The NIJ target backing schematic is shown in Fig. 11.6 in the previous chapter. The target backing consisted of four neoprene sponge layers, a polyethylene foam layer, and two dense rubber layers, respectively, in succession. A Polyart witness sheet was placed on the top, bottom, and middle layers of the neoprene sponge layers. The witness papers indicated the depth of penetration during stab testing. The number of witness papers that were penetrated corresponded to the penetration depths as listed in Table 12.6. Each layer of neoprene sponge is 5.8 mm.

In this section, the focus will be on the results of dynamic stab testing with the S1 knife and the spike. Fig. 12.5(a) is a plot of resistance of neat Kevlar® and STF-impregnated Kevlar® to dynamic knife testing. Noticeably, the neat Kevlar® target had more resistance to knife testing than spike testing. Only two witness layers (5.8–11.6 mm) of the target backing were penetrated at 2.7 J. All five witness layers (>23.2 mm) of the target backing were penetrated at 10.8 J. STF-impregnated Kevlar®

Table 12.5 Energy distribution for dynamic stab testing

Letter designation (spike and knife, respectively)	Energy (J)
A and J	2.7
B and K	5.4
C and L	8.1
D and M	10.8
E and N	13.5
F and O	16.2
G and P	18.9
H and Q	21.6
I and R	24.3

Table 12.6 Explanation of depth measurement technique

Number of layers penetrated	Penetration depth (mm)
0	0
1	0–5.8
2	5.8–11.6
3	11.6–17.4
4	17.4–23.2
5	>23.2

showed better resistance than neat Kevlar® in the knife test. At the lowest energy level (2.7 J) only the first layer of witness paper was penetrated. All five layers were penetrated at 21.6 J, as compared to 10.8 J in case of neat Kevlar®.

Fig. 12.5(b) is a plot of the same combination as the dynamic spike test. Neat Kevlar® had little resistance to spike. At the lowest energy level tested (2.7 J), three witness layers (11.6–17.4 mm) of its target backing were penetrated. The spike then blew through all layers of witness paper at the next tested energy level. STF-impregnated Kevlar® had the best resistance against spike with no layer penetrated at the lowest energy level, and at the highest energy level (24.3 J), only three witness layers were penetrated. Overall, in both the knife and spike tests, STF-impregnated Kevlar® with 12 layers performed better than 15-layer neat Kevlar®.

Fig. 12.6 shows the graphs of dynamic stab resistance comparing correctional Kevlar® with Kevlar®. Correctional Kevlar® has a tight weave, and this property of the fabric has added advantage against spikes. Fig. 12.6(a) is a plot resistance offered

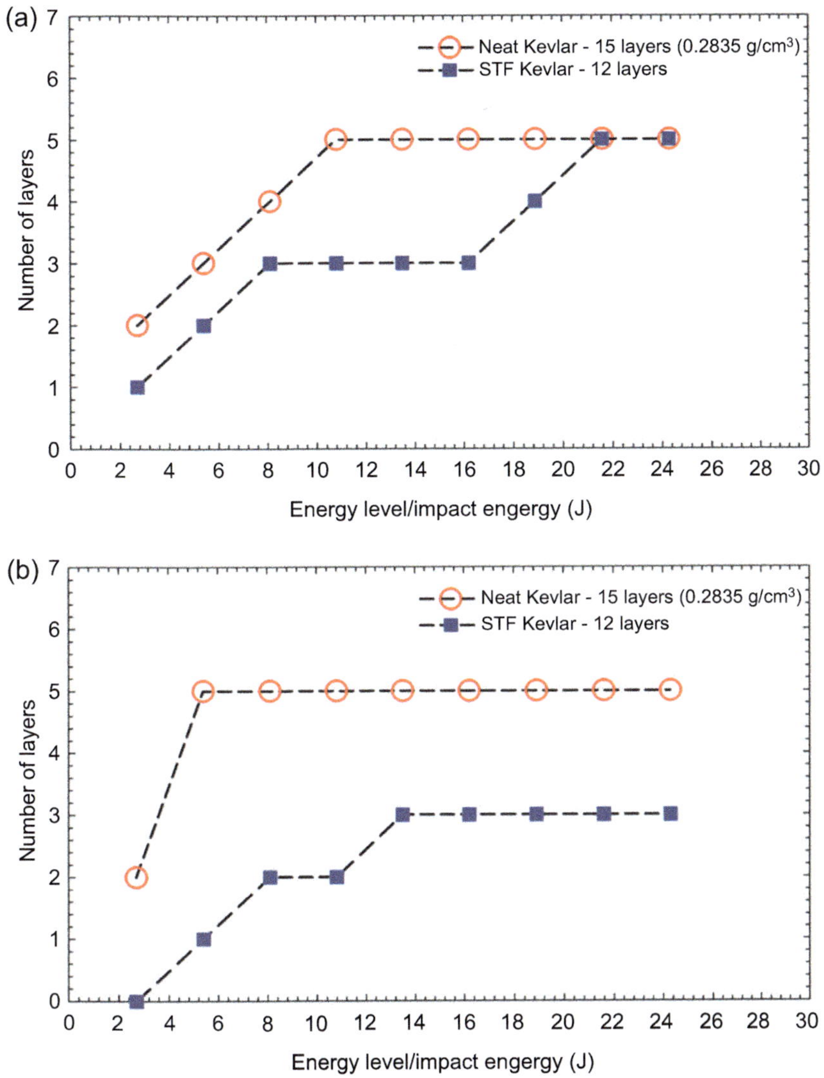

Figure 12.5 Areal density comparison of dynamic stab test results of neat Kevlar and shear thickening fluid Kevlar composite targets: (a) knife and (b) spike.

by correctional Kevlar® against neat Kevlar® to dynamic knife test. At the lowest energy level (2.7 J), one witness layer was penetrated in both the neat as well as the STF-impregnated fabric tests, as against two witness layers in neat Kevlar®. STF correctional Kevlar® had better resistance when compared to neat Kevlar® and correctional Kevlar®. At the highest energy level of 24.3 J, all five layers could be defeated. For a moderate energy level between 8.1 and 16.2 J, only three witness layers could be defeated, as against all five layers in the case of neat Kevlar®. Adding STF to fabric not only reduced the number of layers of the target but increased resistivity as well.

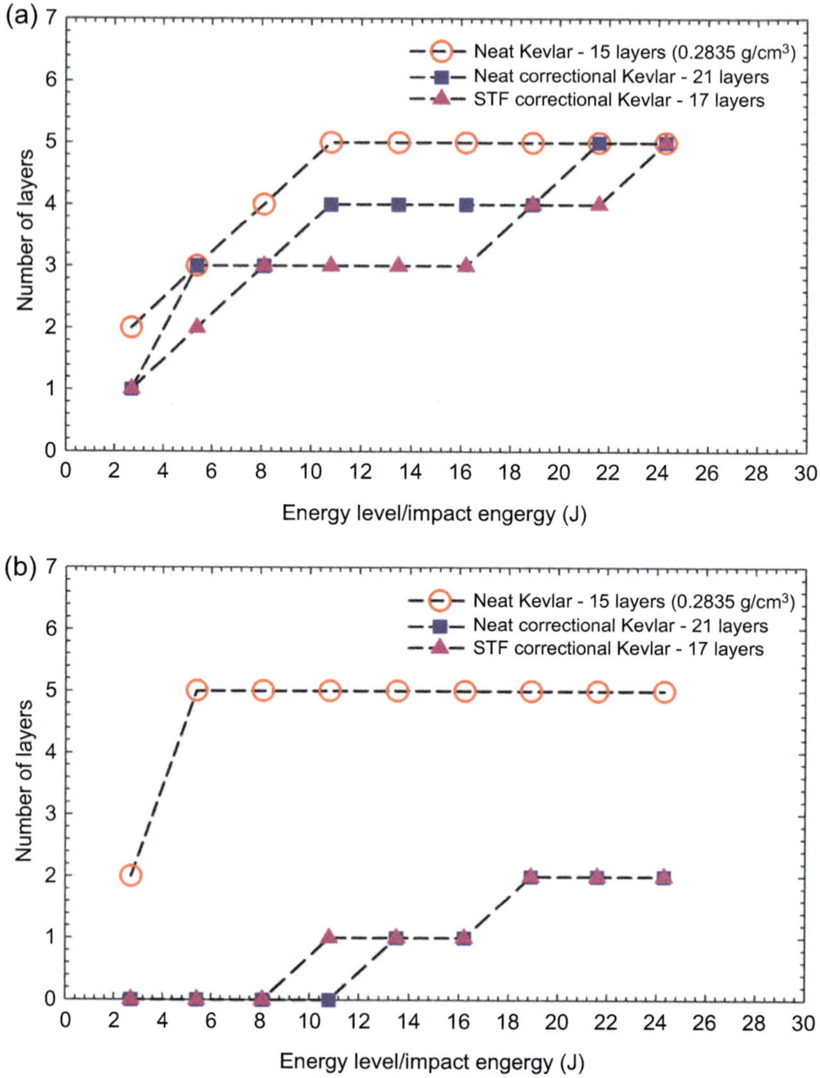

Figure 12.6 Areal density comparison of dynamic stab test results of neat Kevlar and correctional Kevlar composite targets: (a) knife and (b) spike.

Fig. 12.6(b) is a plot of resistance offered by correctional Kevlar® and neat Kevlar® to dynamic spike test. Of all the fabrics tested in the current study, correctional Kevlar® with its tight weave of yarns offered much higher resistance than other fabrics. No witness layers were penetrated until the energy level of 10.8 J, and at the highest energy level of 24.3 J, only two layers were penetrated both in neat as well as STF-impregnated correctional Kevlar®. Neat correctional Kevlar® showed slightly better resistance by having defeated an energy level higher than STF-impregnated correctional Kevlar®. At same time, the number of layers of STF-impregnated correctional Kevlar® was 17 layers compared to neat correctional Kevlar® with 21 layers. Overall, both in the knife and spike tests, STF-impregnated correctional Kevlar® had better resistivity compared to neat correctional Kevlar® and Kevlar®.

Dynamic stab testing of the Kevlar® and correctional Kevlar® composite targets yielded results that were quite interesting. To understand the mechanics behind the results of stab testing, the top layers of the Kevlar® target and the correctional Kevlar® composite target were observed under a scanning electron microscope (SEM) after testing. Damage caused under the highest energy level is discussed. Fig. 12.7 shows the damage area under SEM for neat Kevlar® and STF Kevlar®. Fig. 12.7(a) is picture of the damage area on neat Kevlar®, and Fig. 12.7(b) is the damage area for STF Kevlar®. Here, the damage caused by the knife is apparent. A slit with completely damaged yarns was left by the knife.

Local tearing strength of neat Kevlar® yarns easily exceeded the driving force of the knife. The tiny blade edge in conjunction with the 24.3 J provided enough stress to the examined area to break the fibers of each yarn viewed. Once the surface layer of the neat Kevlar® target was penetrated, the frictional forces of the yarns on the knife did not come into play. The damaged yarns provided little resistance to the knife. In STF Kevlar® fabric, damage was limited to the point or area of impact. Fibers were in place, and it looked like STF acted as an adhesive holding all the yarns together after the impact, thereby offering resistance to further penetration.

Fig. 12.8 shows the damage area under SEM for the dynamic spike test. In the dynamic spike test, the kind of damage was different compared to the dynamic knife test. From the response of the neat Kevlar® during dynamic spike testing, it was obvious the spike penetrated through the neat Kevlar® target with little resistance offered. The damage area was small compared to the dynamic knife test. Here the spike caused a window puncture during dynamic spike testing. The target was easily pierced, and the yarns were spread by the spike instead of getting damaged. The yarns were not irreversibly stretched as in the case of quasistatic testing, which will be discussed later in the chapter.

Figure 12.7 SEM pictures of (a) neat Kevlar and (b) shear thickening fluid Kevlar after dynamic knife testing.

(a) (b)

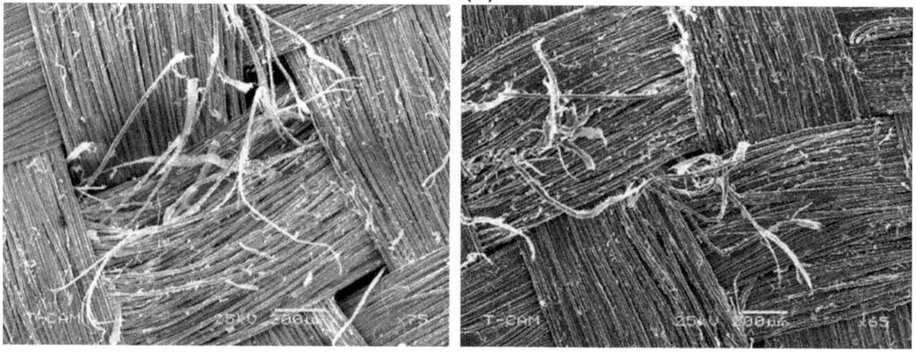

Figure 12.8 SEM pictures of (a) neat Kevlar and (b) shear thickening fluid Kevlar after dynamic spike testing.

For the most part, the yarns returned closely to their original position when the spike was released from the target and target backing. Frictional forces caused by the yarns on the spike were minimal. This is emphasized by the fact that all five witness layers of the neat Kevlar® target backing were penetrated at the 24.3 J level. The local tensile strength of the Kevlar® yarns was not exceeded because the fibers of the yarns did not show much damage. While STF Kevlar® had better resistance, as is evident from Fig. 12.8(b) as no window puncturing was appreciable. Correctional Kevlar® offered high resistance to dynamic spike penetration due to its tight weave and knit.

Heavy damage was seen in neat correctional Kevlar® (Fig. 12.9(a)), where yarns were cut along the length of the knife blade. In the case of STF correctional Kevlar®, not much damage was observed, and again it looked like the STF held the fiber together. Fig. 12.9(b) shows the damage area under SEM for dynamic knife for STF correctional Kevlar®. Fig. 12.10 shows the damage area under SEM for dynamic spike

(a) (b)

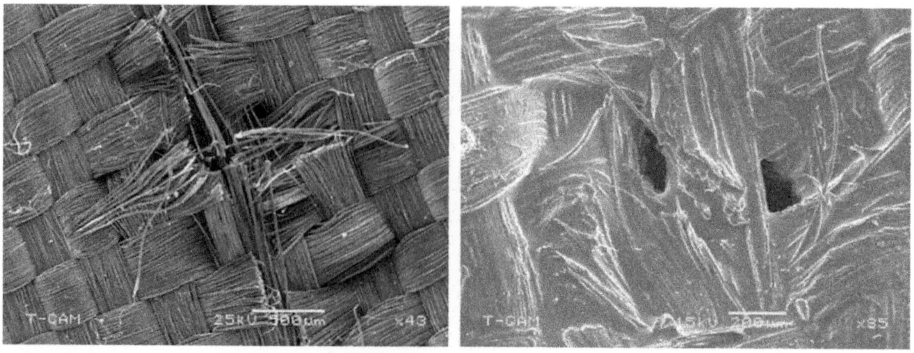

Figure 12.9 SEM pictures of (a) neat correctional Kevlar and (b) shear thickening fluid correctional Kevlar after dynamic knife testing.

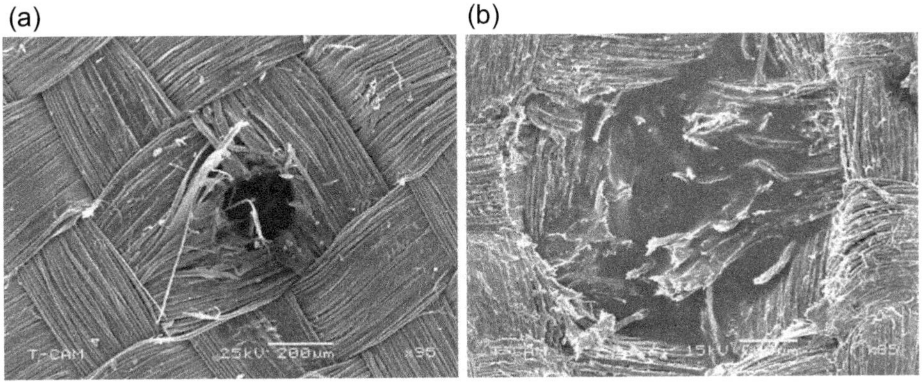

Figure 12.10 SEM pictures of (a) neat correctional Kevlar and (b) shear thickening fluid correctional Kevlar after dynamic spike testing.

test of correctional Kevlar®. Here the damage was in the form of a clear hole in the case of neat correctional Kevlar®. The resistance incurred by the spike after perforation was due to friction on the sides of the spike caused by the yarns. This could be due to tight weave of correctional Kevlar®, where fibers even after been cut offered resistance to the spike. Fig. 12.10(a) shows a puncture or hole, what is called as windowing, which is evident in Fig. 12.10(b) also.

12.4 Stab characterization of TP-Kevlar® hybrid target fabric composites

All the standards followed in the previous section are followed in present section. Single-layer TP-Kevlar® was used as the top and bottom layers of the target. Neat and STF-impregnated fabrics were stacked in between TP-Kevlar® according to equal areal density. In this section, the focus is on the results of dynamic stab testing of TP-Kevlar® hybrid targets.

Fig. 12.11 is a plot of resistance offered by a hybrid target having Kevlar® and TP-Kevlar®. Fig. 12.11(a) is a plot of resistance of the combination to a knife test. The STF-impregnated combination offered better resistance than neat in combination. While at highest energy level 24.3 J, only four layers (17.4—23.2 mm) were of a spike test, where only two layers were penetrated in STF Kevlar® combination at the highest energy level. All five layers were penetrated at an energy level of 10.8 J in the neat Kevlar® combination. This clearly showed the STF-impregnated Kevlar® combination had better performance than neat Kevlar® in combination with TP-Kevlar®.

Fig. 12.12 is a plot of resistance of correctional Kevlar® in combination with TP-Kevlar® in a dynamic test. Both STF correctional Kevlar® and neat correctional Kevlar® had defeated energy of 2.7 J when two witness layers were penetrated in a dynamic knife test, as indicated in Fig. 12.12(a). Only four witness layers were penetrated at the highest energy level (24.3 J) for STF correctional Kevlar® in combination,

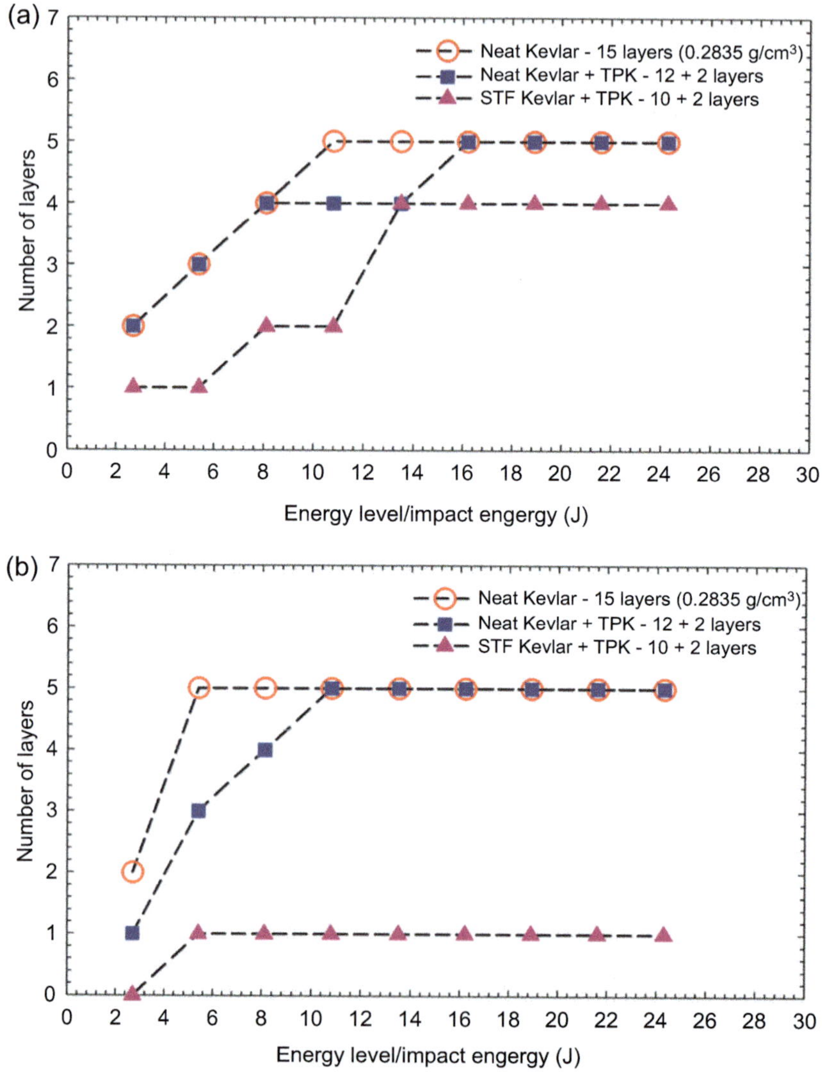

Figure 12.11 Areal density comparison of dynamic stab test results of neat Kevlar and shear thickening fluid Kevlar in thermoplastic Kevlar hybrid composite targets: (a) knife and (b) spike.

and 21.6 J was required for penetration in the neat correctional Kevlar® combination. Fig. 12.12(b) is a plot of resistance in a spike test. STF correctional Kevlar®/TP-Kevlar® hybrid target had defeat energy of 24.3 and 18.9 J for the neat correctional Kevlar®/TP-Kevlar® hybrid target. The first witness layer was penetrated at an energy level of 10.8 J in the STF correctional Kevlar®/TP-Kevlar® hybrid target and 13.5 J for the neat correctional Kevlar® in hybrid target. STF correctional Kevlar® in

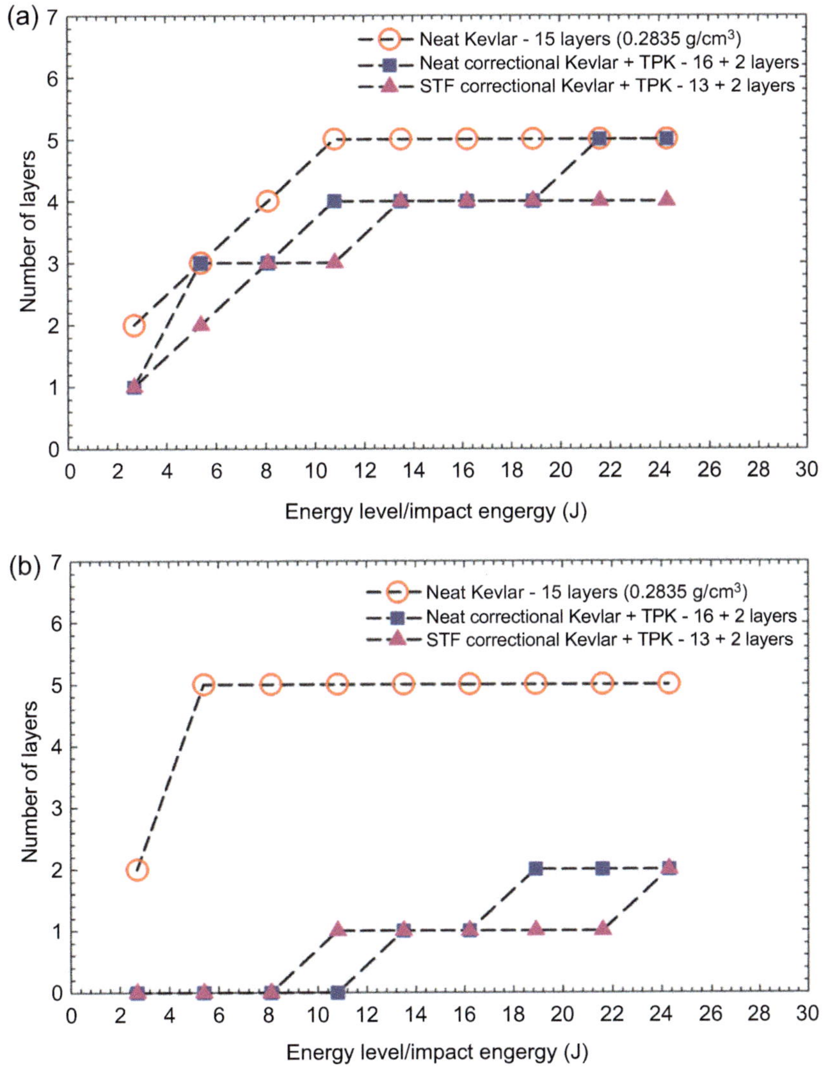

Figure 12.12 Areal density comparison of dynamic stab test results of neat correctional Kevlar and shear thickening fluid correctional Kevlar in thermoplastic Kevlar hybrid composite targets: (a) knife and (b) spike.

combination showed better performance than the neat correctional Kevlar®/TP-Kevlar® hybrid target.

Dynamic stab testing of the Kevlar® and correctional Kevlar® in combination with TP-Kevlar® composite targets yielded results that were quite interesting. To understand the mechanics behind the results of stab testing, the top layers of TP-Kevlar®, Kevlar®, and correctional Kevlar® composite targets were observed under SEM after testing. Damage caused under the highest energy level is discussed.

Fig. 12.13 shows the damage area as seen under SEM of TP-Kevlar® in combination with neat and STF Kevlar® after a dynamic knife test. TP-Kevlar® in Fig. 12.13(a) was used in combination with neat Kevlar®; it can be seen in the figure that the thermoplastic around the impact zone has been pulled. This might be due to lower resistance offered by neat Kevlar® which allowed knife to penetrate further deep into target, thereby pulling the thermoplastic. Though this kind of pullout was not seen in the TP-Kevlar® used in the STF Kevlar® combination, damage to the thermoplastic was seen, as in Fig. 12.13(b). It seemed the thermoplastic completely wetted the fabric and acted as a protecting shield. Fig. 12.14 shows pictures of Kevlar® fabrics used in the hybrid target. Damage on neat Kevlar® was heavy; yarns and fibers were heavily damaged at the point of impact, as seen in Fig. 12.14(a). In the case of STF Kevlar® (Fig. 12.14(b)), there was damage to yarns and fiber cutting, but the magnitude of damage was less compared to neat Kevlar®. STF in the fabric held the fabrics together,

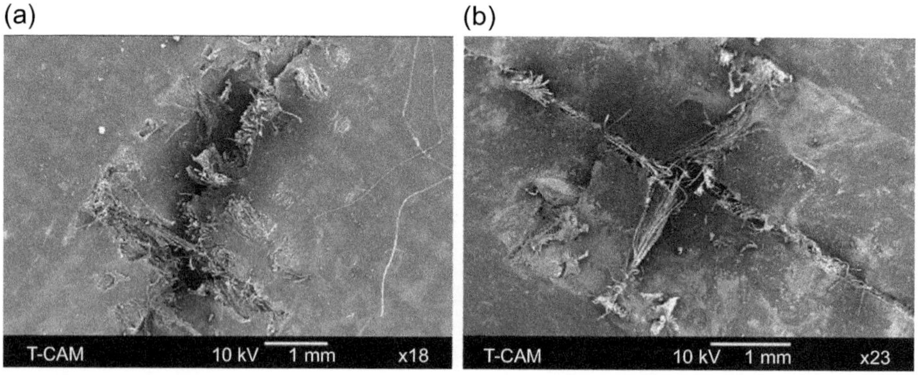

Figure 12.13 SEM pictures of (a) thermoplastic Kevlar in neat Kevlar combination and (b) thermoplastic Kevlar in shear thickening fluid Kevlar combination after dynamic knife testing.

Figure 12.14 SEM pictures of (a) neat Kevlar and (b) shear thickening fluid Kevlar in thermoplastic Kevlar combination after dynamic knife testing.

thereby offering better resistance for further penetration of the knife. Damages in both cases were along the breadth of the knife from one cutting edge to the other cutting edge.

Fig. 12.15 shows the damage area after a dynamic spike test in TP-Kevlar® and Kevlar® combination. Here it can be seen (Fig. 12.15(a)) that the thermoplastic around the impact area has been pulled out from the fabric. This might be due to less resistance offered by neat Kevlar® beneath the top TP-Kevlar®. The surface of the spike rubbed against the thermoplastic and pulled it down along with it as it penetrated through the target. Some of the thermoplastic around the point of impact has been damaged heavily. In Fig. 12.15(b), STF Kevlar® was used in combination, and damage was lower compared to TP-Kevlar® in neat combination. Thermoplastic was not pulled out of the fabric as in previous case, and damage has been local. Fig. 12.16(a) shows the damage area on a neat Kevlar® in a dynamic spike test. Few fibers were damaged, and not

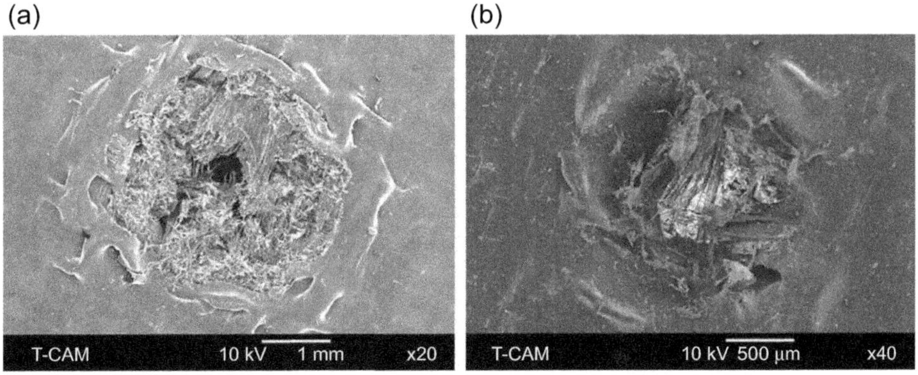

Figure 12.15 SEM pictures of (a) thermoplastic Kevlar in neat Kevlar combination and (b) thermoplastic Kevlar in shear thickening fluid Kevlar combination after dynamic spike testing.

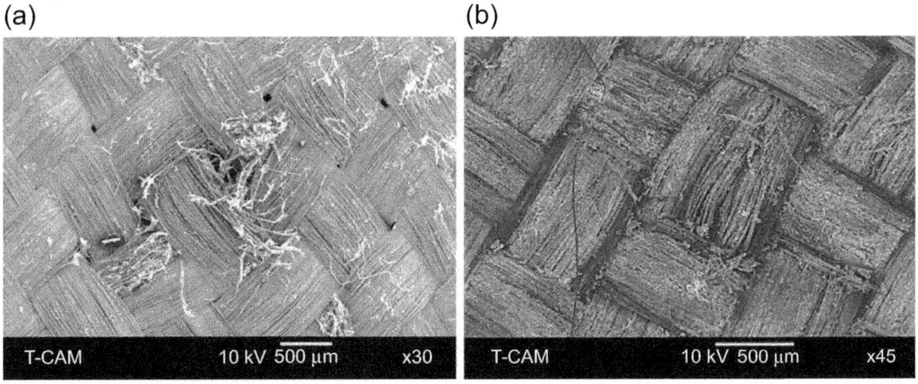

Figure 12.16 SEM pictures of (a) neat Kevlar (b) shear thickening fluid Kevlar in thermoplastic Kevlar combination after dynamic spike testing.

much damage to yarns was seen. The spike made a window and penetrated through the layers of the target. In STF-impregnated Kevlar®, as seen in Fig. 12.16(b), no fibers or yarns were damaged; rather they were pushed aside, and the spike penetrated through.

Fig. 12.17 shows pictures of damage on TP-Kevlar® used in combination with correctional Kevlar®. No pull out of thermoplastic was seen in either case. The knife cut through the TP-Kevlar® fabric along the length of the knife blade and penetrated through the target. Fibers in the yarns were damaged at the point of impact in both the neat and STF fabric combination. Heavy damage to yarns and fibers were seen in neat correctional Kevlar® and can be seen in Fig. 12.18(a). STF correctional Kevlar® had lower damage to yarns and fibers compared to neat correctional Kevlar®. Damage was in the form of slit cutting the fibers in the yarn and the knife making its way through the target layers, as seen in Fig. 12.18(b).

Figure 12.17 SEM pictures of (a) thermoplastic Kevlar in neat correctional Kevlar combination and (b) thermoplastic Kevlar in shear thickening fluid correctional Kevlar combination after dynamic knife testing.

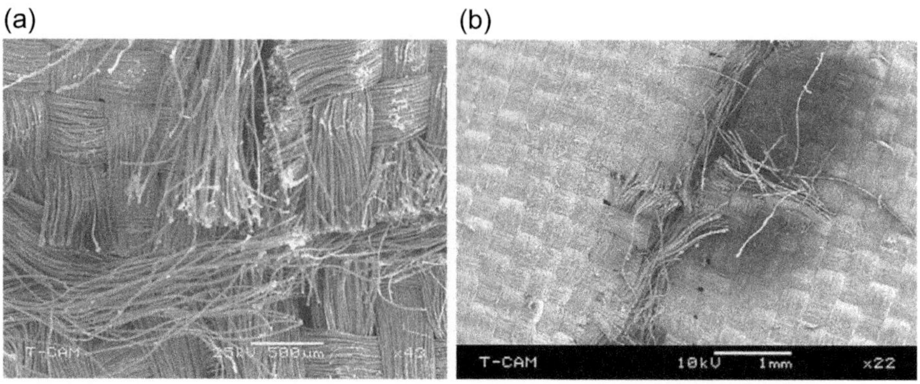

Figure 12.18 SEM pictures of (a) neat correctional Kevlar and (b) shear thickening fluid correctional Kevlar, in thermoplastic Kevlar combination after dynamic knife testing.

TP-Kevlar® used in combination with neat correctional Kevlar® in a dynamic spike test had damage in the form of a hole, Fig. 12.19(a). The sharp tip of the spike made a hole through TP-Kevlar®, and fibers were damaged at the point of impact. Pull out of the thermoplastic was not seen in the neat and STF fabric combination. TP-Kevlar® in combination with STF correctional Kevlar® had a smaller hole compared to TP-Kevlar® in neat combination. Damage was also much lower compared to TP-Kevlar® in neat combination, Fig. 12.19(b).

Of all the fabrics tested, correctional Kevlar® had better performance against a spike threat. In neat correctional Kevlar®, damage was seen in the form of a fiber cut and a window. Fibers covered the hole created by the spike when it was pulled out after the test. Fig. 12.20(a) shows the damage area on neat correctional Kevlar®; yarns and fiber at the point of impact have been damaged. In STF correctional Kevlar®, fibers or yarns were not cut; rather they were pushed to the side, and the spike made its way through

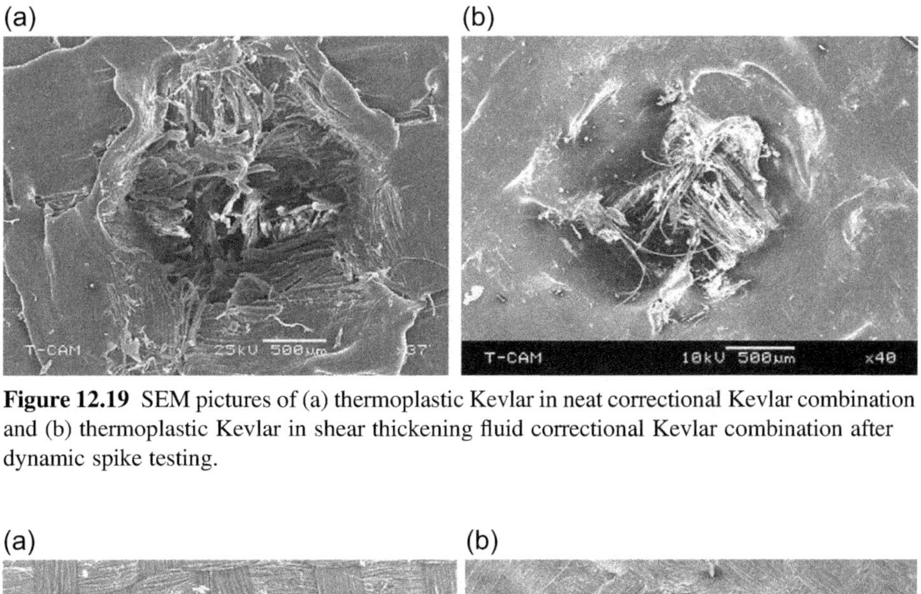

Figure 12.19 SEM pictures of (a) thermoplastic Kevlar in neat correctional Kevlar combination and (b) thermoplastic Kevlar in shear thickening fluid correctional Kevlar combination after dynamic spike testing.

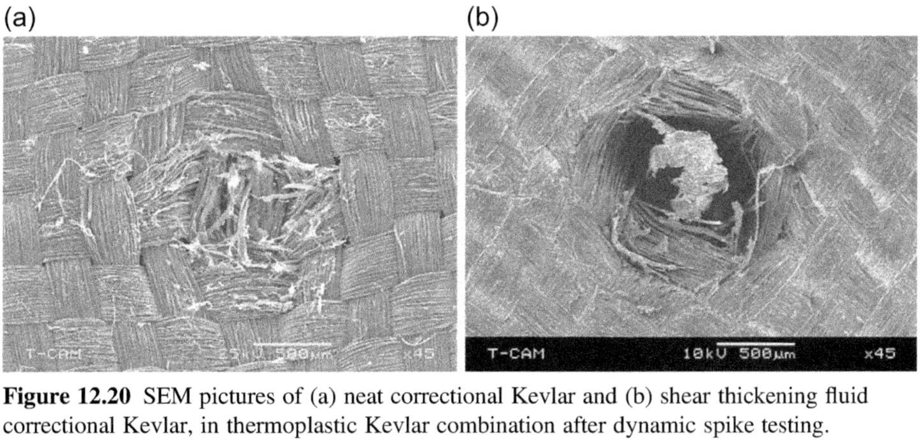

Figure 12.20 SEM pictures of (a) neat correctional Kevlar and (b) shear thickening fluid correctional Kevlar, in thermoplastic Kevlar combination after dynamic spike testing.

the target (Fig. 12.20(b)). A hole of size 1.5 mm was formed, but still the fabric offered higher resistance to the spike. The undamaged fibers offered resistance against the walls of the spike. Impregnating correctional Kevlar® with STF not only increased the resistivity of the fabric but also lowered the fabric damage.

12.5 Conclusions and future trends

This study will aid in the manufacture of complete (multithreat) body armor for different industrial applications. Dynamic stab tests were completed on the neat, STF, and hybrid fabric targets. The STF-impregnated fabrics showed great resistance in that the total wetting through of the fabrics by STF showed greater surface energy due to increased local tearing strengths and greater friction for the individual threads of the fabric. TP-film showed greater surface energy due to increased local tearing strengths and greater friction for the individual threads of the Kevlar® fabric. The greater friction comes from the individual threads being coated by the TP-film. The increase in elasticity was observed during dynamic stab testing in that in some instances the threat weapon bounced off the target instead of sticking upon impact. Use of hydrophilic fabrics in the study aided impregnating the STF onto fabrics easily. Hydrophobic Kevlar® was used in fabrication of thermoplastic Kevlar® fabric. The targets were examined on the resistance to low energy (2.7–24.3 J) threats. More resistance was indicated for the spike as compared to knife threats. This was in place because the spike was stopped by some of the targets while the knife drove through the targets completely. The STF-impregnated targets showed a great amount of resistance. These targets were more rigid and bulky, so they would be well suited for torso and head protection. Of all the nonhybrid targets tested, STF correctional Kevlar® had the highest resistance against dynamic spike, followed by STF Kevlar®. Addition of TP-Kevlar® showed an increase of resistance in all the targets in the study. STF Kevlar®/TP-Kevlar® had the highest resistance in the dynamic knife test, the best performance of all the hybrid and nonhybrid targets studied. TP-Kevlar®/correctional Kevlar® showed improvement in their performance against dynamic spike. At the highest energy level, only two layers of the target were penetrated. Targets with STF fabrics in combination had better performance than neat fabrics in combination. STF correctional Kevlar®/TP-Kevlar® had the best resistance in a quasistatic knife and spike test. Overall, in the TP-Kevlar® hybrid target studied, TP-Kevlar®/STF Kevlar® had the best resistance in dynamic tests. Microscopy of the impact zone after testing shows that damage mechanisms are consistent with the STF reducing fabric mobility. The entrenching of colloidal particles into the filaments may provide enhanced coupling between the STF and fibers, enabling highly efficient load transfer between filaments. The importance of these particle–filament interactions remains to be proven. The thicker Surlyn-Kevlar® composite targets showed great resistance and would be sufficient for body armor that requires rigidity and bulk. Thermoplastic films in most cases protected the fibers and yarns against sharp knife edge and spike. In some cases, it was clearly seen that thermoplastic films were ruptured and the fibers underneath them were undamaged.

Finally, these results show that ballistic fabrics can be modified to provide enhanced stab resistance. Therefore, these results demonstrate that it may be possible to engineer a fabric that is capable of providing meaningful levels of both ballistic and stab protections, properties which are often engineered independently with conventional materials. Here, a proof-of-concept for new protective materials has been demonstrated. TP/STF hybrid fabric composites are worthy of examination for stab protection in multithreat complete body armor. Aging of the fabrics (environmental effects) is a factor that affects the stab resistance.

TP/STF hybrid composites show good resistance against stab threats such as the knife and spike. TP-Kevlar® (Surlyn-Kevlar®) in particular showed increased performance against the knife as well as spike test. Therefore, a suggestion is made to study the resistance with increased use of TP-Kevlar® and STF-impregnated fabrics. This would be complete with stab testing, tear/peel testing, and ballistic resistance testing. Needle studies and other thin threat studies, such as flechettes, are suggested. The nonhybrid and hybrid target composites examined during this study should also be examined for ballistic resistance. Some other suggestions for future work are as follows:

- use of other non-Newtonian fluids with different particle and particle size can be investigated
- preparation of STF by different methods
- importance of particle–filament interactions
- effect of patterned TP coating (rather than just uniform, a way to perhaps get stab resistance while maintaining some flexibility)
- hot press STF-impregnated fabrics and study the resistance
- a systematic study of flexibility of these various armors
- different environmental effects such as moisture exposure and effect of UV radiation on the stab performance can be investigated
- mechanistic study of cutting, at the microscale; how exactly does cutting initiate and progress in neat and STF Kevlar® fabrics?
- investigation into the mechanism as well as direct studies on the effect of particle hardness on STF fabric behavior
- how critical is blade sharpness, and why? Can we use first principles to model these mechanisms?
- other fabrics beyond Kevlar® and correctional Kevlar® have to be studied; TP-correctional Kevlar®, TP-PBO, and Zylon can be used to study the stab performance
- other novel concepts for stab-resistant materials (hard particles, flakes, CNTs), integration of metal wires/flakes/foils, incorporation of ring mesh materials, and bio-inspired designs

Acknowledgments

This work was carried out through funding from the US Army Research Office through grant number W911NF-05-0006-P00004. The authors would also like to acknowledge the technical support of Dr. Eric Wetzel, US Army Research Laboratory, and the late Dr. Jessie Mayo, Jr.

References

[1] Leslie LF, Woods JA, Thacker JG, Morgan RF, McGregor W, Edlich RF. Needle puncture resistance of surgical gloves, finger guards, and glove liners. J Biomed Mat Res 1996; 33(1):41–6.

[2] Verge S. Rapidly deployable structures in collective protection systems. Army Soldier and Biological Chemical Command (Natick, MA) Report, Acc. Number ADA444670. 2006.

[3] Cadogan D, Stein J, Grahne M. Inflatable composite habitat structures for lunar and Mars exploration. Acta Astronaut 1999;44(7–12):399–406.

[4] Henderson JP, Morgan SE, Patel F, Tiplady ME. Patterns of non-firearm homicide. J Clin Forensic Med 2005;12:128–32.

[5] Emerson E. Surviving knife attacks. Police Law Enforc Mag 2004;28(6):56–9.

[6] Taking the Stab Out of Stabbings. National Law Enforcement and Corrections Technology Center (NLECTC). NCJ 211819 2000.

[7] Bleetman A, Watson CH, Horsfall I, Champion SM. Wounding patterns and human performance in knife attacks: optimising the protection provided by knife-resistant body armour. J Clin Forensic Med 2003;10:243–8.

[8] National Institute of Justice. Stab resistance of personal body armor. National Institute of Justice standard 0115.00; 2000.

[9] Walker A, Gray TGF, Nicol AC, Chadwick EKJ. Evaluation of test regimes for stab-resistant body armour. Proc Inst Mech Eng Part L 2004;218(4):355–61.

[10] Shin H-S, Erlich DC, Simons JW, Shockey DA. Cut resistance of high-strength yarns. Text Res J 2006;76(8):607–13.

[11] Shin H-S, Erlich DC, Shockey DA. Test for measuring cut resistance of yarns. J Mat Sci 2003;38(17):3603–10.

[12] Afshari M, Sikkema DJ, Lee K, Bogle M. High performance fibers based on rigid and flexible polymers. Polym Rev 2008;48(2):230–74.

[13] Chae HG, Kumar S. Rigid-rod polymeric fibers. J Appl Polym Sci 2006;100(1):791–802.

[14] Flambard X, Polo J. Stab resistance of multi-layers knitted structures: comparison between para-aramid and PBO fibers. J Adv Mater 2004;36(1):30–5.

[15] Gadow R, von Niessen K. Lightweight ballistic with additional stab protection made of thermally sprayed ceramic and cermet coatings on aramide fabrics. Int J App Ceram Tech 2006;3(4):284.

[16] Anctil M, Keown D, Bourget G, Pageau G, Bolduc M, Shewchenko N. Threat assessment and performance evaluation of multi-threat body armour systems. In: Proceedings of the joint RTO AVT/HFM specialists meeting on "Equipment for personal protection (AVT-097)" and personal protection: bio-mechanical issues and associated physio-pathological risks (HFM-102), Koblenz, Germany; May 19–23, 2003.

[17] Termonia Y. Puncture resistance of fibrous structures. Int J Impact Eng 2006;32(9): 1512–20.

[18] Shedden L, Nash DH, Walker CA. Finite element modelling of stab impact on body armour. In: Proceedings of fourth international conference on modern practice in stress and vibration analysis. UK: University of Nottingham; 2000. p. 209–21.

[19] Wang L, Zhang S, M.Gao W, Wang X. FEM analysis of knife penetration through woven fabrics. Comput Model Eng Sci 2007;20(1):11–20.

[20] Decker MJ, Halbach CJ, Nam CH, Wagner NJ, Wetzel ED. Stab resistance of shear thickening fluid (STF)-treated fabrics. Comp Sci Tech 2007;67(3–4):565–78.

[21] Bartus SD, Vaidya UK. Performance of long fiber reinforced thermoplastics subjected to transverse intermediate velocity blunt object impact. Comp Struct 2005;67(3):263–77.
[22] Dear JP, Lee H, Brown SA. Impact damage processes in composite sheet and sandwich honeycomb materials. Int J Impact Eng 2005;32(1–4):130–54.
[23] Santulli C, Brooks R, Long AC, A.Warrior N, Rudd CD. Impact properties of compression moulded commingled E-glass–polypropylene composites. Plast Rubber Comp 2002; 31(6):270–7.
[24] Putnoki I, Moos E, Karger-Kocsis J. Mechanical performance of stretched knitted fabric glass fibre reinforced poly(ethylene terephthalate) composites produced from commingled yarn. Plast Rubber Comp 1999;28(1):40–6.
[25] Bigg DM. The impact behavior of thermoplastic sheet composites. J Reinf Plast Comp 1994;13(4):339–54.
[26] Alcock B, Cabrera NO, Barkoula N-M, Wan Z, Peijs T. The effect of temperature and strain rate on the impact performance of recyclable all-polypropylene composites. Compos Part B: Eng 2007;67(10):2061–70.
[27] Alcock B, Cabrera NO, Barkoula N-M, Peijs T. Low velocity impact performance of recyclable all-polypropylene composites. Comp Sci Tech 2006;66(11–12):1724–37.
[28] Xu T, Farris RJ. Comparative studies of ultra high molecular weight polyethylene fiber reinforced composites. Polym Eng Sci 2007;4(10):1544–53.
[29] Morye SS, Hine PJ, Duckett RA, Carr DJ, Ward IM. Modelling of the energy absorption by polymer composites upon ballistic impact. Comp Sci Tech 2000;60(14):2631–42.
[30] Karahan M. Comparison of ballistic performance and energy absorption capabilities of woven and unidirectional aramid fabrics. Text Res J 2008;78(8):718–30.
[31] Tan VBC, Khoo KJL. Perforation of flexible laminates by projectiles of different geometry. Int J Impact Eng 2005;31(7):793–810.
[32] Tan VBC, Shim VPW, Tay TE. Experimental and numerical study of the response of flexible laminates to impact loading. Int J Solids Struct 2003;40(23):6245–66.
[33] Walsh SM, Scott BR, Spagnuolo DM. The development of a hybrid thermoplastic ballistic material with application to helmets. U.S. Army Research Laboratory Technical Report ARL-TR-3700. December 2005.
[34] Zivkovic I, Perisic P, Burzic Z, Uskokovic P, Aleksic R. Aramid Fiber Reinf Laminar Thermoplast Compos Mater 2005;37(4):23–31.
[35] Srivastava A, Majumdar A, Butola BS. Improving the impact resistance of textile structures by using shear thickening fluids: a review. Criti Rev Solid State Mater Sci 2012; 37:115–29.
[36] A Hassan T, Rangari V, Jeelani S. Synthesis, processing and characterization of shear thickening fluid (STF) impregnated fabric composites. Mat Sci Eng A 2010;527:2892–9.
[37] Barnes HA. Shear-thickening (dilatancy) in suspensions of nonaggregating solid particles dispersed in Newtonian liquids. J Rheol 1989;33(2):329–66.
[38] Lee Y, Wetzel ED, Wagner NJ. The ballistic impact characteristics of Kevlar woven fabrics impregnated with a colloidal shear thickening fluid. J Mater Sci 2003;38:2825–33.
[39] Majumdar A, Butola BS, Srivastava A. Optimal designing of soft body armour materials using shear thickening fluid. Mater Des 2013;46:191–8.
[40] Mayo J, Wetzel E, Hosur M, Jeelani S. Stab and puncture characterization of thermoplastic-impregnated aramid fabrics. Int J Impact Eng 2009;36(9):1095–105.

13 Polyolefin film–reinforced composites for personal protection

J. Singletary
DuPont Protection Technologies, Richmond, VA, United States

B. Lauke
Leibniz Institute for Polymer Research e. V., Dresden, Germany

13.1 Introduction

Personal protection equipment requires high specific modulus and specific tensile strength. This makes oriented, synthetic polymers attractive for their low density and potential properties in the orientation direction. Historically, most uses of oriented polymer in personal protection materials have been with fibres, dating back to nylon in flak jackets in World War II. This leads to an understanding of the protective article as an assemblage of essentially one-dimensional (1D) structures, which is still a widely held and often useful principle in understanding material design and impact response. (There are many that we offer as exemplary (Roylance and Wang, 1980; Parga-Landa and Hernández-Olivares, 1995; Cunniff, 1999a; Walker, 2001; Phoenix and Porwal, 2003)).

However, an alternative form for oriented polymers in personal protection is oriented films. Efficient utilisation of these materials requires considering them beyond conventional 1D-oriented polymer concepts used previously for protective materials. In terms of practical material design, unlike fibres or yarns, films are self-supporting structures, and so require less effort to convert into a stable, area-covering assembly. Further, their aspect ratios enable higher volume fractions of reinforcement, reducing bulk and needed resin content in resin-impregnated structures. In terms of conceptual understanding and analysis, the off-axis properties of film reinforcements lead to different response under impact. Anticipating such differences in impact response can enable novel new approaches for personal protection.

13.1.1 History of polyolefin film–reinforced personal protection

Film-reinforced personal protection was demonstrated during World War II, responding to not only a need for higher armour mass efficiency,[a] but simply to

[a] Armour mass efficiency, or 'ME', is the ratio of areal densities of a benchmark armour solution to a novel armour solution of equal protection. So ME > 1 indicates the novel solution has higher protection per weight than the benchmark.

Advanced Fibrous Composite Materials for Ballistic Protection. http://dx.doi.org/10.1016/B978-1-78242-461-1.00013-3

augment the availability of protective materials at a time when armour steel and aluminium were in high demand. Working for the DuPont Company, Heckert (1946) obtained a patent on 'synthetic linear superpolymers…formed into fibers or filaments…and cold-drawn or cold-rolled', foreshadowing the solid-state extrusion (SSE) process commonly used today. The patent taught that such materials could be used in many armour articles, including helmets and fragment protective body armour, and as spall liners to protect from fragments that evolve from bullet impact onto armour steel. The film-reinforced armour could be fixed to high hardness elements via bolting or preferably adhesives. Structures could contain combinations of film and fibrous materials and be bonded together with an adhesive throughout the article, or in part of the article (to allow articulated flexibility). Heckert (1946) taught that neighbouring layers of reinforcement should be at or near right angles to each other.

Although the teachings of (Heckert, 1946) were broad, the reductions to practice focused on polyamides. Polyamides are renowned for their toughness, but they have proven historically difficult to orient sufficiently to begin to approach their theoretical, uniaxial properties (Porter and Wang, 1995). It is therefore perhaps not surprising that further development of film-reinforced armour awaited the broader commercial development of polyolefins, which are generally easier to orient and draw.

In the 1970s, Philips Petroleum developed a film-reinforced protective material technology called *XP*. The original patent (Holmes, 1982) taught that a range of polyolefins could be used, including polyethylene and polypropylene. Films would be drawn uniaxially, then either woven as flat strips into fabric or laid up with adjacent tapes, essentially, but joining and neighbouring layers at right angles, then moulded into rigid composites at pressures up to 690 MPa. The polyolefin film component could be combined with high-hardness materials to protect against armour-piercing projectiles, and/or with other fibrous textiles. In several successive reports, XP technology was reduced to practice in personal protection, and preferred embodiments were taught.

Additional publications reduce embodiments of Holmes (1982) to practice, and they may be of interest to material users for practical manufacturing advice, as well as being available online. Alesi et al. (1975) describes cross-plied, oriented film-reinforced helmets, moulded alone and with combinations of glass- and Kevlar®[b]-fibre skins for improved durability. Ballistic *V*50 tests against a range of steel right-circular cylinder fragment-simulating projectiles (FSPs) showed performance generally similar to prototype helmets made with 100% Kevlar® fabric reinforcement in a rubber-modified phenolic resin, despite the fact that the axial strength of the XP films were about 0.34 GPa—one-eighth the specific strength of the Kevlar® prototype. This protection per weight for such a relatively weak material confounds many theories describing the relation of axial properties to ballistic protection in fibre-reinforced armour (noted above), and it is persistent for polyolefin film reinforcements with many structures, including subsequent films made

[b] Kevlar® is a registered trademark of *E*. I. du Pont de Nemours and Company.

from calendared gel-spun PE fibre yarns (Harpell et al., 1992), and later SSE films that are the focus of this chapter. Alesi et al. (1975) identified special precautions for preforming the shear-intolerant film reinforcement and noted that the need for a well-bonded, durable article conflicted with the need for controlled delamination upon impact for high-impact penetration resistance. Moulding at up to 17 Bar and cooling under pressure to near ambient temperature were found to improve durability of monolithic polyolefin film–reinforced articles, although co-moulded skins using polyester and rubber-modified phenolic skins did not adhere well to the oriented polyolefin film–reinforced core. Moulded parts absorbed a trivial amount of water in soak tests.

Appropriate moulding conditions for cross-plied, oriented polyolefin film–reinforced protective composites were elaborated in Prifti et al. (1976). Prifti et al. (1976) used very thin films (30 g/m^2) around 20-cm wide, and taught that it was necessary to mould laminates to a low void content, which they termed *theoretical density*. They used evacuation and high-pressure moulding (17 MPa) at elevated temperatures. Doerr (1979) and Cook (1980) give practical advice on preform evacuation, moulding, quality control and machining, used in panels ranging over 102-mm thick, and suggest co-moulding woven fabric skins to improve abrasion resistance.

So, by the end of the 1970s, working with relatively low-strength oriented polyolefin film armours available at the time, several principles of effective structure design of personal protection materials with polyolefin films had been taught that remain relevant:

- Films should be highly oriented in one direction and have high cut resistance.
- Films should be thin.
- Films can be woven or cross-plied nonwovens. Sequential layers should be oriented at or near right angles to each other.
- Rigid armour articles with film reinforcement should be moulded to 'near theoretical density', through vacuum and/or high pressure. They may have a secondary matrix phase, or be self-reinforced, relying on partial melting of the reinforcement to bond the article. A secondary matrix increases cost of the reinforcement material, but it allows a wider processing window. Self-reinforced film composites are simpler to take to the preform state, but they have a tighter processing window since the article temperature must be uniformly brought to very near but below the final melt temperature.
- Final articles can be combined with other materials, including monolithic or tiled high-hardness elements for protection from armour-piercing rounds and/or fibrous textiles. These other materials may be co-moulded or secondarily bonded.
- Finally, film reinforcements can require different preforming and moulding than fibre-reinforced materials to prevent the reinforcement from shifting.

13.1.2 Solid-state extrusion (SSE)

Significant advances in armour performance came with the practical advent of SSE, in which special polymer powder is compacted into a relatively dense but friable sheet, and then taken through a shearing orifice to orient it. This enables drawing beyond the capacity of melt-extruded materials (on the order of 10–15X, where X is draw ratio (eg, Zachariades et al., 1980) to ~35–50X, improving modulus and strength. A further, critical discovery was the combination of SSE to a relatively low initial draw ratio (enough to provide strength for subsequent drawing, ~5–6X)), followed by tension drawing in multiple steps. This enabled draw to ~75–100X for appropriate commercial PE powders (Kanamoto et al., 1987) and ~250X for PE powders specially prepared under laboratory conditions (Kanamoto et al., 1988). The former gave corresponding strengths of ~1.3–2.0 GPa; the latter up to 5–6 GPa. Beygelzimer and Beloshenko (2004) reviewed SSE processing. Porter and Wang (1995) give a detailed review of SSE and SSE with two-step draw for many semicrystalline polymers. Weedon (2000) describes the reduction of SSE with two-step draw to commercial practice. From this point forward, we will assume 'SSE' to mean a process that includes a separate tension draw, as is practised commercially.

Laboratory production of very high strength SSE-polyethylene (PE) strips uses individual, separate crystals laid out as precipitated mats, and it is discontinuous, thus not suitable for commercial manufacture; however, academic practice informs the potential of the technology. As Porter and Wang (1995) point out, relative molecular simplicity favours unbranched polyolefins and particularly PE for high molecular draw. PE is to date the only polymer that can be drawn after SSE to an axial modulus and strength approaching that predicted for the polymer crystal. Therefore, PE is the most attractive polymer for high-performance protective materials, and it will be the sole focus for the remainder of this chapter. For the interested reader, Ward and Hine (2004) review the principles and some applications of rigid protective articles from oriented polypropylene sheets.

The value of SSE-PE as reinforcement for personal protective articles was recognised early in the technology's development. Narrow strips of lab-prepared SSE-PE up to 4.8 GPa in strength were reported to be valuable in body armour (Anonymous, 1991); fabrics for protective articles and helmets reinforced with commercially prepared SSE-PE were reported in Porter et al. (1994). Body armour from commercially prepared SSE-PE was said to have desirable performance in Weedon (2000).

13.2 Structure of SSE-PE

As noted earlier, oriented polyolefin film–reinforced composites have historically been seen to offer higher protection per weight than would be expected from their mechanical properties, based on axial strength–protection relations established for fibre-reinforced plastics. We will thus discuss the structure of commercial SSE-PE films first, as a basis for subsequently proposing mechanisms for why their performance differs from fibre-reinforced composites.

Commercial SSE-PE films are highly crystalline, highly uniaxially oriented, orthotropic, and have a void structure created during drawing. Melt endotherm by differential scanning calorimetry is preferably above 234 J/g, suggesting crystallinity above about 82%. Fig. 13.1 shows SSE-PE film images in different planes. The rolling and drawing of SSE-PE films produces a uniplanar, 3D structure, with axes of orthotropy parallel to the machine direction/direction of draw (1), the cross direction in the plane of the film (2), and the through-thickness direction (Kanamoto et al., 1987, 1988; Jordan et al., 2002). Individual films can fibrillate into subunits parallel to the 1-direction (Zachariades and Kanamoto, 1988) that are larger in the 2- than the 3-direction (Kanamoto et al., 1988; Jordan et al., 2002). These subunits are highly crystalline, as indicated by differential scanning calorimetry (DSC).

Fibrillar subunits have low adhesion and are poorly adhered to each other, whether as drawn or after moulding. They buckle in compression in kink bands (Fig. 13.1, right), similar to gel-spun PE fibres (DeTeresa et al., 1988). Transverse tensile strength is difficult to determine experimentally; however, our internal testing suggests SSE-PE films have a transverse peel force of 0.35 ± 0.05 N/m, which is similar to gel-spun PE fibres (McGarry and Moalli, 1991). Exposure to temperature near the melt can reveal the underlying shish-kebab structure expected from incompletely drawn PE (Fig. 13.2) (Gibson et al., 1978; Weedon, 2000; Porter and Wang, 1995).

SSE-PE films contain a network of voids creating during drawing. For PE extrudates with lower drawability (up to 36X), Jarecki and Meier (1979) found elliptical voids perpendicular to the draw direction. For PE drawn from precipitated mats of single crystals, drawn to 250–300X, Hu and Schmidt-Rohr (2000) found that voids are highly oriented in the machine direction. Commercial powders have perfection and draw ratios intermediate between these. Fig. 13.3, from Gibson et al. (1978), suggests a structural model for PE chains bridging between crystals that could fit either of the previous observations of void structure, with voids perpendicular to the draw direction when the bridging elements are relatively short, and parallel as the extended chain bridges increase in aspect ratio. We have observed this

Figure 13.1 Scanning electron microscopy of commercial solid-state extrusion polyethylene films. Left: Tensylon™ HS film surface in machine (1)-cross (2) direction plane, showing fibrillar structure. Centre: diamond knife cut in Tensylon™ HT showing through thickness (3)-cross (2) direction plane, showing subunits of thin, wide structures. *(From Jordan, N., Olley, R., Bassett, D., Hine, P., Ward, I., 2002. The development of morphology during hot compaction of Tensylon high-modulus polyethylene tapes and woven cloths. Polymer 43, 3397–3404.* http://dx.doi.org/10.1016/S0032-3861(02)00104-0*; used by permission).* Right: Tensylon™ HS film torn parallel to 1-direction, showing fibrillar structure in 3-1 plane. Note apparent handling-induced kink bands on top (cf. DeTeresa et al., 1988).

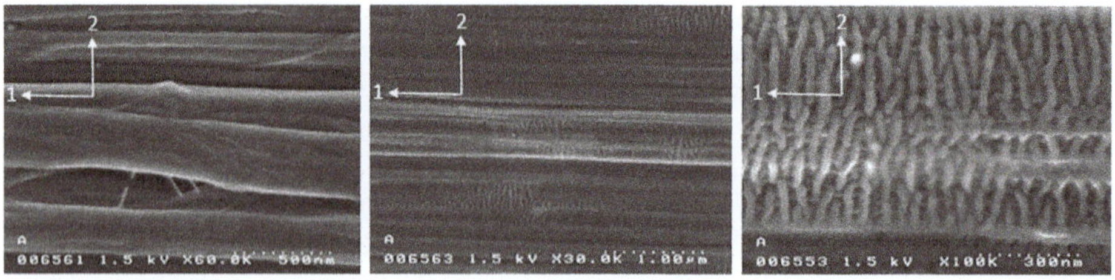

Figure 13.2 Artifacts in solid-state extrusion polyethylene films (Tensylon™ HS). Left: ligands connect neighbouring fibrils. Centre: exposure to high temperature appears to correspond to the onset of lines in the 1-2 plane, parallel to the 2 (cross) direction. Right: higher magnification suggests these lines are ~20-nm thick and connected by fibrils in the 1 (orientation) direction. This is consistent with the classical 'shish-kebab' model of oriented and partially extended UHMWPE (Weedon, 2000; cf. Nakae et al., 2000).

Figure 13.3 Structure transverse to orientation (1) direction.
Proposed by Gibson, A.G., Davies, G.R., Ward, I.M., 1978. Dynamic mechanical behaviour and longitudinal crystal thickness measurements on ultra-high modulus linear polyethylene: a quantitative model for the elastic modulus. Polymer 19 (6), 683—693. http://dx.doi.org/10.1016/0032-3861(78)90123-4, used by permission.

network to be inaccessible to helium during pycnometry. As a result, although DSC would suggest a density of 960–980 kg/m^3, commercial SSE-PE films have an as-drawn specific gravity of around 800–850 kg/m^3, making them the most buoyant commercial reinforcements for high-performance armour. Such films can collapse to near theoretical density when moulded into armour articles (Holmes, 1982; Alesi et al., 1975).

13.2.1 Mechanical properties of SSE-PE: orientation direction

Ballistic protection of oriented polymers is often associated with their axial strength and stiffness. Fig. 13.4 plots the axial specific strength and specific modulus of

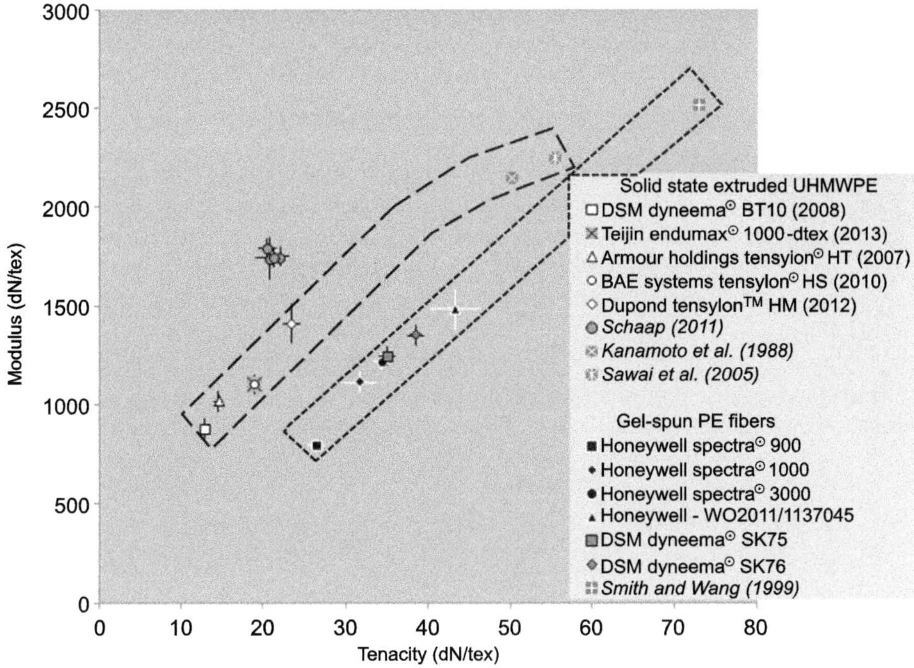

Figure 13.4 Specific modulus versus specific strength (tenacity) for some solid-state extrusion polyethylene films and gel-spun polyethylene fibres. Data on solid-state extrusion polyethylene from Schaap (2011), Kanamoto et al. (1988), and Sawai et al. (2005) and otherwise tested internally at DuPont. Data on gel-spun polyethylene fibres from manufacturers' data sheets (van der Werff et al., 2010; Chocron et al., 2013; Wang and Smith, 1999). Modulus for solid-state extrusion polyethylene films is the low-load modulus defined by ASTM D7744-11.

commercial SSE-PE films and gel-spun PE fibres[c] in textile units. Also shown are highest results (reported in Kanamoto et al., 1988; Sawai et al., 2005; Wang and Smith, 1999), suggesting potentially obtainable properties. Because these results are taken from a variety of sources and different test methods, exact comparison is not appropriate. However, plotted together, the data suggest trends for both SSE-PE films and gel-spun PE fibres. For both materials, the relation between modulus and tenacity are generally linear for commercial materials, with laboratory results suggesting commercial offerings may improve substantially with time. SSE-PE films generally show higher modulus per tenacity than gel-spun PE fibres.

Determining tenacity of SSE-PE film strips free of stress concentrations at the clamps is difficult (Kanamoto et al., 1987, 1988; Sawai et al., 2005). Stipulating that the tenacities in Fig. 13.4 are reasonable representations, it is counterintuitive

[c] Tensylon™ is a trademark of *E*. I. du Pont de Nemours and Company. Endumax® is a registered trademark of Teijin Aramid GmbH. Spectra® is a registered trademark of Honeywell International Inc. Dyneema® is a registered trademark of DSM Dyneema®.

that the modest tenacities shown in Fig. 13.4 for commercial SSE-PE can lead to competitive or even superior ballistic protection compared to PE fibre-reinforced materials. Conventional wisdom from fibre-reinforced composites is that impact perforation resistance increases as (tenacity)$^{2/3}$ (*specific axial strength*)$^{2/3}$ (Cunniff, 1999a; Phoenix and Porwal, 2003).

Fig. 13.5 plots the mean velocity to barely perforate a target, or *V*50, versus a useful scaling measurement in ballistic impact with homologous projectiles, the ratio Γ = (*Target areal density*)/[(*projectile mass*)/(*projectile presented area*)] (Singletary et al., 2008). The projectiles were 0.26-g right-circular steel cylinders, with near unity aspect ratio and a nominal hardness of Rc 30, suggesting projectile deformation is unlikely. The solid trend line shows *V*50 = f(Γ) interpolated from a database of homologous projectile impacts onto a common PE fibre reference material (DSM Dyneema® HB26), and it well represents the trend seen in these tests. As expected, *V*50 for a consistent projectile increases with increasing target areal density, and it is roughly linear in the range presented. Dashed trend lines represent 5% increments or decrements from this trend line. When moulded to low target density, ballistic protection afforded by commercial SSE-PE film-reinforced composites (here, DuPont™

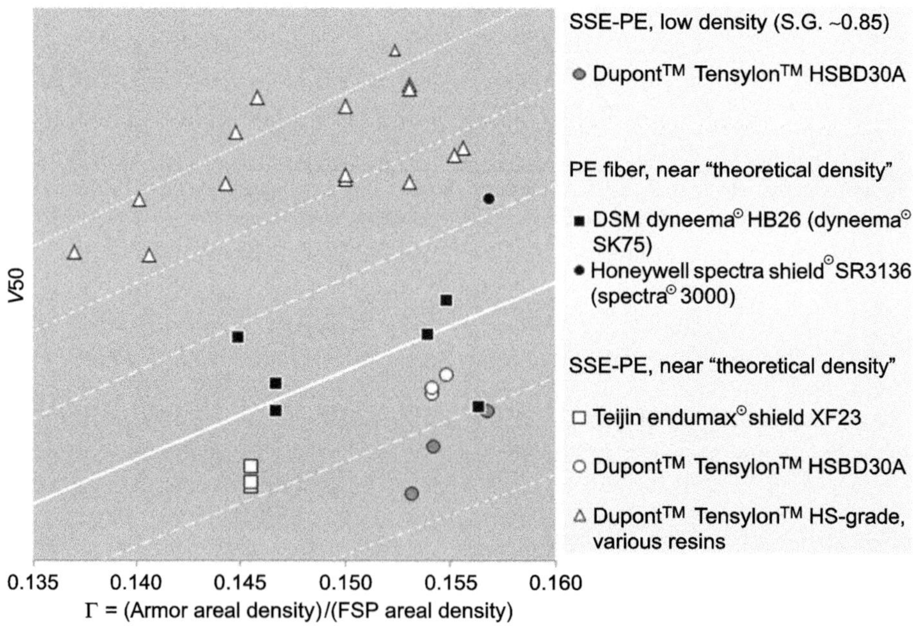

Figure 13.5 Mean velocity to barely perforate, or *V*50, versus the ratio (armour areal density)/(projectile areal density) for impacts with 0.26-g right-circular steel cylinders of near unity aspect ratio, for solid-state extrusion polyethylene film-reinforced and polyethylene fibre-reinforced $[0|90]_n$ composites, moulded to various densities.

Tensylon™ HSBD30A) is around 8% lower than the reference. However, by Fig. 13.4, this compares SSE-PE reinforcements that are only around 55% as strong as the corresponding fibre PE reinforcement. Following the teachings of Holmes (1982) to mould at high pressure (approach a void-free or what Alesi et al. (1975)) called theoretical density composite for polyolefin film reinforcements), the ballistic performance of commercial SSE-PE materials (DuPont™ Tensylon™ HSBD30A, Teijin Endumax® SHIELD XF23) approach within around 4% of the benchmark *V*50. With alternate resins, SSE-PE film-reinforced composites can achieve 10–15% higher *V*50 than the PE fibre-reinforced composite benchmark moulded under similar conditions.

Why SSE-PE film-reinforced composites can demonstrate 10–15% higher *V*50 ballistic protection compared to PE fibre-reinforced armour composites, when conventional wisdom based on fibre composites suggests their *V*50 should be $(55\%)^{2/3} = 67\%$ of benchmark fibre composites, is the subject of current research. In the following section, we suggest three possible explanations:

1. Off-axis mechanical properties of SSE-PE films.
2. Through-thickness mechanical properties of SSE-PE films.
3. Higher reinforcement volume fraction when moulded to near theoretical density.

13.2.2 Off-axis mechanical properties

Ultrasonic testing suggests the in-plane shear modulus, G_{12}, of commercial SSE-PE films is 3 ± 1 GPa. This is 2–20 times the G_{12} reported for gel-spun PE fibres (DeTeresa et al., 1988). The higher shear modulus of SSE-PE films may be due to higher interfibrillar connection than in gel-spun PE (Zachariades and Kanamoto, 1988, cf. Fig. 13.2). When reinforcing a composite panel, the film structure of SSE-PE offers high stiffness and strength on the bias direction because the SSE-PE film reinforcement is essentially continuous and perpendicular to the load, rather than in discrete, 10–20-μm-diameter filaments embedded in a matrix. This fundamental difference between planar film structure of SSE-PE and essentially 1D-oriented fibres used historically may explain why SSE-PE tapes offer ballistic performance far above what would be expected of fibres with similar axial properties.

Fig. 13.6 shows a polar plot of modulus in the plane for a [0|90]-reinforced, balanced composite. Case 1 uses the elasticity of DuPont™ Tensylon™ HS-grade SSE-PE film, determined ultrasonically, and averaged assuming classical laminated plate theory and 10% by volume of a soft matrix. Case 2 uses results of tension tests on-axis (ASTM D3039) and off-axis (ASTM D3518) of the same SSE-PE film in a soft polyolefin resin. Agreement between Cases 1 and 2 is reasonable given the challenges of ultrasonic stiffness measurements off the orientation direction (1) when the reinforcement has such high anisotropy ($E_1/E_2 \approx 50$) and about 10% by volume of a relatively soft resin, which dilutes modulus in the orientation directions. Case 3 considers a

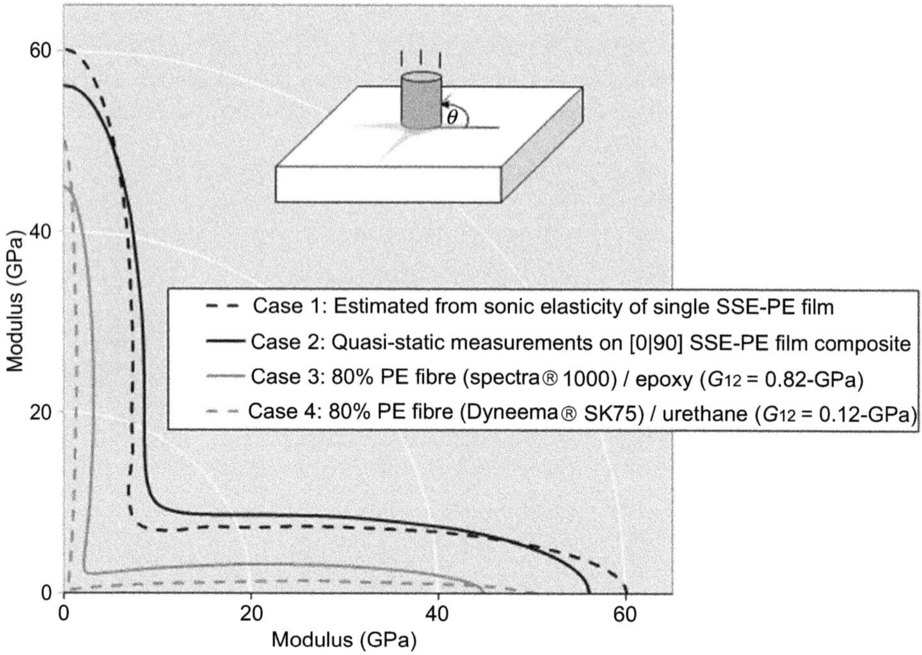

Figure 13.6 Polar plot of normal in-plane modulus for solid-state extrusion polyethylene film-reinforced and polyethylene fibre-reinforced, $[0|90]_n$ composites.

balanced [0|90]-reinforced composite with a high fibre volume fraction typical of armour materials (80%) and a hard resin (epoxy)[d]. Lamina properties are taken from DeTeresa et al. (1996), except E_1 is taken as 80% of the manufacturer's stated yarn modulus ($\sim$108 GPa), to account for the higher volume fraction than DeTeresa reported[e]. Case 4 considers a more realistic composite, with a softer, urethane resin: balanced [0|90]-reinforced with 80% by volume fibre, using again 80% of the manufacturer's stated fibre modulus (122 GPa, see van der Werff et al., 2010) as E_1, and G_{12} taken from $\pm$45 tension tests reported on DSM Dyneema® HB26 in Karthikeyan et al. (2013) (Fig. 13.2, center).

It may be argued that this elastic response based on quasistatic tests reasonably does not represent actual, dynamic response; however, similar reinforcements were seen to have little strain rate sensitivity up to at least $10^3\ s^{-1}$ (Russell et al., 2013). If Fig. 13.6 does reasonably approximate modulus in the plane as a function of direction, $E(\theta)$, then the total area subjected to the advancing strain wave at an arbitrary time after

[d] We avail the estimates of axial tensile properties in part because of the known challenges of correctly performing the test on $[0|90]_n$ PE fibre-reinforced armour composites. Difficulty with grip slippage often leads to spuriously low axial properties (cf. Czechowski et al., 2012).

[e] Peijs et al. (1994) and Stijnman (1995) give similar E_1 and G_{12} properties for unidirectional PE fibre-reinforced composites.

Table 13.1 $\int_0^{\pi/2} \sqrt{E(\theta)d\theta}$ from Fig. 13.6, normalised to Case 1, suggesting relative ability of solid-state extrusion (SSE) polyethylene (PE) film-reinforced and PE fibre-reinforced composites to communicate strain early in the ballistic event

Case	Material	$\int_0^{\frac{\pi}{2}} \sqrt{E(\theta)d\theta}$ normalised to Case 1
1	SSE-PE $[0\|90]_n$ estimated from sonic elasticity of single SSE-PE film, assuming 10% by volume soft resin	1.0
2	SSE-PE $[0\|90]_n$ experimental, 10% soft polyolefin matrix	0.9
3	PE fibre $[0\|90]_n$ experimental, 20% epoxy matrix	0.7
4	PE fibre $[0\|90]_n$ experimental, 20% urethane matrix	0.4

impact should be proportional to $\int_0^{\pi/2} \sqrt{E(\theta)d\theta}$. Table 13.1 lists $\int_0^{\pi/2} \sqrt{E(\theta)d\theta}$ normalised by Case 1. This suggests SSE-PE film-reinforced composites are capable of involving larger volumes of material in resisting impact penetration than composites reinforced with PE fibres, at least early in an impact event.

Higher stiffness and strength in bias direction offers several unique advantages to protective articles made with SSE-PE films, compared to those reinforced with gel-spun fibres:

1. Stress is reacted out of the immediately impacted reinforcement in a shorter distance. This reduces the propensity for a near-perforating partial penetration to cause widespread damage in the laminate, leading to lower dynamic deflection.
2. Likewise, localising deformation around the impact site rather than distributing it in the far field improves a laminate's ability to resist edge shots and to support a brittle strike-face material.
3. Lower anisotropy dramatically increases bolt bearing strength. This is useful in articles that transmit impact loads to a frame by bolted connections, such as vehicle spall liners.

13.2.3 Dynamic deflection

The film form of SSE-PE appears to allow a novel way of overcoming the long observed correlation in fibre-reinforced composites of higher armour mass efficiency coming at the expense of laminate rigidity and dynamic deflection (Walsh et al., 2006; Freitas et al., 2014). Freitas et al. (2014) studied the back-face dynamic deflection response of several composites of potential interest in ballistic protective helmets,

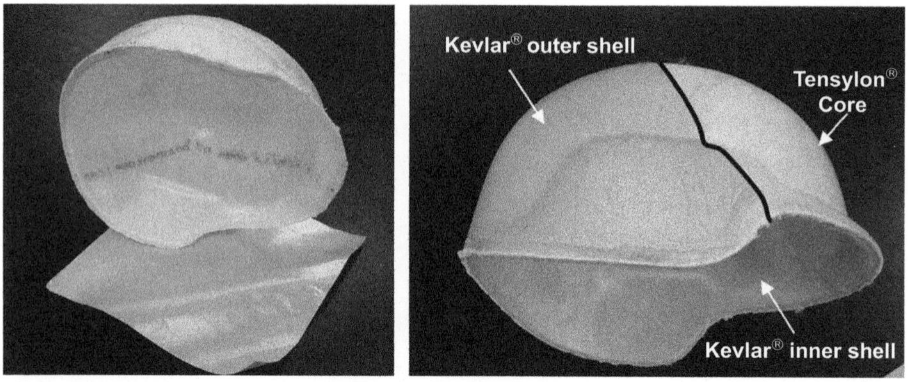

Figure 13.7 Ballistic protective helmets made with solid-state extrusion polyethylene. Left: monolithic. Right: with skins of other materials (here, para-aramid fabric) for abrasion resistance and easier painting.

using 3D image correlation, impacted with FSPs and pistol bullets. They found, as expected, that the average maximum velocity and extent of back-face deflection tend to increase with armour mass efficiency across a wide range of materials. However, they noted the two SSE-PE film-reinforced composites tested (using Tensylon™ HT and HS-grade films, respectively) as exceptions, offering a combination of armour mass efficiency and lower dynamic deflection not seen in fibre-reinforced composites. To reduce the observations of Freitas et al. (2014) to practice, size M 'PASGT' helmet shells were made from the Tensylon™ HS-grade composite and Kevlar® woven fabric skins (Fig. 13.7), and tested for resistance to ballistic penetration per HPW-TP-401.01B. Clay crater depths against 9-mm parabellum bullets averaged 11 mm and had a maximum of 15 mm—comparable or less to conventional fibre-reinforced helmets weighing an up to one-third more.

13.2.4 Rigid support

The ability of SSE-PE film-reinforced composites to distribute impact energy in the plane may explain their unique ability to resist normal impact penetration near target edges. Resistance to penetration decays near unsupported target edges. This leaves an unfortunate compromise between selective edge reinforcement (unusual, expensive), heavy equipment that is overdesigned for protection in the interior or accepting a smaller protected area. Typically, the last option is chosen. This is especially problematic in helmets and torso plates, where even an ineffective distance of 25 mm from edges corresponds to a 10–20% loss in protected area. Fig. 13.8 shows a 4.5-kg/m^2 SSE-PE film-reinforced composite, intended for protection against heavy handgun bullets, impacted with 15.6-g 0.44-magnum semijacketed hollow point pistol bullets at 436 $\pm$ 9 m/s, at varying distances from an unsupported edge. Impacts at $\delta > 1.1$ calibres were stopped repeatedly. Similar edge shots would challenge high-performance fibre-reinforced armour materials, suggesting SSE-PE film

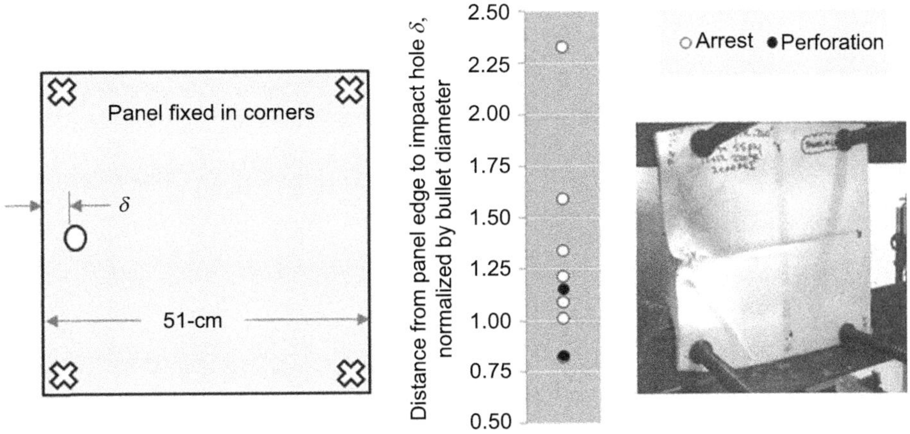

Figure 13.8 Edge shot performance of solid-state extrusion polyethylene film-reinforced composites demonstrated with 15.6-g 0.44-magnum semijacketed hollow point bullets. Left: test set up. Centre: perforation/arrest as function of edge distance δ. Right: panel response to impact.

reinforcements offer an improved compromise between edge protection and protection in the article interior.

13.2.5 Bolt shear-out strength

Armour composite panels are often fixed with bolts in the corners. $[0|90]_n$ reinforcement is desirable for high-impact penetration resistance per weight and preferred for conversion efficiency in large parts, but it gives inherently poor bolt shear-out strength compared to more quasiisotropic lay ups (eg, Duthinh, 2000). Therefore, the cost and weight of fixation to mitigate poor bolt shear-out strength reduce the potential system weight savings of mass-efficient $[0|90]_n$ fibre-reinforced armour materials. Fig. 13.9 suggests an abstraction of the attachment loads on a corner-bolted composite armour panel to bolt pull-out strength on the bias, for instance, ASTM D5961. Table 13.2 compares bolt bearing strengths of an SSE-PE film-reinforced composite with two commercial, PE fibre-reinforced composites, using both a relatively rigid matrix promoted for vehicle armour (polyetherdiol-aliphatic diisocyanate polyurethane) and a more compliant resin intended for highest single-shot mass efficiency (styrene-isoprene-styrene triblock copolymer). Results of this test are sensitive to bolt tightness (Duthinh, 2000); we used minimal torque to simulate how an actual panel reinforced with relatively high creep reinforcement like oriented PE will relieve bolt torque quickly compared to the life of a vehicle. The three materials failed at similar strains ($10 \pm 2\%$), but the much higher G_{12} of the SSE-PE composite gives a bearing strength an order of magnitude or more above the two PE fibre composites. This suggests SSE-PE composites could enable lighter total vehicle armour system design, and/or less fraction of area with compromised protection if bolts are moved further into the panel interior to increase shear-out strength in a composite with lower off-axis properties.

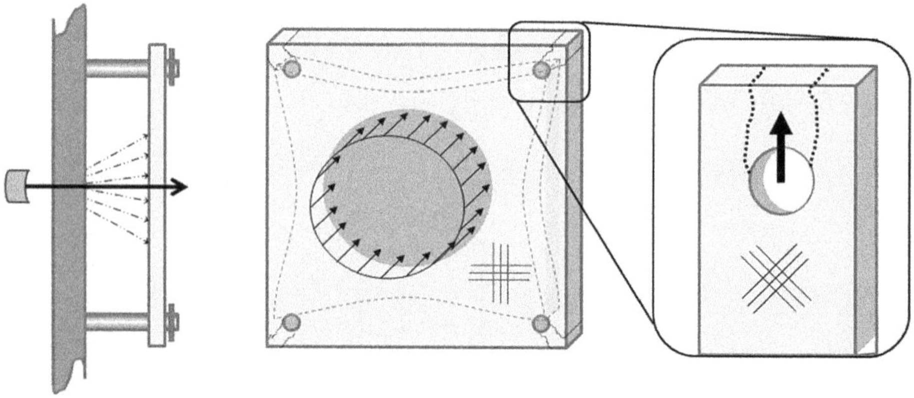

Figure 13.9 Abstraction of spall liner fixation requirements to bolt shear-out test. Left: side view of armour perforated by shaped charge, loading a $[0|90]_n$ reinforced composite spall liner panel attached at standoff with spall. Centre: spall load on composite panel causes it to deflect out of plane and load corner bolt holes. Right: corresponding load is bolt shear-out, simulated here by ASTM D5961.

Table 13.2 Bearing pull-out loads on the bias for $[0|90]_n$ composites with solid-state extrusion (SSE) polyethylene (PE) film and PE fibre reinforcement, per ASTM D5961, Procedure C. Fibre PE composites moulded per manufacturer's recommendation

Material	Width to diameter ratio	Edge distance ratio	Ultimate bearing strength (MPa)	Chord stiffness (MPa)	Ultimate strain (%)
90% SSE-PE film with polyolefin resin	6.10	3.00	58.3	950.8	0.122
80% PE fibre/urethane (DSM Dyneema® HB26)	6.09	3.00	6.0	95.3	0.094
80% PE fibre/S-I-S copolymer resin (DSM Dyneema® HB50)	6.09	3.05	1.5	167.1	0.081

13.2.6 Through-thickness properties of SSE-PE films

Ballistic impact onto laminated targets is generally held to start with transverse shear loading of the first layer contacted, with tensile stresses developing to allow the target to comply out of plane with the projectile. Conventional models of ballistic impact onto fibre-reinforced laminates generally consider the reinforcement as 1D elements (Roylance and Wang, 1980; Parga-Landa and Hernández-Oilvares, 1995; Cunniff, 1999a; Walker, 2001; Phoenix and Porwal, 2003). Consequently, they ignore this transition of the triaxial stress state around the failure locus during the impact event from pure transverse shear to combined transverse shear and axial tension. It is sometimes noted as justification that fibres that offer high ballistic protection are seen to retain a large fraction of their axial strength after transverse loading (Abbott et al., 1974; Cheng et al., 2005; Marissen, 2011), or that the energy required to cut through a layer of ballistic fibrous materials is sufficiently small that it can be neglected, or indeed, that even the specific fibre-reinforced material used on the surface is not important (Cunniff, 1999b). Such assumptions become less reasonable as laminate reinforcement transverse shear strength increases.

The cut resistance of SSE-PE films does indeed appear to be significantly higher than oriented polymer fibres used in protective applications. Para-aramid fibres are generally accepted as offering the highest cut resistance per weight of any all-organic fibre reinforcement, and they were the basis for many current fibrous armour impact models (Roylance and Wang, 1980; Parga-Landa and Hernández-Oilvares, 1995; Cunniff, 1999a; Walker, 2001; Phoenix and Porwal, 2003). Therefore they are a reasonable control. Cut resistance depends in part on textile structure. To minimise the effect of differing structures between SSE-PE film-reinforced composite

Material	Basis weight (g/m^2)	Cutting force F, Mean ± Std. dev. (N)
Kevlar® KM2 woven fabric	129	329 ± 1
Kevlar® KM2 woven fabric	159	325 ± 2
Tensylon™ SSE-PE, woven fabric	146	702 ± 1

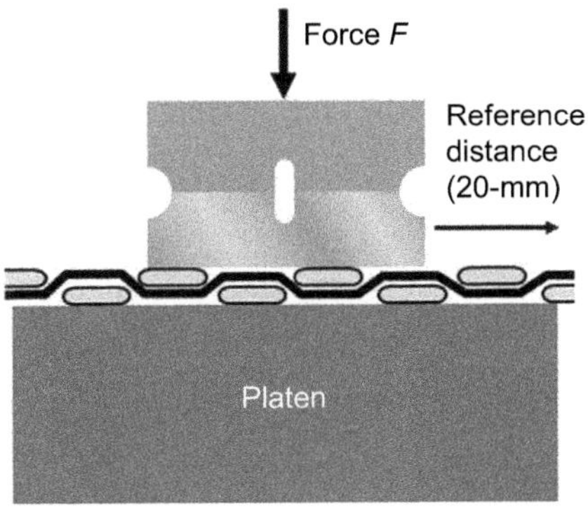

Figure 13.10 Force to cut in reference distance per ASTM F1790-97 as a function of basis weight for fabrics woven from Kevlar® fibre and from narrow strips of Tensylon™ solid-state extrusion polyethylene.

prepregs and fibre-reinforced materials, fabrics were woven from narrow (812-dTex) ribbons of Tensylon™ SSE-PE and compared with similar basis weight fabrics of 660-dTex Kevlar® KM2+, a grade commonly used in armour applications (eg, Cunniff, 1999a,b; Cheng et al., 2005) per ASTM F1790-97. All samples were cut across the bias. Fig. 13.10 compares the results, in terms of cutting force versus basis weight to cut through in the reference distance. The SSE-PE fabric sample offered more than twice the cut resistance of the para-aramid fibre sample, suggesting that the resistance of SSE-PE film-reinforced composites early in the ballistic impact event may offer a unique energy absorption mechanism—particularly early during the ballistic event, when the reinforcement's multiaxial stress state is dominated by transverse shear, and more particularly when, as suggested above, the in-plane properties of SSE-PE have allowed more strain energy to be distributed from the impact point before material failure. The exceptional cut resistance of SSE-PE film-reinforced flexible materials may allow lighter weight personal protection when both bullets and blades must be addressed. Weedon (2000) suggests the high cut resistance of commercial SSE-PE films is due to their high molecular weight and also their relatively large transverse dimensions (~50–65 μm, compared to 10–25 μm for typical PE fibres). A dependence of cut resistance on thickness has been seen in other highly oriented polymers (Knoff, 2001).

13.3 Reinforcement volume fraction of SSE-PE film composites

The film form of SSE reinforcements allows high translation of reinforcement modulus and strength into composites by minimising matrix fraction needed. This is particularly valuable in protective composites, where protection per weight is important. Composites can be 100% self-reinforced and use partial melting under heat and pressure to provide a matrix (Holmes, 1982; Jordan et al., 2002; Ward and Hines, 2004), or they can use a separately applied matrix. In either case, modulus and strength in either reinforced direction in a $[0|90]_n$ composite can exceed 45% of those of the reinforcing film.

13.4 Conclusions

Protective equipment based on film-reinforced composites has a long history, established design principles and performance above predictions based on axial strength ballistic protection design principles established for fibre-reinforced plastics. The commercial realisation of highly oriented and elongated SSE-PE films allow unique opportunities in improving personal protection compared to fibre-reinforced composites. These include buoyancy, light weight, low dynamic and static deflection after impact and corresponding better support for brittle strike faces, improved penetration resistance at composite edges and less cost and weight associated with attachments in vehicle armour systems. Unique characteristics of SSE-PE films and film-reinforced composites may explain why their performance exceeds expectations based on fibre-reinforced composites. Higher film properties demonstrated at laboratory scale suggests these materials offer large opportunities for further improvement.

Acknowledgements

We thank Tom Boyer, Bryce VanArsdalen, Jovan Jovicic, Helga Kuhlmann and Randy Kendzierski for experimental work, and Ken Harding and Leo Carbajal (all of DuPont) for many useful discussions. We also thank Prof. I. M. Ward of the University of Leeds for reviewing this manuscript, and the Leibniz Institute for Polymer Research e.V. for supporting the scientific exchange programme that enabled this work.

References

Abbott, N., Donovan, J., Schoppee, M., 1974. The Effect of Temperature and Strain Rate on the Tensile Properties of Kevlar and PBI Yarns. Technical Report AFML-TR-74–66, Part II. US Air Force Materials Laboratory, Wright Patterson Air Force Base, Dayton, OH, USA. www.dtic.mil/dtic/tr/fulltext/u2/b002047.pdf.

Alesi, A.L., Ames, R.P., Gagne, R.A., Litman, A.M., Prifti, J.J., November 1975. New Materials and Constructions for Improved Helmets. Technical report AMMRC MS 75-9. US Army Materials and Mechanics Research Center, Watertown Arsenal. www.dtic.mil/dtic/tr/fulltext/u2/a018958.pdf.

Anonymous, 1991. Solid-State Extrusion Process Yields Ultra-Strong Polyolefin Fiber. Advanced Materials Newsletter, Smithers-Rapra Publishing (Issue 13).

ASTM D1790-97. Standard Test Method for Measuring Cut Resistance of Materials Used in Protective Clothing, 1997. ASTM International, West Conshohocken, PA, USA.

ASTM D3518-94, Standard Test Method for In-plane Shear Response of Polymer Matrix Composite Materials by Tensile Test of a ±45° Laminate. ASTM International: West Conshohocken, PA, USA (reapproved 2007).

ASTM D3039-08. Standard Test Tensile Properties of Polymer Matrix Composite Materials, 2008. ASTM International, West Conshohocken, PA, USA.

ASTM D7744-11. Standard Test Methods for Tensile Testing of High Performance Polyethylene Tapes, 2011. ASTM International, West Conshohocken, PA, USA.

ASTM D5961-13. Standard Test Method for Bearing Strength of Polymer Matrix Composite Laminates, 2013. ASTM International, West Conshohocken, PA, USA.

Beygelzimer, Y.E., Beloshenko, V.A., 2004. 'Solid State Extrusion'. Encyclopedia of Polymer Science and Technology. John Wiley & Sons, New York. http://dx.doi.org/10.1002/0471440264.pst343.

Cheng, M., Chen, W., Weerasooriya, T., 2005. Mechanical properties of Kevlar® KM2 single fiber. Journal of Engineering Materials Technology 127 (2), 197–203.

Chocron, S., King, N., Bigger, R., Walker, J., Heisserer, U., van der Werff, H., 2013. Impacts and waves in Dyneema® HB80 strips and laminates. Journal of Applied Mechanics 80, 031806, 1–10.

Cook, G.A., September 1980. Molding of Polypropylene Film Armor for Radar Antenna Hardening Applications. Technical report AMMRC-TR80–54. US Army Materials and Mechanics Research Center, Watertown Arsenal. www.dtic.mil/dtic/tr/fulltext/u2/a095726.pdf.

Cunniff, P.M., November 15–19, 1999a. Dimensional parameters for optimization of textile-based body armor systems. In: Proceedings of the 18th International Symposium on Ballistics, pp. 1303–1310. San Antonio, TX.

Cunniff, P.M., November 15–19, 1999b. Decoupled response of textile body armor. In: Proceedings of the 18th International Symposium on Ballistics, pp. 814–821. San Antonio, TX.

Czeckowski, L., Jankowski, J., Kubiak, T., 2012. Experimental tests of a property of composite material assigned for ballistic products. Fibres & Textiles in Eastern Europe 20, 3 (92), 61–66.

DeTeresa, S., Porter, R., Farris, R., 1988. Experimental verification of a microbuckling model for the axial compressive failure of high performance polymer fibres. Journal of Materials Science 23, 1886–1894.

DeTeresa, S., Groves, S., Harwood, P., Sanchez, R., March 1996. Mechanical Characterization of Sonar Window Materials. Lawrence Livermore National Laboratory. Technical Report UCRL-ID-123468.

Doerr, R., March 1979. Manufacturing Methods and Technology Project to Establish Production Techniques to Manufacture Rigid Armor for Radar Antenna Hardening. Contractor Report Accession Nr. ADA069395. US Army Materials and Mechanics Research Center, Watertown Arsenal. www.dtic.mil/dtic/tr/fulltext/u2/a051695.pdf.

Duthinh, D., August 2000. Connections of Fiber-Reinforced Polymer (FRP) Structural Members: A Review of the State of the Art. Technical Report NISTIR 6532. US National Institute of Standards and Technology, Gaithersburg, MD. http://fire.nist.gov/bfrlpubs/build00/PDF/b00043.pdf.

Freitas, C., Bigger, R., Scott, N., LaSala, V., Mackiewicz, J., 2014. Composite materials dynamic back face deflection characteristics. Journal of Composite Materials 48 (12), 1475–1486. http://dx.doi.org/10.1177/0021998313487934.

Gibson, A.G., Davies, G.R., Ward, I.M., 1978. Dynamic mechanical behaviour and longitudinal crystal thickness measurements on ultra-high modulus linear polyethylene: a quantitative model for the elastic modulus. Polymer 19 (6), 683–693. http://dx.doi.org/10.1016/0032-3861(78)90123-4.

Harpell, G., Palley, I., Kavesh, S., Prevorsek, D., March 18, 1992. Ballistic-Resistant Composite Article. European Patent Specification 0191306 B1.

Heckert, W.W., April 30, 1946. Laminated Armor Plate Structure. US Patent 2399184.

Holmes, R.P., January 5, 1982. Laminated Armor. US Patent 4309487.

Hu, W., Schmidt-Rohr, K., 2000. Characterization of ultradrawn polyethylene fibers by NMR: crystallinity, domain sizes and a highly mobile second amorphous phase. Polymer 41, 2979–2987. http://dx.doi.org/10.1016/S0032-3861(99)00429-2.

Jarecki, L., Meier, D.J., 1979. Ultrahigh Modulus Polyethylene. II. Effect of Drawing Temperature on Void Formation and Modulus. Journal of Polymer Science: Polymer Physics Edition 17, 1611–1621. http://dx.doi.org/10.1002/pol.1979.180170913.

Jordan, N., Olley, R., Bassett, D., Hine, P., Ward, I., 2002. The development of morphology during hot compaction of Tensylon high-modulus polyethylene tapes and woven cloths. Polymer 43, 3397–3404. http://dx.doi.org/10.1016/S0032-3861(02)00104-0.

Kanamoto, T., Ohama, T., Tanaka, K., Takeda, M., Porter, R.S., August 1987. Two-stage drawing of ultra-high molecular weight polyethylene reactor powder. Polymer 28, 1517–1520. http://dx.doi.org/10.1016/0032-3861(87)90352-1.

Kanamoto, T., Tsuruta, A., Tanaka, K., Takeda, M., 1988. Superdrawing of ultrahigh molecular weight polyethylene: 1. Effects of techniques on drawing of single crystal mats. Macromolecules 21, 470–477. http://dx.doi.org/10.1021/ma00180a032.

Karthikeyan, K., Russell, B., Fleck, N., Wadley, H., Deshpande, V., 2013. The effect of shear strength on the ballistic response of laminated composite plates. European Journal of Mechanics A/Solids 42, 35–53. http://dx.doi.org/10.1016/j.euromechsol.2013.04.002.

Knoff, W., July 30–August 1, 2001. Fiber transverse deformation work and cut performance of protective apparel. In: Proceedings of the 30th Textile Research Symposium at Mt. Fuji in the New Millennium, Shizuoka, Japan.

Marissen, R., 2011. Design with ultra strong polyethylene fibers. Materials Science and Applications 2, 319–330. http://dx.doi.org/10.4236/msa.2011.25042.

McGarry, F., Moalli, J., 1991. Mechanical behaviour of rigid rod polymer fibres: 1. Measurement of axial compressive and transverse tensile properties. Polymer 32 (10), 1811–1815. http://dx.doi.org/10.1016/0032-3861(91)90368-S.

Nakae, M., Uehara, H., Kanamoto, T., 2000. Structure development upon melt drawing of ultrahigh molecular weight polyethylene: effect of prior thermal history. Macromolecules 33, 2632–2641. http://dx.doi.org/10.1021/ma991330a.

Parga-Landa, B., Hernandez-Olivares, F., 1995. An Analytical Model to Predict Impact Behavior of Soft Armours. International Journal of Impact Engineering 16, 455–466. http://dx.doi.org/10.1016/0734-743X(94)00054-Z.

Peijs, T., Rijsdijk, H., de Kok, J., Lemstra, P., 1994. The role of interface and fibre anisotropy in controlling the performance of polyethylene-fibre-reinforced composites. Composites Science and Technology 52, 449–466. http://dx.doi.org/10.1016/0266-3538(94)90180-5.

Phoenix, S.L., Porwal, P.K., 2003. A new membrane model for the ballistic impact response and V_{50} performance of multi-ply fibrous systems. International Journal of Solids and Structures 40, 6723–6765. http://dx.doi.org/10.1016/S0020-7683(03)00329-9.

Porter, R.S., Wang, L.-H., 1995. Uniaxial extension and order development in flexible chain polymers. Journal of Macromolecular Science, Part C: Polymer Reviews 35 (1), 63–115. http://dx.doi.org/10.1080/15321799508014590.

Porter, R.S., Kanamoto, T., Zachariades, A.E., 1994. Property opportunities with polyolefins: a review. Preparations and applications of high stiffness and strength by uniaxial draw. Polymer 35 (23), 4979–4984. http://dx.doi.org/10.1016/0032-3861(94)90652-1.

Prifti, J.J., DeLuca, E., Alesi, A.L., 1976. Hardened Tuned-Wall Plastic Radomes for Military Radars: Fabrication of XP Plastic Radome Panels. U.S. Army Materials and Mechanics Research Center, Watertown, Massachusetts. http://handle.dtic.mil/100.2/ADA026146.

Roylance, D., Wang, S.S., 1980. Ballistic materials and penetration mechanics. In: Laible, R.C. (Ed.), Ballistic Materials and Penetration Mechanics. In: Wolsky, S.P., Czanderna, A.W. (Eds.), Methods and Phenomena, vol. 5. Elsevier Scientific Publishing Co.

Russell, B., Karthikeyan, K., Deshpande, V., Fleck, N., 2013. The high strain rate response of ultra high molecular-weight polyethylene: from fibre to laminate. International Journal of Impact Engineering 60, 1–9. http://dx.doi.org/10.1016/i.ijimpeng.2013.03.010.

Sawai, D., Nagai, K., Kubota, M., Ohama, T., Kanamoto, T., 2005. Maximum tensile properties of oriented polyethylene, achieved by uniaxial drawing of solution-grown crystal mats: effects of molecular weight and molecular weight distribution. Journal of Polymer Science: Part B: Polymer Physics 44, 153–161. http://dx.doi.org/10.1002/polb.20682.

Schaap, A., 2011. Interlaboratory Study to Establish Precision Statements for ASTM D7744-11, Test Methods for Tensile Testing of High Performance Polyethylene Tapes. Research Report D13:1131. ASTM International, West Conshohocken, PA, USA.

Singletary, J., Carbajal, L., Boogh, L., October 6–10, 2008. Fragment simulating projectile V_{50} scaling rules. In: Proceedings of the Personal Armour Systems Symposium, Brussels, Belgium.

Stijnman, P., 1995. Determination of the elastic constants of some composites by using ultrasonic velocity measurements. Composites 26, 597–604.

Test Procedure HPW-TP-401.01B: Bullet Resistant Helmet, October 1995. H.P. White Laboratory, Inc., Street, MD, USA. In: http://www.bodyarmornews.com/wp-content/uploads/2013/HP-White-helmet-401-01b_TestingProcedure.pdf.

van der Werff, H., Heisserer, U., Phoenix, S., September 13–17, 2010. Modelling of ballistic impact on fiber composites. In: Proceedings of the Personal Armour Systems Symposium, Quebec City, CA.

Walker, J.D., May 7–11, 2001. Ballistic limit of fabrics with resin. In: Proceedings of the 29th International Symposium of Ballistics, pp. 1409–1414. Interlaken, Switzerland.

Walsh, S., Scott, B., Spagnuolo, D., Wolbert, J., November 30, 2006. Hybridizing thermoplastic aramids: enabling material technology for future force headgear. In: Proceedings of the US Army Science Conference. www.dtic.mil/cgi-bin/GetTRDoc?AD=ADA481016.

Wang, J., Smith Jr., K., 1999. The breaking strength of ultra-high molecular weight polyethylene fibers. Polymer 40, 7261–7274. http://dx.doi.org/10.1016/S0032-3861(99)00034-8.

Ward, I.M., Hine, P.J., 2004. The science and technology of hot compaction. Polymer 45, 1413–1427. http://dx.doi.org/10.1016/j.polymer.2003.11.050.

Weedon, G., 2000. Solid-state extrusion high molecular weight polyethylene fibres (Chapter 4.5). In: Hearle, J.W.S. (Ed.), High Performance Fibres. Woodhead Publishing, Cambridge.

Zachariades, A., Kanamoto, T., 1988. New model for the high modulus and strength performance of ultradrawn polyethylenes. Journal of Applied Polymer Science 35, 1265–1281.

Zachariades, A.E., Watts, M.P.C., Porter, R.S., May 1980. Solid state extrusion of ultra high molecular weight polyethylene. Processing and properties. Polymer Engineering and Science 20 (8), 555–561.

Ballistic performance evaluation of woven fabrics based on experimental and numerical approaches

14

D. Sun
School of Textiles and Design, Heriot-Watt University, Galashiels, United Kingdom

14.1 Introduction

Body armour can be classified into two categories: hard and soft body armours. Ceramic, plastic and steel were materials used for hard body armours. Many different materials, such as leather (animal skins), linen, cotton, silk and nylon have been used as soft body armour materials for antiballistic applications in history [1,2]. The revolutionisation of the modern generation of soft body armour occurred in the late 1960s, when novel synthetic fibrous materials with high-performance antiballistic performance were developed. In early 1965, a novel aramid fibre, Kevlar®, was discovered by the polymer research group in Dupont. Since then, more high-performance aramid fibres such as Twaron®, Technora® and ultra-high molecular weight polyethylene (eg, Spectra®, Dyneema®) were developed to be used as commercially available body armour materials. Cunniff evaluated the ballistic property of Zylon® (PBO) and believed that it is lighter, more comfortable and stronger than aramid body armour [3]. However, its physical properties degraded when exposed to moisture [4]. M5 is the most recently known fibre showing great ballistic impact potential [5]. The fibre is still in its development stage, not commercially available. These fibrous materials have been proved stronger than steel and have better ballistic resistance. Some of these ballistic materials, such as Kevlar® and Dyneema® have been widely used as today's soft body armour materials.

Ballistic impact is normally a low-mass high-velocity impact by a projectile propelled by a source onto a target. Since the ballistic impact is a high-velocity event, the effects on the target can be only near the location of impact. The technical approaches for stopping penetrating missiles and mitigating nonpenetrating impacts are different. The principles of minimising the effects of energy transfer from a projectile are as follows [6]:

1. absorbing energy in armours by making it do work on materials before it gains access to the body, such as breaking, stretching and compressing materials, or extend the time over which it is applied to the body; and
2. redistributing the energy so that other materials or the body wall are more able to withstand the total energy.

Advanced Fibrous Composite Materials for Ballistic Protection. http://dx.doi.org/10.1016/B978-1-78242-461-1.00014-5

A flexible armour system defeats the projectile by absorbing the kinetic energy of the projectile and spreading it over a larger area before the projectile has a chance to perforate. The energy absorbed by the fabric goes into (1) strain energy via the stretching of yarns, (2) kinetic energy of the fabric due to transverse deflection of the fabric, (3) movement of material towards the impact point and (4) yarn breakage.

When a fabric sample is impacted, two wave fronts (transverse and longitudinal wave) are generated, shown in Fig. 14.1. The longitudinal wave travels outward along the fibre axis at the sound velocity of the material from the point of impact. This wave also causes the yarn to be stretched and have in-plane movement.

The longitudinal wave velocity a_0 for isotropic solids is expressed as follows according to Lyons [8].

$$a_0 = \sqrt{\frac{E}{\rho}} \qquad [14.1]$$

where E is young's modulus, and ρ is material density.

Porwal and Phoenix [9] discovered the cone wave velocity C_i in the ith layer can be written as follows:

$$C_i = \text{const} \times a_{0i}^{1/3}\left(\frac{v}{\sqrt{2}}\right)^{2/3} \qquad [14.2]$$

where a_{0i} is the longitudinal strain wave velocity in the ith layer, which is constant for each layer; const is constant and v is projectile velocity.

During the ballistic impact event, cone formation takes place on the exit side of the target just below the point of impact [10–12]; this is attributed to the transverse wave propagation. The radius of the formed cone increases with the increase of impact time. The cone moves along with the projectile in the transverse direction; as a result, the height of the cone increases with impact time. The projectile displacement at any instant of time and the cone height formed would be the same, reported by Naik [13]. Initially, the transverse wave velocity of the impacted objective increases significantly; then it remains nearly constant. The cone height variation is nonlinear to impact time. The rate of the increase of cone height or depth of projectile penetration decreases with impact time; this is caused by the velocity variation of the projectile

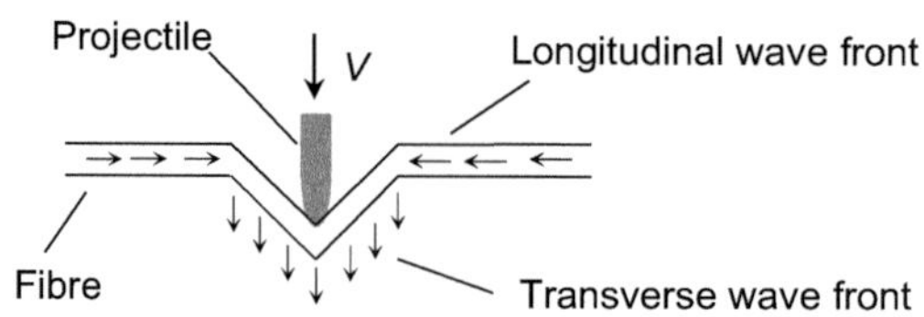

Figure 14.1 A yarn impacted by a projectile [7].

during ballistic impact event. Roylance [7] investigated ballistic impact of textile structure and concluded that the vast majority of ballistic energy was deposited in the orthogonal fibres/yarns passing through the impact points, while other fibres/yarns are essentially ineffective. The finding suggested possible improvement in the design of textile structures intended for dynamic impact applications. During the development of body armour, it has always been desirable to have lighter and stronger materials so that the advancement of performance could be achieved at a reduced weight. Due to their unique molecular structures, the previously listed fibres offer outstanding high performance over traditionally used synthetic fibres and fulfil the design requirement of light weight, flexibility and covert body armour for its users.

Apart from fibre/yarn properties, the same as traditional garment fabric construction, fabric construction, finishing processes, armour panel design and construction also play important roles in the determination of body armour performance. It has been revealed that nonwoven structure is effective in only capturing low-speed fragments by having some of the fibres prealigned along the projectile trajectory, and it is more likely to cause larger trauma [14]. Similar to nonwoven fabric, knitted fabrics are avoided from being used for body amour application caused by greater trauma due to the low modulus [15]. Nonwovens consisting of unidirectional layers are also made for the use of antiballistic applications, according to a US patent [16]. Basic 2D plain weave (shown in Fig. 14.2) is the most used structure currently for antiballistic body armours; it exhibits the highest level of yarn interlacing density, which is therefore the best dimensional stability of plain weave among the three 2D basic weave structures. Previous studies [17] on the effect of fabric type suggest

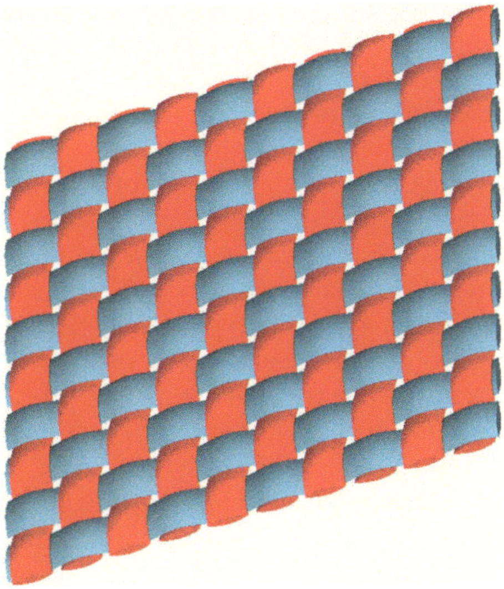

Figure 14.2 Plain-weave structure.

that the structures with fewer interlacing yarns show better ballistic performance due to the reduced interference of the strain wave propagation upon ballistic impact. Most of today's fabrics for the ballistic impact application are woven because they provide high area density at the lowest possible weight and are extremely stable. Besides, fabric construction parameters such as yarn count and density also affect the ballistic performance. Cork and Foster [18] revealed that square fabrics which possess the same thread density and yarn linear density provided better ballistic impact performance than others under the same fabric areal density. Three-dimensional fabrics have attracted attention in recent years. The main advantage of 3D fabric structure is in through-the-thickness reinforcement; as a result, the dimensional stability of 3D structures is much better compared to 2D structures. 3D composites exhibit excellent damage tolerance upon ballistic impact [19].

As all the high-performance fibres used currently for body armour have very high molecule crystallinity, their surfaces are very smooth, and the coefficient of friction of the surface is very low. It has been reported that increasing interyarn friction improves the ballistic resistance of woven fabrics [20,21]. Duan's finite element (FE) model revealed that fabric possessing a high level of friction absorbed more energy than that with no friction [22]. The projectile—fabric friction delays yarn breakage by distributing the maximum stress along the periphery of the projectile—fabric contact zone. The delay of yarn breakage substantially increases the fabric energy absorption during the later stages of the impact. The yarn—yarn friction hinders the relative motion between yarns and thus resists decrimping of fabric weave tightness. To improve the surface friction between fabric and projectile, techniques have been investigated to increase the energy absorption during a ballistic impact event. Hogenboom's patent [23] describes a method by introducing a second yarn possessing a high surface friction to cover the surface of the ballistic yarn through core spinning. The yarn with high tensile strength but low surface friction is used as a core. Coating is another technique that has been used to add a foreign layer with higher surface friction onto the ballistic fabric [24]. However, the performance to weight ratio was affected due to the add-on of the coated layer.

Furthermore, the layering out and construction of an armour panel also play an important role in the improvement of ballistic performance. A combined structure has been created to improve the protection against projectiles. Steeghs et al. [16] invented a type of ballistic vest containing a stack of flexible woven fabrics and a stack of flexible unidirectional layers, where two types of ballistic fibres are used for the two stacks of fabrics. Plasma as a surface modification method has been investigated to modify the surface of body armour materials; it only modifies the outermost surface of the treated substrates without altering its bulk inherent property [25].

Research in ballistic impact on fabric body armour is mainly based on three approaches: analytical, experimental and numerical studies. Analytical models are believed to be at the expense of accuracy, although they require less time than numerical models. This chapter reports ballistic performance of woven fabrics based on experimental and numerical approaches.

14.2 Ballistic testing principles and equipment

Methods have been developed to evaluate ballistic performance of armour panels and a single or a few layers of fabrics. Fig. 14.3 shows the principle of a ballistic test.

Ballistic energy loss test is used to define energy absorption of a single or a few layers of fabric by measuring the striking velocity v_1 $\left(\frac{s_1}{t_1}\right)$ and the residual velocity v_2 $\left(\frac{s_2}{t_2}\right)$. The result is presented by the energy absorption of the impacted fabric. The kinetic energy loss or energy absorbed by fabric ΔE after a projectile perforates the fabric can be calculated by Eq. [14.3]. This method is mainly used for research purposes to evaluate the ballistic performance of any newly developed fabric for body armour. In this test, the projectile passes through the first pair of detectors before perforating the fabric. Then it passes through the second pair of detectors, as shown in Fig. 14.3: (1) energy lose test.

$$\Delta E = \frac{1}{2}m\left[\left(\frac{s_1}{t_1}\right)^2 - \left(\frac{s_2}{t_2}\right)^2\right] = \frac{1}{2}m\left(v_1^2 - v_2^2\right) \quad [14.3]$$

where, m is the mass of projectile; s_1 and s_2 are the distance of the front and back pair detectors, respectively; t_1 and t_2 are the time spent for the projectile to pass through the front and back pair of detectors.

Ballistic resistance test is performed based on HOSDB body armour standards for UK police and NIJ Standard 0101.04 by measuring the overall ballistic performance of the armour according to pass/fail criteria and the back-face signature

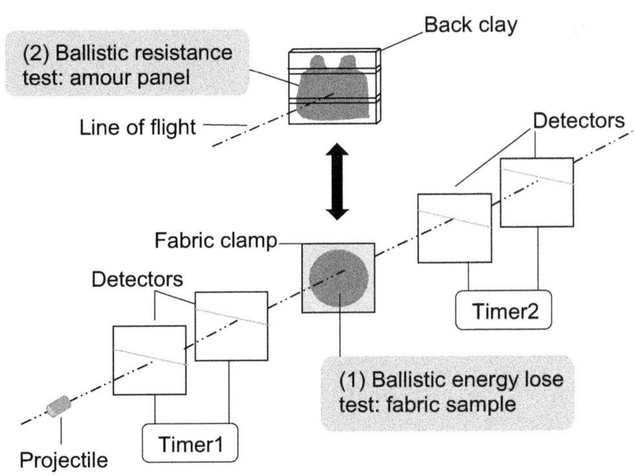

Figure 14.3 Principle of a ballistic test: (1) energy loss test, (2) ballistic resistance test.

(the indentation depth) created by the impact of projectile on the back clay during a nonperforating threat round impact to statistically measure penetration performance. This is shown in Fig. 14.3: (2) ballistic resistance test. The testing method has been used for both research and commercial use following the Standards and Testing Program. Apart from the back-face signature, the amount of materials used to stop a projectile can also be discovered. Areal density of the panel to stop a projectile from penetration has been used for research purposes to compare the ballistic performance of different materials. In this method, material ballistic performance is tested by determining the number of layers of fabric required to stop a projectile. The areal density needed to stop a projectile from penetration can then be calculated based on the number of layers required to stop the projectile. It is a discretised measuring method, as assembly weight could only be increased or decreased by adding or subtracting a single layer, but it is more relevant to end-use applications where preventing penetration is the goal.

Ballistic Limit Test V_{50} is the average of an equal number of highest partial penetration velocities and the lowest complete penetration velocities which occur within a specified velocity speed. It determines the limit of performance and the probability of penetration of ballistic resistant material. MIL-STD-662F is the standard for measuring ballistic limit V_{50} [26]. Certain approaches lead to approximation of V_{50}, the velocity at which complete penetration and incomplete penetration are equally likely to occur.

In 1998, Czarnecki studied ballistic limit V_{50} using a semiempirical procedure to estimate the ballistic limit [27]. The theoretical kinetic energy relationship versus the typical empirical energy relationship is shown in Fig. 14.4. The assumption was

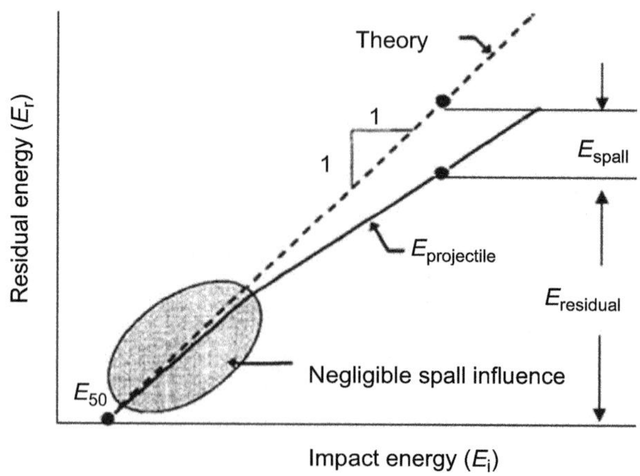

Figure 14.4 Theoretical kinetic energy relationship versus typical empirical energy relationship.
Adapted from G.J. Czarnecki, Estimation of the V_{50} using semi-empirical (1-point) procedures, Composites Part B 29 (3) (1998) 321–329.

made that at penetration velocities near V_{50}, spall becomes insignificant and negligible. E_{50} and V_{50} can be estimated according to the following equations:

$$E_{50} = E_i - E_r \quad [14.4]$$

$$\frac{1}{2}mV_{50}^2 = \frac{1}{2}mv_i^2 - \frac{1}{2}mv_r^2 \quad [14.5]$$

$$V_{50} = \sqrt{v_i^2 - v_r^2} \quad [14.6]$$

E_{50} is effectively constant regardless of the mass of projectile for a given impact target; therefore V_{50} can be estimated by the following energy equation:

$$V_{50} = \sqrt{\frac{2E_{50}}{m}} \quad [14.7]$$

where, E_{50} and V_{50} are ballistic limit energy and velocity, respectively; E_i and v_i are impact energy and velocity, respectively; E_r and v_r are residual energy and velocity, respectively; m is mass of the projectile.

Cunniff studied V_{50} performance of body armour and found that V_{50} is closely associated to projectile impact obliquity at high areal density of an armour panel [28].

14.3 Finite element simulation of ballistic impact on woven fabrics

14.3.1 Three-dimensional continuum finite element models

Ballistic energy absorption of woven fabrics has been studied by many researchers [29–32]. In general, there are two approaches in the theoretical study of ballistic impact. One is numerical, the other analytical. Analytical models are established using general continuum mechanics equations. However, in many problems, analytical methods seem to fail to find solutions. An analytical method has limitations in the case of nonlinear problems, and in many cases, the problems are simplified from the reality. This has significant influence when dealing with very complex geometry and loading conditions. Analytical methods have been used to study ballistic impact on soft body armour by researchers [33,34]. However, the analytical models only calculated the breakage of principal yarns. No considerations were taken to the secondary yarns in absorbing projectile impact energy and fabric kinetic energy. This has affected the accuracy of the results.

The FE method was developed by Argyris and his coworkers. Initially, it was used in the aerospace industry. Then it was applied to structural and solid mechanics. It is a very useful tool in solving engineering problems with regard to complex geometry, loading and material properties. The FE method involves modelling the structure using small interconnected elements (finite elements). The method consists of three main steps: (1) preprocessing to create the FE geometrical model, apply loads, constraints, boundary conditions, etc.; (2) processing which conducts numerical calculation by a computer and (3) postprocessing to analyse the results based on calculation.

A 3D continuum element model is formed by a 3D continuum element, which enables the simulation to take account of the discrete nature of woven fabrics. ABAQUS, ANSYS, DYNA3D and LS-DYNA are the softwares that have been developed for FE analysis. Ballistic performance of woven fabrics modelled with 3D continuum elements have been studied and reported by many [22,32,35]. This method enables the simulation of woven fabric impacted by a projectile while accounting for fabric geometry and material properties of the fabric described by such parameters as crimp, yarn geometry at a crossover and transverse compression of yarns. Ballistic impact and the damage behaviour of the impacted target would be predicted in a more accurate manner while taking into consideration the yarn–yarn, layer–layer contact and friction properties. FE simulation of high-speed projectile–fabric impact can be performed with little difficulty, but the important issue in numerical studies of impact on and perforation through ballistic materials and structures is the accuracy in material and structure definition. The benefit of FE modelling of ballistic impact lies in the fact that some of the very important properties of a projectile and the impacted fabric can be predicted based on the validated model using experimental results. The provided information can be used for ballistic material and panel design and engineering.

14.3.2 Geometric model generation

Fabric geometry can be generated directly in any of the commercial FE softwares mentioned earlier. There are also specific softwares for fabric geometry generation; the generated fabric geometry is then imported into FE softwares. The accuracy of the fabric geometry development is very important to ensure the model simulates the actual fabric.

To work out a woven fabric geometry, these fabric parameters are needed: warp and weft yarn cross-sectional shape, width of yarn (W_{warp} and W_{weft}), yarn spacing (S_{warp} and S_{weft}) and fabric thickness. The surface of the fabric and yarn cross-section can be obtained through SEM observation, shown in Fig. 14.5. Fabric thickness T_{fabric} can be measured using textile testing devices, yarn cross-sectional shape and yarn width can be analysed based on the SEM images, and yarn spacing can be calculated by equations as follows:

$$S_{\text{warp}} = 1/D_{\text{warp}} \qquad [14.8]$$

$$S_{\text{weft}} = 1/D_{\text{weft}} \qquad [14.9]$$

where, D_{warp} and D_{weft} are warp and weft yarn density, respectively.

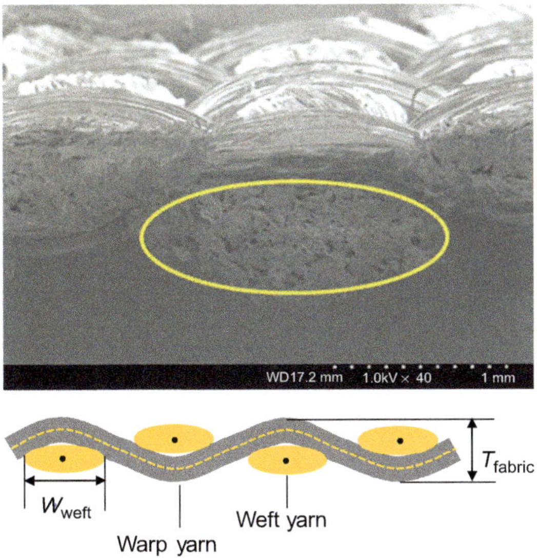

Figure 14.5 SEM observations: fabric surface and yarn cross-section of a woven fabric.

Figure 14.6 A geometric model of projectile and fabric.

Fig. 14.6 shows a ballistic geometric model consisting of a cylindrical-shaped steel projectile with both diameter and height of 5.56 mm, a plain-woven fabric made of 36 Kevlar 49 yarns in both warp and weft directions (7.5 ends and picks per cm). Fig. 14.7 shows a unit yarn path and the cross-section of a yarn in the fabric model (Fig. 14.6) generated in Abaqus®.

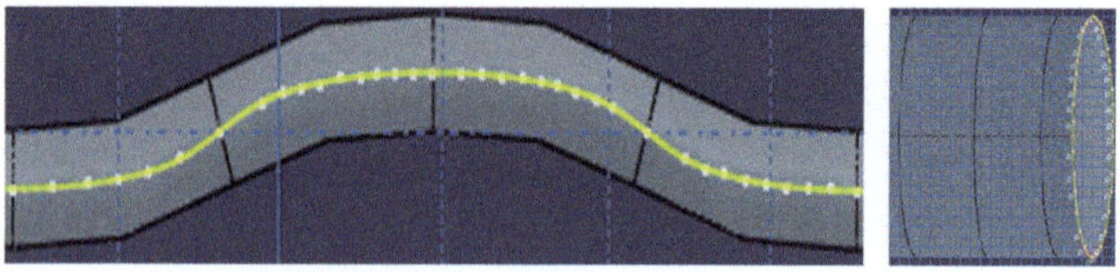

Figure 14.7 Yarn path and cross-section made in Abaqus®.

14.3.3 FE simulation of projectile–fabric impact event

In Fig. 14.6, there are three parts in the finite element model: warp yarns, weft yarns and projectile. Contacts are made between projectile, warp and weft yarns. The projectile is defined as the master surface, and the warp and weft yarns are defined as slave surfaces, as described by Sun and Chen [32]. General contact (explicit) is selected for calculations. The analytical rigid surface of the projectile is used in the general contact algorithm, which provides a smoother surface description that reduces contact noise and is a better approximation of the physical contact constraint. This is because the projectile used in ballistic tests does not deform significantly during penetration into one layer (or more layers) of fabric [36,37].

Parts mesh scheme: a general rule for ballistic impact numerical simulation is that the computer never has enough memory. Therefore we often utilise crude mesh or at least crude in some region. The FE mesh subdivides the geometry into elements, upon which are found nodes. The nodes, which are just point locations in space, are generally located at the element corners. For a 3D solid analysis the elements have physical thickness in all three dimensions.

Sun and Chen [32] designed a type of plain-woven fabric constructed of Kevlar 49 yarn with linear density 158 tex and yarn sett 7.6 threads/cm in both warp and weft directions of the fabric. A ballistic perforation test to evaluate the energy absorbed by the impacted fabric sample had been carried out. Meanwhile, a FE model was developed with a quarter of the projectile and fabric specimen (containing eight warp and eight weft yarns). Yarn density and the structural parameters of the fabric and the size of projectile are the same as the actual dimension for experimental test. Fabric warp and weft yarns and projectile are all meshed by a 3D 4-node linear tetrahedron solid element C3D4 (stress/displacement element in Abaqus), shown in Fig. 14.8. The tetrahedron element is very convenient to mesh a complex shape like the lenticular-shaped yarns with presence of crimps.

Boundary conditions are defined on each part of the boundary by specifying components with respect to a local coordinate system defined by the orthogonal basis. For a ballistic perforation test, boundary conditions have significant influence on fabric absorption while fabric samples are clamped. Cunniff [38] revealed that the ballistic performance of impacted fabric depended on the aperture dimension at impact velocity near the ballistic limit of the fabric. Research work has been done to investigate the clamping pressure and position in the influence of ballistic energy absorption of the impacted fabric [39]. It was discovered that the amount of energy absorbed by fabric

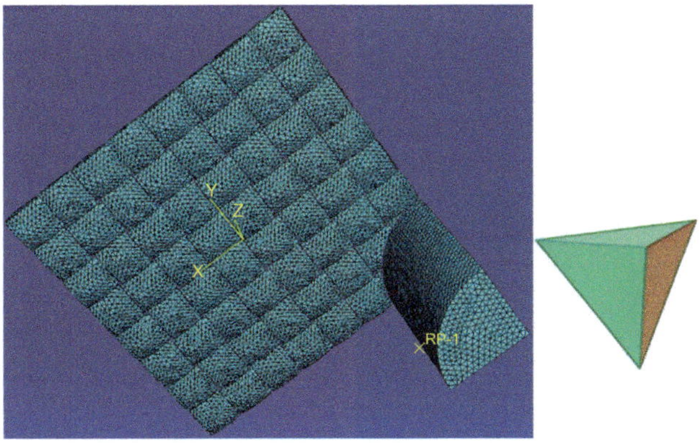

Figure 14.8 Three-dimensional 4-node linear tetrahedron solid element C3D4.

is affected by clamping pressure. Fabric samples with two of the edges clamped absorbed more impact energy compared to those clamped on all four edges. Zeng et al. studied ballistic fabric energy absorption affected by three different boundary conditions: two clamped edges, four clamped edges at 0 degree and four clamped edges at 45 degree. The results are shown in Fig. 14.9 [40]. It is clear that in the low-speed penetration regime, clamping of fabric at two edges is superior to clamping at all four edges. Four edges clamping at 45 degree to the yarn direction facilitates greater energy absorption compared to clamping along 0 degree. Their numerical results show that relaxation of completely ideal clamping at boundaries results in much higher energy absorption.

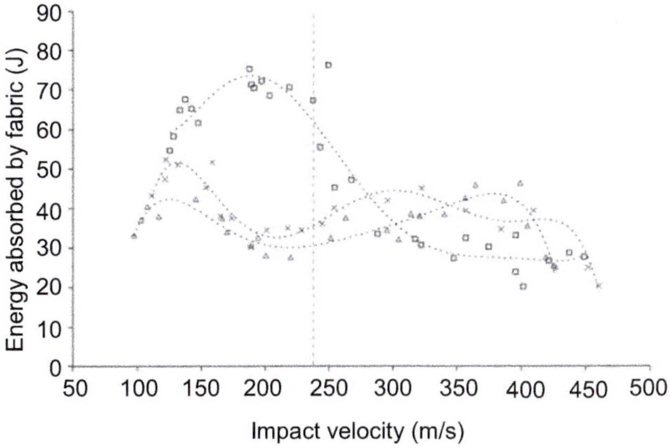

Figure 14.9 Experimental results on energy absorption characteristics for fabric targets with different boundary conditions.
Adapted from X.S. Zeng, V.P.W. Shim, V.B.C. Tan, Influence of boundary conditions on the ballistic performance of high-strength fabric targets, Int. J. Impact Eng. 32 (2006) 631–642.

Displacement boundary conditions (fixed surface) are used for a fabric when modelling a ballistic perforation test. The initial conditions, which describe the state of a body at the start of an analysis specifying the position and velocity at the initial time, are taken as zero. Fig. 14.10 shows the ballistic impact model (quarter of the whole model) with boundary conditions defined for the projectile and fabric/yarns [32]. For a projectile, the translational freedom along the X and Y axes and the rotational freedom along the Z axis are constrained and set to zero, $v_x = v_y = 0$. The translational freedom perpendicular to the symmetrical plane and the rotational freedom in the symmetrical plane are constrained and set to zero, namely $U_x = UR_y = UR_z = 0$ and $U_y = UR_x = UR_z = 0$. One-quarter of the fabric circumference is fixed, $U_x = U_y = U_z = UR_x = UR_y = UR_z = 0$.

Model validation is made by comparing the residual velocities of the projectile between results from FE analysis and the actual ballistic test with the same impact velocities. The results are shown in Fig. 14.11. The theoretical residual velocities are invariably higher (with a maximum value of 4.3%) than the experimental results, as attributed by the assumption that yarns in the FE model are considered as solid. Practically, there is friction between fibres (956 filaments in a 158 tex Kevlar 49 yarn) and yarns during a ballistic impact event. Therefore fabrics impacted by a projectile through a shooting test absorb more energy compared to those of the FE simulated fabric. However, the residual velocities of a projectile between FE simulation and

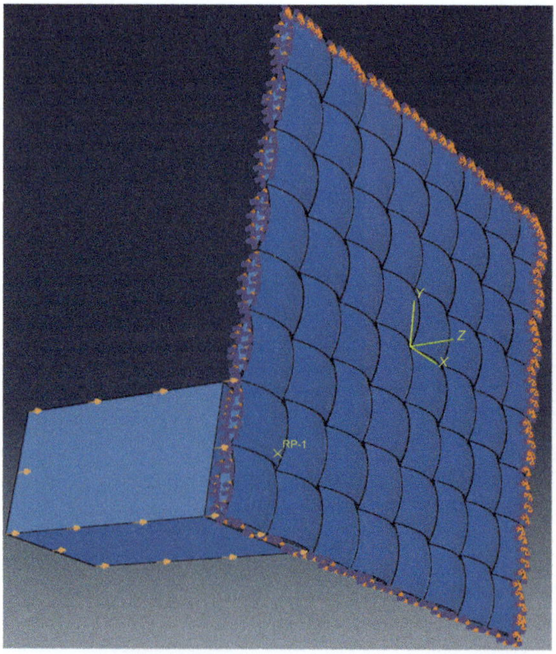

Figure 14.10 Boundary conditions shown in the numerical model of ballistic impact on a single-layer woven fabric.

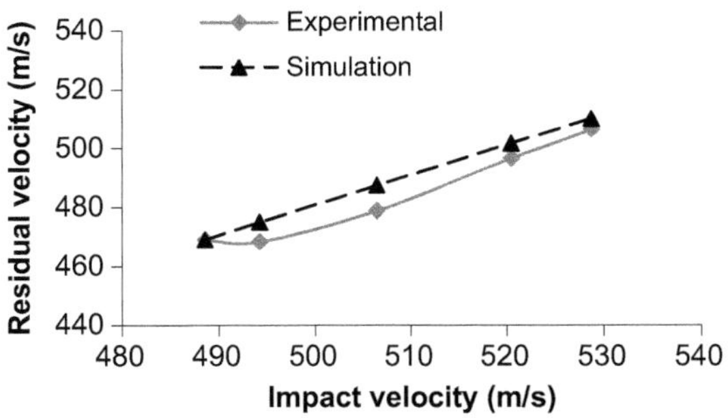

Figure 14.11 Comparison of residual velocity of the projectile between FE simulation and experiment.
Adapted from D. Sun, X. Chen, E. Lewis, G. Wells, Finite element simulation of projectile perforation through a ballistic fabric, Text. Res. J. 83 (14) (2013) 1489–1499.

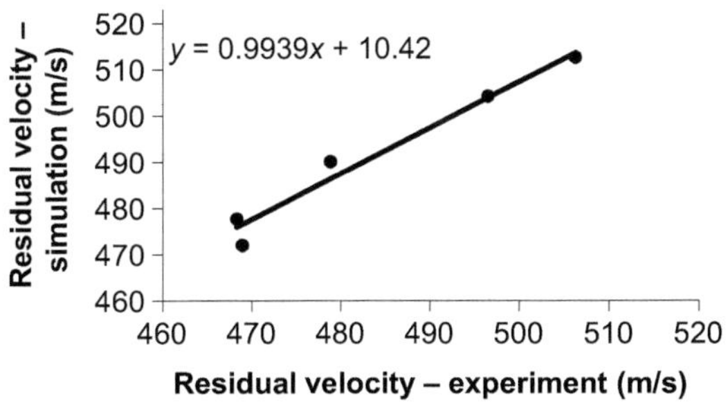

Figure 14.12 Correlation of velocities between results from a simulation and an experiment [32].

an experiment have a significant linear relationship, shown in Fig. 14.12. Fig. 14.13 shows the deformation of the target fabric before and after a ballistic impact as a comparison between FE simulation and an experiment.

14.4 Comparisons and discussions

Fig. 14.14 shows deformation of the target fabric during ballistic impact at different time points in the whole impact process, under a projectile impact velocity of $v_0 = 480$ m/s. FE simulation made it possible to view the impacted fabric deformation and stress distribution in each time step frame. A shock wave is created and centred at the projectile impact location on the fabric, reflected as a stress wave from the distal surface, as seen in Fig. 14.14(b). A longitudinal strain is formed from the impacted area, and the wave is propagated outward to the edge of the fabric model along the warp and weft yarns through the impact vicinity, shown in Fig. 14.14(c) and (d). A pit which is a cone from the back side of the fabric view is formed in the impacted

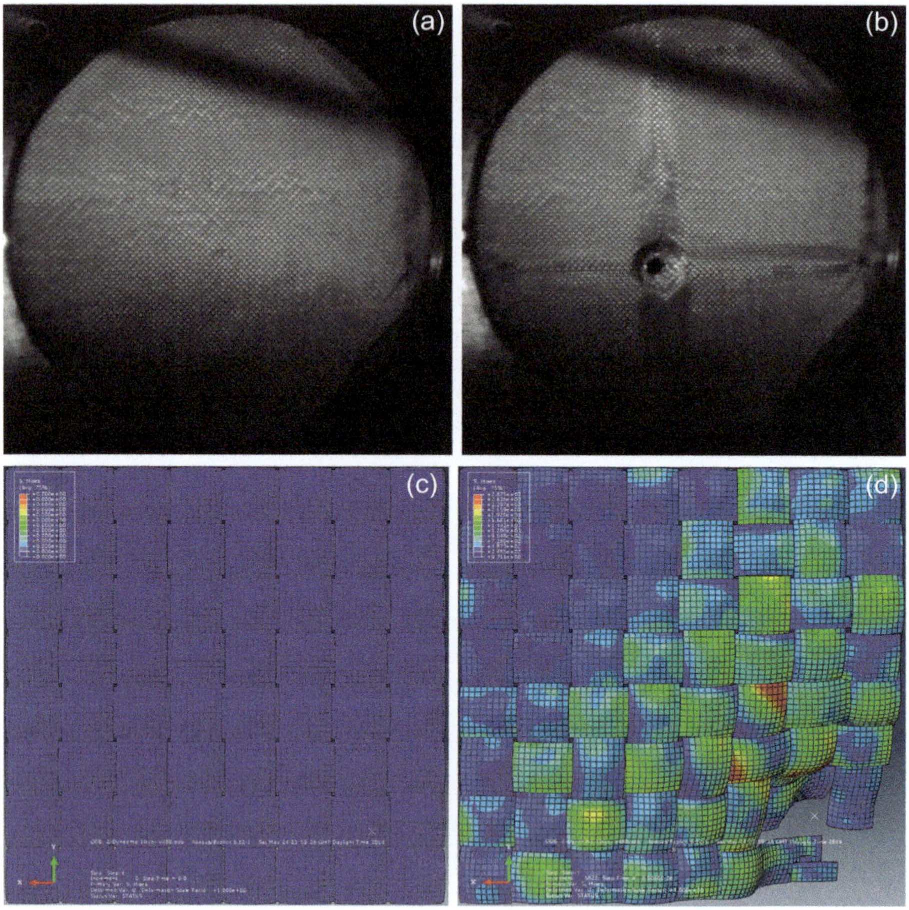

Figure 14.13 Fabric before and after projectile impact from an experiment: (a) and (b), and FE simulation: (c) and (d).

side of the fabric; it shows that the pit area increases with the increase of the impact time. Stress spreads in a larger area through interlacing points between warp and weft yarns. The strain in the yarn starts to exceed the dynamic breaking strain, leading to partial failure of the yarns, shown in Fig. 14.14(b), and the further development of the strain in the principal yarns around the pit area, shown in (c) and (d). In the end, the projectile perforates the fabric with a reduced (residual) velocity.

Based on the validated FE model, Sun and Chen [32] studied the energy absorbed by the fabric. Fig. 14.15 shows various energies absorbed by fabric during an impact event. Immediately after the impact of projectile onto the fabric target, fabric kinetic energy increases rapidly, similar to strain energy. It remains more or less the same after impact and starts to decrease after the projectile penetrates (at about 7 μs), while the plastic energy keeps increasing during the whole impact process. The latter is an indication of the existence of increased plastic deformation in yarns and fibres because of

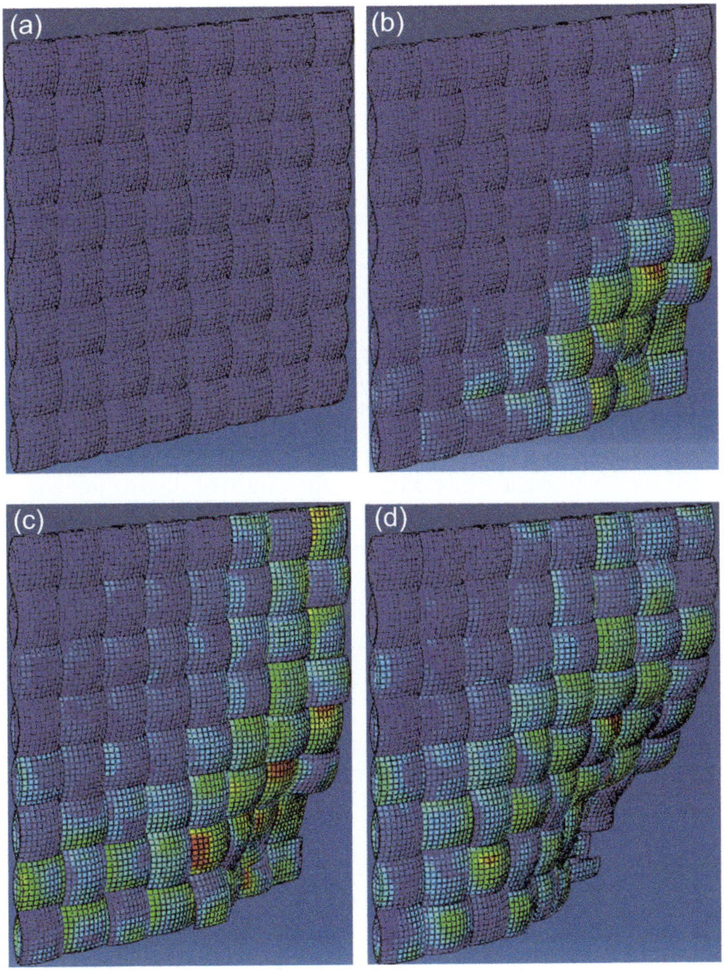

Figure 14.14 Fabric perforated by the projectile and stress distribution at different step time points (a) $t = 0$ μs, (b) $t = 1.33$ μs, (c) $t = 3.33$ μs and (d) $t = 6.67$ μs.

the cone development. As part of the strain energy, the elastic energy absorbed by the fabric, mainly caused by the longitudinal stress, increases at the early stage of the impact and keeps virtually constant for the rest of the impact process. Energy absorbed by fibre and yarn breakage is not in existence until about 3.5 μs into the impact, and this energy reaches its maximum when the projectile penetrates the fabric at around 7 μs. The important information provided by the FE simulation is that the energy absorbed by fabric fracture (ie, damage energy) is relatively small, about only 5% of the total energy absorption of the fabric.

The transverse wave propagation caused cone generation in the back face of the impacted fabric target. The formed cone moves along with the projectile and causes some energy absorption. The cone height and base radius of the formed cone are

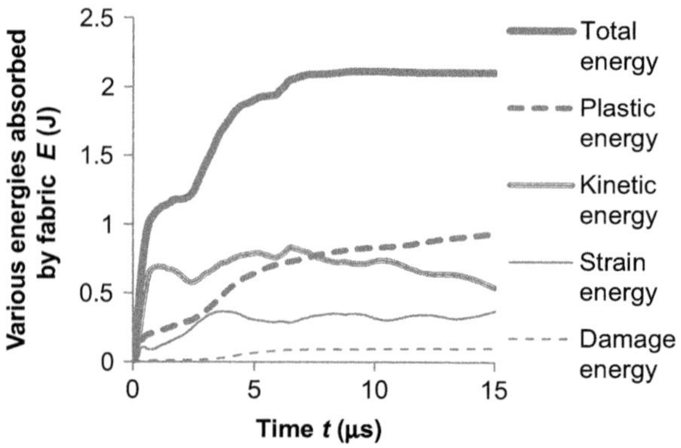

Figure 14.15 Composition of the absorbed energy by the fabric ($v_p = 494.217$ m/s) [32].

important parameters indicating ballistic performance of a fabric. A large cone base radius formed on the impacted fabric means the effective dissipation and absorption of the impact energy, and a small cone height indicates small blunt trauma. The relationship between the radius and height of the cone at different projectile impact velocities is illustrated in Fig. 14.16. It shows that the cone height development is virtually linear with respect to impact time due to the thin fabric layer (0.4 mm).

Its height and base radius of the cone increase as time goes on during the impact process. Basically, a good ballistic fabric should be associated with a large ratio of the cone base radius to the cone height. The cone radius increases rapidly at the very beginning of the impact at about 1 μs of the projectile impact. The change of the cone radius with respect to time becomes much less significant afterwards. It also shows that the final cone base radius after the projectile perforates the target fabric is smaller when the impact velocity is higher due to the fact that there is less time to allow the yarns and fabric to stretch and deform.

Jin et al. [41] studied the ballistic property of 3D angle-interlock fabric through experimental and FE simulation. Fig. 14.17 shows the FE model set up and ballistic impact effect comparison between experimental and FE simulation. In their study, the acceleration fluctuation record of the projectile and the stress wave propagation in fabric obtained from the validated FE model were investigated to further understand the impact damage mechanisms of the 3D angle-interlock fabric. It was believed that the fluctuation of the projectile acceleration was caused by the increased contact area

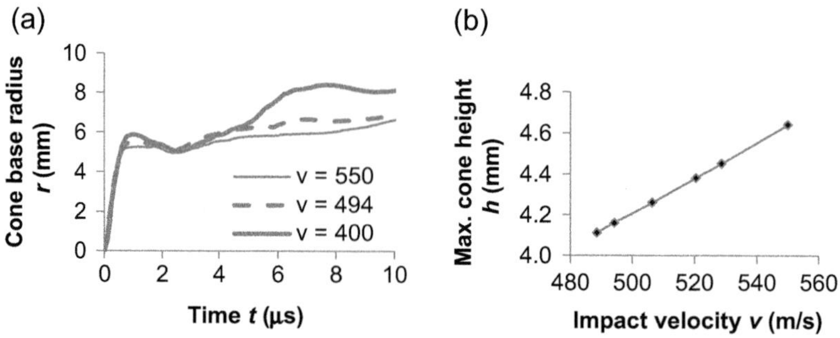

Figure 14.16 Cone radius and maximum height at various impact velocities.

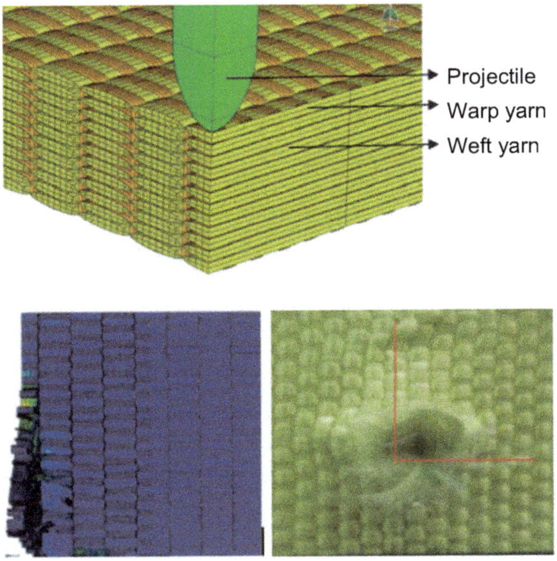

Figure 14.17 FE model set up and ballistic impact effect comparison between experimental and FE simulation of 3D angle-interlock fabric.
Adapted from L. Jin, B. Sun, B. Gu, Finite element simulation of three-dimensional angle-interlock woven fabric undergoing ballistic impact, J. Text. Inst. 102 (11) (2011) 982–993.

between projectile and fabric during ballistic penetration, the number of yarns that contacted with projectile. It has been found that stress waves in fabric were mainly propagated along warp and weft directions, and yarn tensile breakage was the main impact damage modus. As the ballistic penetration process goes, the damage area developed rapidly until the generation of the ballistic hole. Through FE analysis, a valuable fact was found that the projectile was captured by the fabric target at a specific velocity of 248 m/s, and it was believed that it is important for lightweight body armour design to protect against a specific type of threat.

The ballistic impact performance of 3D woven fabrics has also been investigated by many other researchers through the FE method [42–44]. Experimental work has been done in investigating ballistic impact performance of 3D angle-interlock fabric with various construction parameters (shown in Table 14.1). The results were compared to the most used plain-woven fabrics [45], shown in Fig. 14.18. It shows that the four joint-layered 3D honeycomb fabric performs better than the assembly with same layer 2D plain-woven fabric of similar yarn density. The four-layer 2D plain-woven fabric assembly absorbs more projectile impact energy compared to the three 3D angle-interlock fabrics. This may be due to the difference of number of crossover points contained in the fabrics. There are more yarn crossovers in a 2D woven fabric compared to a 3D angle-interlock fabric; the crossover points help to transfer energy to the adjacent yarns. The stress wave propagates in a larger area for the 2D woven fabric. As a result, it absorbs more projectile impact energy. Ballistic panels

Table 14.1 **Fabric specifications**

Fabrics	Weave structure	Fibre type	Yarn density in fabric		Yarn counts	
			Warp (ends/cm)	Weft (picks/cm)	Warp (tex)	Weft (tex)
4PW	4 separate layer plain-weave fabric	Kevlar 49	7.5	7.5	158	158
4LHC	4 jointed-layer 3D honeycomb weave					
4LAI12 × 26	Angle interlock		12	26		
4LAI12 × 28	Angle interlock			28		
4LAI12 × 30	Angle interlock			30		

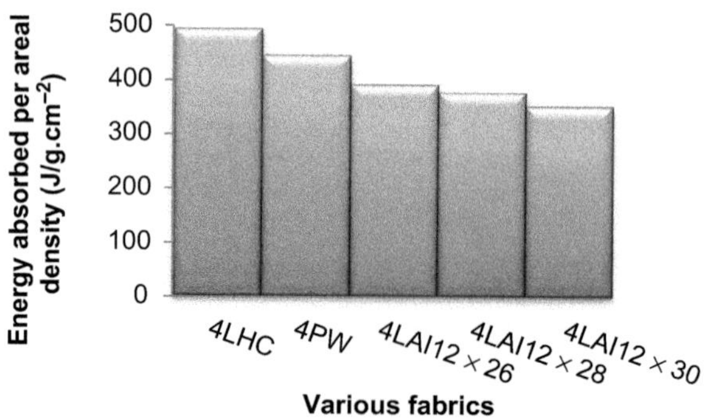

Figure 14.18 Normalised energy absorption of fabrics with different structures.

made of multiple-layer 2D woven fabrics have been studied experimentally and numerically by many researchers. Chen et al. [35] studied the ballistic property of a hybrid fabric panel consisting of 2D plain-weave fabric and unidirectional (UD) fabric made of UHMWPE fibre material. It was found that the woven structure exhibited better shear resistance, and the UD structure gave better tensile resistance and wider transverse deflection upon ballistic impact.

Cork [18] revealed that narrow fabrics with selvedges have improved ballistic performance over full-width fabric of the same structure. This is in line with results reported elsewhere [46,47]. The inclusion of a selvedge to the fabric strip further increases performance. However, when they were tested in the form of an off-set ballistic panel, the performance was disappointing because of yarn discontinuity across the fabric panel [48]. In addition, the narrow fabrics are also disadvantageous in body armour manufacturing, which is both tedious and time-consuming. Sun et al. [49] and Zhou [50] believed that leno structure can provide a similar yarn gripping effect to the fabric selvedges and affect ballistic energy absorption of the fabric.

Table 14.2 shows the fabrics used for their research.

Fig. 14.19 shows the energy absorption of plain-woven fabric and plain-woven fabrics with various bands of leno insertions from an experimental shooting test. Fabrics with leno insertions PW02, PW04 and PW06 perform better than fabric with entire plain-weave PW. It is concluded from this set of tests that fabrics with gripping yarns through leno weave structure contribute more effectively in absorbing impact energy. Zhou [50] simulated the ballistic impact effect of plain fabric with leno insertions and comparison made to plain, only woven fabric.

Fig. 14.20 shows part of the FE geometrical model of plain-woven fabric with leno insertion at 3-cm interval and the results of its ballistic energy absorption at different impact velocities as compared to simply plain-woven fabric. Fabric PWL03 exhibits higher energy absorption than PW, and the differences decrease with the increase of impact velocity.

Table 14.2 Fabric specifications of fabric with plain weave and fabrics with a combination of plain and leno weaves

			Yarn densities in fabric		Yarn counts	
Fabrics	**Weaving structure**	**Fibre type**	**Warp (ends/cm)**	**Weft (picks/cm)**	**Warp (tex)**	**Weft (tex)**
PW	Plain	Kevlar 49	7.5	7.5	158	158
PW02	Plain with leno insertion in every 2 cm intervals					
PW03	Plain with leno insertion in every 3 cm intervals					
PW04	Plain with leno insertion in every 4 cm intervals					
PW06	Plain with leno insertion in every 6 cm intervals					

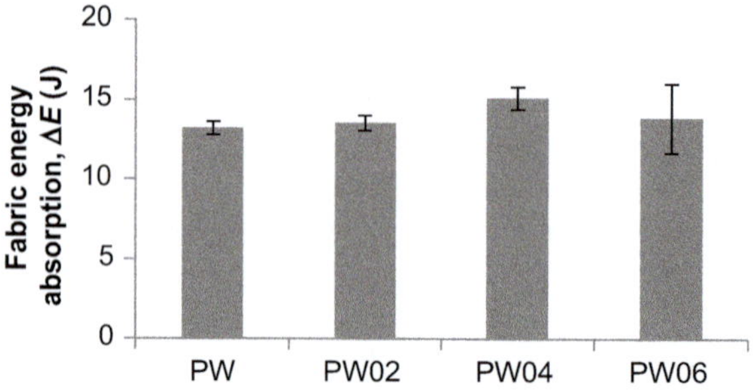

Figure 14.19 Energy absorption of fabrics with plain weave and plain weave with leno insertions.

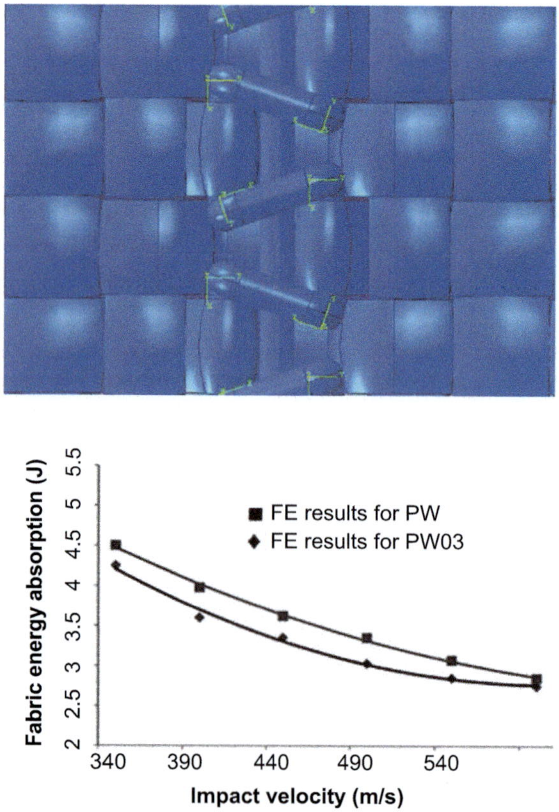

Figure 14.20 FE geometrical model with leno and plain-weave structure and fabric energy absorption as a function of impact velocities.
Adapted from Y. Zhou, Development of Lightweight Soft Body Armour for Ballistic Protection (Ph.D. thesis), University of Manchester, Manchester, 2013.

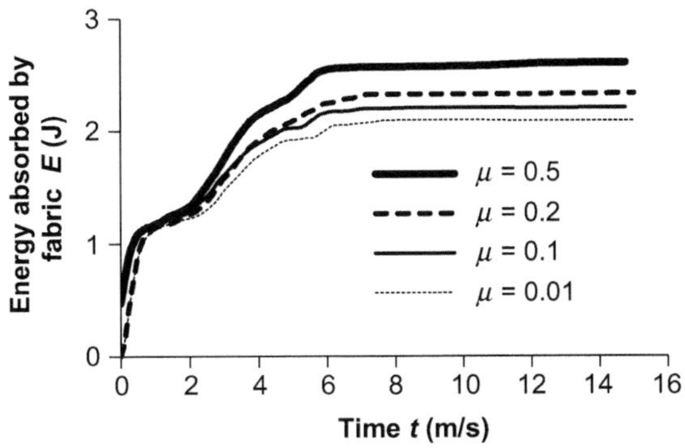

Figure 14.21 Energy absorbed by fabric affected by the surface coefficient of friction.

The idea of designing fabrics with gripping yarn is to increase the friction between fibres and yarns for improved ballistic performance. To study how the coefficient of friction influences the ballistic performance of a fabric, the validated model (shown in Fig. 14.14) was analysed further. Fig. 14.21 shows the energy absorbed by fabric with different coefficients of friction of yarn material versus impact time. It is clear that more energy is absorbed by fabric with a higher coefficient of friction. This indicates that ballistic performance can be improved by increasing the surface friction of the impacted fabric and yarns in terms of ballistic energy absorption.

Fig. 14.22 shows how the height of back-face deformation is affected by the yarn coefficient of friction. It is obvious that a higher coefficient of friction associates to a smaller height of back-face deformation. This phenomenon indicates that a rougher surface of Kevlar fabric and yarn would be beneficial to prevent from trauma. However, consideration must be taken that surface modification of Kevlar fabric and yarn does not affect its weight so as to ensure the weight to performance ratio.

Plasma treatment has been introduced for modification of textiles from fibres to fabrics for various types of applications in recent years. The advantage of the technique is that the surface properties can be changed significantly both physically and chemically

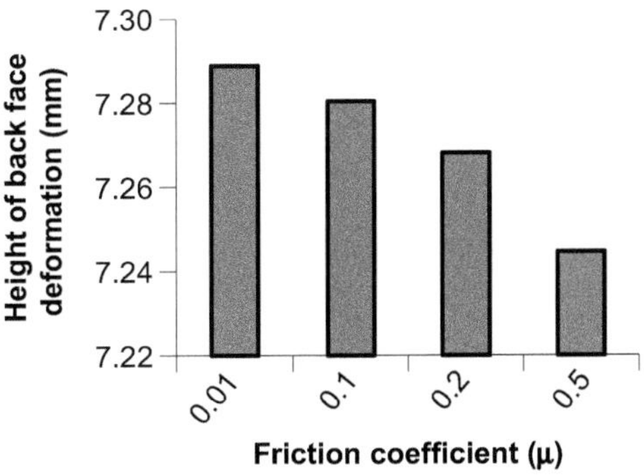

Figure 14.22 Height of back-face deformation at various coefficients of friction.

within a very short treatment time without altering the bulky property of the treated material. Fig. 14.23 shows an alternative current plasma treatment system which Sun and Chen [25] used for treatment of Kevlar fabrics. Nitrogen and dichlorodimethysilane plasma gas were employed.

Because the system can only allow treat a fabric with a small dimension and is not big enough for a ballistic impact test, a yarn pull-out test followed. The load needed to pull out yarns from plasma-treated fabrics was compared with untreated fabric; the results are presented in Fig. 14.24. The load indicates the resistance to the yarns pulling from fabrics. It has been proven that low-temperature non-polymerising reactive plasma treatment caused etching and ablation action on the treated fabric substrate surface, resulting in a roughening effect on the fabric surface [51,52], reflected by the higher value in surface roughness for the plasma-treated fabrics as a comparison to that of the untreated. The rougher surface might impart

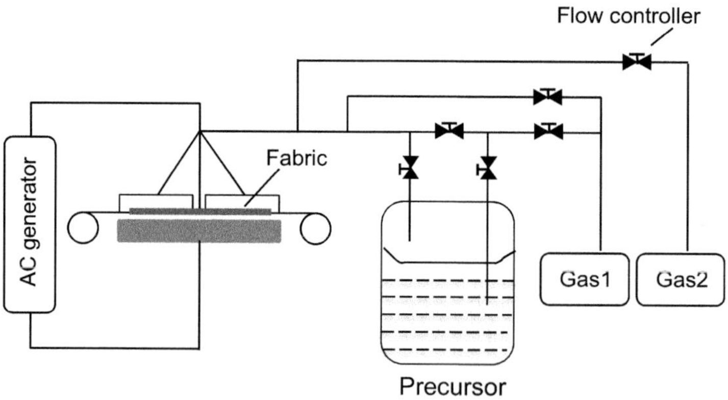

Figure 14.23 An alternating current plasma treatment system.

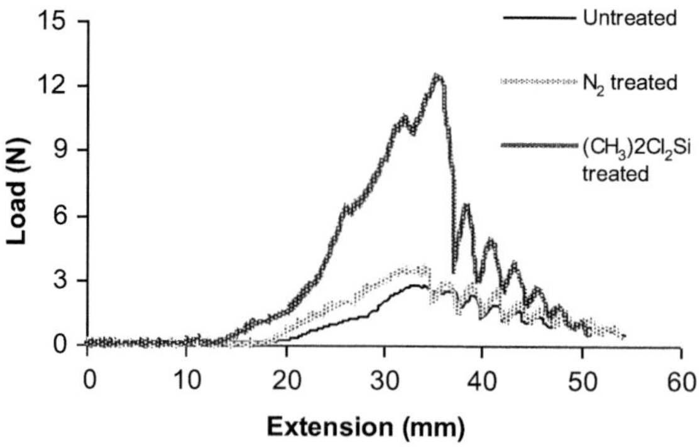

Figure 14.24 Resistance of pulling out a yarn from fabrics.
Adapted from D. Sun, X. Chen, Plasma modification of kevlar fabrics for ballistic applications, Text. Res. J. 82 (18) (2012) 1928–1934.

more contact points within the fibres and yarns, resulting in the enhanced interyarn and fibre friction [53], so that a larger amount of load is required during pulling out a yarn from nitrogen plasma-treated fabric. It is also believed that the increased friction among fibres and yarns produces a greater cohesive force between the yarns and the fibres; this may increase the tensile strength of the treated yarns and fabrics. The load of pulling out a yarn from diclorodimethysilane-treated plain-woven Kevlar fabric is 3 times the load needed to pull out a yarn from the untreated fabric. This may be caused by the added plasma polymer layer on the treated fabric surface which sticks fibres of the fabric surface and thin plasma polymer layer together. As a result, a much higher load is required to overcome the resistance from the adjacent fibres and the plasma polymer layer to pull out a yarn from diclorodimethysilane-treated Kevlar fabric than that of the untreated. Chu et al. [54] also tried diclorodimethysilane using the same plasma treatment system to modify Twaron 930 dtex yarn instead of fabric. The research concluded that both the static and kinetic frictional coefficient were increased after plasma treatment, and the treated yarn strength and Young's modulus were also affected positively by plasma surface modification.

Apart from woven fabric structure and surface modification, research has also been extended to the construction of armour panels for better ballistic protection based on the engineered fabrics. Chen et al. reported panel formation by angle-layering the fabric and by using different types of layer materials. Fig. 14.25 shows the examples of selected fabric assemblies [55]. ABAQUS was used in the research to predict the ballistic impact performance of the fabric assemblies based on the model validated by

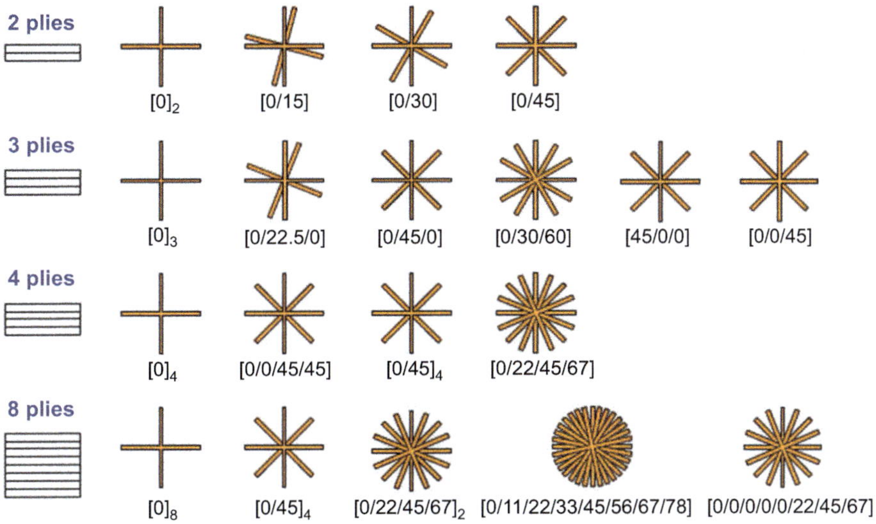

Figure 14.25 Examples of selected fabric assemblies.
Adapted from X. Chen, Y. Wang, Y. Zhou, D. Sun, 2D/3D woven fabrics for ballistic protection, JEC Compos. Mag. (79) (March 2013) 89–96.

experimental results. The research confirmed that angle-layering improves the extent of isotropy of the panel and caused higher energy absorption. The study on hybrid design of panels showed the possibility of using fabrics with different structures for achieving improved performance of the ballistic panel.

14.5 Conclusions

Ballistic performance evaluation of woven fabrics based on experimental and numerical approaches has been reported. The following aspects have been discussed:

1. Techniques and principles to evaluate woven fabric and panel ballistic performance lie in three aspects: ballistic resistance to study the trauma and material required to stop a projectile; energy absorption to examine the energy absorbed by fabric or panel after a projectile has perforated; and ballistic limit to estimate the projectile velocity at which a given projectile will perforate a given target 50% of the time.
2. FE modelling method involves techniques for geometry generation, parts meshing schemes and interaction, as well as definition of boundary conditions. The advantage of the FE method is that some of the properties which are unable to be tested by experiment can be predicted by the validated FE models.
3. Some of the interesting results predicted by the validated FE models have been discussed, such as the surface frictions in the effective improvement of ballistic performance, etc. They are meaningful for fabric design, engineering and armour panel assembly.

14.6 Future trends

Protection is often required against a variety of threats, projectiles, fragments and improvised weapons. This has led to a rapid evolution of the technology in both protection level and comfort. Therefore body armour research has two main aspects: one is how to improve the ballistic performance to weight ratio, the other is to improve the comfortability.

One of the drawbacks of currently used body armour materials, especially high performance polyethylene, is the low surface coefficient of friction. To develop techniques and modify the surface of currently used body armour materials would be beneficial to ballistic protection. Plasma treatment could be further explored by optimising gas(es) and other treatment parameters. It is also necessary to test the ballistic performance of treated fabrics.

Although the currently used materials for body armour are much stronger than steel, they are synthetic and not good with wear comfort. NATICK soldier research made a decision to take a second look at wool, to develop, test and evaluate wool fabrics for combat clothing since 2012. Wool has very good resilience, significant fire-resistant qualities, good strength and durablility due to its molecular structure. To find the specific type of wool source for a desired fibre diameter and length is the key because it directly affects the antiballistic properties. It looks promising to make a fabric that is much stronger with the mixing of long-staple (four-inch) worsted wool yarns and high-performance fibrous materials. Due to the scales on the wool fibre surface, shrink-resistant treatment is an important issue so the treatment should neither affect the bulk property nor the weight of the wool. Regarding ballistic materials, nanofibre, due to its enormous surface area, has been investigated for ballistic applications. Nonwoven fabrics can be made by staple nanofibres. However, to make woven fabrics, it needs continuous filament yarns with filaments in nano scale. Will nano yarns be commercially available in the near future?

References

[1] X. Chen, I. Chaudhry, Ballistic protection, in: R. Scott (Ed.), Textiles for Protection, Woodhead Publishing Limited, Cambridge, 2005, p. 229.
[2] M. Saxtorph, Warriors and Weapons of Early Times, Macmillan Co., New York, 1972.
[3] P. Cunniff, The performance of poly (para-phenylene benzobizoxazole) (PBO) fabric for fragmentation protective body armor, in: 18th International Symposium on Ballistics, San Antonio, 1999.
[4] P. Walsh, X. Hu, P. Cunniff, A. Lesser, Environmental effects on poly-p-phenylenebenzobisoxazole fibers. I. Mechanisms of degradation, J. Appl. Polym. Sci. 102 (2006) 3517–3525.
[5] P. Cunniff, M. Auerbach, E. Vetter, D. Sikkema, High performance "M5" fibre for ballistics/structural composites, in: Proceedings of the 23rd Army Science Conference, Orlando, 2002.
[6] G. Cooper, P. Gotts, Ballistic protection, in: Ballistic Trauma, Springer-Verlay London Limited, London, 2005, pp. 67–84.
[7] D. Roylance, A. Wilde, G.C. Tocci, Ballistic impact of textile structures, Text. Res. J. 43 (1973) 34–41.
[8] W. Lyons, Impact Phenomena in Textiles, The M.I.T. Press, Cambridge, 1963.
[9] P.K. Porwal, S. Phoenix, Modelling system effects in ballistic impact into multi-layered fibrous materials for soft body armour, Int. J. Fract. 135 (2007) 217–249.
[10] S. Chocron, A. Pintor, F. Galvez, C. Rosello, D. Cendón, V. SánchezGálvez, Lightweight polyethylene non-woven felts for ballistic impact applications: material characterization, Composites Part B 39 (7) (2008) 1240–1246.
[11] G. Zhu, W. Goldsmith, C.K. Dharan, Penetration of laminated kevlar by projectiles: I: experimental investigation, Int. J. Solids Struct. 29 (1992) 399–420.
[12] S.S. Morye, P.J. Hine, R.A. Duckett, D.J. Carr, I.M. Ward, Modelling of the energy absorption by polymer composites upon ballistic impact, Compos. Sci. Technol. 60 (14) (2000) 2631–2642.
[13] N.K. Naik, P. Shrirao, B.C.K. Reddy, Ballistic impact behaviour of woven fabric composites: parametric studies, Mater. Sci. Eng. 412 (2005) 104–116.
[14] B.R. Scott, New ballistic products and technologies, in: A. Bhatnagar (Ed.), Lightweight Ballistic Composites, Woodhead Publishing Limited, Cambridge, 2006, p. 348.
[15] J.W.S. Hearle, Research on a Basic Study of the High Speed Penetration Dynamics of Textile Materials, University of Manchester Institute of Science and Technology, Manchester, 1974, p. 51.
[16] P.H.W. Steeghs, M. Blaauw, W.A.R.M. Pessers, J.L. Lindemulder, Ballistic Resistant Body Armor Articles, USA Patent US 2005/0097647 A1, May 12, 2005.

[17] J.G. Donovan, B. Kirkwood, F. Figucid, Development of Lower Cost Ballistic Protection, US Army Natick RD&E Centre, Natick, MA, 1985. Technical Report Natick/TR-85/019L.
[18] C.R. Cork, P.W. Foster, The ballistic performance of narrow fabrics, Int. J. Impact Eng. 34 (2007) 495–508.
[19] S.T. Jeng, J.T. Kuo, L.T. Sheu, Ballistic impact response of 3D four-step braided glass/epox composites, Key Eng. Mater. 141 (1998) 349–366.
[20] B.J. Briscoe, F. Motamedi, The ballistic impact characteristics of aramid fabrics: the influence of interface friction, Wear 158 (1992) 229–247.
[21] R.A. Prosser, Penetration of nylon ballistic panels by fragment-simulating projectiles, Part II: mechanism of penetration, Text. Res. J. 58 (3) (1988) 161–165.
[22] Y. Duan, M. Keefe, T.A. Bogetti, B.A. Cheeseman, B. Powers, A numerical investigation of the influence of friction on energy absorption by a high-strength fabric subjected to ballistic impact, Int. J. Impact Eng. 32 (2006) 1299–1312.
[23] H.M. Eric, P.B. Hogenboom, Combinations of Polymer Filaments or Yarns Having a Low Coefficient of Friction and Filaments or Yarns Having a High Coefficient of Friction, and Use Thereof, USA Patent US5035111 A, July 30, 1991.
[24] Y. Termonia, Increased Ballistic Performance of Fabrics Coated with Polymer Stripes, Patent EP1913330 A2, April 23, 2008.
[25] D. Sun, X. Chen, Plasma modification of kevlar fabrics for ballistic applications, Text. Res. J. 82 (18) (2012) 1928–1934.
[26] Department of Defense Test Method Standard V_{50} Ballistic Test for Armor, MIL-STD-662F, Department of Defence, United State of America, 1997.
[27] G.J. Czarnecki, Estimation of the V_{50} using semi-empirical (1-point) procedures, Composites Part B 29 (3) (1998) 321–329.
[28] P.M. Cunniff, The V50 performance of body armour under oblique impact, in: Proceedings of the 18th International Symposium on Ballistics, San Antonio, 1999.
[29] M. Grujicic, G. Arakere, T. He, M. Gogulapati, A numerical investigation of the influence of yarn-level finite-element model on energy absorption by a flexible -fabric armour during ballistic impact, in: Proceedings of the Institution of Mechanical Engineers Part I: Journal of Materials Design and Applications, 2008.
[30] M. Ethan, N. Parsosn, M.J. King, S. Socrate, Modeling yarn slip in woven fabric at the continuum level: simulations of ballistic impact, J. Mech. Phys. Solids 61 (2013) 265–292.
[31] M. Tran, T. Ngo, E.C. Yang, P. Mendis, W. Humphries, Effects of architecture on ballistic resistance of textile fabrics: numerical study, Int. J. Damage Mech. 23 (3) (2014) 359–376.
[32] D. Sun, X. Chen, E. Lewis, G. Wells, Finite element simulation of projectile perforation through a ballistic fabric, Text. Res. J. 83 (14) (2013) 1489–1499.
[33] I.S. Chocron-Benloulo, J. Rodríguez, V. Sánchez-Gálvez, A simple analytical model to simulate textile fabric ballistic behavior, Text. Res. J. 67 (7) (1997) 520–528.
[34] B. Parga-Landa, F. Hernandez-Olivares, An analytical model to predict impact behaviour of soft armors, Int. J. Impact Eng. 16 (3) (1995) 455–466.
[35] X. Chen, Y. Zhou, G. Wells, Numerical and experimental investigations into ballistic performance of hybrid fabric panels, Composites Part B 58 (2014) 35–42.
[36] C.H. Cheong, C.T. Lim, V.B.C. Tan, Perforation of high-strength fabric of projectiles of different geometry, Int. J. Impact Eng. 28 (2) (2003) 207–222.
[37] C.T. Lim, Y.H. Ng, V.P.W. Shim, Finite element modeling of the ballistic impact into fabric armor, Int. J. Impact Eng. 28 (1) (2003) 13–31.
[38] P.M. Cunniff, An analysis of the system effects of woven fabrics under ballistic impact, Text. Res. J. 62 (1992) 495–509.

[39] B.L. Lee, T.F. Walsh, S.T. Won, H.M. Patts, J.W. Song, A.H. Mayer, Penetration failure mechanisms of armour-grade fibre composite under impact, J. Compos. Mater. 35 (2001) 1605–1633.
[40] X.S. Zeng, V.P.W. Shim, V.B.C. Tan, Influence of boundary conditions on the ballistic performance of high-strength fabric targets, Int. J. Impact Eng. 32 (2006) 631–642.
[41] L. Jin, B. Sun, B. Gu, Finite element simulation of three-dimensional angle-interlock woven fabric undergoing ballistic impact, J. Text. Inst. 102 (11) (2011) 982–993.
[42] F. Cui, B. Sun, B. Gu, Fibre inclination model for finite element analysis of three-dimensional angle interlock woven composite under ballistic penetration, J. Compos. Mater. 45 (14) (2011) 1499–1509.
[43] X. Jia, B. Sun, B. Gu, A numerical simulation on ballistic penetration damage of 3D orthogonal woven fabric at microstructure level, Int. J. Damage Mech. 21 (2) (2012) 237–266.
[44] C. Ha-Minh, F. Boussu, T. Kanit, D. Crepin, A. Imad, Analysis on failure mechanisms of an interlock woven fabric under ballistic impact, Eng. Fail. Anal. 18 (8) (2011) 2179–2187.
[45] X. Chen, D. Sun, G.M. Wells, Effect of inter-yarn friction on ballistic performance of woven fabrics, in: B.J. McCarthy (Ed.), Polymeric Protective Technical Textiles, Smither Rapra, Shawbury, UK, 2013, pp. 41–60.
[46] K. Greenwood, C.R. Cork, Ballistic Penetration of Textile Fabrics – Phase V, Final report to the United Kingdom Ministry of Defence, 1990.
[47] L. Dischler, Method for Improving the Energy Absorption of a High Tenacity Fabric During a Ballistic Event, Patent US5466503, November 14, 1995.
[48] D. Sun, X. Chen, G. Wells, Engineering and analysis of gripping fabrics for improved ballistic performance, J. Compos. Mater. 48 (11) (2014) 1355–1364.
[49] D. Sun, X. Chen, M.M. Mrango, Investigating ballistic impact on fabric targets with gripping yarns, Fibres Polym. 14 (7) (2013) 1184–1189.
[50] Y. Zhou, Development of Lightweight Soft Body Armour for Ballistic Protection (Ph.D. thesis), University of Manchester, Manchester, 2013.
[51] D. Sun, G. Stylios, Surface roughness investigation of low temperature plasma treated wool and cotton fabrics, in: Proceedings of International Textile Design and Engineering Conference, INTEDEC, Edinburgh, 2013.
[52] R. d'Agostino, Plasma Deposition, Treatment and Etching of Polymers, Academic Press, New Jersey, 1990.
[53] C.W. Kan, K. Chan, C.W.M. Yuen, Plasma modification of wool: the effect of plasma gas on the properties of the wool fabric, J. China Text. Univ. 15 (1998) 5–10.
[54] Y. Chu, X. Chen, D.W. Sheel, J.L. Hodgkinson, Surface modification of aramid fibres by atmospheric pressure plasma-enhanced vapor deposition, Text. Res. J. 84 (12) (2014) 1288–1297.
[55] X. Chen, Y. Wang, Y. Zhou, D. Sun, 2D/3D woven fabrics for ballistic protection, JEC Compos. Mag. (79) (March 2013) 89–96.

Thermoplastic matrix combat helmet with carbon-epoxy skin for ballistic performance

15

F. Folgar
INTER Materials, LLC, Midlothian, VA, United States

15.1 Introduction

Thermoplastic matrix advanced composites have long held the potential to reduce the weight of personnel body armor carried by a soldier, including ballistic vests and helmets. When combined with a thermoplastic matrix, high-performance fibers such as ultra-high molecular weight polyethylene (UHMWPE) have demonstrated the capacity to provide both weight reduction and superior ballistic protection beyond that afforded by the incumbent para-aramid fiber phenolic thermoset resin systems. As a result, hybrid designs using the new thermoplastic matrix ballistic materials with structural carbon composite skins can reduce the blunt trauma and durability limitations inherent with these compliant materials.

Since these materials require unique processing conditions to achieve optimal performance, a cost-effective manufacturing process for thermoplastic matrix composite helmets is required with rapid temperature cycling times to compete with the current compression molding process used for para-aramid fiber thermoset matrix helmets and high molding pressures to provide superior ballistic performance and meet all structural requirements.

A brief description of the Personnel Armor System Ground Troop (PASGT) helmet system is included in Section 15.2 to illustrate that the technology known since 1940 of using pinwheel patterns to assemble a safety helmet and the compression molding process with match-metal dies for making thermoset resin composite parts have not changed for the past 70 years. Most of the limited ballistic improvements and weight reduction for combat helmets have come from using tougher and finer denier para-aramid fibers and lighter fabrics.

This article also describes INTER Materials' efforts for the fabrication of thermoplastic matrix combat helmets using para-aramid fibers to assess the potential to improve the ballistic performance and structural characteristics of the US Army Advanced Combat Helmet (ACH) using a finer para-aramid yarn denier, a greater ply count with a lighter fabric, and a thermoplastic matrix. The results of these efforts demonstrated that, when using para-aramid fibers, the opportunities to further improve the ballistic performance and structural characteristics of the ACH helmet were limited without making them thicker and heavier.

Advanced Fibrous Composite Materials for Ballistic Protection. http://dx.doi.org/10.1016/B978-1-78242-461-1.00015-7

Since those limitations were attributed to the fibers and fabric architecture, Section 15.4 is included to describe a more relevant class of materials for fabricating high ballistic performance helmets known as unidirectional 0 degree/90 degree cross-plied construction. The UHMWPE fibers when combined with a thermoplastic matrix in a two-layer cross-plied sheet-like prepreg have improved the ballistic protection beyond that afforded by the incumbent para-aramid fiber thermoset phenolic resin composite system. Because published ballistic performance of helmets made with thermoplastic matrix composite materials is very limited, all the references made in this section are ballistic results from testing flat panels.

In the last two sections, INTER Materials describes the results from testing thermoplastic matrix combat helmets made with its own technology (patent pending) including the following:

1. a method of designing a helmet to eliminate the overlap of the pinwheel legs, to control the fiber orientation between neighboring cross-plied layers, and to increase the overall ballistic performance of thermoplastic matrix helmets by 70–80%; and
2. a low-cost and high-volume manufacturing process for making thermoplastic matrix combat helmets with or without a carbon-epoxy structural reinforcement at a molding rate of less than 20 min without compromising the ballistic performance.

15.2 PASGT combat helmet

The PASGT helmet was developed by research scientists at the US Army Natick Research Center, Natick, Massachusetts, in the late 1970s [1]. The PASGT helmet has been the most popular combat helmet for the past 30 years and is still being manufactured using a Kevlar® fabric coated with a thermoset resin. The PASGT is a helmet with an areal density of approximately 11.2 kg/m^2 (2.30 lbs/ft^2) and made using 1500-denier Kevlar® 29 yarn in a 2 × 2 basket fabric construction [2]. The fabric is impregnated with 16–18% by weight of polyvinyl butyral (PVB)-phenolic resin. The helmets are fabricated by assembling a helmet preform using 19 equivalent layers of prepreg and then compression molded. Because of the thermoset nature of the PVB-phenolic resin matrix, molding of the standard PASGT helmets requires constant temperatures between 160°C and 180°C (320–355°F) and pressures well over 3.45 MPa (500 psi) for roughly 10–15 min on average for the resin to fully cure.

In the case of a medium-size PASGT helmet, 19 layers were required to deliver a helmet with the target weight or areal density of 11.2 kg/m^2 (2.3 lbs/ft^2). To obtain 19 equivalent layers during the preform assembly, the helmet preform was put together using 16 pinwheels/rosettes, Fig. 15.1(a), and three crowns, Fig. 15.1(b).

During the preform assembly, the overlap of the pinwheel/rosette legs provides an increase in the actual number of layers on the side of the helmet walls higher than the crown area. Therefore, crown patches of preform material are placed in the crown area of the helmet to compensate and balance the number of layers.

The use of pinwheel preform patterns to facilitate the fabrication of helmets has been mentioned in the prior art as early as 1942 on US patents issued to Le Grand Daly [3] and to H. Goldsmith [4]. Daly's patents disclose the use of the pinwheel

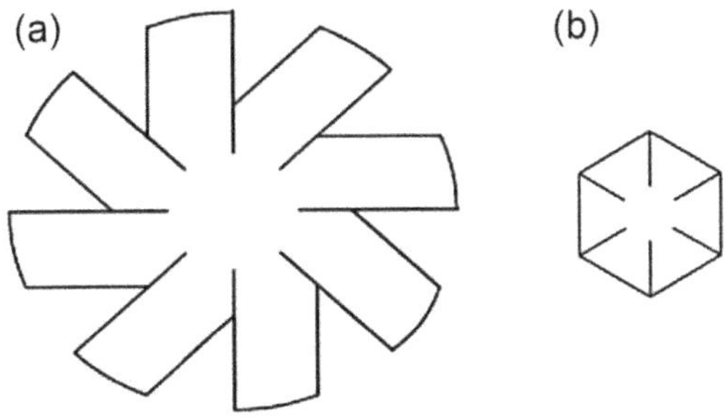

Figure 15.1 Helmet preform patterns: (a) pinwheel/rosette, (b) crown.

helmet preform pattern as a conventional practice of making safety football helmets from flat sheets of resin-impregnated fabrics in a more convenient manner. Daly's patents claim the shape of the individual pieces already referred to as *preforms* and their arrangement going into the molding dies in a way that the seams are staggered without bending and prevent the formation of wrinkles and folds. Fig. 15.2(a) is a view illustrating the preform pattern, and Fig. 15.2(b) shows the pattern being folded together and ready to be assembled for molding.

Forty years later in the early 1980s, with the availability of Kevlar® as a new high-performance fiber better suited for the fabrication of ballistic helmets, the cost of materials became a pressing issue since a large number of plies were required for a specific level of ballistic protection. Although using Daly's preform patterns to be laid up in a mold and pressed into a helmet resulted in a high-quality method of making a helmet, it did not make efficient use of the prepreg material because of the waste during the tracing and cutting process of the pinwheel patterns.

In 1983, S.J. Grick [5] filed for a patent addressing a more efficient use of prepreg materials by introducing a preform pattern design which substantially reduced the amount of prepreg needed to make a helmet shell. Grick's helmet pinwheel pattern increased the efficiency of the prepreg material by the method used in putting together the pinwheel, as shown in Fig. 15.3(a). The patent discloses cutting pinwheel segments, Fig. 15.3(b), and jointing them together by stitching or by a heat-sensitive

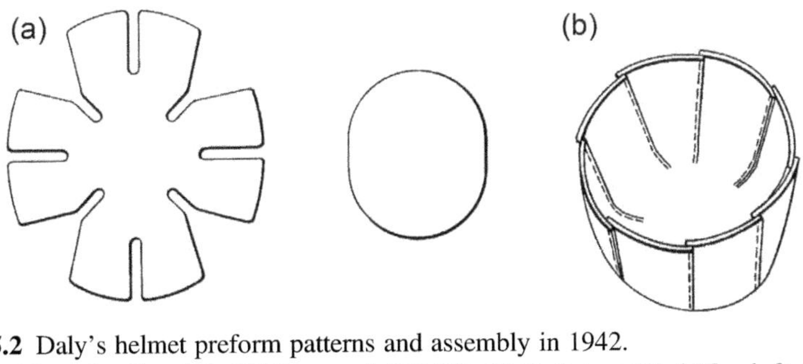

Figure 15.2 Daly's helmet preform patterns and assembly in 1942.
Reproduced from Le Grand D. US Patents 2,420,522, 1947, Patent Filed March 9, 1942 and Granted in May 13, 1947.

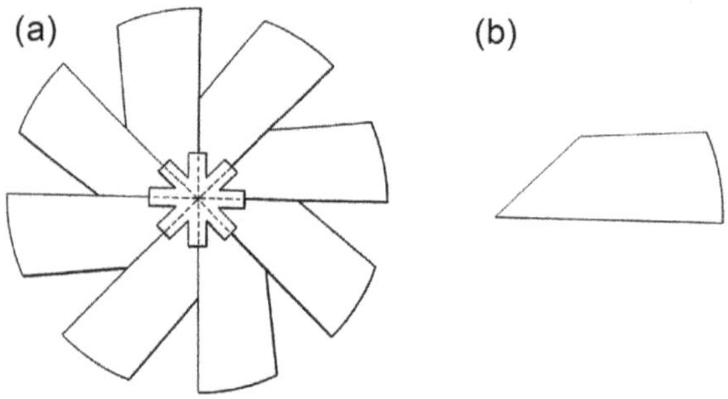

Figure 15.3 Grick's helmet preform pattern: (a) pinwheel preform pattern assembly, (b) pinwheel segment.
Reproduced from Grick SJ. US Patent 4,596,056, 1986.

tape at the center. Cutting pinwheel segments of the same shape and size reduced the waste during the cutting process from a roll of fabric prepreg.

Since the assembly of a preform into a molded helmet left one ballistically weak spot on top of the helmet, the pattern design was later improved by using preform patterns of hexagonal shape slit from the corners of the hexagon up to a certain distance from the center, as shown in Fig. 15.4 [6]. This was an improved method of making a ballistic helmet with an efficient use of the prepreg fabric material and without sacrificing any degree of ballistic protection around the whole helmet.

The efficiency of the prepreg fabric usage came from the reduction of waste during cutting the hexagonal shape of the preform patterns and the staggering of the hexagonal patterns of different sizes during the assembly of the preform before being laid into the compression molding dies.

From the early 1980s through the 1990s, very little changed about the molding methods known for making PASGT helmets, including the use of matched metal dies during compression molding. In addition, improvements on ballistic performance and weight reduction of the PASGT helmet system during that period were limited. One helmet design of particular interest is the Kevlar® KM2 fiber system developed by DuPont in 1990 [8], as described in Table 15.1.

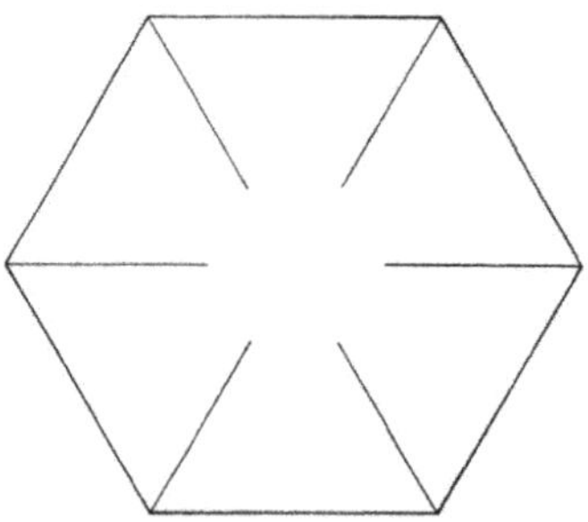

Figure 15.4 New hexagonal helmet preform pattern design [7].
Reproduced from White MR. US Patent 4,778,638, 1988.

Table 15.1 PVB-phenolic matrix PASGT and Kevlar® KM2 helmet systems

Properties	PASGT helmet	Kevlar® KM2 helmet
Areal density, kg/m^2 (psf)	11.24 (2.3)	9.53 (1.95)
Yarn denier/fiber	1500-denier Kevlar® 29	850-denier Kevlar® KM2
Fabric construction/weight	2 × 2 basket, 475 g/m^2	31 × 31 plain weave, 234 g/m^2
Matrix resin/weight fraction	PVB-phenolic, 16–20%	PVB-phenolic, 16–18%
Molding process	Compression molding	Compression molding
Molding cycle time, min	10–15 min	10–15 min
V_{50} (f/s), 17-grain FSP	2100 f/s	2200 f/s

The KM2 helmet innovation produced a 15% lighter helmet shell with equal or slightly superior ballistic performance to the incumbent Kevlar® 29 PASGT helmet system. This improvement was achieved mostly by using a higher toughness 850-denier KM2 Kevlar® fiber than the 1500-denier Kevlar® 29 used in the standard PASGT, finer Kevlar® yarn denier, greater ply count, and a special fiber surface treatment to control the adhesion strength between the PVB-phenolic resin and the fiber surface. The KM2 Kevlar® helmet development illustrated the concept of improved ballistic performance of a helmet by using a greater number of Kevlar® prepreg layers (ply count) with lighter fabrics at the same helmet weight or the same ballistic protection at lighter weight than conventional helmets.

15.3 Para-aramid fiber thermoplastic matrix composite combat helmets

In 2005, INTER Materials [9] started working on a low-cost manufacturing molding process for thermoplastic matrix combat helmets with superior ballistic performance and lighter weight. To facilitate the comparison of any improvements on the ballistic performance and structural characteristics of the new thermoplastic matrix combat helmet, the 850-denier Kevlar® KM2 fiber-PVB/phenolic helmet previously developed by DuPont in 1990 [8] was selected as the control composite system.

Since the geometry of the ACH design changed very little when compared to the PASGT geometry, Fig. 15.5, the thermoplastic matrix helmet ballistic performance goal was to meet or exceed the minimum ballistic limits, V_{50} (ft/s), described on the solicitation first article testing requirements of the ACH Purchase Description [10].

To achieve the required ballistic performance, three different thermoplastic matrix composite systems with para-aramid fiber reinforcement were selected, a 600-denier Kevlar® KM2 woven fabric coated with a DuPont proprietary thermoplastic resin, an 850-denier Kevlar® KM2 fabric coated with an INTER Materials proprietary

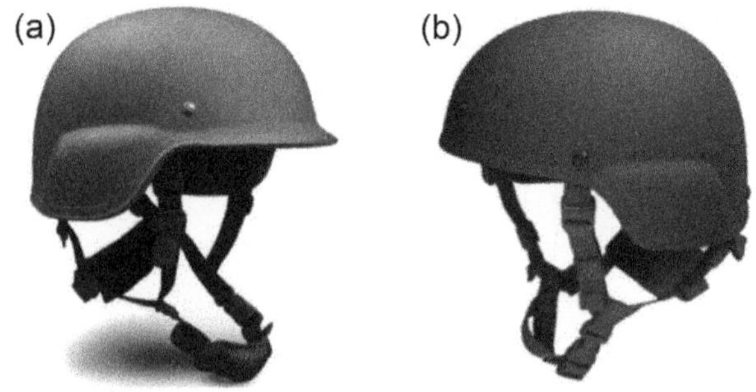

Figure 15.5 US Army combat helmet designs: (a) PASGT, (b) ACH.
Public source/internet www.google.com/PASGTballistichelmetpicturegallery.

thermoplastic resin, and Gold Shield® from Honeywell, a 0 degree/90 degree unidirectional cross-plied prepreg system made with Twaron® para-aramid yarn and a proprietary thermoplastic resin.

For the structural reinforcement, we used the carbon-epoxy skin as a sandwich of two-ply 0 degree/90 degree IM7 carbon-epoxy prepreg made by Cytec [11] that weighs 448 g/m^2 (13.2 oz/yd^2) between two layers of 850-denier Kevlar® KM2 fabric coated with PVB-phenolic resin for better chemical compatibility and for easier surface release during the forming and molding steps. The carbon-epoxy skin with the two layers of 850-denier KM2/PVB-phenolic weighs 1.0 kg/m^2 (30.0 oz/yd^2).

Once the area density of 8.55 kg/m^2 (1.75 lbs/ft^2) was estimated to obtain a 25% helmet weight reduction, the wall thickness of the helmet was calculated at 7 mm (0.28 inches). Molding the new Kevlar® thermoplastic matrix combat helmet still required matched steel dies. However, the molding cycle of a thermoplastic matrix helmet takes 90 min compared to the 10–15 min it takes for molding a PASGT helmet, due to the fact that a thermoplastic matrix helmet has to be under pressure during the heating and the cooling phase of the molding cycle.

Ballistic testing results against a 17-grain fragment simulator projectile (FSP) using 15 × 15-inch flat panels and molded helmets with an areal density of 8.55 kg/m^2 (1.75 lbs/ft^2) are summarized in Fig. 15.6. This figure shows the normalized ballistic limit, V_{50} (m/s), for all three materials systems, where the value of 100 corresponds to the V_{50} (m/s) of 850-denier KM2 PVB-phenolic flat panels with 8.55 kg/m^2 (1.75 lbs/ft^2) areal density [8].

Within the statistical distribution of the ballistic limit V_{50} (m/s) results against the 17-grain FSP, the three composite materials with thermoplastic matrix using para-aramid fibers in both unidirectional and woven configurations met the ballistic requirements in panel form. Since these tests used the same compression molding conditions of temperature, pressure, and time for all the thermoplastic matrix materials, it was expected that a common molding process might not be optimum for each and all of these thermoplastic matrix composite systems.

As shown in Fig. 15.6, there is a difference in ballistic performance between the groups of flat panels and molded helmet shells. The two woven fabric systems have different basis weight and ply count with two different thermoplastic matrices. Lower

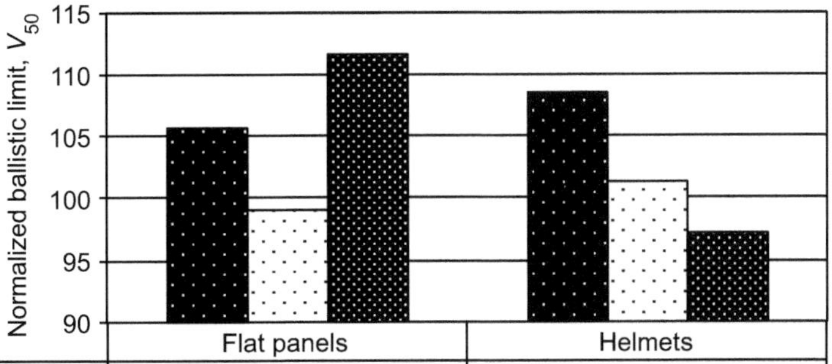

	Flat panels	Helmets
600d KM2	106	109
850d KM2	99	101
Goldshield	112	97

Figure 15.6 Ballistic performance comparison between para-aramid fiber thermoplastic matrix composite flat panels and molded helmets [9].

ply basis weights and greater ply count for the same final areal density seem to suggest better ballistic efficiency for the fragment sizes and velocities considered.

In addition, it was observed that the two Kevlar® KM2® fabrics not only performed as well as a flat panel, but their ballistic performance increased as a molded helmet. However, the Gold Shield® thermoplastic matrix composite system that performed the best as a flat panel, performed close to the 850-d Kevlar® KM2 system with PVB-phenolic matrix against the 17-grain FSP fragment. Of particular interest to our research was how to modify the helmet preform design and the helmet molding conditions of the thermoplastic matrix Gold Shield® material to achieve in the molded helmets the ballistic performance demonstrated in flat panels taking into consideration the unidirectional nature of the reinforcement architecture.

Before more thermoplastic matrix helmets were further evaluated ballistically, they were first subjected to the side-to-side compression test as the structural screening procedure. The results from testing the structural side-to-side compression properties were not consistent. At the end of the program, we concluded that para-aramid fabrics impregnated with PVB-phenolic thermoset resin or thermoplastic matrix resins have little potential left for additional ballistic optimization without making thicker and heavier helmets. In addition, by the late 1990s, newer UHMWPE fibers impregnated with thermoplastic matrix materials have demonstrated potential to provide both weight reduction and enhanced protection against small arms threats [12].

15.4 Ballistic performance of unidirectional thermoplastic matrix composites

As demonstrated in the previous section, some of the limitations for further ballistic improvements in a helmet when using a finer para-aramid yarn denier, a greater ply count with a lighter fabric, and a thermoplastic matrix come from the fabrics themselves.

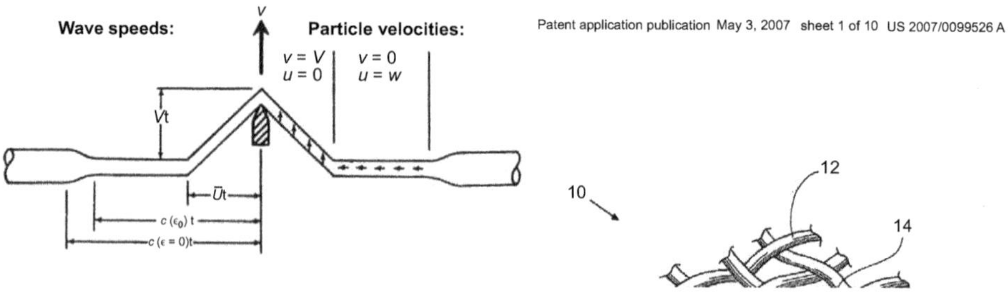

Figure 15.7 Ballistic impact on a woven fabric.
Reproduced from Roylance D, Wang Su-Su. US Army Natick Technical Report TR-80/021, Penetration Mechanics of Textiles Structures, MIT, Contract No DAAG 17-76-C-0013, 1979.

During a ballistic impact [13], the individual yarns in a woven fabric are constrained at the crossover points (Fig. 15.7). The tensile wave propagated along the longitudinal direction of the yarns is reflected at the crossover points breaking the yarns much before the maximum amount of energy can be absorbed along its length. Therefore one way to improve the ballistic performance of woven fabrics was to reduce the crossover points by reducing the fabric crimp or spreading the crossover points farther apart by weaving fabrics in a more open construction.

A more open fabric construction resulted in a lighter fabric. Therefore, more fabric layers for the same yarn denier is needed to meet given ballistic requirements. The physics behind the improvement in ballistic performance due to more layers of a lighter fabric for the same total areal density is attributed to a more uniform distribution of the impact energy throughout more fabric layers and a reduced density of crossover points [14].

Since there is a limit to how open a woven fabric can be before it loses its ballistic properties due to becoming like an open net, the next step was to reduce the yarn crimp. Modern textile weaving equipment allowed us to spread the filaments of the yarns in a flat and oriented fashion on a loom, giving rise to what is currently known as unidirectional fabric constructions. Since unidirectional fabric constructions have very little crimp, two of the most common methods to hold together two or more layers of unidirectional fabrics are by using a low modulus thermoplastic film to bond the fabric layers together or using fine-denier yarns of low-tensile modulus and low-strength fibers to stitch the layers together. Karl-Mayer offers warp-knit looms to weave unidirectional multiple-layer fabric structures that are popular in the composites industry with glass, carbon, and para-aramid fibers [15].

A common unidirectional fabric construction known as 0 degree/90 degree cross-plied construction has two layers with their yarns at 90 degrees from each other bonded together by a thermoplastic film or stitched together by a fine-denier yarn, Fig. 15.8. The higher ballistic efficiency of unidirectional fabrics is then attributed to the fact that the yarns can dissipate more impact energy along their length resulting in the impact energy being transmitted to a much larger area within each unidirectional fabric layer away from the place of impact [14].

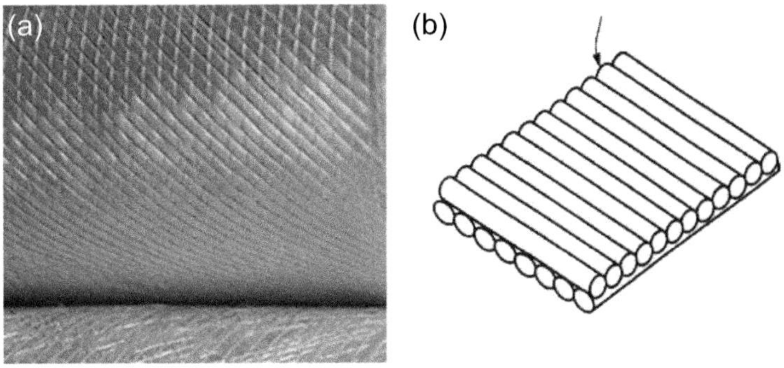

Figure 15.8 Unidirectional fabrics: (a) stich-bonded unidirectional fabric, (b) unidirectional 0 degree/90 degree cross-plied fabric construction.
Adapted from Karl-Mayer Technical Textiles, http://www.karlmayer.com/internet/.

Some examples of commercially available unidirectional fabric constructions made with para-aramid fiber yarns for ballistic applications are disclosed on patents issued to Andrew Park [16,17] and patents issued to Barrday Corporation [18,19]. Park's patents disclose an unidirectional fabric construction of two layers of high-performance fibers 0 degree/90 degree cross-plied and bonded together by a thermoplastic film or scrim, where it claims that the film or scrim, once laminated by heat and pressure, does not penetrate the fiber filaments inside the yarn. Barrday's patents also disclose a two-layer 0 degree/90 degree unidirectional fabric construction where the yarns are held together by stitching with a finer, lower modulus yarn that Barrday refers to as an encapsulating yarn.

A more relevant class of unidirectional materials for fabricating high-performance ballistic helmets are commercially available with the trade names of Gold Shield® and Spectra Shield® from Honeywell International and Dyneema® from DSM Dyneema® Corporation. Both of these brands are a two-layer sheet-like prepreg where an array of high-performance para-aramid or UHMWPE fiber filaments are aligned parallel to each other, coated or impregnated with a thermoplastic elastomer matrix, and then cross-plied together at a 90 degree angle to produce a single sheet 0 degree/90 degree construction of ballistic material.

High ballistic efficiency fibers such as UHMWPE, when combined with a thermoplastic matrix, have improved the ballistic protection beyond that afforded by the incumbent para-aramid fiber thermoset phenolic resin composite system. References for ballistic performance of ballistic helmets made with these new materials are very limited at best. Most of the references found for ballistic performance of these new materials are results from testing flat panels.

For example, Van der Loo from DSM NV [11] disclosed a ballistic-resistant flat panel made by a compressed stack of layers containing 0 degree/90 degree unidirectionally oriented reinforcing UHMWPE fibers with 30 wt.% of a thermoplastic matrix material. Van der Loo achieved a high ballistic performance against a 7.62×39 Mild Steel Core PS Ball M1943 when the stack of unidirectional UHMWPE material reaches at least 99.5% of the theoretical maximum density by compressing the stack at a pressure of more than 15 MPa (2175 psi) during heating at elevated temperature and during cooling down to room temperature. He suggested that the required compression time and compression temperature depend on the kind of fiber, matrix resin, and on the thickness of the molded panel.

Similar findings were reported by Ashok Bhatnagar [20] from Honeywell International, where he demonstrated that the ballistic performance and structural characteristics of unidirectional composite materials with thermoplastic matrix increase with increasing the molding pressure.

The patent disclosed that unidirectional para-aramid fibers (Twaron® T2000 from Teijin) were coated with 16% by weight of a polyurethane resin to make a uni-tape. The uni-tapes were cross-plied at 90 degree, and multilayer composite flat panels were pressed under different molding pressures. Multiple layers of the two-ply para-aramid fiber construction were molded separately at 115.6°C (240°F) at molding pressures of 3.4 MPa (500 psi) and 10.3 MPa (2500 psi) for a period of 20 min and allowed to cool to room temperature. Ballistic testing using a 7.62 × 51 mm M80 NATO bullet demonstrated that the ballistic resistance was substantially higher when molding at 10.3 MPa (2500 psi) of pressure than using the same matrix resin but molding at a low pressure.

Bhatnagar also demonstrated that the increase in ballistic performance and structural characteristics of the unidirectional para-aramid composite material is not guaranteed by increasing the molding pressure alone if the resin matrix is changed. Bhatnagar illustrated this behavior by using two different matrix resins: one thermoplastic and one thermoset. The thermoplastic was Kraton® D1107 styrene-isoprenestyrene block copolymer thermoplastic elastomer at 20% by weight and molded at 121.1°C (250°F) for 30 min. The thermoset resin was an epoxy vinylester resin (Derkane 411) at also 20% by weight and molded at 93.3°C (200°F) for 30 min. In both cases, the ballistic improvement by increasing the molding pressure was not observed. In addition, ballistic improvements of those unidirectional para-aramid composite materials with either Kraton® or epoxy vinylester resins were not observed when testing against a 9-mm full metal jacket (FMJ) hand gun bullet either.

15.5 INTER Materials unidirectional UHMWPE fiber thermoplastic matrix composite combat helmet

15.5.1 Unidirectional 0 degree/90 degree cross-plied ballistic materials

For most of our R&D efforts, the ballistic material used for the molding of helmet shell prototypes was a 0 degree/90 degree cross-plied unidirectionally oriented high-tenacity ultra-high molecular weight polyethylene (UHMWPE) fiber coated with a *thermoplastic polyurethane* resin. The 0 degree/90 degree layup consists of two individual plies with unidirectionally oriented fibers at 90 degree from each other, consolidated together by a solution-coating process with a polyurethane thermoplastic matrix resin where the individual filaments are impregnated by the continuous matrix resin.

Table 15.2 Unidirectional UHMWPE thermoplastic matrix composite materials

Thermoplastic matrix composite helmet 0 degree/90 degree prepreg materials	Spectra Shield® SR-3130	Dyneema® HB-26	Dyneema® HB-80
Helmet area density, kg/m^2 (lbs/ft^2)	11.24 (2.30)	11.24 (2.30)	11.24 (2.30)
Helmet wall thickness, mm (inches)	10.92 (0.430)	10.92 (0.430)	10.92 (0.430)
Fiber	UHMWPE	UHMWPE	UHMWPE
Fiber construction/prepreg	0 degree/ 90 degree	0 degree/ 90 degree	0 degree/ 90 degree
Prepreg weight, g/m^2 (oz/yd^2)	263 (7.75)	276 (8.14)	142 (4.18)
Thermoplastic prepreg matrix resin	PU	PU	PU
Number of layers/flat panel	42.7	40.7	79.2

Some commercially available unidirectional materials with the polyethylene (PE) fibers in a 0 degree/90 degree cross-plied prepreg configuration included Dyneema® HB-80 and HB-26 from DSM Dyneema® Corporation and Spectra Shield® SR-3130, SR-3136, or SR-3137 from Honeywell International, to name a few. Table 15.2 lists the properties of some of the 0 degree/90 degree cross-plied UHMWPE prepreg materials used.

15.5.2 Carbon-epoxy structural skin materials

Because all the thermoplastic matrix ballistic composite materials have low flex strength and stiffness, a carbon-epoxy structural reinforcement has been incorporated to meet the side-to-side compression and the blunt impact requirements. The commercially available carbon-epoxy prepreg materials that can be used as structural reinforcements for helmets are listed in Table 15.3. They have compatible low cure temperature (250°F) requirements and provide very high flexural stiffness once cured.

15.5.3 Helmet preform design and assembly

As a standard practice, a preform is assembled before molding a helmet. INTER Materials developed a patent-pending pinwheel/rosette pattern design called Plus/Plus that increases the overall ballistic performance of thermoplastic matrix helmets by reducing the number of cuts or seams, eliminating the pinwheel leg overlap, and

Table 15.3 Carbon-epoxy 0 degree/90 degree prepreg materials for structural reinforcement

Structural composite materials	IM7/381	C24T700150
Fiber	Carbon	Carbon
Fiber construction/prepreg	0/90/0/90	0 tape
Prepreg weight, g/m^2 (oz/yd^2)	448 (13.0)	117 (3.44)
Thermoset matrix resin	Epoxy	Epoxy
Manufacturer	Cytec [11]	Park Aerospace [21]

adjusting the fiber orientation between adjacent or neighboring cross-plied layers during the preform assembly. By eliminating the pinwheel leg overlap, the Plus/Plus pinwheel/rosette pattern provided a uniform areal density in the molded helmet and closely duplicated the same number of layers as a flat panel of the same areal density.

15.5.4 Molding and finishing

INTER Materials has used a set of compression molding dies to make medium-size ACH helmet prototypes for most of the R&D efforts. For ACH helmets with 11.24-kg/m^2 (2.30 lbs/ft^2) areal density made with the unidirectional UHMWPE composite materials listed in Table 15.2, the tooling for the compression molding dies has to be redesigned for a helmet wall thickness estimated to be between 10.16 mm (0.400 inches) and 11.43 mm (0.450 inches). The tooling was machined for a 10.92-mm (0.430 inches) helmet wall thickness. During the troubleshooting phase of molding thermoplastic matrix composite helmets, the number of layers listed in Table 15.2 corresponding to a flat panel with 11.24 kg/m^2 (2.30 lbs/ft^2) areal density was used as the starting point. The number of layers was later adjusted for the addition of the carbon-epoxy reinforcement and the quality of the molded helmet.

During the molding of thermoplastic matrix helmets, some of the efforts were focused on reducing the molding cycle time from 1.5 h down to just under 1 h by designing a more uniform distribution of the heating and cooling inside the die cavities, as shown in Fig. 15.9. To fine-tune the quality of the preform design and the molding process, helmet shells with no carbon-epoxy skin were molded, as shown in Fig. 15.10(a). The molded helmet shells were exposed to an incandescent light bulb in a dark room as shown in Fig. 15.10(b) to observe the level and uniformity of the translucency of the molded shell.

Adjusting the preform design, the preform assembly, and the pressure molding steps, INTER Materials has been able to achieve a uniform high level of translucency, Fig. 15.10(b), both in the crown area and all around the helmet side walls. The uniformity and high level of translucency in a molded helmet shell translate into an overall ballistic performance improvement of *a thermoplastic matrix composite helmet.*

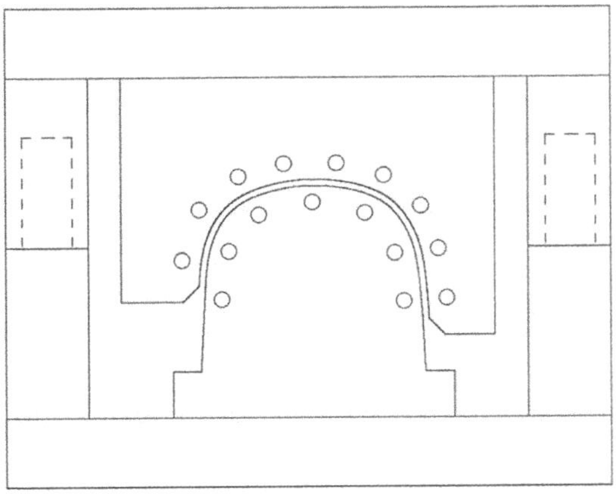

Figure 15.9 Compression molding match-metal dies for combat helmets.

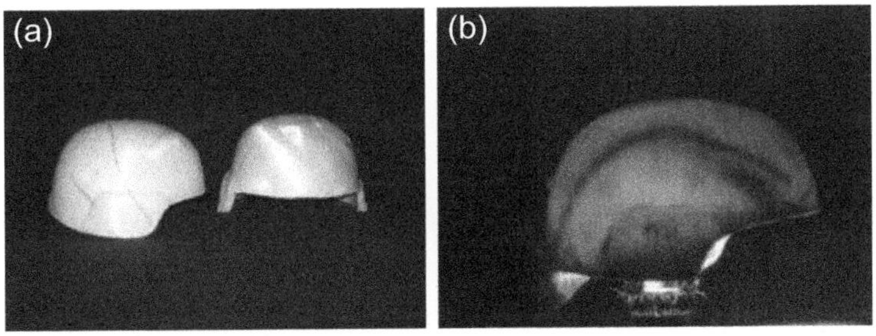

Figure 15.10 UHMWPE thermoplastic matrix composite molded helmets with no carbon-epoxy skin (a) exposed to incandescent light (b).

After a helmet has been molded, it is trimmed by waterjet for accurate dimensions, Fig. 15.11. INTER Materials has designed the tooling shown in Fig. 15.12 for holding the helmet during the water jet trimming.

The final finishing steps of drilling the holes, painting, rubber edging, and installation of the suspension system are standard practice for manufacturing helmets, Fig. 15.13.

15.5.5 Ballistic performance of thermoplastic matrix composite helmets

Example #1—ballistic performance as a function of 0 degree/90 degree cross-plied UHMWPE ballistic materials

Thermoplastic matrix composite ballistic helmets were molded using the thermoplastic matrix 0 degree/90 degree cross-plied prepreg materials listed in Table 15.2, Spectra Shield® SR 3130, Dyneema® HB-26, and Dyneema® HB-80. For each 0 degree/90 degree prepreg material, Table 15.2 also includes the number of ballistic prepreg layers per panel to produce a helmet shell with 10.92 mm (0.430 inches) of

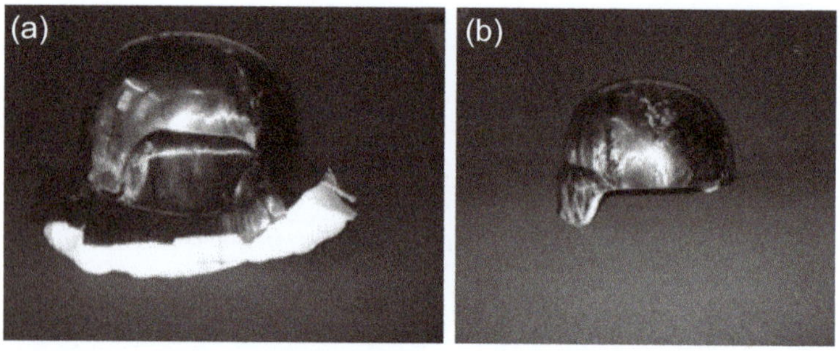

Figure 15.11 Helmet molding: (a) thermoplastic matrix helmet molded with a carbon-epoxy skin, (b) helmet trimmed by waterjet.

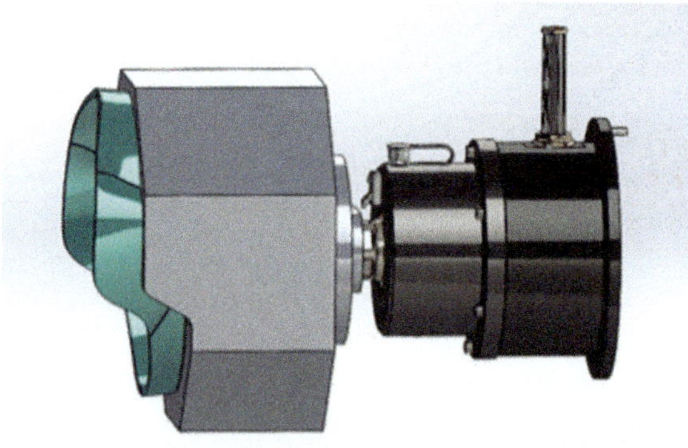

Figure 15.12 INTER Materials tooling design for helmet trimming by waterjet.

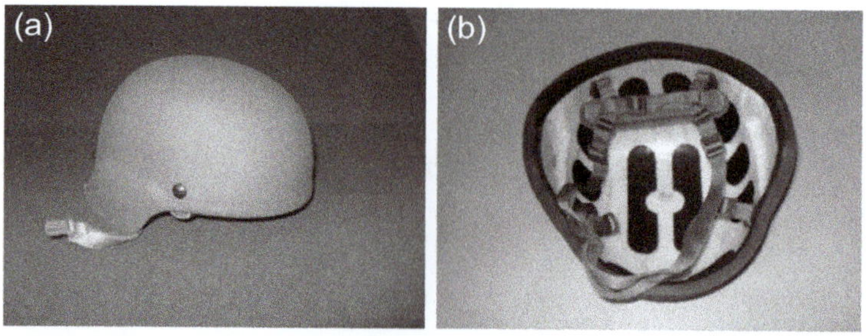

Figure 15.13 Finished ACH helmet: (a) painting, rubber edging, and hardware, (b) suspension system and pads installation.

wall thickness and 11.24 kg/m^2 (2.30 lbs/ft^2) of total area density, not including the carbon-epoxy skin. For helmets, the final number of the 0 degree/90 degree ballistic prepreg layers is adjusted to include the carbon-epoxy reinforcement and to meet the weight requirements. In addition, a fine tuning takes place during compression molding helmets to adjust the molding quality for a smooth surface and uniform consolidation.

All helmets were made using the Plus/Plus pinwheel/rosette pattern design with the ballistic 0 degree/90 degree prepreg layers. The structural skin was made using two 0 degree/90 degree carbon-epoxy prepreg C24T700150 with a total area density of 468 g/m^2 (13.76 oz/yd^2). The carbon-epoxy preform was incorporated on the outside surface of the helmet during the compression molding cycle. All helmets with carbon-epoxy skin were molded in a 10.92-mm (0.430 inches) die cavity using a standard compression molding method. Before molding, the helmet preform was preheated between 200°F and 240°F. The actual mold temperatures of the male and female steel dies were kept at about 260–270°F. All helmets where cooled down to 140°F under pressure before being removed from the dies. The ballistic data from testing at an NIJ-certified laboratory [22] is summarized in Table 15.4. This table shows the ballistic limit, V_{50} (m/s), as a percentage increase over the US Army ACH ballistic requirement against each threat for all three UHMWPE cross-plied prepreg systems.

As illustrated by Table 15.4, the ballistic performance of UHMWPE helmets made with the Plus/Plus pinwheel/rosette pattern design is far superior to the currently manufactured para-aramid ACH. The Plus/Plus pinwheel/rosette pattern design controls the fiber orientation between neighboring layers of cross-plied unidirectional UHMWPE materials. Depending upon the fragment size, the ballistic performance against right-circular cylinder (RCC) and FSP increased between 41% and 54% when compared to the current para-aramid ACH helmet requirements.

Example #2—thermoplastic matrix composite helmet ballistic performance as a function of helmet preform design

From the experience during the development of the Plus/Plus helmet preform design, INTER Materials developed a proprietary technology labeled the *Solid Design* to make a helmet with no cuts, no seam, and accurate control of the relative fiber orientation between neighboring layers throughout the whole molded helmet shell.

Table 15.4 Percentage increase of ballistic performance of helmets made with three 0 degree/90 degree UHMWPE cross-plied prepreg materials

Ballistic threat fragments (grains)	Spectra Shield® SR-3130	Dyneema® HB-26	Dyneema® HB-80
4-grain RCC	54%	54%	N/A
16-grain RCC	41%	47%	42%
64-grain RCC	46%	46%	42%
17-grain FSP	48%	46%	49%

In Example #2, we used 0 degree/90 degree cross-plied UHMWPE Dyneema® HB-80 material and the new *Solid Design*. As before, all helmet shells were made with 10.92 mm (0.430 inches) of wall thickness, 11.24 kg/m^2 (2.30 lbs/ft^2) of total area density, and molded by the standard compression molding method with a carbon-epoxy skin on the outside. The ballistic data is summarized in Table 15.5.

As the data in Table 15.5 illustrate, the ballistic performance of the helmets molded with Dyneema® HB-80 using the Solid Design is far superior to the ballistic performance of helmets made with the same cross-plied layers of Dyneema® HB-80 using the Plus/Plus pinwheel/rosette pattern design. Depending upon the fragment size, the ballistic performance against RCC and FSP increased between 72% and 77% when compared to the current para-aramid ACH helmet requirements.

Comparing the V_{50} (ft/s) values of Dyneema® HB-80 helmets using the Plus/Plus pinwheel/rosette pattern design with the *Solid Design*, we demonstrated that the *Solid Design* is not only a far superior design than the Plus/Plus, but it also has provided HB-80 helmets with exceptionally superior ballistic performance against small rifle fire as demonstrated by helmets stopping 7.62 × 39-mm FMJLC at muzzle velocities.

To further illustrate the benefits of the helmet material design, Fig. 15.14 compares the average ballistic limit, V_{50} (ft/s), against 17-grain FSP of helmets made with Dyneema® HB-80 and molded by conventional compression molding process as a function of three helmet material configurations: the pinwheel/rosette preform design with overlaps, the Plus/Plus preform with no overlaps and less seams, and the Solid Design with no cuts or seams and uniform fiber orientation. Fig. 15.14 shows the normalized ballistic limit, V_{50} (m/s), against 17-grain FSP for all three helmet designs, where the value of 100 corresponds to the V_{50} (m/s) of the standard US Army ACH requirement [10].

Example #3—thermoplastic matrix composite helmet ballistic performance as a function of lighter weight and short molding cycle.

In the quest for a lighter weight helmet, ACH helmets were made with 7.11 mm (0.280 inches) of wall thickness to produce a 25% lighter helmet with an approximated area density of 1.65 lbs/ft^2 including the carbon-epoxy skin. These helmets were molded with and without a carbon-epoxy skin on the outside by two molding methods, the standard compression molding method and a new molding process labeled *Short Cycle*.

Table 15.5 Ballistic performance percentage increase of ACH helmets made with the Plus/Plus helmet preform pattern design and the *Solid Design*

Ballistic threat fragments (grains)	Dyneema® HB-80 Plus/Plus Design	Dyneema® HB-80 Solid Design
4-grain RCC	N/A	N/A
16-grain RCC	42%	77%
64-grain RCC	42%	72%
17-grain FSP	49%	76%

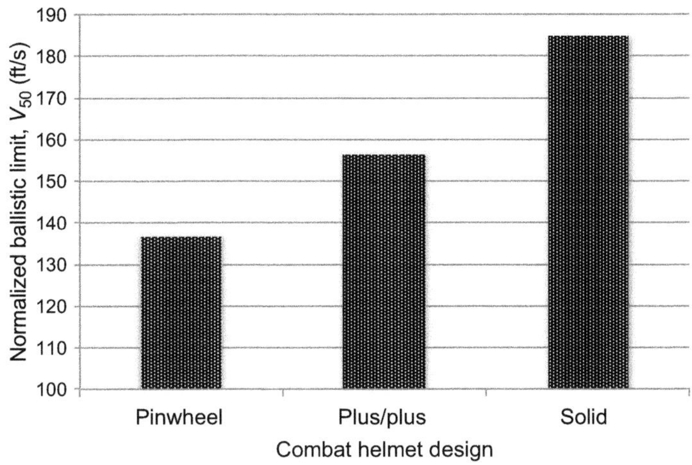

Figure 15.14 Normalized ballistic limit, V_{50} (ft/s), against 17-grain FSP of helmets made with Dyneema® HB-80 and molded by conventional compression molding process as function of the helmet materials design.

All the lightweight helmets were made with Dyneema® HB-80 using the *Solid Design* construction. They were tested against 17-grain FSP, and their ballistic performance is displayed in Fig. 15.15. This Figure illustrates that, on average, the same ballistic performance is obtained for Dyneema® HB-80 when molded with either

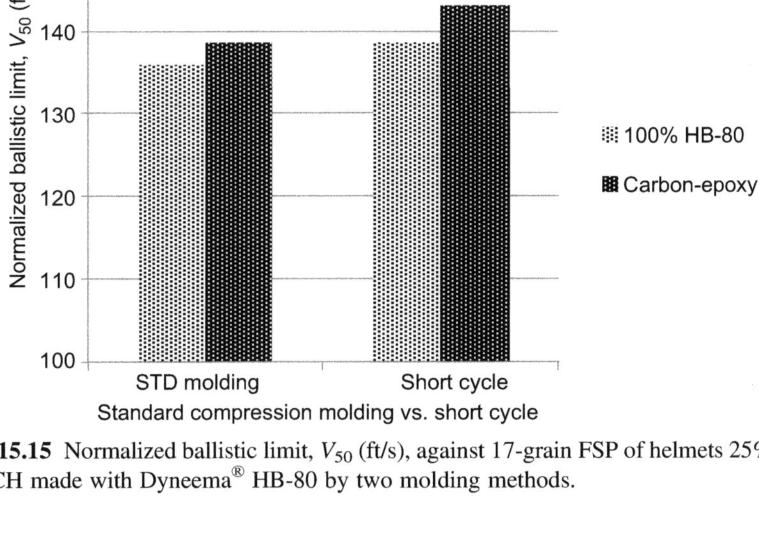

Figure 15.15 Normalized ballistic limit, V_{50} (ft/s), against 17-grain FSP of helmets 25% lighter than ACH made with Dyneema® HB-80 by two molding methods.

molding method. In addition, using the carbon-epoxy skin on the outside of the molded thermoplastic matrix helmets not only improves the structural characteristics of the thermoplastic matrix composite helmet, but it also slightly improves their ballistic performance against small fragments including 2-grain, 4-grain, and 16-grain RCC and 17-grain FSPs.

INTER Materials has developed the Short Cycle molding process as a low-cost and high-volume manufacturing process for making combat helmets with cross-plied unidirectional para-aramid or UHMWPE fiber thermoplastic matrix resin composite materials. The patent-pending innovation includes the reduction of the molding cycle time from 1.5 h down to less than 20 min for making thermoplastic matrix composite helmets with or without a carbon-epoxy skin, as illustrated in Fig. 15.15, without compromising their ballistic performance.

Fig. 15.15 shows the normalized ballistic limit, V_{50} (m/s), against 17-grain FSP for helmets 25% lighter than ACH, where the value of 100 corresponds to the V_{50} (m/s) of the standard US Army ACH requirement against 17-grain FSP [10].

15.6 Structural requirements of thermoplastic matrix composite combat helmets

Although in Example 3, Section 15.5, we achieved a ballistic improvement of 36–43% against 17-grain FSP with ACH helmets that were 25% lighter, we were not able to meet the structural requirements of back-face deformation (BFD) against 9-mm FMJ. The BFD data clearly showed that, for a thermoplastic matrix helmet with a carbon-epoxy skin, 7.11 mm (0.280 inches) of wall thickness and 8.07 kg/m^2 (1.65 lbs/ft^2) of area density, the structural requirements were still a challenge, even when using the best cross-plied unidirectional UHMWPE materials available on the market.

In the quest for a lighter weight helmet, INTER Materials has optimized the weight reduction for the ACH helmet between 10% and 15% with the most advanced UHMWPE available materials that meet all the structural requirements.

INTER Materials now offers two basic designs, a lightweight helmet and a helmet with protection against rifle fire. The lightweight helmet is 10–15% lighter than the ACH with 35–40% superior ballistics performance. The helmet with protection against small arms has the same weight as the ACH but with 70–75% superior ballistics to the ACH and capable of stopping some rifle rounds. Both designs meet all the structural requirements including BFD against 9-mm FMJ, side-to-side compression, and blunt trauma.

For the lightweight helmet that is 10–15% lighter than the ACH, the BFD results from testing with the FARO arm against 9-mm FMJ at an NIJ-certified laboratory [22] are shown in Table 15.6. All BFD ACH requirements were met, including less than 16 mm on the crown and left and right sides; and less than 25.4 mm on the front and back of the helmet.

In addition, the lightweight helmet that is 10–15% lighter than the ACH also met the ACH requirements for blunt impact and side-to-side compression. The maximum

Table 15.6 Back-face deformation (BFD) with FARO arm against 9-mm FMJ

Average BFD against 9 mm FMJ–FARO arm				
Crown	**Left**	**Right**	**Back**	**Front**
7.213	6.732	7.876	15.177	9.830

peak acceleration at room temperature was less than 100 G (gravitational constant) after a mass of 4.969 kg is dropped from a 19.6 inch height, thus meeting the ACH requirement of no individual acceleration exceeding 150 G's.

The side-to-side maximum compression at room temperature was less than 0.040 inches immediately after 25 cycles (within 5 min) and less than 0.030 inches after 24 ± 1 h. The side-to-side ASTM D-76 Compression Test procedure includes placing the helmet in a 2.5″ flat-loading anvil, centering the helmet on the widest part of the helmet shell, compressing the shell at a rate of 12 inches/min until 300 lbs are reached, and releasing the load down to 5 lbs and repeating this loading cycle for a total of 25 cycles. After the loading cycle, the change in dimension was measured to be less than 0.200 inches immediately after the 25 loading cycles (within 5 min) and to be less than 0.125 inches after 24 ± 1 h. These tests were performed on shells conditioned at room temperature and humidity.

15.7 Discussion and future trends

Some of the great challenges to the military and law enforcement communities will always be to enhance ballistic protection and reduce weight of the personal body armor carried by a soldier and law enforcement personnel, including ballistic vests and helmets. From the history of composite armor systems, improvements in ballistic protection have been associated with the advent of new and superior ballistic materials. In the late 1970s, Kevlar® revolutionized the personal ballistic protection and composite armor markets that have relied mainly on steel, fiberglass, and nylon [1]. During the past 10 years, with the advent of UHMWPE fibers and cross-plied thermoplastic matrix composites [23], progress is being made to design a thermoplastic matrix composite lightweight helmet that not only has superior ballistic performance, but also meets all the structural requirements and is made with a cost-effective manufacturing process.

It has been demonstrated that high-performance ballistic fibers, when combined with thermoplastic matrices, improve the ballistic protection beyond that afforded by the incumbent para-aramid fiber PVB-phenolic thermoset resin systems. Helmets made with cross-plied UHMWPE fiber thermoplastic matrix composites have reached a ballistic performance 70–75% superior to the current para-aramid PVB-phenolic ACH helmet, and they are capable of stopping some rifle rounds at the same weight or areal density.

However, when helmet weight reductions higher than 15% have been attempted with hybrid designs using a carbon-epoxy composite as the structural reinforcement, meeting the structural requirements, including the back-face deformation against 9-mm FMJ, side-to-side compression, and blunt impact, still remains a challenge.

In addition, the new military testing protocol for the first article test will require extensive testing to meet the acceptance criteria. The new acceptance criteria will rely on the statistical analysis of ballistic testing data from four helmet sizes combined (small, medium, large, and extra-large) at three different temperatures.

Therefore, the overriding factor for introducing a new lightweight helmet technology using the new thermoplastic matrix composite materials is not the ballistic performance, but rather meeting all the structural requirements and having a cost-effective manufacturing process.

References

[1] Laible RC. Fibrous Armor, Ballistic Materials and Penetration Mechanism. Elsevier Scientific Publishing Company; 1980. p. 73–115.
[2] MIL-H44099-A. Military Specification-Helmet, Ground Troops and Parachutists. U.S. Government Printing Office; December 22, 1986.
[3] Le Grand D. US Patents 2,532,442 and 2,420,522, 1947.
[4] Goldsmith H. US Patent 2,451,483, 1946.
[5] Grick SJ. US Patent 4,596,056, 1986.
[6] White MR. US Patent 4,778,638, 1988.
[7] White MR. US Patent 4,908,877, 1990.
[8] Riewald PG, Folgar F, Yang HH, Shaughnessy WF. Light weight helmet from a new aramid fiber. Kiamesha Lake, New York. In: 23rd International SAMPE Technical Conference, vol. 23; October 21–24, 1991. p. 684–95.
[9] Folgar F, Scott BR, Walsh SM, Wolbert J. Thermoplastic matrix combat helmet with graphite-epoxy skin. In: 23rd International Symposium on Ballistics, Tarragona, Spain; April 16–20, 2007.
[10] ACH Purchase Description, AR/PD 10-02, December 14, 2010.
[11] Cytec Engineer Materials, Anaheim, CA. (800-256-1090), www.cytec.com.
[12] van der Loo LH. US Patent 6,183,834 B1, 1997.
[13] Roylance D, Wang Su-Su. US Army Natick Technical Report TR-80/021, Penetration Mechanics of Textiles Structures. MIT Contract No DAAG 17-76-C-0013. 1979.
[14] Cunniff PM. An analysis of the system effect in woven fabrics under ballistic impact. Text. Res. J. 1992;62:495–509.
[15] Karl-Mayer Technical Textiles, http://www.karlmayer.com/internet/.
[16] Park AD. US Patent 5,952,078, 1998.
[17] Park AD. US Patent 7,148,162 B2, 2004.
[18] van Heerden J, et al. US Patent 2007/0099526 A1, Barrday, 2006.
[19] Cunningham DV. US Patent 7,407,900 B2, Barrday, 2008.
[20] Bhatnagar A, et al. Honeywell international, US Patent 2007/0194490 A1, 2007.
[21] Park Aircraft Technologies, Newton, Kansas, http://www.parkelectro.com.
[22] Chesapeake Testing Services, Belcamp, Maryland, http://chesapeaketesting.com.
[23] Bhatnagar A. Lightweight Ballistic Composites Military and Law-Enforcement Applications. England: Woodhead Publishing Ltd.; 2006.

Numerical analysis of the ballistic performance of textile fabrics

16

C. Ha-Minh
R&D Center, Duy Tan University, Da Nang, Viet Nam; LMT, ENS Cachan, CNRS, Université Paris Saclay, Cachan, France

16.1 Introduction

Textile 3D warp interlock structures show a high performance in ballistic protection with high flexibility and light weight [1]. The advantages of this material over 2D structures have been confirmed [1–5]. We can summarize two main advantages of 3D textile structures in comparison with the 2D ones: (1) higher resistance to multiimpacts (less damage for an impact) and (2) easier and cheaper achievement of complex shape structures. However, as of 2015, numerical and analytical studies are still at a modest level and usually limited to 2D woven fabrics.

To describe the impact behavior of 2D woven fabrics, many papers have been published [6]. In fact, analytical models were developed from the simplest case (impact behavior of a single yarn without boundary conditions [7]) to the complex one (a multilayer 2D plain-woven fabric submitted to a ballistic impact [8]). The analytical model in the case of impact behavior of a single yarn of Smith et al. [7] was based on experimental data previously published in 1956 [9]. In this model, constant impact velocity and an experimental strain–stress curve were used as input parameters. Single yarns have been considered infinite (no reflection), and the contact between projectile and yarn has been assumed sharp. However, the reality is more complex with cases where yarns are fixed at two ends. Carr [10] carried out ballistic tests between a steel sphere of 0.68 g and the para-aramid and ultra-high molecular weight polyethylene (UHMWPE) yarns. Fifty ballistic tests were performed on five different types of yarn with impact velocities varying between 346 and 720 m/s. Results showed that yarns failed by two fracture modes, shearing and tension, depending on impact energies. Moreover, interactions among reflection and incident waves can lead to premature yarn failures due to stress concentration [5].

Developing from the model of Smith et al. [7], many authors [11–16] later proposed analytical modeling of impact behavior of woven fabrics but were limited to the linear and elastic behavior of yarns. All of these works neglected frictions and undulations of yarns that were proven numerically important in the case of 3D woven fabrics [17]. Moreover, reflection of strain waves on yarns at interlacement points is another complex phenomenon that has not been considered yet. On the other hand, analytical models of Naik et al. [13,14] and Mamivand and Liaghat [15] calculated stress distribution on primary yarns (the ones in direct contact with a projectile) based

Advanced Fibrous Composite Materials for Ballistic Protection. http://dx.doi.org/10.1016/B978-1-78242-461-1.00016-9

on experimental data of laminated composite. Thus, these works are not very reliable because behavior of a laminated composite is different from dry woven fabrics without a matrix. Recently, we proposed an analytical model that permits taking this into account [8]. Indeed, reflection of strain waves causes a sharp increase of yarn deformation at impact point that results in premature failure of principal yarns.

Roylance et al. [18] proposed a numerical model since the 1970s that approximated woven layers of fabric as a network of nodes with a certain mass connected together by 1D elements. This approach is conceptually similar to that used by other workers, for example, Shim et al. and Joo et al. [19–21]. These models use fabric material property data together with a failure criterion and a constitutive equation as inputs of 1D elements. Nodal mass is determined such that the surface density of numerical fabric is equal to the real one. One-dimensional elements, which do not have mass, can only suffer tension or compression. The ballistic limit and the residual velocity after impact may be determined as outputs, and the influence of the boundaries of the fabric on these values is accounted for. Shim et al. [19] included fiber viscoelasticity in the formulation of an identical model to analyze the impact between a small spherical projectile and a 2D plain-woven fabric made by poly-PPTA (*p*-phenylene terephthalamide) fibers. Since crimp is a specific character of yarns in a plain-woven structure, they performed a static tensile test on Twaron fabric. The decrimping contributes 1.5% to the total deformation of the fabric without increase of yarns stress. Therefore, they took account for this effect in calculating the real strain of yarns in their model by using an adaptive formulation. Joo and Kang [20] developed the model of Roylance et al. [18] to predict the impact of a rigid, steel sphere onto a target composed of several layers of 2D plain-woven fabrics. In this model, the curvature of the crossing points of yarns is explicitly presented. Based on this model, in another work, Joo et al. [20,21] have characterized the different energies during the impact on the fabric. These energies correspond to deformation of yarns and interactions among yarns in a layer or at the interface between layers. Particularly, this work also permitted the calculation of energy of interactions between yarns with clamp cadre.

In general, the main difficulty of the method of Roylance et al. [18] lies in a simple mapping of 2D woven fabrics that cannot take into account the influence of its architecture. The extension to 3D woven fabric is even more difficult. Moreover, this mapping leads to an overall deformation of the fabric having a rectangular form with edges parallel to those of clamp cadre, which is different from the reality [5]. According to the work of Ivanov and Tabiei [22], Vinson et al. (1975) proposed a macroscopic model where the fabric is described schematically by a homogeneous plate. This continued modeling of textile materials allows the use of a loose mesh and therefore low computation time. In the model of Vinson et al. (1975) and Taylor et al. (1990), the material is isotropic and leads to an identical behavior of fabric in all directions of the plate. Such a result is far from the experimental observations of Ivanov and Tabiei [22]. In another work, the macroscopic model of Lim et al. [23] incorporated the viscoelastic behavior of fibers. However, numerical results showed a deformation of the plate-shaped cone, whereas the experimental tests show a pyramid deformation.

Ivanov and Tabiei [22] have attempted to integrate crimp and sliding of yarns in the macroscopic model of the fabric. The ballistic behavior of the fabric is closer to reality with a global dynamic deformation. Unit cell—based approaches have been extensively used to derive the equivalent (smeared) continuum-level material models of woven composites from the knowledge of the mesoscale yarn properties, fabric architecture, and interyarn and inter-ply frictional characteristics. Many discussions and extensions of this approach have been mentioned in recent literature [24—31].

Gu [32], Duan et al. [33,34], and Rao et al. [35] proposed mesoscopic models in which yarns are geometrically modeled and meshed with 3D solid elements. This approach has permitted the description of complex phenomena such as slipping, crimp and fracture of yarns, and delamination between the layers. Ballistic response of the fabric is more sophisticatedly predicted. Barauskas and Abraitiene [36] used 3D shell elements to construct their mesoscopic model, thus significantly reducing the computation time. On the other hand, we found in the works of Duan et al. [33—34,37] a study on the effects of yarn—yarn and yarn—projectile frictions where the friction coefficient is expressed as a function of the relative velocity of the contacting surfaces and the exponential decay coefficient representing the transition from static friction to the stable dynamic one. This expression of the friction coefficient is currently widely used by several authors. Barauskas and Abraitiene [36] have adjusted their model assessing the value ranges of static and dynamic friction coefficients such that the number of broken yarns predicted is equal to the experimental one in the case of an impact on the 2D plain-woven fabric made by 709 Twaron fibers. Rao et al. [35] have carried out an experimental study on a quasistatic coefficient of friction for the case of Kevlar®KM2 fabric. These authors have shown the important role of friction coefficient as well as Young's modulus and strength of yarns on the ballistic performance of fabrics.

In general, it can be noted that the prediction of full mesoscopic models is more realistic than macroscopic ones for the ballistic behavior of the fabric but with a very high cost. The multiscale model is an effective solution [35,38—39]. The idea is that the macroscopic model is applied to a zone far from the impact region, and the mesoscopic one is used for the contact zone with a projectile. The multiscale model presents an interesting approach to solving impact problems with a gain of computing time.

This chapter presents a full review on the numerical modeling of the ballistic behavior of 2D and 3D textile woven fabrics. In the first part, macroscopic and mesoscopic models will be described, and results are validated with experimental data. In the second part, a multiscale strategy will be discussed for gain of computation time. Last, prediction of impact behavior of 3D woven fabric is assessed using a numerical tool recently developed.

16.2 Numerical macro-mesoscopic simulation of dynamic behavior of a 2D plain-woven fabric

16.2.1 Macroscopic model

A fabric is made of interlaced warp and weft yarns. Under ballistic impact, fabric response is similar to a thin homogeneous plate. This phenomenon can be easily observed at low velocities where fabrics have a global deformation or a cone is formed with the top being the impact point. In this model, the plain-weave fabric is considered as a homogenous plate (Fig. 16.1). Therefore, shell elements can be used to represent this homogenous plate. Since cross sections of warp and weft yarns are the same and parallel yarns are located side by side in the fabric, the size of the shell elements should be equal to the distance between yarns. This choice will give the ballistic impact response of the model close to the reality where the link between yarns is much weaker than the one between fiber components in a yarn. Therefore, the shear modulus is taken close to zero as reported by Rao et al. [40] in the macroscopic model. Certainly, this model cannot describe detailed deformations of each yarn, but it can reduce the time of computation because of the large size of elements. Thus, the model can be applied to larger structures such as complete ballistic armor. Logically, the fabric thickness is used as an element thickness, and the mass density of each element is calculated to ensure the mass of the whole plate, which is equal to the fabric [35]. However, since this technique does not account for the porosity of the fabric and the undulation of yarns, the calculated density for the plate ends up being a little lower than the density of the yarn material. In this case, the 2D plain-woven structure of the fabric is like a plane plate, and yarn undulation is negligible; therefore, similar to other simple analytical models [12,14–16], this density approximation of the macroscopic model is acceptable. The number of integration points throughout the thickness of each element is equal to 2 to have a good approximation of deformation of plate in this direction. Since the mechanical properties of the fabric are similar to each other in both warp and weft directions, the plate can be considered as orthotropic. Similar to the previous density approximation, the elastic moduli of the plate in warp and weft directions are assumed to be equal to the elastic modulus of a real yarn (see other analytical models [12,14–16]).

With a fabric held at two sides, yarn reorientation is an important phenomenon that changes the angle between warp and weft yarns in the fabric [22,25]. This reorientation results in a nonlinear in-plane shear modulus for the shell elements that initially is very

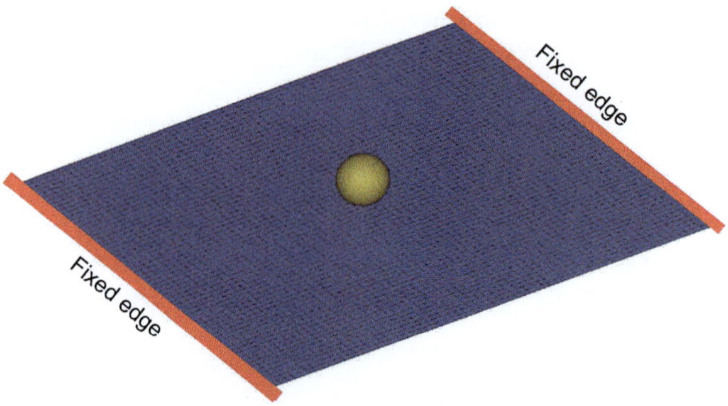

Figure 16.1 Macroscopic model.

low and then grows larger after yarn locking. Therefore, using a nonlinear value of shear modulus that varies with yarn angle will give a better prediction of the impact behavior of the fabric in this case. However, a constant value of shear modulus is used in this study for simplification.

16.2.2 Mesoscopic model

The macroscopic model does not take into account actual geometry of the fabric, for which modeling needs to be proposed at the mesoscopic scale. In fact, this model describes fabric across the yarn (Fig. 16.2). A cross section of a yarn may be assumed to have an elliptical shape. In our case, the dimensions of a yarn section are the height, $h = 0.115$ mm, and the width, $b = 0.59$ mm (Fig. 16.3(a)). This section is described by shell elements with different thicknesses (Fig. 16.3(b) and (c)). It should be noted that the number of elements per cross section is always even due to the symmetry. Indeed, in our study, we chose two configurations, four and eight elements, to check the influence of the number of elements. In the case of shell elements, four elements are the minimum to describe the elliptical shape of the cross section. Therefore, both cases (four and eight elements) for a section are checked assuming that the area of the section is constant (Fig. 16.3). The thicknesses of the elements on a section in both cases are calculated to ensure this condition. In this model, yarn is assumed homogeneous and orthotropic. The mechanical properties are constant throughout the length of yarn. The symmetry planes of a yarn cross section are taken as orthotropic plans.

If the orthogonal directions of the yarn are numbered as shown in Fig. 16.4, the behavior of the orthotropic material, the relationship between the deformations

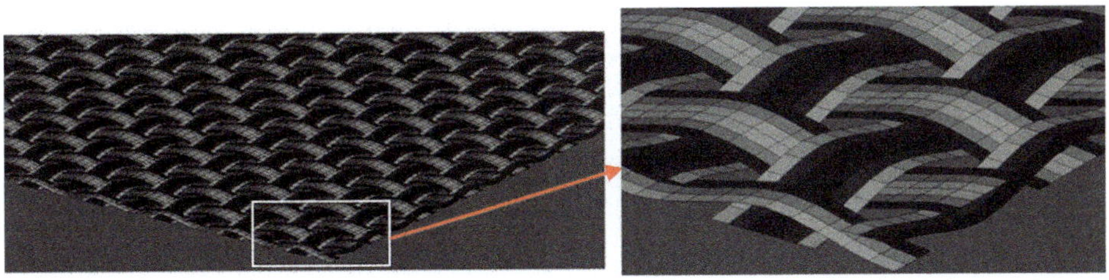

Figure 16.2 Mesoscopic modeling of the plain-weave fabric using shell elements in Radioss.

Figure 16.3 Modeling of a yarn cross section by shell elements: (a) dimensions of a cross section of yarn; (b) model of four elements; and (c) model of eight elements.

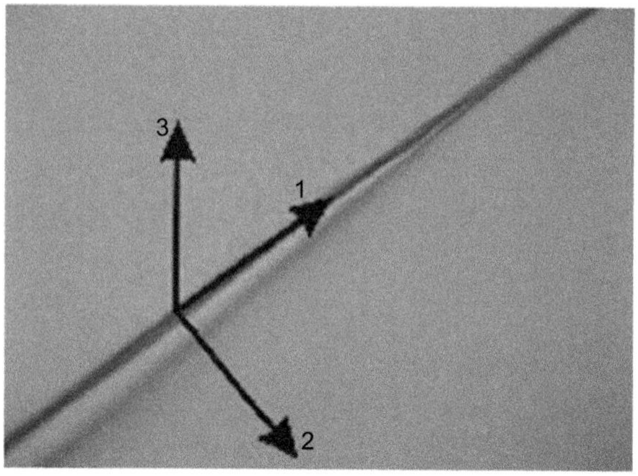

Figure 16.4 Directions of yarn.

(ε_{11}, ε_{22}, ε_{12}, ε_{23}, ε_{31}) and the constraints (σ_{11}, σ_{22}, σ_{12}, σ_{23}, σ_{31}) of a yarn, is modeled by the following equation:

$$\begin{pmatrix} \varepsilon_{11} \\ \varepsilon_{22} \\ 2\varepsilon_{12} \\ 2\varepsilon_{23} \\ 2\varepsilon_{31} \end{pmatrix} = \begin{pmatrix} \frac{1}{E_{11}} & \frac{-\upsilon_{21}}{E_{22}} & 0 & 0 & 0 \\ \frac{-\upsilon_{12}}{E_{11}} & \frac{1}{E_{22}} & 0 & 0 & 0 \\ 0 & 0 & \frac{1}{G_{12}} & 0 & 0 \\ 0 & 0 & 0 & \frac{1}{G_{23}} & 0 \\ 0 & 0 & 0 & 0 & \frac{1}{G_{31}} \end{pmatrix} \begin{pmatrix} \sigma_{11} \\ \sigma_{22} \\ \sigma_{12} \\ \sigma_{23} \\ \sigma_{31} \end{pmatrix}$$

where E_{11} is the longitudinal Young's modulus of a bundle; E_{22} is the transversal Young's modulus of a bundle; υ_{12} and υ_{21} are Poisson's ratios of a bundle; G_{12}, G_{23}, and G_{31} are the shear moduli of a bundle. In this formulation, we excluded the strain and stress in yarn cross section height ε_{33}, σ_{33} due to a very small ratio between the height and the width of yarn cross section. For a simplification, we can assume $G_{12} = G_{31}$ (shearing in the direction perpendicular to the bundle). Moreover, we can consider $G_{23} \approx 0$ (shearing in the direction parallel to the bundle) because a bundle is composed of thousands of separate single fibers (Fig. 16.3). Moreover, we know that the compliant matrix is symmetric, so $\upsilon_{12}/E_{11} = \upsilon_{21}/E_{22}$. In general, we need four experimental values for material elastic constants of a bundle: E_{11}, E_{22}, υ_{12}, and G_{12}.

The number of integration points through the thickness of elements should be equal to 2 for calculating strain in this direction and for predicting well the elastic behavior of yarns. Barauskas and Abraitiene [36] used only one integration point for shell elements in thickness to eliminate the bending modulus of yarns. In real conditions of application, the bending modulus of yarns is very small but not equal to zero. Therefore, with two integration points through the thickness of orthotropic shell elements, prediction of bending resistance of yarns is better than with only one point of integration. Moreover, if yarns sections are modeled by only one element in thickness, physical fields are constant when using one integration point in calculations. It follows that yarns only undergo tensile strain and no shear strain at all. This is not compatible with physical laws, as shear resistance of yarns, which cannot be negligible according to the works of Cheng and Chen [41]. In this study, the optimal number of integration points through the thickness of a shell element is found and equal to 2. When this value increases beyond 2, the behavior of yarns does not change, and the computation time is longer.

16.2.3 Criteria of rupture

The material used in this study is a Kevlar® KM2 fabric. The behavior of yarn is elastic until failure [41]. The breaking stress for this material is equal to 3.4 GPa [41]. This value is equivalent to elastic break deformation equal to 5.48%. In our modeling, the failure criterion is of type FLD (forming limit diagram) where the rupture zone is defined in a plane of principal strains (Fig. 16.5).

This area must be over the curve of maximum deformation in function of minimum deformation. Indeed, in our case, the maximum deformation should be less than the breaking strain measured, or $\varepsilon_R = 5.48\%$.

16.2.4 Calculation conditions

The material used is a dry, plain-weave fabric of 50.6 × 50.6 mm (ie, without resin) (Fig. 16.6(b)) with end and pick densities of 13.4 yarns/cm, which is equivalent to a distance of 1.49 mm between yarns. Each yarn is composed of thousands of single Kevlar® KM2 fibers. Therefore, the target in the impact system of this study is named

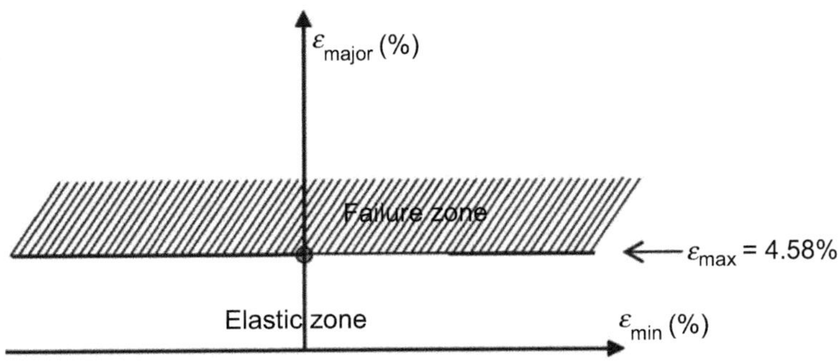

Figure 16.5 Schematization of a forming limit diagram (FLD) for rupture criterion of Kevlar® KM2 yarns.

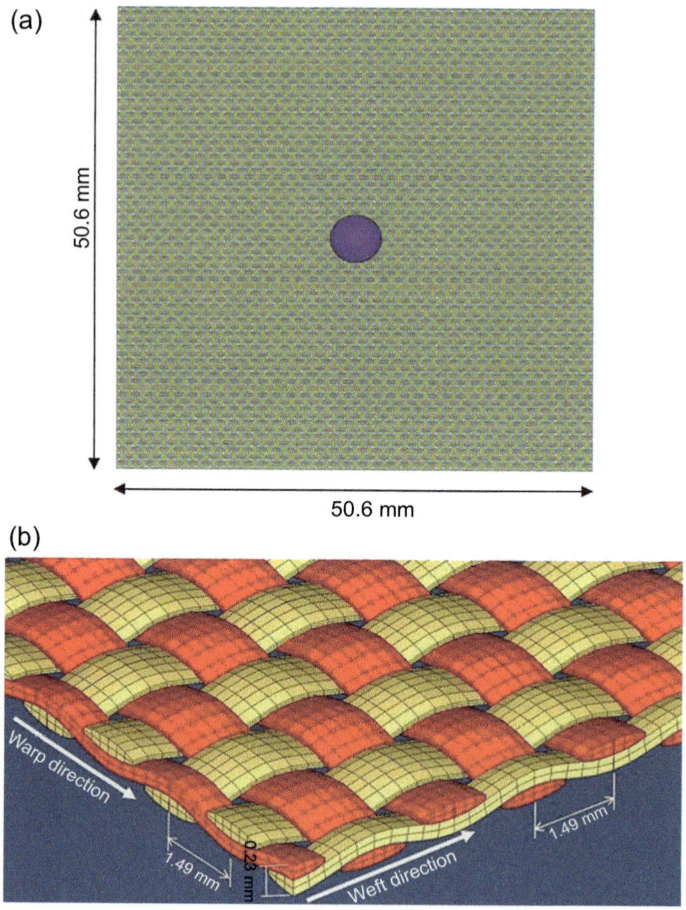

Figure 16.6 (a) Initial configuration of the ballistic impact system simulated in the present study; (b) detailed illustration of Kevlar® KM2 plain-woven fabric.

as dry 2D Kevlar® KM2 plain-weave fabric. The spherical projectile is supposed to be rigid, with a diameter of 5.35 mm and a mass of 6.25×10^{-4} kg (Fig. 16.6(a)).

The fabric is fixed at the two edges; the other two are considered free (Fig. 16.7(a)). It is assumed that the point of contact between the fabric and the projectile is a point of intersection between a warp and a weft yarn in the center of the fabric. Only a quarter of the model is calculated because of the symmetry of this impact (Fig. 16.7(b)) to optimize the computation time. In the case of a single layer of plain-weave fabric, the deformation of the projectile after impact is considered negligible. Thus, the projectile is assumed to be infinitely rigid.

For simplicity, only one static Coulomb's coefficient ($\mu_s = 0.20$) is used to describe two friction types (yarn–yarn and yarn–projectile) in this study. According to the work of Rao et al. [40], Kevlar® KM2 yarns have a density of 1310 kg/m^3 with a

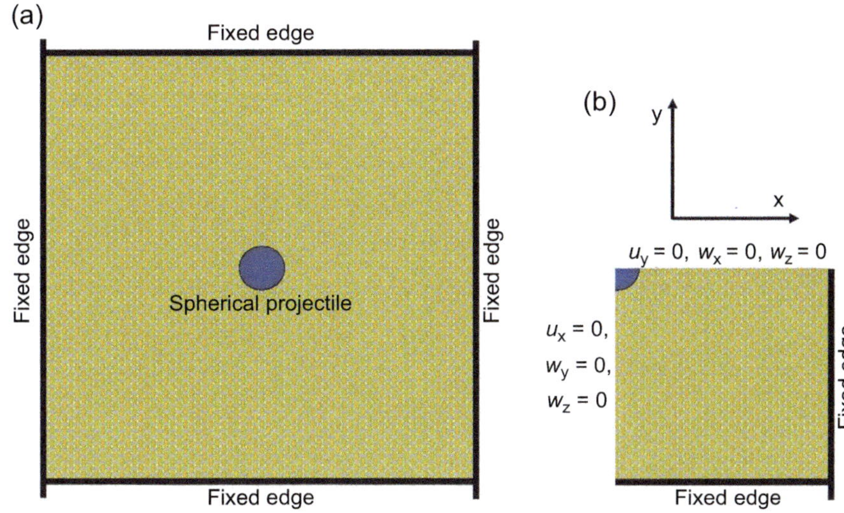

Figure 16.7 Boundary conditions of the model: (a) full model, (b) a quarter of the model.

longitudinal Young's modulus taken equal to 62.0 GPa. The transverse and shear moduli are assumed to be 100 to 1000 times lower than the longitudinal modulus [41]:

- a transverse modulus equal to 0.62 GPa
- a shear modulus equal to 0.126 GPa

When yarn failure criterion is programmed by the maximum deformation, the shell elements are sensitive to the low value of the shear modulus. In addition, Cheng and Chen [41] presented a technique to measure the shear modulus of a single fiber. The shear modulus of Kevlar® KM2 R fibers is determined to be equal to 24.4 GPa by this technique. Therefore, a shear modulus of 7.56 GPa has been proposed for the mesoscopic model. In the case of the macroscopic model, the shear modulus is always set equal to 0.126 GPa for the homogenous plate because of the porosity of the fabric. The importance of yarn shear modulus in the numerical modeling is detailed and discussed in Ref. [42].

16.2.5 Results and discussions

16.2.5.1 Mesh optimization

Fig. 16.8 shows the evolution of the projectile speed in terms of time, respectively, at the impact velocities of 60.6 m/s and 245.0 m/s. Before appearance of yarn failure, there is an insignificant difference between the two models: four elements and eight elements. This result is similar to the one of Boisse et al. [43], who executed a numerical simulation without failure.

Yarn failure appears almost at the same time for both cases of four and eight elements (at the moment 165 μs with impact velocity 60.6 m/s and 21 μs with impact

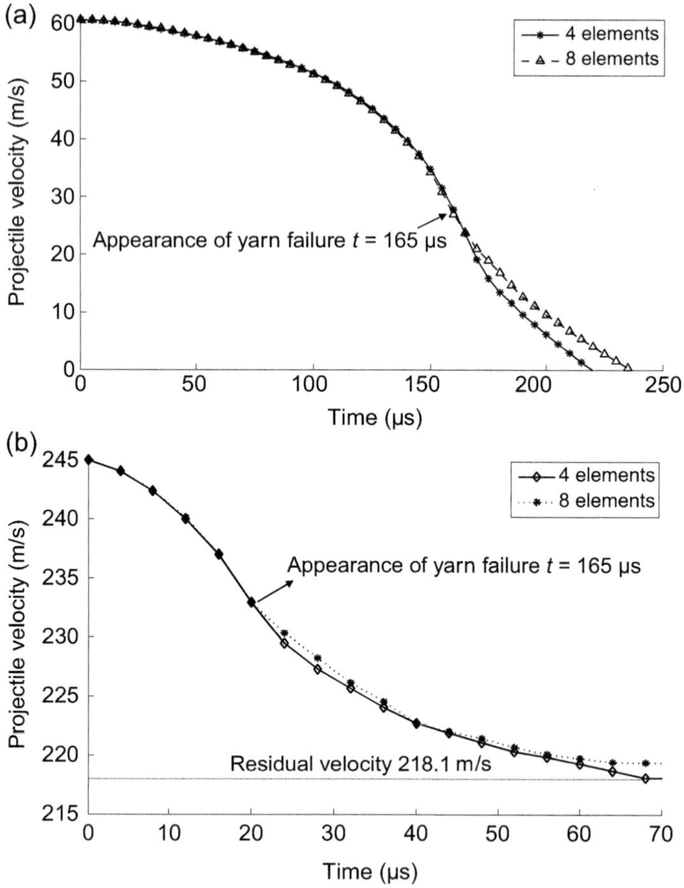

Figure 16.8 Evolution of the projectile velocity for the impacts: (a) 60.6 m/s; (b) 245.0 m/s.

velocity 245.0 m/s) (Fig. 16.8). This indicates that main modes of yarn failure, tension, and transverse shearing in the fabric subjected to ballistic impact do not change when the number of elements in cross section enhances doubly. In fact, changing the number of elements without variation of the area of cross section of yarns cannot influence yarn tensile and transverse shear resistances.

The first appearance of yarn failure initiates distortions by subsequent ruptures of yarns in the zone impact between the fabric and the projectile. This causes a little larger difference on the projectile velocity–time curves between two models, four and eight elements, than the time before yarn failure (Fig. 16.8). Therefore, the time of projectile stopped at impact velocity 60.6 m/s is 220 μs for four elements and 235 μs for eight elements. Residual velocities (velocities of the projectile after impact) for the impact of 245 m/s are also distinguished for two cases: 218.1 m/s with four elements and 219.5 m/s with eight elements.

These differences on calculated values are insignificant, but the computation time of a model of eight elements is always doubly higher than for four elements (Fig. 16.9).

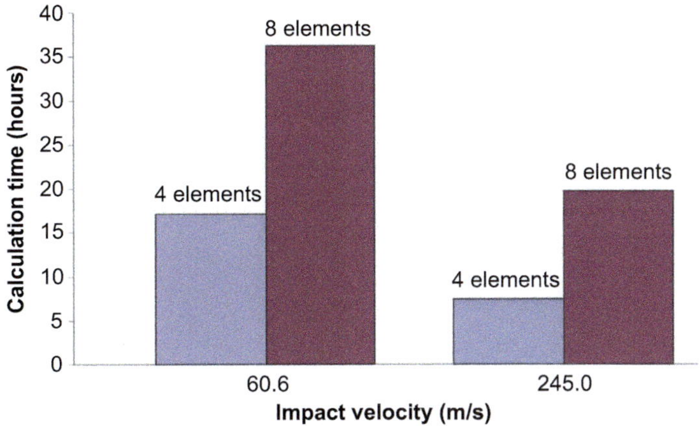

Figure 16.9 Calculation time (comparison between four and eight elements).

We can use experimental results as the reference to determine the exact number of elements on a section following residual velocity. In both cases of four and eight elements, the calculation gives a nonperforation of the fabric with the 60 m/s impact and the range of the ballistic limit from 60 m/s to 92 m/s as in reality. However, in experience, with the impact velocity of 245 m/s, the residual velocity is 207 m/s; hence, the one with four elements on a yarn cross section is more convenient than eight elements. Thus, in the next sections, we only study and continue to validate the mesoscopic model of four elements on a yarn section based on the global behavior and the damage phases of the fabric subjected to ballistic impact.

16.2.5.2 Numerical analysis of impact behavior of a 2D plain-woven fabric

Fig. 16.10 shows the state of fabric after impact in the case of a macroscopic model (a) and a mesoscopic model (b). Overall, the areas affected by the impact constitute a zone in the form of a cross, centered at the impact point of fabric. These phenomena were also observed during a ballistic impact event in both cases of macroscopic and mesoscopic models. In the case of the mesoscopic model, Fig. 16.10(b) shows the extension of the deformation and failure modes of a fabric under impact with perforation. Indeed, in the impact zone of the fabric, yarns slide over each other on the projectile surface leading to local failure of yarn. Similarly, the central yarns whose ends are free are displaced with the projectile due to the lack of efforts at two edges. These findings are consistent with those observed experimentally (Fig. 16.11). Indeed, the experimental results show that during the penetration time of the projectile, the response of the fabric is divided into different mechanisms or damage phases (Fig. 16.11):

- formation of a pyramid with the apex being the head of the projectile;
- displacement of primary yarns (the yarns crossing the contact zone with the projectile and perpendicular to the free edge) from the fabric plane due to their free ends;
- narrowing of the fabric and yarns burst at two free edges due to impact with the projectile;

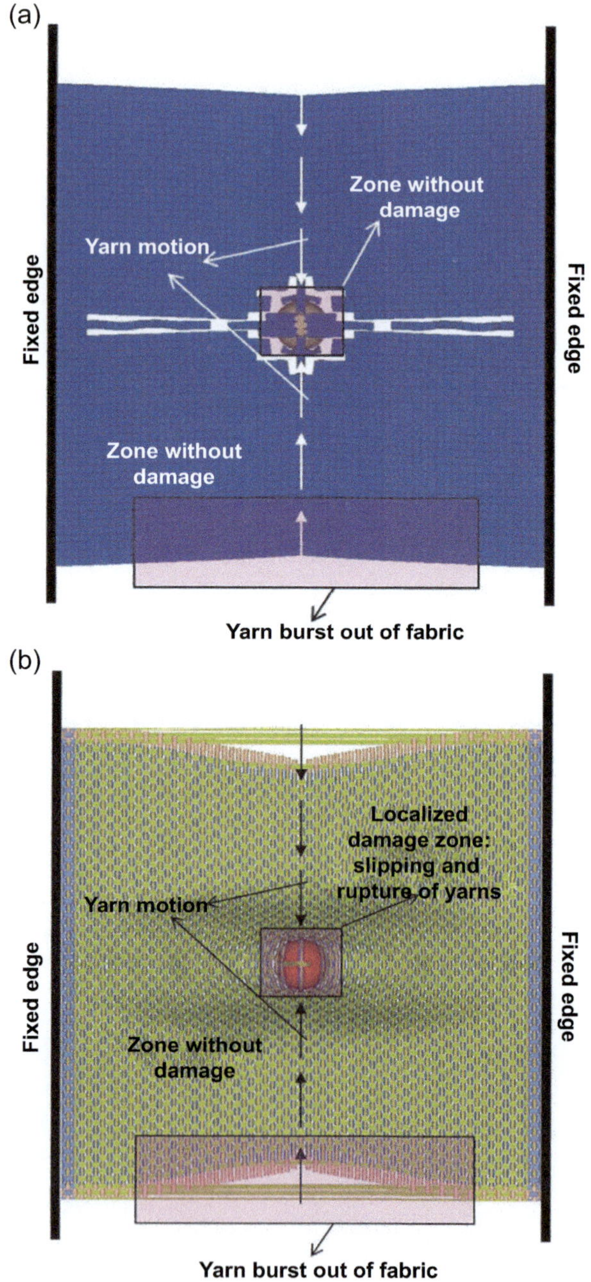

Figure 16.10 Different damage zones in the case of a perforation impact with velocity of 245.0 m/s: (a) macroscopic model; (b) mesoscopic model.

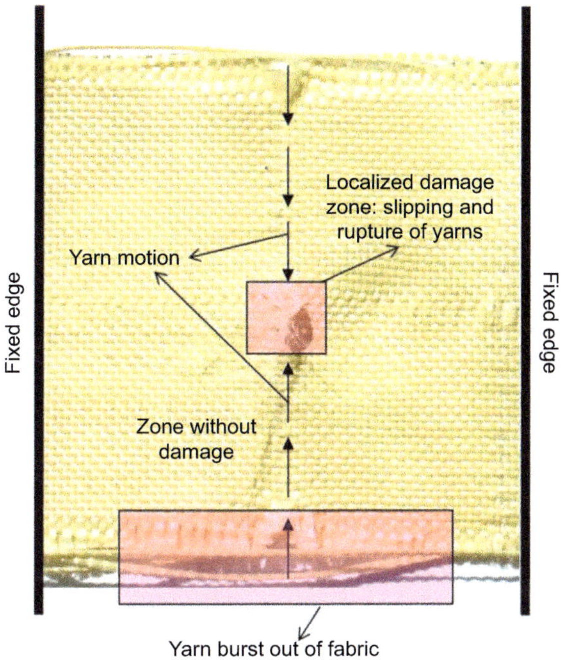

Figure 16.11 State of a plain-woven fabric with free surface of 5 × 5 cm after a 245 m/s impact.

- a strongly damaged and localized area (contact area between yarns and projectile) where primary yarns perpendicular to the free edges essentially slid on the surface of the projectile, and the ones parallel to these edges are essentially broken in tension due to the fixation at both ends; and
- four nondamaged areas far from the point of contact.

16.2.5.3 Comparison between macroscopic and mesoscopic models

Fig. 16.12 shows the variation of the projectile velocity at impacts of 60.6, 92.1, and 245 m/s in both the macroscopic and mesoscopic models. These curves can be divided into two different zones:

- The first zone starts from the initial instant of impact (at $t = 0$ μs) and ends as soon as yarn rupture initiates. This zone can be considered as the phase before rupture of yarn. The first zones in all numerical models presented in this work are parabolic. It is deduced that, in this period, the response consists of cone formation without yarn failure. The kinetic energy of the projectile is absorbed by the deformation of primary and secondary yarns, where primary yarns are those that pass through the contact zone with the projectile, and secondary yarns are those that lie beyond the contact zone, that is, the rest of the yarns. Initially, the velocity of projectile is dissipated slowly because only the decrimping or straightening of primary yarns contributes to energy absorption. Then, the deformation constituted in primary yarns causes the propagation of strain waves to the neighboring secondary yarns and the deceleration rate

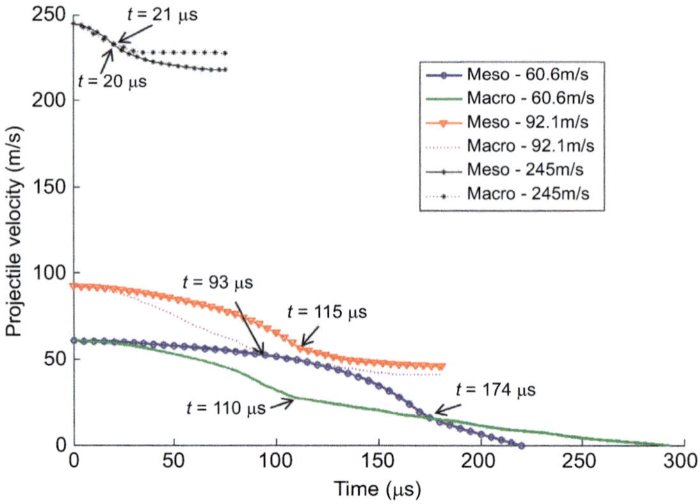

Figure 16.12 Projectile velocity evolution versus time for macroscopic and mesoscopic models at impact velocities of 60.6, 92.1, and 245 m/s.

of projectile increases. At the end of the parabolic zone, the deceleration of projectile velocity reaches its maximal peak because this is the point where the decrimping of primary yarns terminates and these yarns become completely tense. It can be also observed that at the parabolic period, strong decrease of the projectile velocity of the macroscopic model is earlier than the mesoscopic. In fact, there is no crimping of yarns in the macroscopic model. Hence, the fabric is submitted to tensile stress as soon as the presence of contact; the propagation of strain waves occurs immediately in this model. Comparing the couples of meso–macro curves in Fig. 16.12 corresponding to different impact velocities, it can be found that when the projectile velocity increases, in the parabolic period, the curve of the macroscopic model is closer to the one of the mesoscopic model. The difference between the two models on the first time of the appearance of yarn failure is also reduced (from the impact velocity 60.6–245.0 m/s). It means that for higher impact velocity, the mesoscopic model behavior is close to the macroscopic one. This result is in good agreement with experimental observations. In fact, in the higher impact velocity case, the time for strain wave propagation in the fabric decreases. With high impact velocity like that, only principle yarns contribute essentially to stop the projectile; the contribution of secondary yarns can be negligible. We can better understand this physical phenomenon with Fig. 16.13. Behaviors of both mesoscopic and macroscopic models come mostly from principle yarns in a zone of cross form.

- The second zone of velocity variation is a quasilinear curve. It begins with the initial rupture of elements or initial failure of numerically modeled primary yarns and slip of several primary yarns out of the contact zone. Due to being clamped at the two edges, primary yarns perpendicular to these edges are tensile and failed. When primary yarns fail completely, their impact resistance and contribution for the propagation of strain waves toward corresponding secondary yarns (the yarns perpendicular to them that are not fixed at both ends) are insignificant. That is why the deceleration of the projectile in this zone is slower than at the end of the first zone. This phenomenon can also be verified by making observations on the curve.

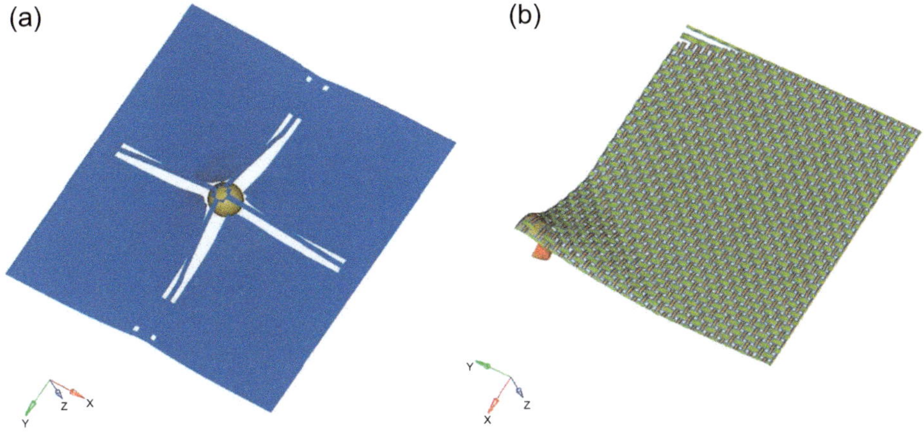

Figure 16.13 The deformed fabric configurations with high impact velocity of 245 m/s: (a) macroscopic model; (b) mesoscopic model.

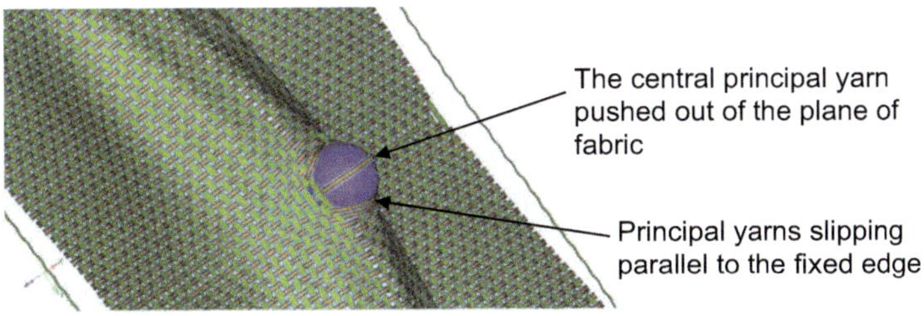

Figure 16.14 Localized observation on contact zone of the mesoscopic model with impact velocity of 60.6 m/s.

The curve has a lower slope in the second zone. The successive penetrations of the projectile are determined by two mechanisms: yarn slip out of contact zone and rupture of primary yarns. The rupture mechanism occurs essentially for primary yarns perpendicular to the two fixed edges, while the slippage is observed in other primary yarns parallel to these edges. This represents the physical behavior of yarns correctly as the yarns, which are under tension due to their boundary conditions, fail easily. At the impact velocity of 60.6 m/s, the projectile cannot perforate the fabric. The impact ends after continuous development of slippage and rupture of a few primary yarns. In this period, the projectile is in contact with the central primary yarn parallel to fixed edges (Fig. 16.14). The motion of primary yarns is depicted in Fig. 16.13. The central principal yarn has not slipped out of the contact zone. However, due to projectile penetration, this yarn is pulled out from the fabric plane (Fig. 16.14).

Calculated results of both models, macroscopic and mesoscopic, show a good prediction of residual velocity of the projectile. The fabric models can stop the projectile at 60.6 m/s and not at 92.1 and 245 m/s. Both models can determine residual velocities of the projectile after impact. Fig. 16.15 depicts calculated results for an impact velocity of 245 m/s to compare them with experimental work in [33].

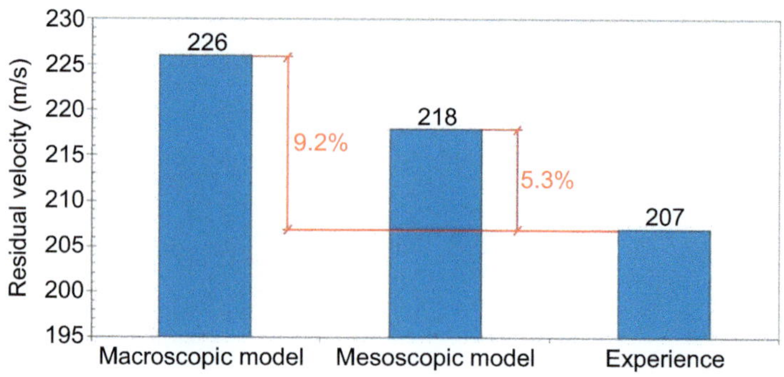

Figure 16.15 Comparison of the residual velocity between our numerical results and experimental data in Ref. [34].

16.3 Multiscale modeling for the cases of 2D woven fabrics

On the one hand, a mesoscopic model of the ballistic impact on a 2D fabric needs a large number of elements to detail crimps and elliptic cross sections of bundles. On the other hand, a macroscopic model that considers 2D fabrics as a homogenized plate can reduce the computation time, but it cannot present damage mechanisms of fabrics at the impact point. A combination of both models was used to develop a multiscale model to minimize the computation time. Fig. 16.16 shows the distribution of stresses of the mesoscopic model at different moments to be able to better understand the cross shape of the damaged zone. We can consider this cross as the principal work zone of a 2D woven fabric during the ballistic impact. Therefore, in the multiscale model, this zone needs to be retained at the mesoscopic level where bundles are described in detail (Fig. 16.17(b) and (c)) by a fine mesh. For the remaining areas where the bundle deformation is not major, it is convenient to replace bundles in detail by a homogeneous plate: a macroscopic zone using a large mesh. The multiscale model proposed in this study considers two zones of calculation: (a) a mesoscopic zone corresponding to the localized damage zone—strong slipping and rupture of yarns (Fig. 16.17(a)) and (b) a macroscopic zone corresponding to the zone without damage (Fig. 16.17(a)). Fig. 16.17(b) and (c) shows clearly the architecture of the proposed model. Fig. 16.18(d) describes the connection between the mesh of the macroscopic zone and the one of the mesoscopic zone. Note that, in this study, we used 3D shell elements for describing both the mesoscopic and macroscopic models. This element type has a certain thickness. However, in this figure, these 3D shell elements were presented by their neutral surfaces, and their thickness is not visible. The neutral surface of the macroscopic zone is placed at the center of the fabric thickness. Hence, due to the symmetry of the fabric, this surface cuts yarns in the mesoscopic zone to create boundary lines at the meso–macro interface (Fig. 16.17(d)). Therefore, at this interface, all of

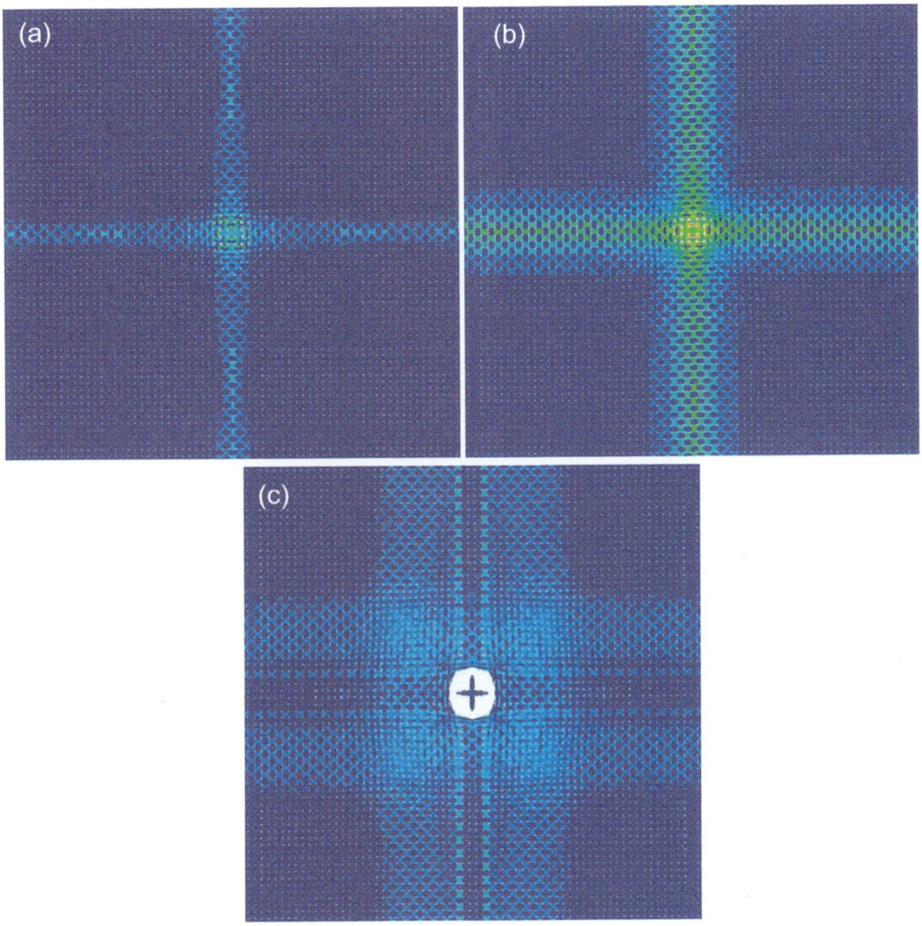

Figure 16.16 von-Mises stresses distribution on the fabric subjected to 245 m/s impact at (a) 9 μs; (b) 17 μs; and (c) 33 μs.

nodes of the mesoscopic zone can be pasted perfectly on the boundary of the macroscopic zone. It means that this bond cannot fail.

According to Drumheller (1998) [5,38], the transmission and reflection of a wave across an interface between two different linear elastic materials is governed by the ratio of their acoustic impedance, K. If $K = Z_1/Z_2 = 1$ where Z_1 and Z_2 are acoustic impedances of two different linear elastic materials, then the wave is transmitted across the interface; otherwise both reflected and transmitted waves are generated. In the case where there is no difference on contact area of two dissimilar elastic materials at the interface, Drumheller (1998) [5,38] proposed the generalized formulation of acoustic impedance for any elastic material:

$$Z = \rho c$$

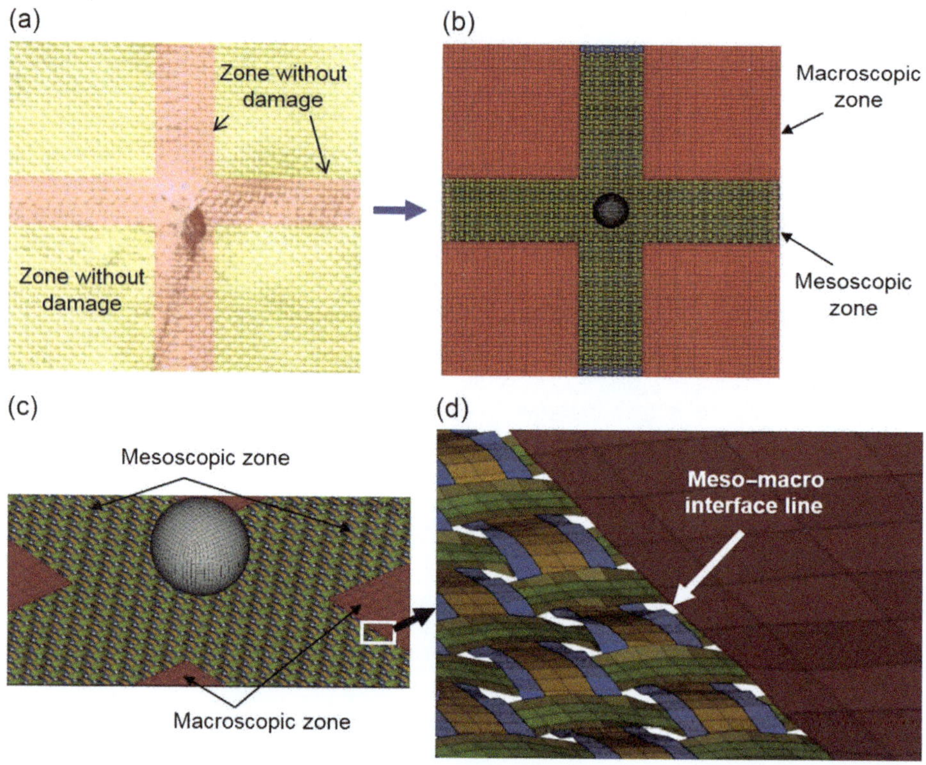

Figure 16.17 Definition of a multiscale model: (a) an image of the fabric after impact [23]; (b) a global view of a multiscale model; (c) a local view of a multiscale model at the impact point; (d) connection between mesoscopic and macroscopic zones.

where Z is the acoustic impedance of the cross section of material, and c is deformation wave velocity in the material. In a yarn, the deformation wave is propagated in the yarn longitudinal direction with the velocity [5,38]:

$$c = \sqrt{\frac{E}{\rho}}$$

where c is the deformation wave velocity in the longitudinal direction of a yarn; E and ρ are the Young's modulus and volume mass of the material, respectively. Hence, we have the following:

$$Z = \sqrt{E\rho}$$

In this study, at the meso–macro interface, the area of the mesoscopic part is different from the macroscopic one (Fig. 16.17(d)). The reason is that at the side of the mesoscopic zone, yarn undulation and gaps between yarns exit; while at the macroscopic side, it is a flat, continual, homogenous plate (Fig. 16.17(d)). Rao et al. (2009) [35]

proposed a new formulation of acoustic impedance through a cross section of a material to take account for this difference:

$$Z = A\sqrt{E\rho}$$

where A is the section area. With this new formulation, the smooth transition of the transverse wave across the meso–macro interface can be modeled by equating the acoustic impedances of the mesoscopic and macroscopic zones. Basing on this condition, an expression for the elastic modulus of the macroscopic zone is deduced. This influences the impact behavior of multiscale models that are discussed [38]. When the acoustic impedance is continuous across a junction between two macroscopic and mesoscopic areas, reflection of strain waves will be minimized at these joints. Therefore, the bundles will not fail early due to strain wave reflections at the macro–meso junctions.

In this investigation, we vary the ratio between macroscopic and mesoscopic areas to study the variation of the behavior of the fabric against impact and find out the optimal ratio. We study three ratios: 56.3–43.7%, 65.5–34.5%, and 75.3–24.7% with two impact velocities of 60 and 245 m/s (nonperforation and perforation). The mesoscopic model (ie, the 0.0–100.0% multiscale model) that showed a good agreement with experimental ballistic results is considered the reference to compare the precision of multiscale models with different macro-mesoscopic area ratios. Indeed, the multiscale models are a reduction of the mesoscopic model; therefore, in following, we need to verify their accuracy by a few important criteria.

16.3.1 Computation time of models

Fig. 16.18(a) and (b) illustrate the computation time of all models. We can observe that the calculation time is strongly reduced when comparing the multiscale models with the mesoscopic model. Among the multiscale models, the reduction of calculation time is slighter when the macroscopic area increases. The reason is that with a large macroscopic zone, the multiscale model can use a large mesh instead of a fine mesh. In that zone, yarn undulation is not presented; hence, the number of elements also can be decreased more regarding the mesoscopic model. Moreover, with only a homogenous plate in the macroscopic, the time to calculate contact between yarns does not exist; thus, the total calculation time of the multiscale models can reduce in comparison with the mesoscopic model. Therefore, absolutely, the multiscale models can save much time compared to the mesoscopic model. These figures also indicate that the computation time increases when the impact velocity diminishes. The reason is that impact phenomena take place faster when the velocity increases.

16.3.2 Validation of macro–meso interface continuity

To check the continuity between meso and macro areas, we assess the displacement at the nodes along a diagonal line from the center of the fabric (Fig. 16.19(a)).

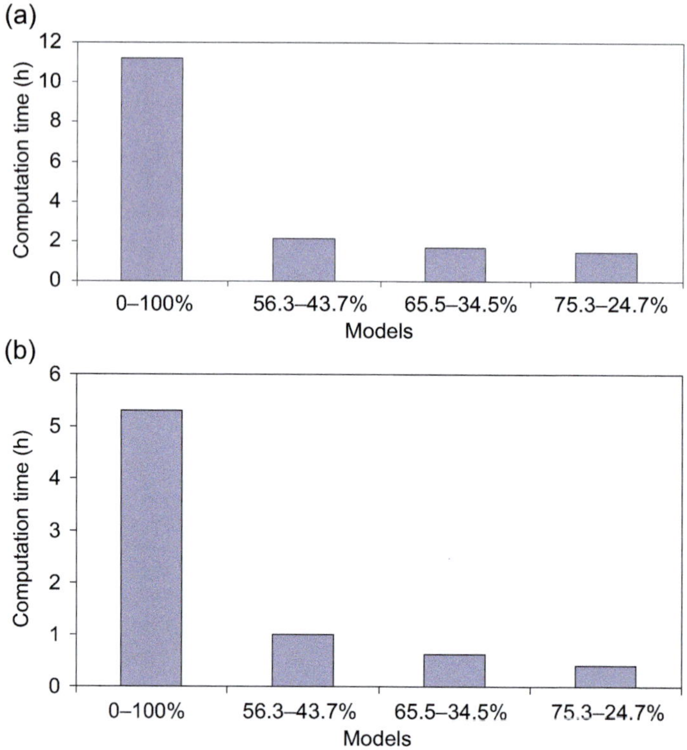

Figure 16.18 Computation time of models: (a) the impact velocity case 60 m/s; (b) the impact velocity case 245 m/s.

Fig. 16.19(b) shows the overall evolution of the displacement of nodes on the diagonal line for two different times in the event of an impact velocity of 245 m/s: before the rupture of bundles (17 μs) and just after the complete perforation (33 μs). Fig. 16.19(b) clearly shows the continuity of displacements at the macro–meso interface. These results validate the multiscale model chosen in this study. Globally, all curves have a similar global configuration at both instances 17 and 33 μs. This means that the multiscale models also assure the precision in computing displacement of the fabric and the continuity across the macro–meso junction.

In Fig. 16.19(b), we found a slight difference of the 75.3–24.7% multiscale model regarding the others at 33 μs in the impact velocity case 245 m/s. This difference comes from the failure of yarns, which we will discuss in the next section. We also can observe the two parallel parts of curves at associated moments $t_1 = 17$ μs to $t_2 = 33$ μs with a horizontal distance called $d = 7$ mm. It shows the propagation of the same displacement between two nodes on the diagonal line from t_1 to t_2. In the wave theory, this is the propagation of a transverse strain wave in the fabric. Therefore, it is possible to evaluate the propagation velocity of a transverse strain wave on the

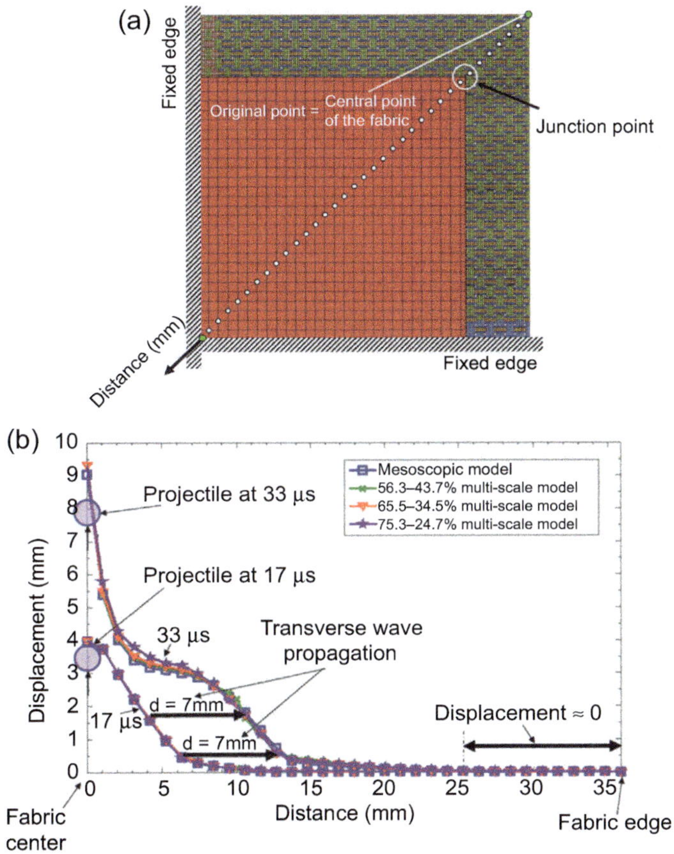

Figure 16.19 (a) Nodes on the diagonal line of a quarter of the fabric; (b) displacement of the nodes on the diagonal line of a quarter of the fabric impacted by the projectile of velocity of 245 m/s at 17 and 33 μs.

diagonal line of the fabric called u_d using the following formulation: $u_d = \frac{\text{distance}}{\text{time}} = \frac{d}{(t_2 - t_1)} = \frac{7\times10^{-3}\ \text{m}}{(33-17)\times10^{-6}\ \text{s}} = 437.5\ \text{m/s}$. This value can be compared with the results obtained by Barauskas and Abraitiene [36] for values of propagation velocity varying from 180 to 325 m/s. They calculated for a fabric of Twaron textiles CT709 with cross section 0.952 and 0.15 mm. The material density is 1440 kg/m^3, and the stiffness modulus in extension is 90 GPa. Their fabric is impacted by a 9-mm bullet at velocities in the range 270−300 m/s. On the other hand, this propagation velocity is certainly much slower than the one on the two warp and weft yarn directions. Hence, approximately 10 mm from the fabric edge (zone between 25 and 35 mm), the displacement of the fabric is always equal to zero, as shown by Fig. 16.19(b). It means that the transverse strain wave cannot reach this part of the fabric during the whole impact event.

16.3.3 Evolution of the projectile velocity

Figs. 16.20 and 16.21 show the evolution of the projectile velocity as a function of time obtained by the multiscale models and the mesoscopic model for two impact velocities 60 and 245 m/s. For the impact velocity case of 60 m/s (Fig. 16.20), curves of all multi-scale models give similar results to those obtained by the mesoscopic model. Generally, we can note that the evolution of the projectile velocity is slightly affected during the process of impact velocity case. The projectile is completely arrested by the fabric and then begins to rebound at around 150 μs for all models (Fig. 16.20). When the transverse strain wave is contained in the local zone (mesoscopic area), the multiscale models give the same evolution of the impact velocity corresponding to the first phase before around 80 μs (phase 1 in Fig. 16.20).

This figure shows also a slight difference between the results of the mesoscopic model and the three configurations of the multiscale model in the phase from 80 to 150 μs. In this phase (phase 2 in Fig. 16.20), the projectile velocities obtained by the mesoscopic model are always higher, and stopping of the projectile occurs later

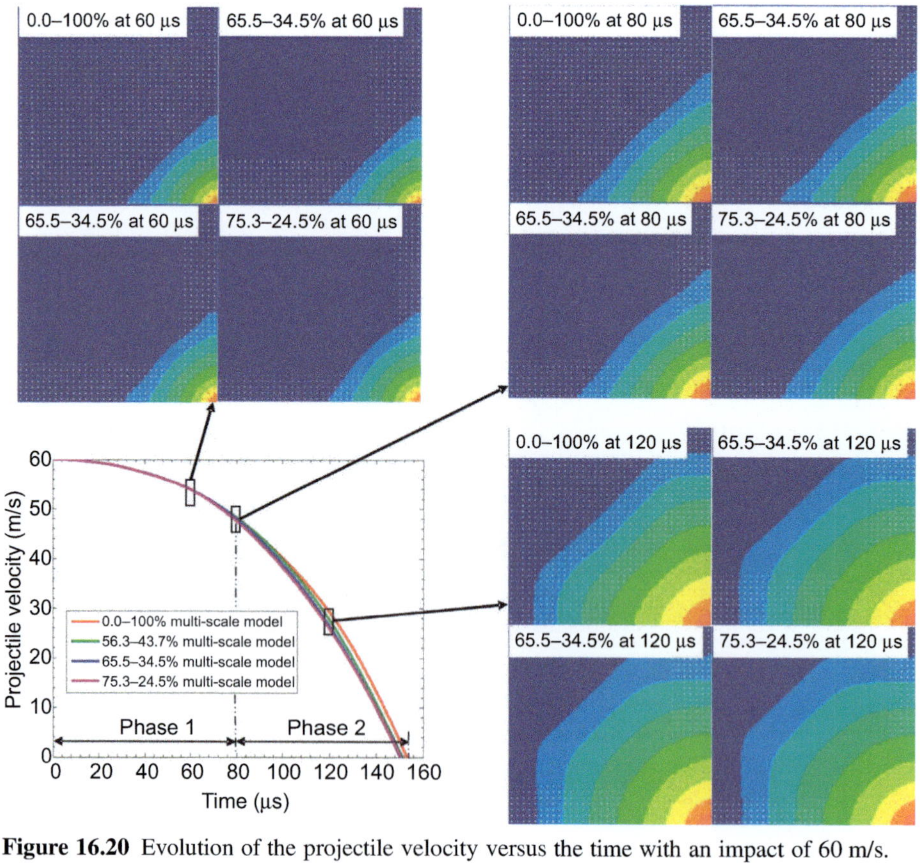

Figure 16.20 Evolution of the projectile velocity versus the time with an impact of 60 m/s.

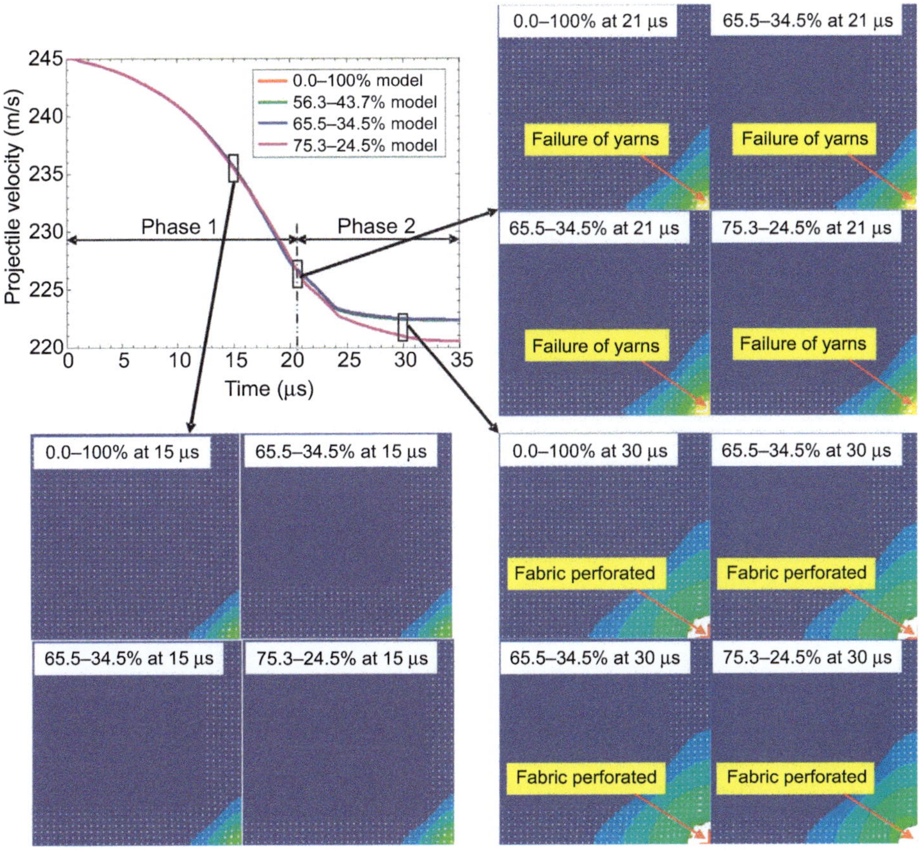

Figure 16.21 Evolution of the projectile velocity versus the time with an impact of 245 m/s.

compared with predictions of multiscale modeling. We know that the mesoscopic model where all of the bundles in the fabric are detailed or the area of localized mesoscopic zone is 100% over the fabric and the macroscopic one is 0%. In contrast, the multiscale models have a certain area where the detailed bundles are replaced by a homogeneous macroscopic plate. It is clear that the homogenous structure of the plate is more rigid than the one of the mesoscopic model with separated bundles. This explains the slight difference in behavior between the two multiscale and mesoscopic models. Hence, we also understand why the deceleration of the projectile increases with the development of the size of the macroscopic area as observed in Figs. 16.20 and 16.21. Fabric resistance increases, and its ballistic performance is improved.

Fig. 16.21 shows that the evolution of all curves in the case of velocity impact 245 m/s can be divided into two phases: before and after the rupture of yarns. In the first phase (phase 1, 0–21 μs), all of these curves seem to be the same because the strain wave front is only developed in the mesoscopic area. In this phase, the effect of the macroscopic zone is therefore very small (Fig. 16.21). We can also observe here that with a high impact velocity like 245 m/s, the projectile can fail yarns and perforate the fabric so rapidly that the transverse strain wave cannot yet considerably propagate on the macroscopic area. Note that for the impact of 245 m/s, the result of the multiscale model of 75.3−24.7% is significantly different from other multiscale models after around 21 μs corresponding to phase 2 (Fig. 16.21). It can indicate that when the rupture of bundles appears due to high impact velocity, the effect of the macroscopic zone becomes more important for the accuracy of multiscale models. The reason is that in the case of a high-speed impact like 245 m/s, bundles begin to fail at the impact point (ie, elements are removed here) and a local perturbation is initiated. In this case, the initial time of the rupture of the bundle is different from this model to other ones because each multiscale model has its own macroscopic area, and therefore, it has an associated resistance. The initial moment of rupture of bundles is very important; it decides the following failure processes of bundles. Therefore, the deviation on the initial moment of rupture causes the difference on the evolution of projectile velocity among multiscale models. This difference is greater when the macroscopic area of the multiscale model increases.

Globally, all configurations of transverse strain wave propagations are identical for the multiscale and mesoscopic models in both impact velocity cases: 60 and 245 m/s (Figs. 16.20 and 16.21). Again, this indicates that macro−meso interface is well modeled; the reflection of strain waves seem to be insignificant here.

16.4 FEM modeling for the cases of 3D woven fabrics

16.4.1 Numerical tool for geometrical modeling of 3D woven fabrics

Numerical studies for the design and production of the 3D weave structures and the associated composite have been limited. The structural complexity of 3D interlock woven fabrics can explain this limitation. Therefore, there has not yet been any software that permits correctly representing the geometry of all 3D interlock woven fabrics.

In fact, there were a lot of works and associated programs or softwares to model geometrically woven fabrics [44]. TexGen and WiseTex were the best known softwares in these works. TexGen software was developed by a group of researchers of the University of Notthingham. This software supposes that the cross section of yarns has a constant form in certain fabrics; section forms can be circular, elliptic, lenticular, etc. Actually, it has passed three versions: v1, v2, and v3. The third version was much improved versus the two preceding versions. In version 3, yarn paths are defined by an interpolation of a series of points measured from analyzing micrographic images. In theory, TexGen permits us to construct any woven fabric in two ways. In the first way, we can use a graphic interface to model fabric geometry in TexGen, but the user can only introduce a few parameters in the model: number of warp yarns, number of weft yarns in a unit cell, thickness of the fabric, width of cross section of yarn, and distance between yarns. Sometimes, even with input dimensions experimentally

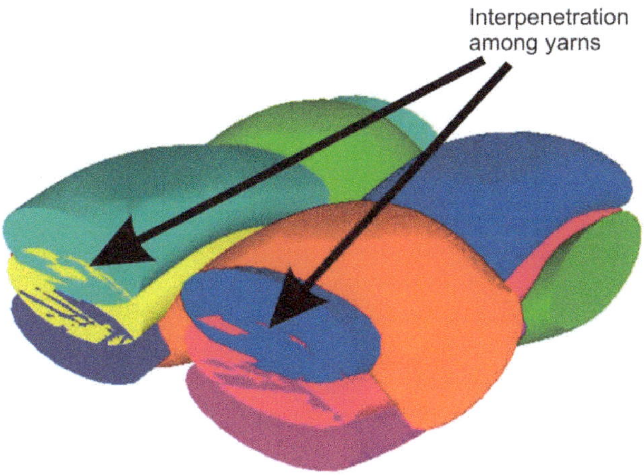

Figure 16.22 A geometrical model of a 3D orthogonal fabric of three layers constructed by TexGen.

measured, interpenetration appears between yarns (Fig. 16.22) that the user cannot modify any more.

In the second way, we can also use this software as a programmer; algorithms can be modified for correcting errors such as interpenetration between yarns. However, certainly, this work is not easy, and it demands a good knowledge of the Python language. This software is in progress of development; nowadays, it is essentially used for representing the geometry of textile fabrics. Applying this software to finite element calculations has not been verified yet.

WiseTex is the product of a group researchers led by Stephan Lomov, professor at Leuven University. This software was created for modeling the geometry of 2D and 3D woven fabrics, UD performs, 2D braids with and without inlays, and multiaxial multi-ply warp-knit stitched performs [44–49]. Similar to TexGen, the cross section of yarns is assumed to be constant along the length of yarns and to have a certain form with characteristic dimensions (height and width, etc.). The input parameters of this software are structure type, yarn densities in warp and weft directions, and characteristics of yarns (geometry, yarn mechanical properties in tension, flexion, compression, etc.). The stable state of a fabric is considered as balanced; hence, the internal energy of a fabric is at the minimum point. Therefore, yarn paths are determined by contact conditions between yarns and minimizing of the internal energy. According to this theory, WiseTex can create correct geometric models of any woven textile fabric without interpenetration between yarns; however, this problem still can happen (Fig. 16.23).

In general, we can summarize that the current softwares in the literature still have certain different limitations:

- only for a few types of fabric structures
- difficulties in use because users cannot correct the errors of the geometrical models immediately on the graphic interface of these softwares

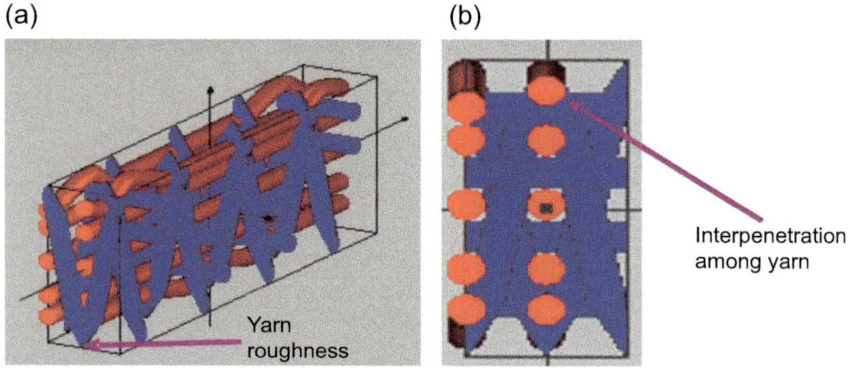

Figure 16.23 A geometrical model of a 3D orthogonal fabric of five layers constructed by WiseTex: (a) 3D view; (b) side view.

- interpenetration between yarns of a fabric in the geometrical models
- incompatibility between these softwares and the finite element codes

This section presents a tool named Ktex_pattern that permits the generation of a geometry model of any woven fabric. Then it allows the meshing of this model in shell or solid elements and applying mechanic properties of yarns to them. These operations can be easily done on a graphic interface in a few steps. This tool has many advantages over the current softwares:

Tool applied for all textile woven fabrics

- smoothness of yarns and exclusion of interpenetration among yarns
- compatibility with commercial finite element codes: Radioss, Ansys, Abaqus, etc.
- facility in use thanks to a control by graphic interface and menu with numbering and coloring yarns

This tool uses two assumptions: weft threads are straight in fabric structures, and yarn cross section is constant along the central geometrical line of yarns (Figs. 16.24–16.26).

Input parameters of this tool consist of the architecture schema, the yarn density in weft and warp directions, the number of layers, and the shape and dimensions of yarn

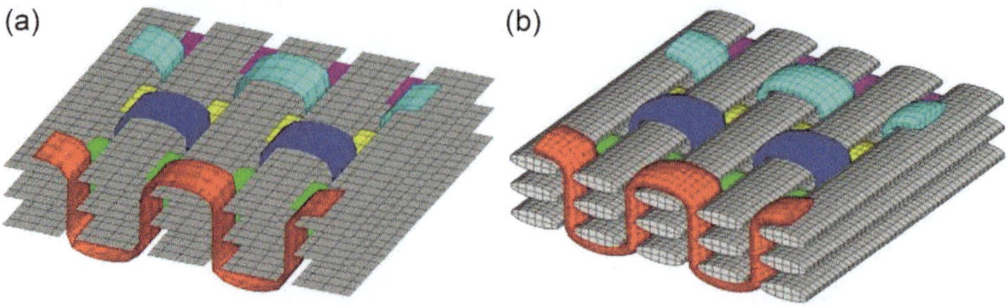

Figure 16.24 Orthogonal interlock woven fabric meshed with (a) shell elements; (b) solid elements.

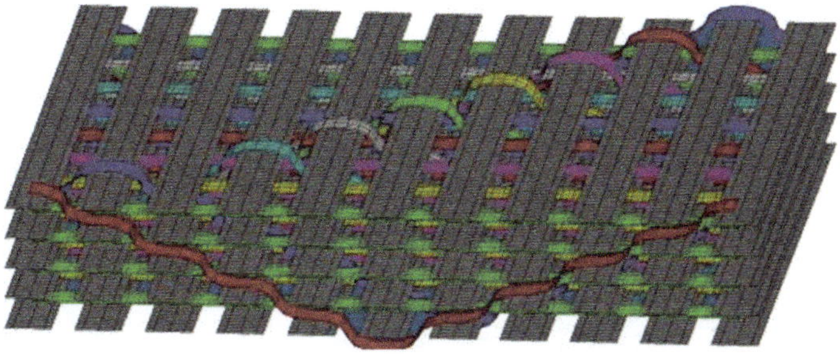

Figure 16.25 Angle−through the thickness interlock woven fabric meshed with shell elements.

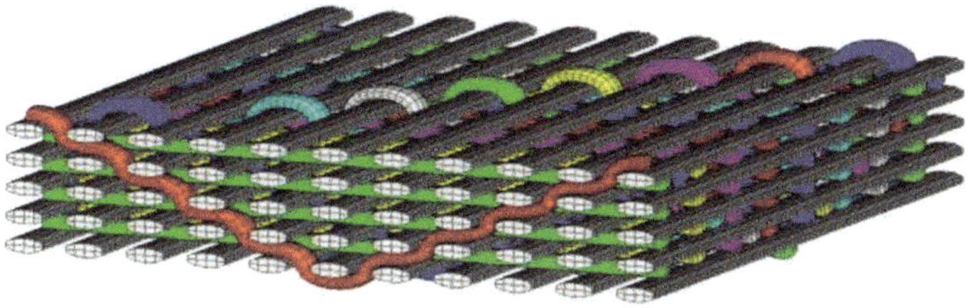

Figure 16.26 Angle−through the thickness interlock woven fabric meshed with solid elements.

cross section. In this tool, different cross section shapes of yarns are available, such as elliptic, circular, and lenticular. Based on the two assumptions previously mentioned and input parameters, this tool employs a special algorithm to calculate the position and path of yarns in any woven fabric. This algorithm is constructed to ensure that yarn paths are regularly smooth. With this algorithm, interpenetration at the contact between yarns does not exist anymore (Figs. 16.24−16.26).

Using this tool, yarns are created in a geometrical model where users can arbitrarily select the dimension of the fabric. However, a fabric is periodic; hence, the user is advised to create a geometrical model only for a basic cell. This technique helps the user to easily control the geometrical models at the level of a basic cell. Once the geometrical model of a basic cell is correct, then this basic cell can be multiplied many times until the user obtains the desired complete fabric.

After the geometrical modeling of fabric, each yarn is discretized into shell or solid elements whose size users can decide for finite element analyses. Elements are created and arranged in groups or different components corresponding to types of yarn and characteristics of element and material. These components are named and numbered logically (Fig. 16.27). This advantage helps users to control models easily in applying loads and contacts for yarns in popular finite element softwares such as Ansys, Abaqus, and Hypermesh. In particular, elements are created so that the local coordinate axes of elements are oriented immediately along the centerlines of yarns (Fig. 16.28). It favors defining a material law for elements of a yarn because the material properties of yarn are constant along yarn length.

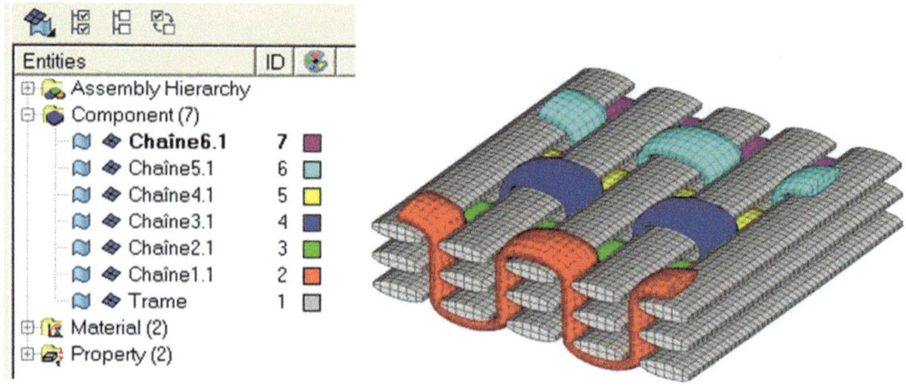

Figure 16.27 Organization of groups of yarns and elements with the new tool.

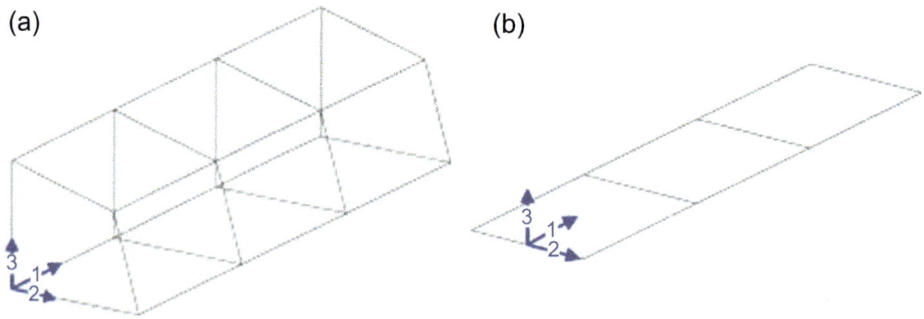

Figure 16.28 (a) Directions of local axes of a solid element in a yarn; (b) directions of local axes of a shell element in a yarn.

This tool is written in Tcl (Tool Command Language). Hypermesh is the actual work environment of this tool. However, with a few modifications of this tool in Tcl, the user can work with it on the graphic interface of another finite element software such as Abaqus or Ansys. Hence, this helps to avoid errors concerning incompatibility between softwares as has been seen when geometric models are imported from other softwares (for example, WiseTex or TexGen). The reason is that current tools miss the unification from creating a geometrical model to constructing the associated finite element one. The new tool was programmed based on this unification. With Hypermesh and Radioss, this tool created a convenient unified numerical program to construct the finite element models of woven fabrics. After meshing geometric models, users can modify normally the properties of finite elements, the material laws for the components, by using functions of Hypermesh. From Hypermesh, a finite element model of fabric can be exported to a file under different formats compatible with other popular codes such as Ansys, Abaqus, and Radioss. A contact between yarns will be automatically created in these exported files.

16.4.2 Numerical modeling of impact behavior of a 3D woven fabric

The last section used the K_tex pattern tool to create and to mesh a geometrical model of a 3D four-layer angle-interlock structure with two orthogonal yarns, weft and warp yarns, (Fig. 16.29) at mesoscale. In the fabric plane, the density of warp yarns is 20 ends/cm and 27.8 picks/cm for weft yarns. All yarns are Twaron 3360 dTex that induces a surface density of 1.66 kg/m^2 for this fabric.

In general, the shape of a yarn cross section can be considered elliptical (Fig. 16.29). A section of weft yarns has a greater height but a smaller width in comparison with warp ones. In fact, even between only warp yarns or between only weft ones, the difference of cross sections is clearly visible. To simplify, in the modeling, we consider that warp yarns have the same elliptical cross section whose height and width are taken equal to the average dimensions of all sections of warp yarns. With the same method, we can obtain the average size for a cross section of weft yarns.

In this case, as discussed earlier, 3D shell elements are used to model the 3D fabric (Fig. 16.30). A cross section of yarns is divided into four elements. The width b and the height h of each section are taken as in reality for assuring physical contacts between yarns.

In Fig. 16.29, we note that weft yarns are arranged in regular columns. These yarns also seem to be right in Fig. 16.29(b). However, warp yarns approach from each other to fill gaps. This is the result of weaving process; yarns are moved to reach the steady

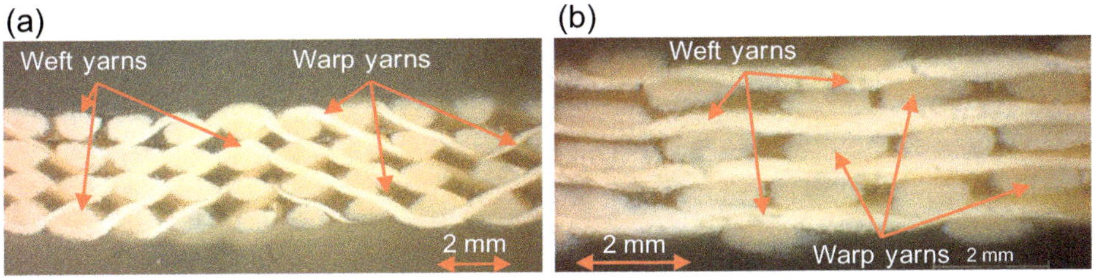

Figure 16.29 Microscopic images of sections perpendicular to (a) weft yarns; (b) warp yarns.

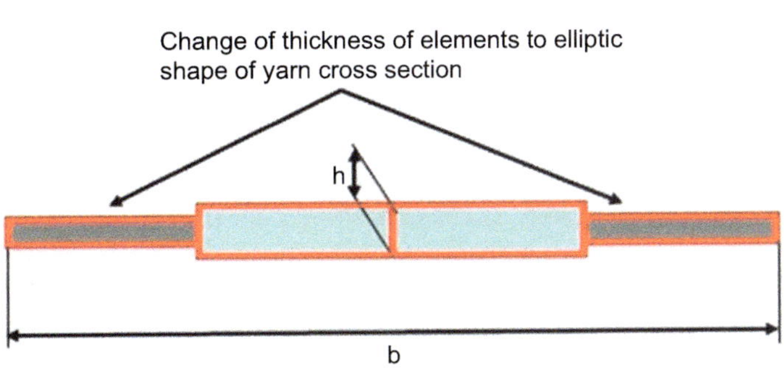

Figure 16.30 Modeling of the cross section of yarns in a 3D woven fabric.

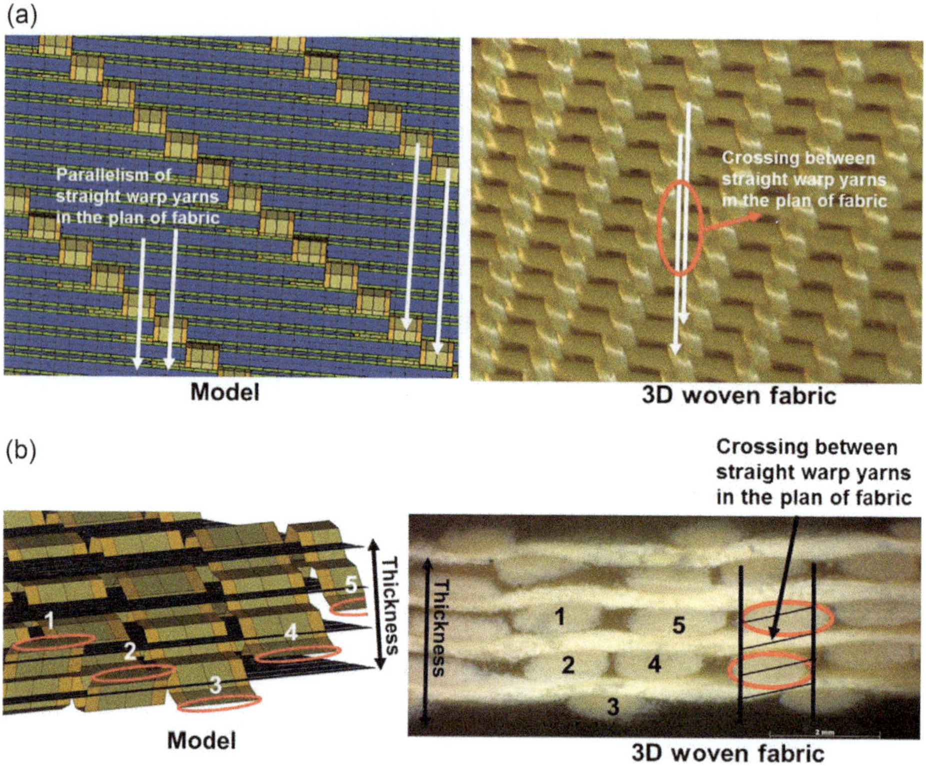

Figure 16.31 Comparison between the geometry of numerical model and reality: (a) in the plane of the fabric, (b) in the cross section orthogonal to warp yarns.

state. Fig. 16.31 compares the geometry between the numerical model and reality. In fact, the model geometry is ideal; weft yarns are always right and do not intersect each other in the plane of the fabric. So, with weft yarns, the difference between the model and reality is slight, but it is considerable for warp yarns. In the model, although the distance between warp yarns is close to zero, the density of these yarns is still significantly lower than that of reality (6.25 yarns/cm regarding to 20 yarns/cm). However, when taking the density of a yarn equal to the one of a fiber (1440 kg/m^3), the model has a significantly higher surface density than the reality (2.8 kg/m^3 compared to 1.6 kg/m^3). The reason is that the model considers a yarn as a homogeneous material. Therefore, the gaps in a yarn are taken into account in the fabric weight modeled. To solve these problems, we take the density of shell elements equal to that of fibers to ensure the physical velocity of the longitudinal wave of fibers. The area of transverse sections of warp and weft yarns is changed so that the surface density of the fabric is the same between the model and reality. We only modified thicknesses of the two outer elements to suit the equal condition of surface density of the fabric in comparison with reality (Fig. 16.30).

Fig. 16.32 shows the numerical model to validate the ballistic tests on 3D fabric. The size of the fabric is identical to reality: 24 × 26 cm. Since the geometry of this

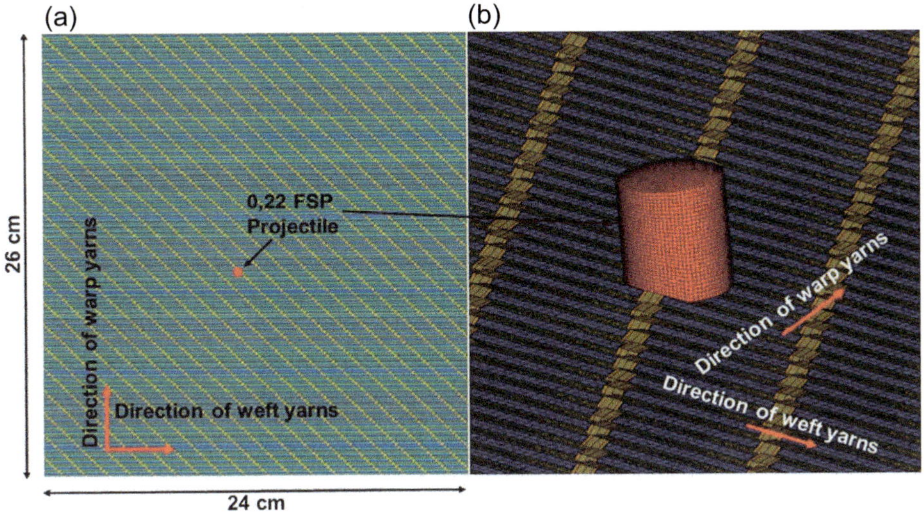

Figure 16.32 Configuration of the numerical model of a 3D fabric subjected to ballistic impact: (a) overview; (b) mesh detail.

Table 16.1 Mechanical properties of 3360 dtex Twaron yarn

State	Longitudinal elastic modulus E_{11} (GPa)	Failure strain ε_R (%)	Transversal elastic modulus E_{22} (GPa)	Shear moduli (GPa)			Poisson's coefficient
				G_{12}	G_{23}	G_{13}	
Static	47.7	4.7	1.34	24.4	0.001	0.001	0.6
Dynamic	225.0	2.2	1.34	24.4	0.001	0.001	0.6

fabric is not symmetrical, we have to calculate a complete model. We assume the following:

- The fixation of fabric is ideal.
- The projectile is infinitely rigid.

The friction coefficient is taken equal to 0.25 for the contact between yarns, and 0.2 for those between yarns and projectile. Based on the experimental work [5], the static and dynamic mechanical properties of yarn are incorporated into the model (Table 16.1). Other mechanical components as the transverse modulus, the shear moduli, and Poisson's ratio are taken equal to those of experimentation over Kevlar® KM2 yarns [5] due to lack of available values.

16.4.2.1 No perforation impact

Fig. 16.33 illustrates the comparison between the two numerical models of impact at 306 m/s and the experimental observations. The results show that the projectile cannot

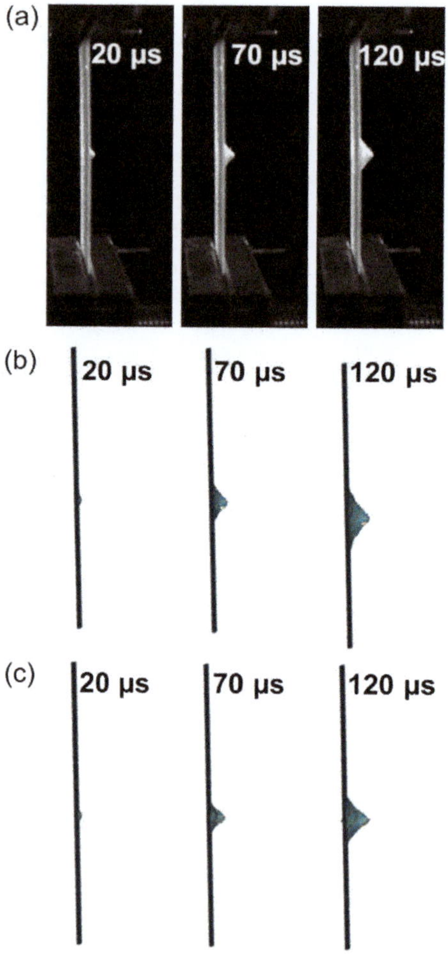

Figure 16.33 Comparison in the impact case of 306 m/s with a side view between: (a) experiment, (b) model using dynamic properties, (c) model using static properties.

penetrate the fabric up to 120 μs. Formation of a deformation pyramid is the unique mechanism observed between 0 and 120 μs. It seems that the developments of the pyramid to the two edges of the numerical model using the dynamic properties and the experimental test are identical and larger than the case using the static values. We observe that the pyramid top of three results has almost the same distance from the plane of fabric. This is why the pyramid of the numerical model using the static constants sloped more slightly than the others.

Fig. 16.34 shows a view from the back surface of the fabric during experimentation and the two numerical models corresponding to the use of dynamic and static constants in the impact case of 306 m/s. The two numerical models are presented under a velocity contour to show more clearly the propagation of a longitudinal deformation wave

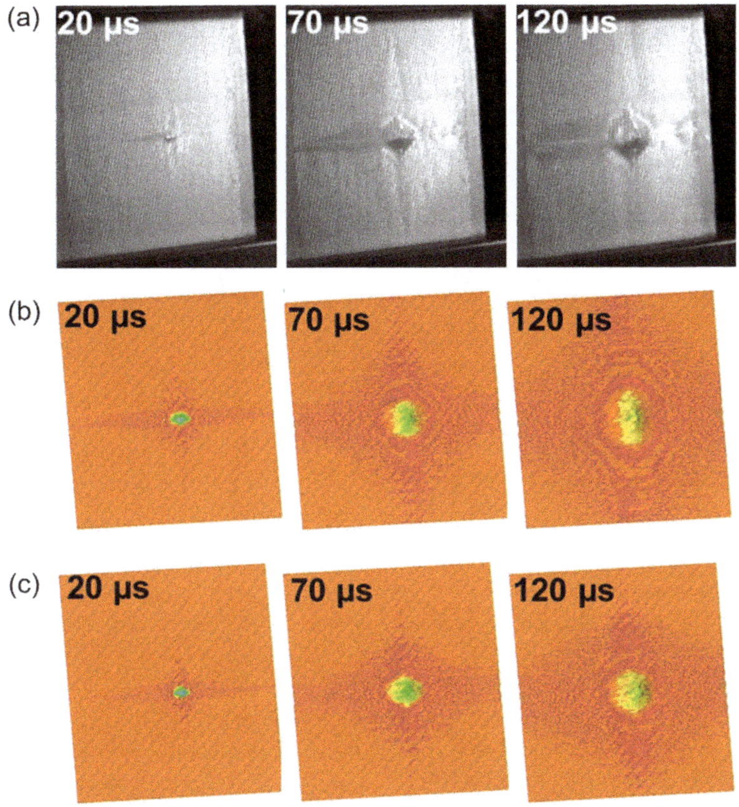

Figure 16.34 Comparison in the impact case of 306 m/s with an opposite view from behind: (a) experiment, (b) velocity contour of the numerical model using dynamic properties, (c) velocity contour of the numerical model using static properties.

on the 3D warp interlock fabric during impact. It can be noted that the propagation of longitudinal deformation waves is always faster in the weft direction than in the warp one for all three results. The reason is that yarns are more crimped in the warp direction. Therefore, strain waves in the case of the experiment and the numerical model using the dynamic properties can reach the two edges at 70 μs in warp direction. On the other hand, in the case of the numerical model using the static values, these waves propagate more slowly.

At 120 μs, for the experiment and the numerical model using dynamic constants, the front of the longitudinal wave is being spread along the edges, while for the model using static constants, the longitudinal wave front has just reached the two edges. Prediction of propagation of deformation waves is considerably improved in the case of the numerical model using the dynamic properties. It should be noted that the dynamic Young's modulus is equal to 225 GPa, and the static one is 49.7 GPa.

Fig. 16.35 shows the evolution of experimental displacement of the top of a deformation pyramid and those of models using dynamic and static constants with the same

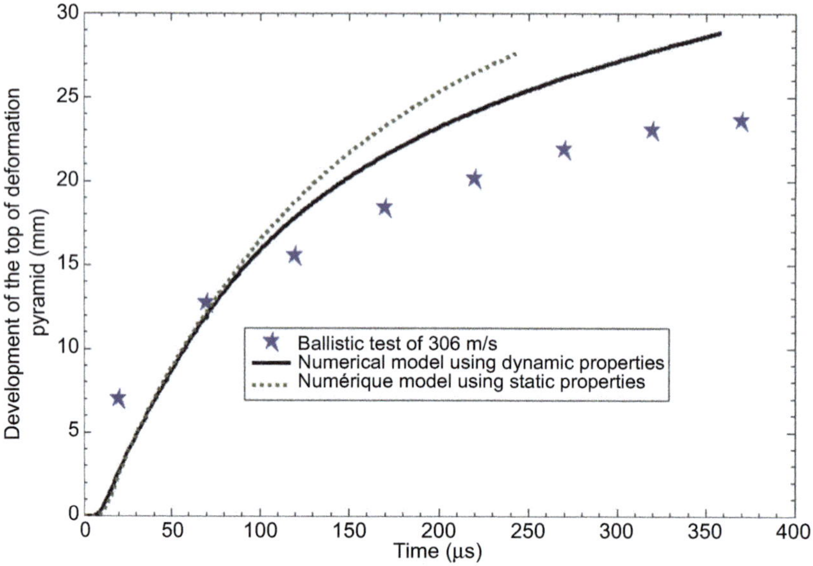

Figure 16.35 Comparison in the case of nonperforating impact on the development of the top of a deformation pyramid between experience, the model using dynamic properties, and the model using static properties.

impact velocity. In this case, the projectile cannot penetrate the fabric; therefore, the motion of this top is equivalent to that of the projectile. It should be noted that all three results are identical during the first 100 μs. After this period, the projectile moves faster in the case of the numerical model using the static constants.

Note that during the first 100 μs, there is a difference on propagation velocity of deformation between the numerical models and the experiment. However, the effect of this difference on the projectile displacement is still low because of the area where the waves have not yet propagated significantly in this period.

After 120 μs, the deformed zone becomes larger, and the effect of the propagation velocity of waves on the projectile displacement is more visible, as shown in Fig. 16.35. On the other hand, we can also observe a difference between the numerical model using the dynamic parameters and the experience at about 370 μs. The reason can be attributed to the geometry of the numerical model that is not perfectly realistic.

Figs. 16.36 and 16.37 show developments of a deformation pyramid in the warp and weft directions for the three cases: the experiment and models using dynamic and static parameters with an impact velocity of 306 m/s. Fig. 16.36 indicates that the numerical model using the dynamic properties is in good agreement with the experimental result describing the evolution of pyramid deformation in the warp direction. On the other hand, both results are significantly different in comparison with the numerical model using the static constants. Again, we emphasize that the development of pyramid deformation of the numerical model using the static mechanical properties is always slower in the warp direction regarding reality. Dependence of the propagation velocity on the Young's modulus of yarn can explain this phenomenon.

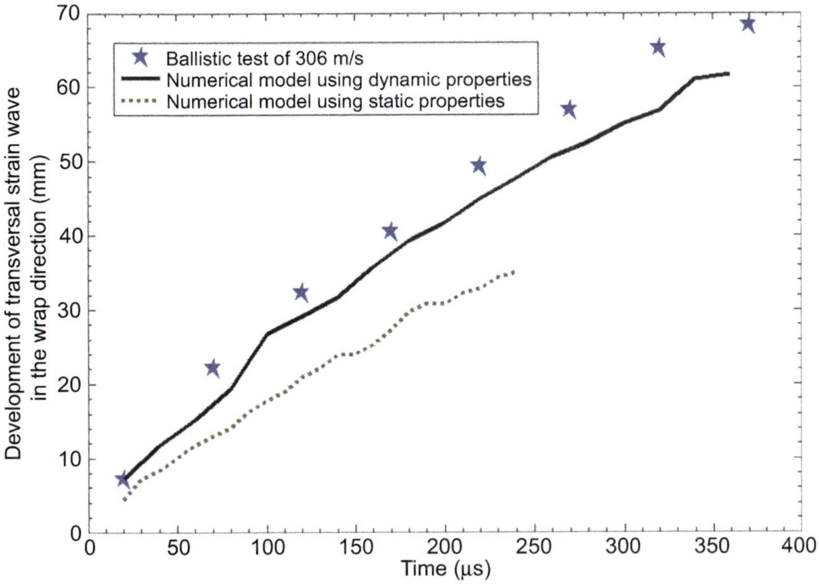

Figure 16.36 Comparison in the case of nonperforating impact on the development of a transversal strain wave in the warp direction between experience, the model using dynamic properties, and the model using static properties.

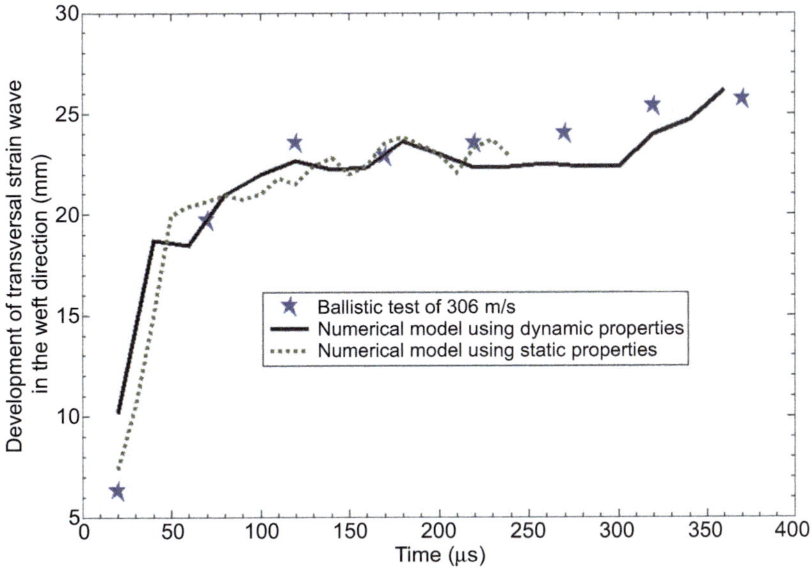

Figure 16.37 Comparison in the case of nonperforating impact on the development of a transversal strain wave in the weft direction between experience, the model using dynamic properties, and the model using static properties.

Regarding the spread of the transverse wave in the weft direction, Fig. 16.37 shows that the difference between the three results is low. A deformation pyramid develops inevitably in the weft direction during the first 50 μs. After 50 μs, the size of the pyramid in this direction does not vary significantly. Note that weft yarns are not fixed at two of their edges. Therefore, they are pulled to the impact point with advancing motion of projectile. This restricts the development of the pyramid in the weft direction.

Comparing Figs. 16.36 and 16.37, it is interesting to note that the transverse strain wave travels faster in the warp direction than in the weft one, although warp yarns are curved and the weft ones are quasi-straight. In fact, the two ends of warp yarns are clamped, while free for the weft ones. Therefore, materials can move more easily in the weft direction with a projectile. This phenomenon causes a certain delay of the transversal wave in this direction compared with warp direction.

16.4.2.2 Perforation impact

Fig. 16.38 illustrates the comparison between the two numerical models of 400 m/s impact and images taken by a camera in a side view. Note that the exact moment of impact is difficult to determine exactly from data of the high-speed camera because the time step of 16 μs is not fine enough. So, we calibrate numerical results with experimental ones for a common time of impact. We consider this instant as the reference (0 μs). In general, we observe that the projectile completely perforates the fabric at 70 μs for two cases: the model using the dynamic properties and the experimental test. However, the projectile is still stuck between yarns for the model using the static properties. Thus, to 120 μs, the projectile cannot displace as far as the other cases. Also, the size of the deformation pyramid in the warp direction is greater for this case. The reason is that the projectile has more time to penetrate the fabric.

Fig. 16.39 shows the impact at 400 m/s from both numerical models and experience in a view on the back face of the fabric. In this figure, the numerical results are described under contour of displacement. We can observe fully the development of the deformation pyramid. From 0 to 70 μs, this pyramid propagates mainly in weft direction. The reason is that the undulation of weft yarns is significantly low. Thus, these yarns more easily follow projectile motion because they are not fixed at the two edges. This displacement leads to damages in the area near the free edges.

Fig. 16.40 details evolution of projectile velocity of both numerical models and experience in the case of 400 m/s velocity impact. We can observe that the two curves related to models using dynamic and static properties are identical during the first 15 μs. After this period, in the case of dynamic properties, a projectile could pass through the fabric by breaking primary yarns (which are in contact with projectile), and the projectile begins to decelerate slowly. Perforation occurs in 50 μs with a residual velocity equal to 305 m/s, close to the experimental value (303 m/s). If using static properties, the projectile velocity continues to decrease after 15 μs because the projectile is still stuck within some unbroken yarns. In fact, with a large failure strain ($\varepsilon_R = 4.7\%$ higher than the dynamic case where $\varepsilon_R = 2.21\%$), yarns seem to be more difficult to fail for 400 m/s impact, which leads to a residual velocity of 232 m/s, very low compared to the experimentally determined value.

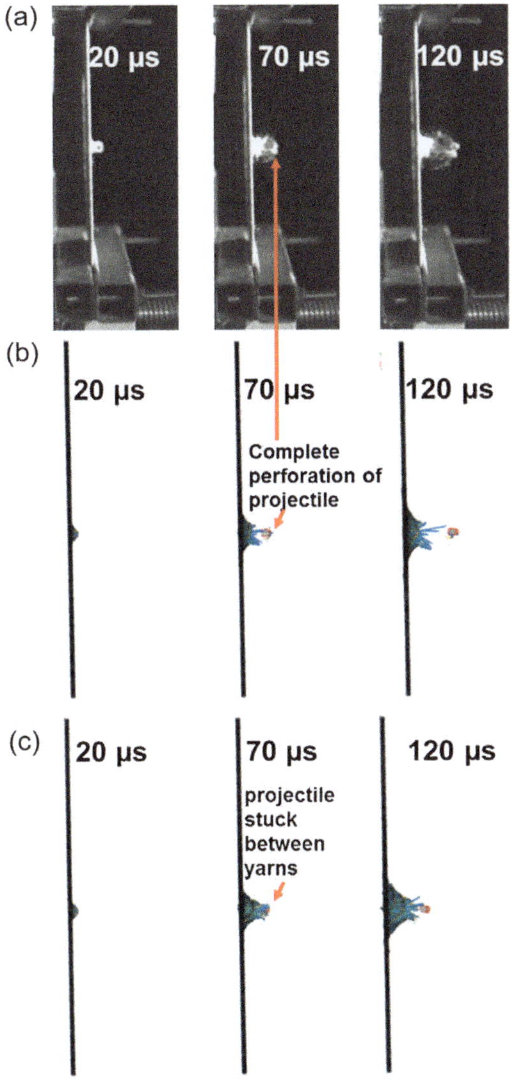

Figure 16.38 Comparison in the impact case of 400 m/s with a side view between (a) experiment, (b) model using dynamic properties, and (c) model using static properties.

Fig. 16.41 shows configurations at the impact point in the cases of the experimental test and numerical models at 54 μs. Fig. 16.41(a) indicates that all primary warp and weft yarns have failed. In fact, weft yarns are free at the two edges; they are only held by intersections with the others. Due to these constraints, a high-velocity impact can break them. In Fig. 16.41(b) and (c), we observe the failure of all primary yarns in the case of the model using the dynamic constants, while the projectile is still prevented by two warp yarns in the case using the static constants.

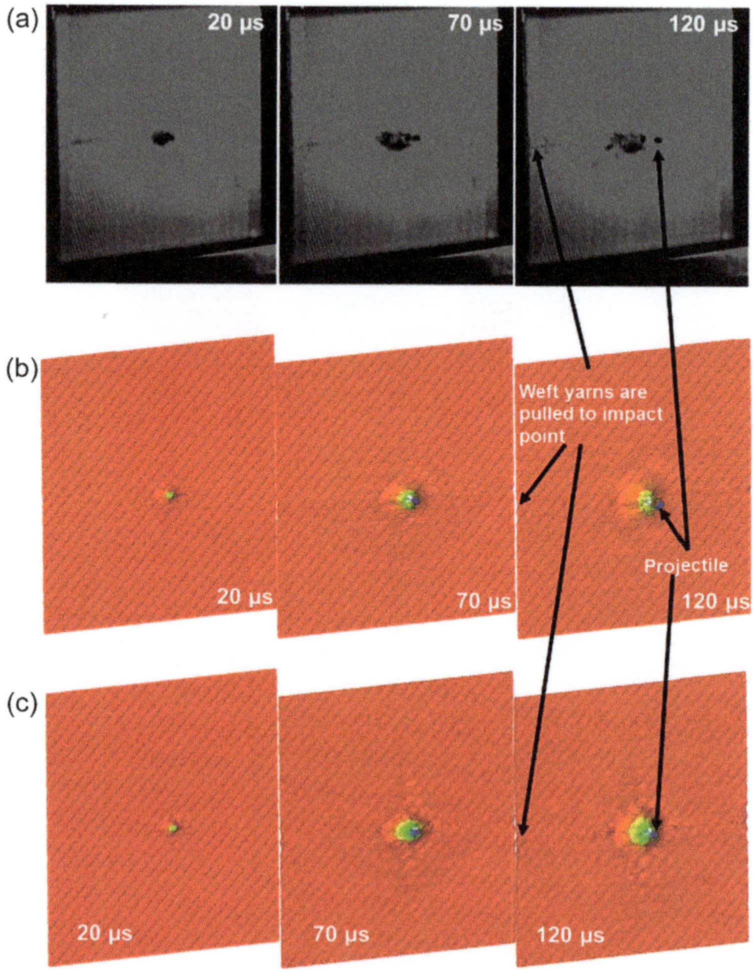

Figure 16.39 Comparison in the impact case of 306 m/s with an opposite view from behind: (a) experiment, (b) velocity contour of the numerical model using dynamic properties, (c) velocity contour of the numerical model using static properties.

Figs. 16.42 and 16.43 indicate the numerical and experimental results concerning the evolution of the formation of the pyramid in both warp and weft directions. Overall, we note that the model using the dynamic properties seem in agreement with experimental results for both figures. In the case of the numerical model using the static constants, development of the pyramid is significantly greater from 74 μs. It should be noted that the projectile perforates the fabric about 50 μs for the experimental test and model using the dynamic constants. Therefore, after 50 μs, the formation of a deformation pyramid is no longer supported by the projectile for these two cases. However, during this period, the projectile can still act on the top of a pyramid in the case using static parameters.

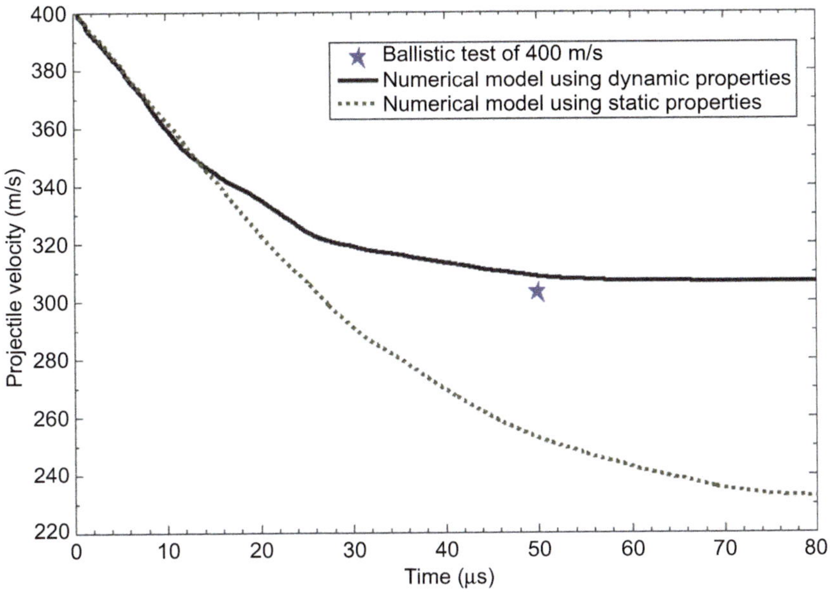

Figure 16.40 Comparison in the case of perforation impact (400 m/s) on the evolution of projectile velocity between experience, the model using dynamic properties, and the model using static properties.

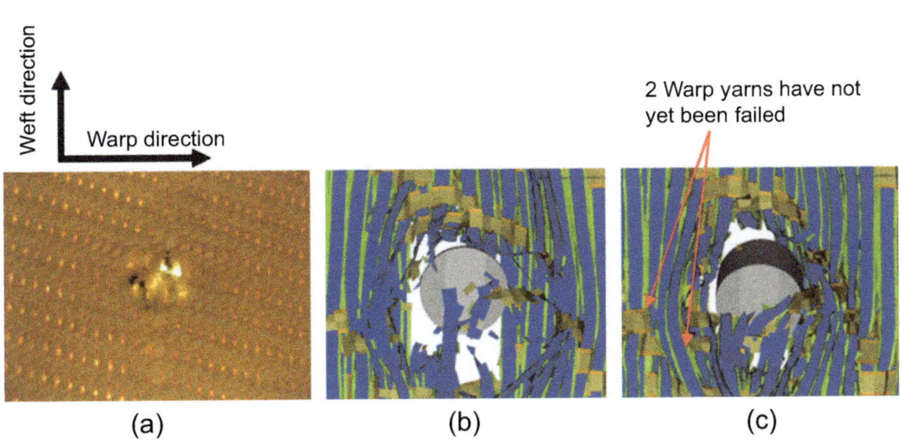

Figure 16.41 Comparison for the case of impact velocity of 306 m/s at the impact zone: (a) experiment, (b) result with digital dynamic properties, (c) result with digital static properties.

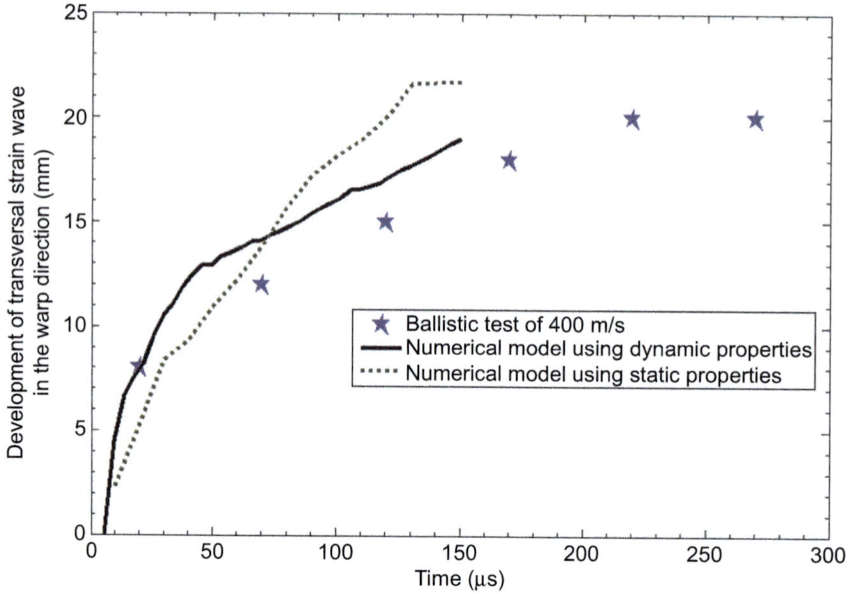

Figure 16.42 Comparison in the case of 400 m/s impact velocity on the development of a transversal strain wave in the warp direction between experience, the model using dynamic properties, and the model using static properties.

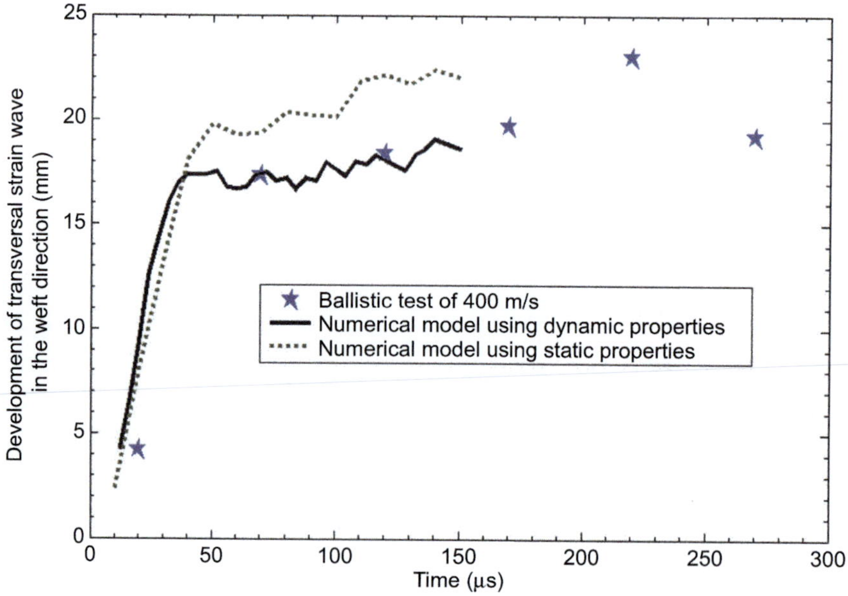

Figure 16.43 Comparison in the case of 400 m/s impact velocity on the development of a transversal strain wave in the weft direction between experience, the model using dynamic properties, and the model using static properties.

16.5 Conclusions

The chapter showed that, actually, a mesoscopic model can predict well the impact behavior of 2D and 3D woven fabrics. However, this model should be combined with a macroscopic model in a multiscale model to permit engineers to calculate large textile structures in different industrial applications. This strategy is indispensable in the case of a 3D woven fabric with so many geometrical complexities. Advanced numerical methods, for example, domain decomposition and isogeometry analysis, should also be performed to resolve these complexities for reducing time computation and regularizing high nonlinearities.

On the other hand, the model using the dynamic properties appears closest to the experimental results. The reason is that with the elastic modulus and failure strain in dynamic state, the numerical model has a good prediction of the propagation velocity of longitudinal and transverse waves which significantly influence behavior and damage phases of 3D fabric. Therefore, development of experimental techniques for measuring dynamic properties yarns is a practical need in the future to obtain a better accuracy of numerical models.

Another difficulty of numerical models is to have the initial state of woven fabrics as in reality (eg, residual stress exits in these types of material after a weaving process, or the real geometry of 3D woven fabrics is not ideal due to automatic arrangement of yarns to reach a steady state during the weaving process). New experimental measure methods should be performed to determine these important input parameters.

References

[1] Hu J. 3D fibrous assemblies, properties applications and modelling of three dimensional textile structure. 80 High Street, Sawton, Cambridge, CB223HJ: Woodhead Publishing Limited; 2008.

[2] Chen X, Sun D, Wang Y, Zhou Y. In: Proceedings to 4th world conference on 3D fabrics and their applications: 4th world conference on 3D fabrics and their applications; Sep 10, 2012–Sep 11, 2012. Aachen, Germany. Manchester: TexEng/RWTH Aachen; 2012.

[3] Ha-Minh C, Boussu F, Thoral-Pierre K. Ktex_Pattern: Numerical tool for textile fabrics subjected to ballistic impact. In: International conference of textile composite, Texcomp10, Lille, France; October, 2010.

[4] Lefebvre M, Boussu F. High energy absorption of warp interlock fabrics: application to high speed impact of fragments. In: 19th conference on the mechanical and physical behaviour of materials under dynamic loading, Belgium; 2009.

[5] Ha-Minh C. Comportement mécanique des matériaux tissés soumis à un impact balistique: Approches expérimentale, numérique et analytique [Ph.D. thesis]. France: University of Lille 1; 2011.

[6] Ha-Minh C, Imad A, Boussu F, Kanit T. Experimental and numerical investigation of a 3D woven fabric subjected to a ballistic impact. Int J Impact Eng 2016;88:91–101.

[7] Smith JC, McCrackin FL, Scniefer HF. Stress-strain relationships in yarns subjected to rapid impact loading. Part V. wave propagation in long textile yarns impacted transversely. Text Res J 1958;28(4):288−302.
[8] Ha-Minh C, Imad A, Boussu F, Kanit T. On analytical modelling to predict of the ballistic impact behaviour of textile multi-layer woven fabric. Compos Struct 2012. http://dx.doi.org/10.1016/j.compstruct.2012.10.011.
[9] Smith JC, McCrackin FL, Schiefer HF, Stone WK, Towne KM. Stress-strain relationships in yarns subjected to rapid impact loading: 4 transverse impact tests. J Res Natl Bur Stand 1956;57(2):83−9.
[10] Carr DJ. Failure mechanisms of yarns subjected to ballistic impact. J Mater Sci Lett 1999; 18:585−8.
[11] Billon HH, Robinson DJ. Models for the ballistic impact of fabric armour. Int J Impact Eng 2001;25:411−22.
[12] Gu B. Analytical modeling for the ballistic perforation of planar plain-woven fabric target by projectile. Compos Part B 2003;34:361−71.
[13] Naik NK, Shrirao P. Composite structures under ballistic impact. Compos Struct 2004;66: 579−90.
[14] Naik NK, Shrirao P, Reddy BCK. Ballistic impact behaviour of woven fabric composites : Formulation. Int J Impact Eng 2006;32:1521−52.
[15] Mamivand M, Liaghat GH. A model for ballistic impact on multi-layer fabric targets. Int J Impact Eng 2010;37(7):806−12.
[16] Chocron S, Figueroa E, King N, Kirchdoerfer T, Nicholls AE, Sagebiel E, Weiss C, Freitas CJ. Modeling and validation of full fabric targets under ballistic impact. Compos Sci Technol 2010;70:2012−22.
[17] Ha-Minh C, Kanit T, Boussu F, Crépin D, Imad A. Effect of yarns friction on the ballistic performance of a 3D warp interlock fabric: numerical analysis. Appl Compos Mater 2011. http://dx.doi.org/10.1007/s10443-011-9202-2.
[18] Roylance DK, Wilde AF, Tocci GC. Ballistic impact of textile structures. Text Res J 1973; 43:34−41.
[19] Shim VPW, Tan VBC, Tay TE. Modelling deformation and damage characteristics of woven fabric under small projectile impact. Int J Impact Eng 1995;16(4):585−605.
[20] Joo K, Kang TJ. Numerical analysis of multi-ply fabric impacts. Text Res J 2007;77: 359−68.
[21] Joo K, Kang TJ. Numerical analysis of energy absorption mechanism in multi-ply fabric impacts. Text Res J 2008;78:561−76.
[22] Ivanov I, Tabiei A. Loosely woven fabric model with viscoelastic crimped fibres for ballistic impact simulations. Int J Numer Methods Eng 2004;61:1565−83.
[23] Lim CT, Ng YH, Shim VPW. Finite-element modeling of the ballistic impact of fabric armor. Int J Impact Eng 2003;28(1):13−31.
[24] King M, Jearanaisilawong P, Socrate S. A continuum constitutive model for the mechanical behavior of woven fabrics. Int J Solids Struct 2005;42(13):3867−96.
[25] Shahkarami A, Vaziri R. A continuum shell finite element model for impact simulation of woven fabrics. Int J Impact Eng 2007;34(1):104−19.
[26] Zhu D, Vaidya A, Mobasher B, Rajan SD. Finite element modeling of ballistic impact on multi-layer Kevlar 49 fabrics. Compos Part B 2014;56:254−62.
[27] Tan V, Shim V, Zeng X. Modelling crimp in woven fabrics subjected to ballistic impact. Int J Impact Eng 2005;32(1−4):561−74.

[28] Stahlecker Z, Mobasher B, Rajan S, Pereira J. Development of reliable modeling methodologies for engine fan blade out containment analysis. Part II: finite element analysis. Int J Impact Eng 2009;36(3):447−59.
[29] Bansal S, Mobasher B, Rajan SD, Vintilescu I. Development of fabric constitutive behavior for use in modeling engine fan blade-out events. J Aerosp Eng 2009;22(3): 249−59.
[30] Naik D, Sankaran S, Mobasher B, Rajan S, Pereira J. Development of reliable modeling methodologies for fan blade out containment analysis. Part I: experimental studies. Int J Impact Eng 2009;36(1):1−11.
[31] Zhu D, Mobasher B, Vaidya A, Rajan SD. Mechanical behaviors of Kevlar 49 fabric subjected to uniaxial, biaxial tension and in-plane large shear deformation. Compos Sci Technol 2013;74(1):121−30.
[32] Gu B. Ballistic penetration of conically cylindrical steel projectile into plain-woven fabric target − a finite element simulation. J Compos Mater 2004;38:2049−74.
[33] Duan Y, Keefe M, Wetzel ED, Bogetti TA, Powers B, Kirkwood JE, Kirkwood KM. Effects of friction on the ballistic performance of a high-strength fabric structure. In: International conference on impact loading of lightweight structure Forianopolis, Brazil. May 8−12, 2005; 2005.
[34] Duan Y, Keefe M, Bogettic TA, Powers B. Finite element modeling of transverse impact on a ballistic fabric. Int J Mech Sci 2006;48:33−43.
[35] Rao MP, Nilakantan G, Keefe M, Powers BM, Bogetti TA. Global/local modeling of ballistic impact onto woven fabrics. Compos Mater 2009;43:445−67.
[36] Barauskas R, Abraitiene A. Computational analysis of impact of a bullet against the multilayer fabrics in LS-DYNA. Int J Impact Eng 2007;34:1286−305.
[37] Duan Y, Keefe M, Bogetti TA, Cheeseman BA, Powers B. A numerical investigation of the influence of friction on energy absorption by a high-strength fabric subjected to ballistic impact. Int J Impact Eng 2006;32:1299−312.
[38] Ha-Minh C, Kanit T, Boussu F, Imad A. Numerical multi-scale modeling for textile woven fabric against ballistic impact. Comput Mater Sci 2011;50:2172−84.
[39] Ha-Minh C, Imad A, Boussu F, Kanit T. Multi-scale model to predict the ballistic impact behavior of multi-layer plain-woven fabrics. Int J Comput Method 2014;11(3):12 pages.
[40] Rao MP, Duan Y, Keefe M, Powers BM, Bogetti TA. Modeling the effects of yarn material properties and friction on the ballistic impact of a plain-weave fabric. Compos Struct 2009; 89:556−66.
[41] Cheng M, Chen W. Mechanical properties of Kevlar® KM2® single fiber. J Eng Mater Technol 2005;123:197−203.
[42] Ha-Minh C, Boussu F, Kanit T, Crépin D, Imad A. Numerical study on the effects of yarn mechanical transverse properties on the ballistic impact behavior of textile fabric. J Strain Anal 2012. http://dx.doi.org/10.1177/0309324712457901.
[43] Ha-Minh C, Boussu F, Kanit T, Crépin D, Imad A. Numerical analysis of a ballistic impact on textile fabric. Int J Mech Sci 2013. http://dx.doi.org/10.1016/j.ijmecsci.2013.01.014.
[44] Sherburn M. Geometric and mechanical modelling of textiles [Ph.D. thesis]. The University of Nottingham; 2007.
[45] Searles K, Odegard G, Kumosa M. Micro- and mesomechanics of 8-harness satin woven fabric composites: I − evaluation of elastic behavior. Compos Part A 2001;32:1627−55.
[46] Hivet G, Boisse P. Consistent 3D geometrical model of fabric elementary cell: application to a meshing preprocessor for 3D finite element analysis. Finite Elem Anal Des 2005;42: 25−49.

[47] Lomov S, Verpoest I, Robitaille F. Manufacturing and internal geometry of textiles. In: Long AC, editor. Design and manufacture of textile composites. Woodhead Publishing Limited and CRC Press LLC; 2005. p. 1−61.
[48] Verpoest I, Lomov S. Virtual textile composites software WiseTex: integration with micro-mechanical, permeability and structural analysis. Compos Sci Technol 2005;65:2563−74.
[49] Birkefeld K, Röder M, Reden TV, Bulat M, Drechsler K. Characterization of biaxial and triaxial braids: fiber architecture and mechanical properties. Int J Appl Compos Mater 2011. http://dx.doi.org/10.1007/s10443-011-9190-2.

Damage modeling of ballistic impact in woven fabrics

17

S.D. Rajan, B. Mobasher
School of Sustainable Engineering & the Built Environment, Arizona State University, Tempe, AZ, United States

17.1 Introduction

A number of engineered systems are subject to low- and high-velocity impacts. Examples include concealable armor worn under street clothing, body armor worn on the outside, containment systems used with aircraft engines, protection systems for land, air, and sea-based military vehicles, and transparent barrier walls for VIPs. The ballistic mitigation systems designed for use in these different scenarios are different since the threats to defeat are different and because weight, cost, size, thickness, and manufacturability play important roles in the design of the final solution.

High-strength woven fabrics are ideal candidate materials for use in structural systems where high-energy absorption is required. Their high strength-to-weight ratio and the ability to resist high-speed fragment impacts enable them to be very efficient compared to metals. One of the more widely used applications for woven fabrics is in propulsion engine containment systems. The engine containment system is typically constructed by wrapping multiple layers of an aramid material, for example, Kevlar® or Twaron®, around a thin aluminum encasement (Fig. 17.1). The fabric is then covered with a protective layer. Fig. 17.2 shows a tactical vest worn by law enforcement personnel where the underlying material is an aramid woven fabric.

Fabrics are also used as a part of composite systems consisting of the woven fabric embedded in an epoxy. Ultra-concealable multithreat body armor used by law enforcement is a multipurpose armor that protects against attacks from knife, spikes, and small caliber rounds. The design of this type of armor involves fiber-resin composite materials that are flexible and light, are not unduly affected by environmental conditions, and perform as required. Fig. 17.3(a)–(d) shows two different commercially available systems where Kevlar® fabric is reinforced with a specially designed resin system (Vokshi, 2011).

In the rest of this chapter, we will examine the steps needed to use explicit finite element analysis as a predictive tool in modeling structural systems constructed using dry woven fabrics subjected to ballistic impact. First, we will discuss the finite element modeling philosophy, followed by details of the experimental tests to characterize the fabric properties needed for finite element modeling and finite element modeling

Advanced Fibrous Composite Materials for Ballistic Protection. http://dx.doi.org/10.1016/B978-1-78242-461-1.00017-0

Figure 17.1 Aircraft engine.
Honeywell Aerospace, 2015. https://commerce.honeywell.com/webapp/wcs/stores/servlet/eSystemDisplay?catalogId=10251&storeId=10651&categoryId=14233&langId=-1 (accessed 27.03.15.).

Figure 17.2 Soft body armor.
Dupont, 2015. http://www.dupont.com/products-and-services/personal-protective-equipment/body-armor/uses-and-applications/tactical-vests-law-enforcement.html (accessed 27.03.15.).

guidelines. This is followed by numerical examples showing how the developed constitutive model implemented in LS-DYNA explicit finite element analysis program as MAT_214 is used (LSTC, 2014). Finally, in the last part of the chapter, sources for further information and future trends are identified.

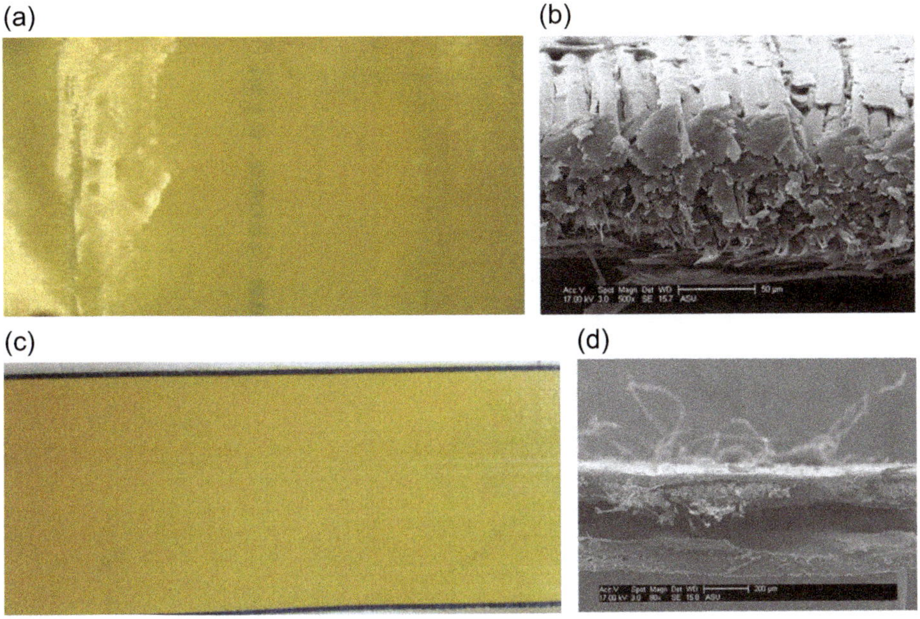

Figure 17.3 Fabric-based composite systems for low-level threats: (a) Honeywell GN2118® sheet, (b) SEM image of GN2118®, (c) Saati AS299® sheet, and (d) SEM image of AS299®.

17.2 Development of constitutive model for dry fabrics

Creating finite element models of dry fabrics that include yarn geometry details at a mesoscale level for use in the analysis of ballistic events is not practical. A more practical approach is to create an equivalent continuum model at a macroscale level, as shown in Fig. 17.4 for a simple weave fabric. Typically, the thickness of the fabric is small (at most a fraction of an inch), and the use of a thin-shell finite element is appropriate.

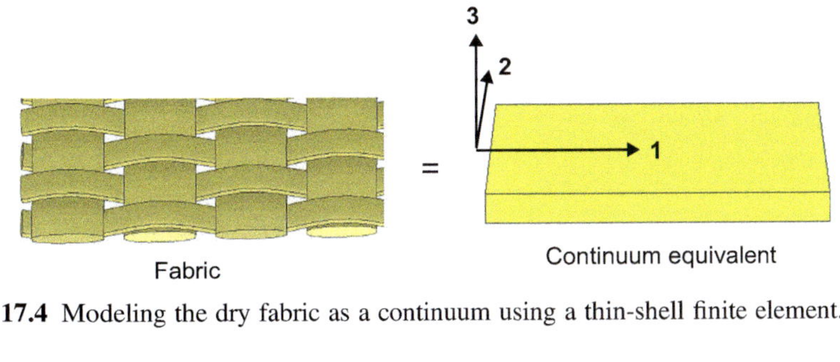

Figure 17.4 Modeling the dry fabric as a continuum using a thin-shell finite element.

Determining the effective or macromechanical properties of a woven fabric can be a challenging task and is usually carried out using appropriate experimental techniques. A typical fabric yarn (eg, Kevlar® yarn) is made up of hundreds of fibers (or filaments), and several yarns in the warp and fill directions make up a fabric swatch. The fabric swatch forms the basic building block of the material model. The orthotropic material behavior (strain—stress relationship) can be expressed as shown in Eq. [17.1]:

$$\begin{bmatrix} \varepsilon_{11} \\ \varepsilon_{22} \\ \varepsilon_{33} \\ \varepsilon_{23} \\ \varepsilon_{31} \\ \varepsilon_{12} \end{bmatrix} = \begin{bmatrix} \frac{1}{E_1} & -\frac{\upsilon_{12}}{E_1} & -\frac{\upsilon_{13}}{E_1} & 0 & 0 & 0 \\ -\frac{\upsilon_{12}}{E_1} & \frac{1}{E_1} & -\frac{\upsilon_{13}}{E_1} & 0 & 0 & 0 \\ -\frac{\upsilon_{13}}{E_1} & -\frac{\upsilon_{13}}{E_1} & \frac{1}{E_3} & 0 & 0 & 0 \\ 0 & 0 & 0 & \frac{1}{2G_{13}} & 0 & 0 \\ 0 & 0 & 0 & 0 & \frac{1}{2G_{13}} & 0 \\ 0 & 0 & 0 & 0 & 0 & \frac{1}{E_1} - \frac{\upsilon_{12}}{E_1} \end{bmatrix} \begin{bmatrix} \sigma_{11} \\ \sigma_{22} \\ \sigma_{33} \\ \sigma_{23} \\ \sigma_{31} \\ \sigma_{12} \end{bmatrix} \quad [17.1]$$

In the case of a fabric swatch, we take material direction 11 as the main longitudinal direction of the fabric (warp direction), direction 22 as the direction along the width of the fabric (fill direction), and direction 33 refers to the direction perpendicular to both warp and fill directions (Fig. 17.4). Experimental and numerical evidence show that the coupling between different directions is weak, and the constitutive behavior suitable for use in an explicit finite element analysis in stiffness incremental form can be expressed as follows:

$$\begin{bmatrix} \Delta\sigma_{11} \\ \Delta\sigma_{22} \\ \Delta\sigma_{33} \\ \Delta\sigma_{12} \\ \Delta\sigma_{31} \\ \Delta\sigma_{23} \end{bmatrix} = \begin{bmatrix} E_{11} & 0 & 0 & 0 & 0 & 0 \\ 0 & E_{22} & 0 & 0 & 0 & 0 \\ 0 & 0 & 0 & 0 & 0 & 0 \\ 0 & 0 & 0 & 2G_{12} & 0 & 0 \\ 0 & 0 & 0 & 0 & 2G_{31} & 0 \\ 0 & 0 & 0 & 0 & 0 & 2G_{23} \end{bmatrix} \begin{bmatrix} \Delta\varepsilon_{11} \\ \Delta\varepsilon_{22} \\ \Delta\varepsilon_{33} \\ \Delta\varepsilon_{12} \\ \Delta\varepsilon_{31} \\ \Delta\varepsilon_{23} \end{bmatrix} \quad [17.2]$$

The shell element formulation, Belytschko-Lin-Tsay (Hallquist, 2006), does not consider out-of-plane normal strains and stresses: $\Delta\varepsilon_{33}$ and $\Delta\sigma_{33}$ are both zero. The values for E_{11}, E_{22}, G_{12}, G_{31}, and G_{23} are functions of several factors, including the current stress and strain, the stress and strain history and the strain rate. The determination of these material properties is discussed next.

17.2.1 Experimental procedures

In this section, details of the experimental procedure to obtain the equivalent continuum material properties are discussed (Naik et al., 2009). The tension test setup is shown in Fig. 17.5(a), and the shear test setup is shown in Fig. 17.5(b). The tests are conducted until failure or until the test assumptions are no longer valid.

17.2.1.1 Tension tests (E_{11} and E_{22})

Typical tension stress–strain curves for a dry fabric are shown in Fig. 17.6(a) and (b).

For use in the constitutive model, these curves are approximated as shown in Fig. 17.7(a).

Initially, the tension test should be carried out to ascertain the quasistatic behavior of the fabric swatch and should include loading, unloading, and reloading in the prepeak as well as the postpeak regions. If appropriate, the tests should be conducted at higher rates of loading to find the rate-dependent behavior, if any (Zhu et al., 2011a,b, 2013a).

17.2.1.2 Picture frame tests (G_{12})

The shear stress–strain relationship should be determined based on picture frame shear tests (Naik et al., 2009). A typical response is shown in Fig. 17.7(b). The shear

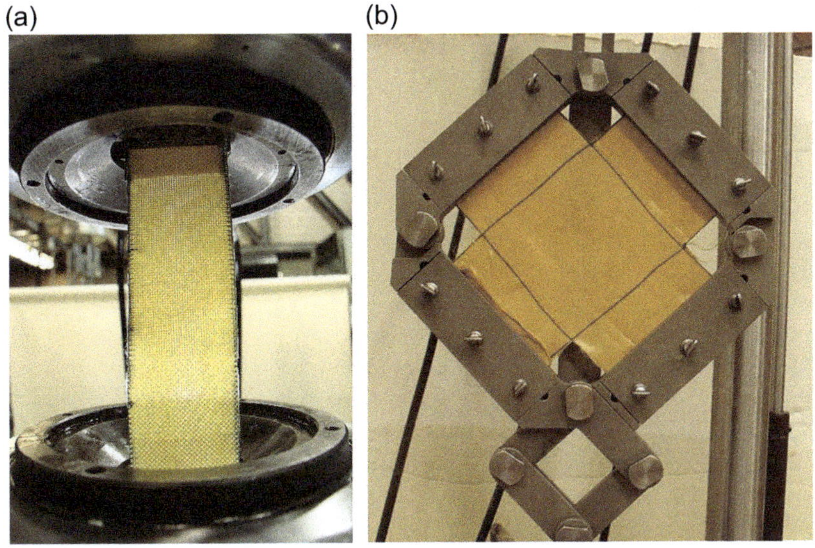

Figure 17.5 Test setups: (a) tension test and (b) picture frame test.

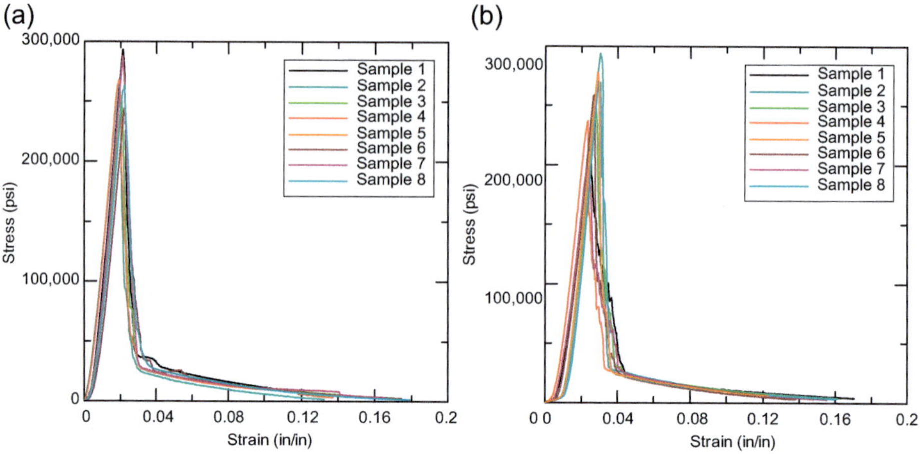

Figure 17.6 Stress–strain curve for woven Kevlar® fabric: (a) true stress versus true strain–warp direction and (b) true stress versus true strain–fill direction.

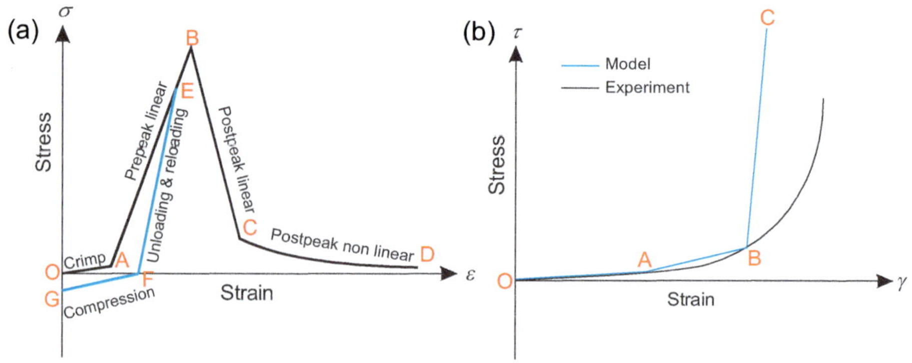

Figure 17.7 Equivalent stress–strain curve for use in frictional energy (FE) analysis: (a) tension stress–strain curve and (b) shear stress–strain curve.

resistance increases with an increase in shear strain. At low-shear strains the fabric has little resistance to shear deformation. The yarns rotate, and the warp and fill directions are no longer orthogonal. At some point, there is a very rapid increase in the shear stress value. This is caused by the reorientation and packing of the fabric yarns as the shear strain increases. One should be careful in looking at the fabric's deformation during the picture frame tests. Wrinkling may occur at the edges during the initial stages of loading and the fabric may buckle during the later stages of loading. Hence, the shear stress–strain curve should be corrected to include only the behavior captured by yarn reorientation. In the material model, a piecewise linear approximation of the corrected results is used (see Fig. 17.7(b)). The fabric is assumed to unload and reload along the same path.

17.2.1.3 Friction tests

Friction tests should be conducted to calculate the coefficients of static and dynamic friction (Naik et al., 2009). These values can then be used to specify the fabric-to-fabric behavior when multiple layers of fabrics are used. Similar tests should also be carried out to characterize the frictional behavior between fabric and other parts used in the finite element model, for example, fabric and steel.

17.2.1.4 Equivalent fabric density

Since a continuum model is used, the equivalent fabric density should be computed. For example, the following is the procedure for 17 × 17 denier Kevlar®49 simple weave fabric. The actual measured density of Kevlar®49 is 0.052 lb/in^3 (weight density), or 1.44 g/cm^3 (mass density). Since the material model's properties are computed based on the measured fabric thickness of 0.011 in, the actual density needed to be adjusted in the model. This is done by first measuring the mass of a 1 × 1-in fabric sample, which is approximately 0.144 g. To obtain the mass density of the fabric in the model, the actual mass is divided by the volume of material assumed in the model, or (1 in)(1 in)(0.011 in) = 0.011 in^3. Thus the fabric mass density used in the material model is 0.80 g/cm^3, or 7.48(10^{-5}) lb − s^2/in^4.

17.2.1.5 Other material constants

Some material values are either difficult to obtain or are known to have minimal effect on the simulations. These material values are to be found through numerical experimentation.

Compressive modulus

Dry fabrics typically have negligible compressive stiffness. If a zero (or numerically tiny) compressive stiffness is used, the model behavior in an explicit finite element analysis is unrealistic: the projectile simply cuts through the fabric. To avoid this problem, a very small stiffness should be assumed, for example, the compressive stiffness can be taken as a certain percentage of the prepeak longitudinal stiffness.

Out-of-plane shear modulus (G_{31},G_{23})

Dry fabrics typically do not experience noticeable shear deformations in the out-of-plane directions of the fabric (31 and 23 directions) when loaded. Hence a conservatively low value can be assumed for G_{31} and G_{23}, and numerical experiments should be carried out to ascertain the adequacy of frictional energy (FE) models with those values.

17.2.2 Finite element modeling guidelines

As with any other material model, the finite element model should be calibrated before use (Stahlecker et al., 2009). The fabric should be modeled using the

Belytschko-Lin-Tsay shell element that is a computationally efficient element and is the default shell element for LS-DYNA explicit finite element analysis (Hallquist, 2006). The corotational portion of the formulation avoids the complexities of nonlinear mechanics by embedding a coordinate system in the element. The choice of velocity strain or rate of deformation in the formulation facilitates the constitutive evaluation, since the conjugate stress is the physical Cauchy stress. During the analysis, the element tracks the principal material directions and updates the strains and stresses in the principal material directions. A one-point reduced integration scheme is used with this element. We also use the shear correction factor which scales the transverse shear stresses (to compensate for not satisfying the zero traction condition on top and bottom surfaces of the shell). Since it is likely that element rotations and strains may be large especially in the vicinity of the impact, we improve the accuracy of the solution by turning on the computation of second-order objective stress updates. This increases the computation time but provides results that are more accurate. The user should also turn on the option of computing hourglass energy, stonewall energy dissipation, sliding interface energy dissipation and Rayleigh (or damping) energy dissipation and include them in the energy check. Additionally, the option of monitoring the warpage of the shell elements flagging elements should be turned on. To suppress the hourglass deformation modes resulting from the use of reduced integration elements, the hourglass control through Flanagan-Belytschko stiffness formulation should be invoked. Our numerical experience shows that the stiffness formulation hourglass control is preferable to any form of viscous control.

To ensure that spurious results are not obtained, energy checks (Rajan et al., 2010) are carried out as a part of the postprocessing step. At any instant during the analysis, the sum of energies in the model must be equal to the sum of initial energies as given below[a]:

$$E_{\mathrm{K}} + E_{\mathrm{I}} + E_{\mathrm{S}} + E_{\mathrm{H}} + E_{\mathrm{rw}} + E_{\mathrm{damp}} = E_{\mathrm{K}}^{0} + E_{\mathrm{I}}^{0} + W_{\mathrm{Ext}} \qquad [17.3]$$

where E_{K} is the kinetic energy, E_{I} is the internal energy, E_{S} is the sliding interface (contact) energy, E_{H} is the hourglass energy, E_{T} is the total energy, E_{rw} is the rigid wall energy, E_{damp} is the damping energy, E_{K}^{0} is the initial kinetic energy, E_{i}^{0} is the initial internal energy and W_{Ext} is external work. The total energy is the sum of the terms on the left-hand side:

$$E_{\mathrm{T}} = E_{\mathrm{K}} + E_{\mathrm{I}} + E_{\mathrm{S}} + E_{\mathrm{H}} + E_{\mathrm{rw}} + E_{\mathrm{damp}} \qquad [17.4]$$

17.2.2.1 Internal energy

Internal energy is associated with elastic strain energy and work done in permanent deformation.

[a] http://www.dynasupport.com/tutorial/ls-dyna-users-guide/energy-data.

17.2.2.2 Kinetic energy

Kinetic energy is work done due to the motion of the nodes/elements with a certain velocity.

17.2.2.3 External work

External work is done by the applied forces and pressure as well as work done by velocity, displacement, or acceleration boundary conditions.

17.2.2.4 Sliding energy

It is the work done by sliding interfaces and is the sum of slave energy (SE), master energy (ME), and frictional energy (FE), where slave and master energy are associated with the sliding energy of the slave and master parts of the model during the impact. Sliding energy is expected to be positive when friction between the surfaces is defined. Negative contact energy sometimes is generated when parts slide relative to each other. When a penetrated node slides from its original master segment to an adjacent though unconnected master segment and a penetration is immediately detected, negative contact energy is the result. Abrupt increases in negative contact energy may be caused by undetected initial penetrations.

17.2.2.5 Hourglass energy

Underintegrated elements are used mainly to increase the computational efficiency and accuracy. However, in certain problems, spurious modes of deformations may result that are associated with the zero-energy modes of deformation (zero strain and no stress). To combat this problem, hourglass stabilization techniques are used. LS-DYNA provides several hourglass control options, and the energy associated with these stabilization techniques can be computed. This nonphysical hourglass energy should be relatively small compared to peak internal energy for each part of the model.

17.2.2.6 Energy ratio (ER)

It is the ratio of total energy to the initial total energy and external work and is given below. The energy balance is perfect if the ratio is equal to 1.

$$\mathrm{ER} = \frac{E_{\mathrm{T}}}{E_{\mathrm{I}}^{0} + E_{\mathrm{K}}^{0} + W_{\mathrm{Ext}}} \qquad [17.5]$$

If the E_{T} rises above the right-hand side, energy is being introduced artificially, for example, by numerical instability or the sudden detection of artificial penetration through a contact surface. The latter condition is often shown by sudden jumps in the total energy. If the left-hand side falls below the right-hand side, energy is being

Table 17.1 **Energy checks**

Description	Acceptable limit
Energy ratio, ER	>0.9 and <1.1
Max. sliding energy ratio, SER (sliding energy/total energy)	<0.1
Max. kinetic energy ratio, KER (kinetic energy/total energy)	<1.0
Max. internal energy ratio, IER (internal energy/total energy)	<1.0
Max. hourglass energy ratio, HER (hourglass energy/total energy)	<0.1

absorbed artificially, perhaps by excessive hourglassing or by stonewalls or overcompliant contact surfaces. In Table 17.1 the values used in the energy checks are listed.

17.2.2.7 Element verification tests

In the following paragraphs, we discuss results from element verification tests. Mesh convergence check is carried out using a modified form of FORTRAN code[b] that contains Richardson's extrapolation technique. Methods for examining the spatial and temporal convergence of computational fluid dynamics (CFD) simulations are presented by Roache (Roache, 1998) including a convergence analysis technique for studying the convergence behavior of the FE models. Roache suggests a grid convergence index (GCI) to provide a consistent manner in reporting the results of grid convergence studies and perhaps provide an error band on the grid convergence. It is suggested that one start with simple tests such as the one-element test and tension tests (Stahlecker et al., 2009) before moving on to more sophisticated tests. These simple tests can be used to check if the constitutive model can be verified and for finding the effect of element size (mesh density) on the final results.

17.3 Numerical modeling of high-speed impacts

In this section, we show the predictive capability (validation) of the developed fabric model by conducting ballistic tests and correlating the results with finite element predictions. Ballistic experimental data were generated at NASA-GRC. In the test, a steel projectile was shot through multiple fabric layers wrapped around a steel ring. Fig. 17.8(a) and (b) shows the test setup. The steel ring, 40-in diameter and 10.5-in

[b] http://www.grc.nasa.gov/WWW/wind/valid/tutorial/spatconv.html.

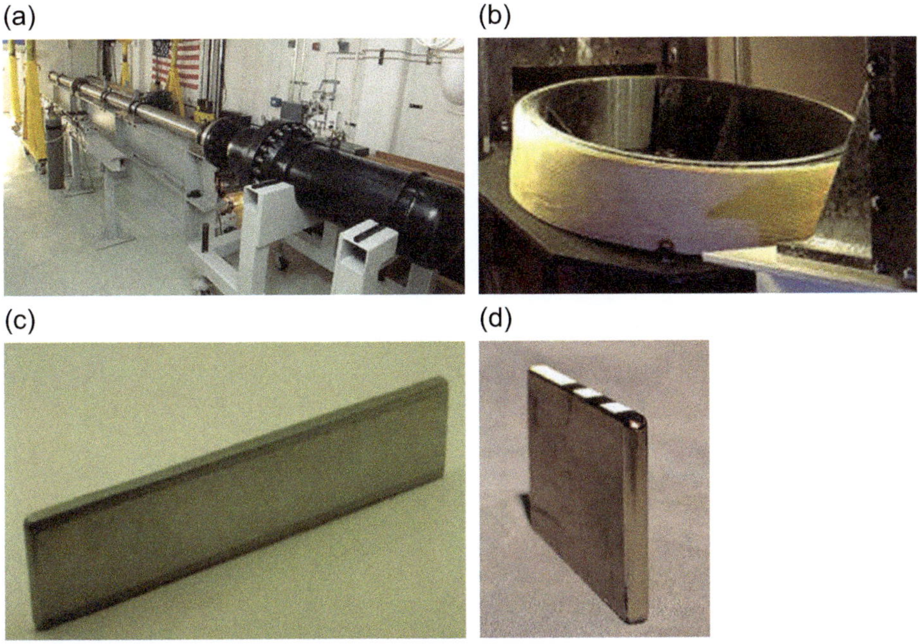

Figure 17.8 NASA-GRC ballistic tests on Kevlar®-wound steel ring: (a) gas gun used to shoot the projectile, (b) Kevlar® 49 wraps as target, (c) thin, long projectile, and (d) thick, short projectile.

width, was placed at a 15-degree angle with the horizontal. There is a 10-in-wide cut out at one end. A 10-in-wide Kevlar® 49 was wrapped around this ring starting from the point diametrically opposite the cut-out end. Glue was used to bond the first layer to the steel ring.

After the required numbers of layers were wrapped around the ring, a portion of the last layer was glued to the layer under it. The number of layers wrapped around the ring varies from 4 to 32. A gas gun shown in Fig. 17.8(a) was placed in line with the open end of the ring. Two different types of steel projectile were used. The thicker, shorter projectile is $4 \times 2 \times 0.312$ in (Fig. 17.8(d)), while the thinner longer projectile is $7 \times 1.5 \times 0.235$ in (Fig. 17.8(c)). The projectiles have rounded edges, and the tests were designed to replicate some key features of the engine fan blade-out tests. The following test parameters were varied: projectile geometry, projectile orientation, projectile velocity, number of fabric layers and projectile type. The variations helped in covering a wide spectrum of possible fan blade-out events. For each test, the input velocity of the projectile and output velocity of the projectile after penetrating the fabric layers were measured using high-speed cameras.

The FE model predictions are compared against experimental data in Fig. 17.9.

Further details of this study can be found in Deivanayagam et al. (2013) and Zhu et al. (2013b).

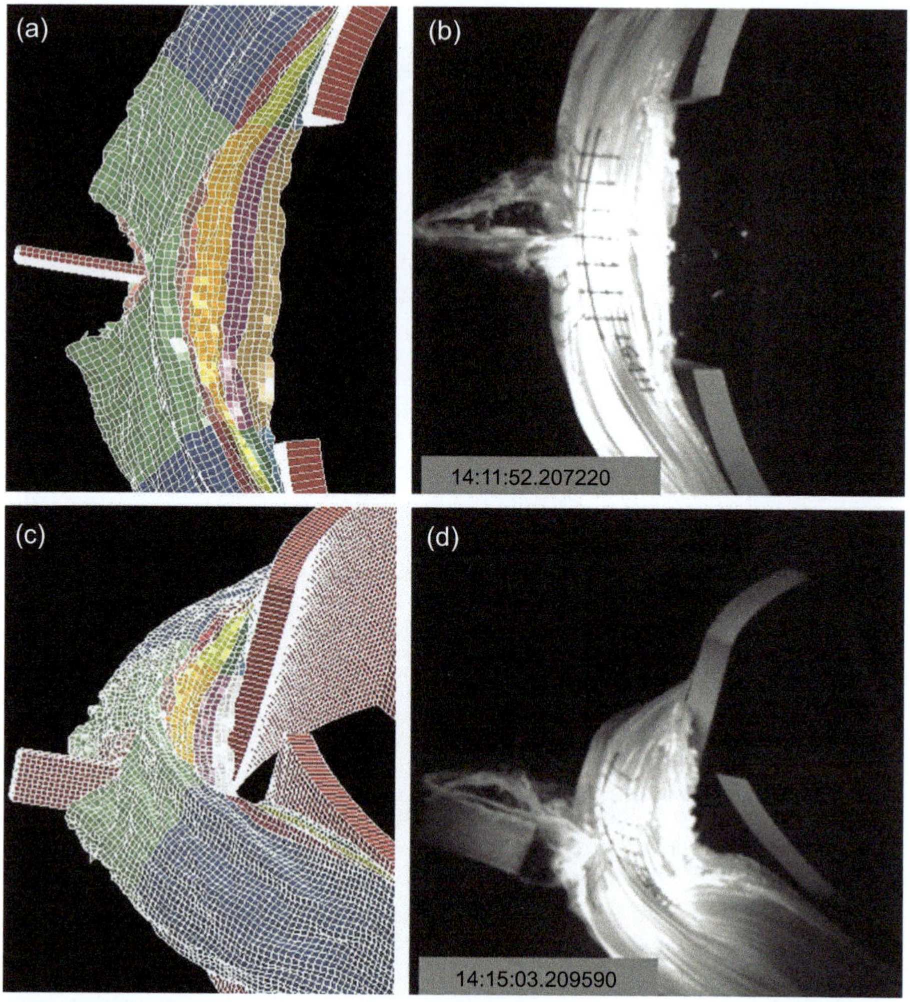

Figure 17.9 Finite element prediction versus experimental data: (a) FE model prediction at time t1, (b) high-speed camera image at time t1, (c) FE model prediction at time t2, and (d) high-speed camera image at time t2.

17.4 Conclusions

In this chapter, we have presented a framework to develop and use dry, woven fabrics that are a part of systems subjected to ballistic impacts. When used correctly and effectively, the modeling system can be a valuable predictive tool. Innovations are constantly taking place to make such systems lighter, more economical, and increasingly applicable.

17.4.1 Materials

There are other materials used to make woven fabrics such as high-modulus polypropylene (HPPE), carbon fibers, S2-glass, etc. Clearly, high-strength fabric-based systems provide attractive solutions due to their flexibility, high strength-to-weight ratios and adaptability.

17.4.2 Experimental techniques

One of the more important developments in the last few years is the ready availability of hardware and software for measuring full-field displacements and strains called digital image correlation (DIC). An example use of DIC is shown in Fig. 17.10.

High-speed camera images are used to track the orientation and velocity of the projectiles as well as to obtain the full-displacement field and surface strains of the fabric. In addition, DIC is also being used in the quasistatic and rate-dependent material characterization tests to yield very high-quality data negating the use of extensometers or strain gages for most tests.

17.4.3 Finite element modeling

The need for accurate material models to simulate the deformation, damage, and failure response of polymer matrix composites under impact conditions is becoming more critical. In general, the existing models often require significant a priori knowledge of the damage and failure such that their use as predictive tools can be limited.

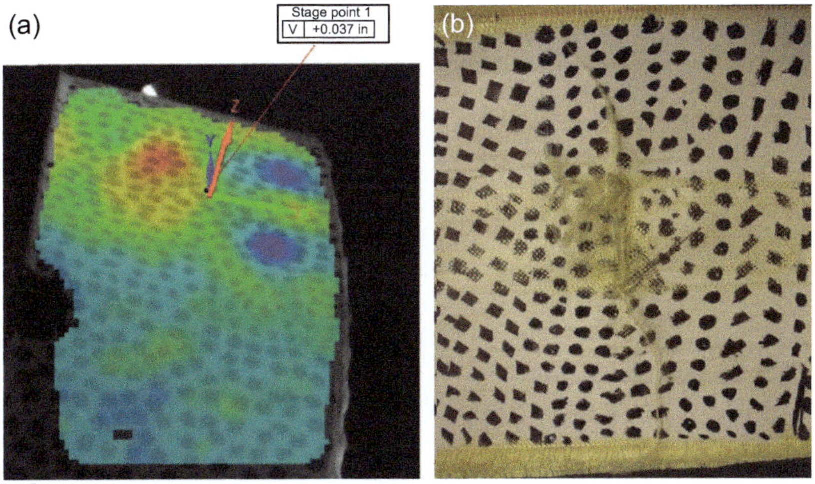

Figure 17.10 Use of DIC during ballistic impact testing using images from high-speed cameras and DIC software: (a) full-field displacement and strains from Kevlar®-wound steel ring test and (b) speckled pattern on the Kevlar® fabric to enable the use of DIC.

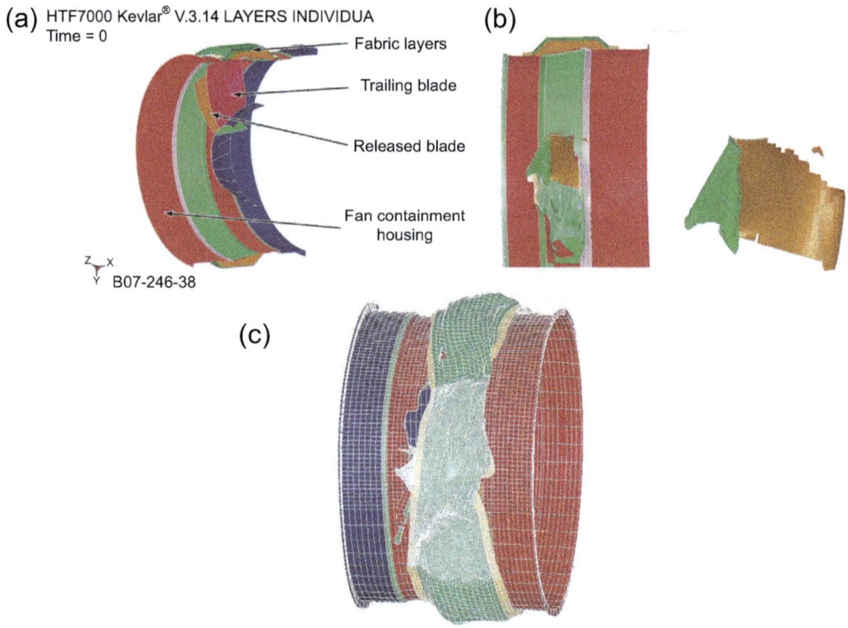

Figure 17.11 (a) LS-DYNA model for fan blade-out containment analysis, (b) engine blade orientation and deformed blade prediction, and (c) final state of the containment system.

Development of more general composite material models that overcome most of the deficiencies of existing models are underway and are likely to be available for public use shortly (Goldberg et al., 2014; Hoffarth et al., 2014).

17.4.4 Engine containment analysis

Design by analysis helps manufacturers cut down on design cost by using finite element analysis as a predictive tool and reducing the number of full-scale tests.

Fig. 17.11 shows simulation of an actual engine fan blade-out test using the dry fabric material model and LS-DYNA (Gomuc, 2004; Bansal et al., 2009). In the actual full-scale containment test, a fan blade was intentionally released while the engine operated for 15 s after the release. Complete fan blade-out containment was achieved. The LS-DYNA simulation correctly predicted that the fan blades would be contained. The fabric damage from the simulation was slightly less than the damage noticed from the actual fan blade-out test.

References

Bansal, S., Mobasher, B., Rajan, S.D., Vintilescu, I., 2009. Numerical modeling of engine fan blade-out events. ASCE Journal of Aerospace Engineering 22, 249–259.

Deivanayagam, A., Vaidya, A., Rajan, S.D., May 9, 2013. Enhancements to modeling of dry fabrics for impact analysis. ASCE Journal of Aerospace Engineering. http://dx.doi.org/10.1061/(ASCE)AS.1943-5525.0000350.

Dupont, 2015. http://www.dupont.com/products-and-services/personal-protective-equipment/body-armor/uses-and-applications/tactical-vests-law-enforcement.html (accessed 27.03.15.).

Goldberg, R., Carney, K., Du Bois, P., Hoffarth, C., Harrington, J., Rajan, S.D., Blankenhorn, G., 2014. Theoretical development of an orthotropic elasto-plastic generalized composite material model. In: 13th International LS-DYNA User's Conference, Detroit, MI, June 2014.

Gomuc, R., November 2004. Explicit Finite Element Modeling of Multilayer Composite Fabric for Gas Turbine Engine Containment Systems. Part 4: Model Simulation for Ballistic Tests, Engine Fan Blade-Out, and Generic Engine. DOT/FAA/AR-04/40,P4.

Hallquist, J., 2006. LS-DYNA Theory Manual. LSTC, Livermore, CA.

Hoffarth, C., Harrington, J., Rajan, S.D., Goldberg, R., Carney, K., Du Bois, P., Blankenhorn, G., 2014. Verification and validation of a three-dimensional generalized composite material model. In: 13th International LS-DYNA User's Conference, Detroit, MI, June 2014.

Honeywell Aerospace, 2015. https://commerce.honeywell.com/webapp/wcs/stores/servlet/eSystemDisplay?catalogId=10251&storeId=10651&categoryId=14233&langId=-1 (accessed 27.03.15.).

LSTC, 2014. LS-DYNA Keyword User's Manual. In: Material Models, LS-DYNA R7.1, vol. II. Livermore, CA.

Naik, D., Sankaran, S., Mobasher, B., Rajan, S.D., Pereira, J.M., 2009. Development of reliable modeling methodologies for fan blade-out containment analysis. Part I: experimental studies. Journal of Impact Engineering 36 (1), 1−11.

Rajan, S.D., Mobasher, B., Stahlecker, Z., Bansal, S., Zhu, D., Morea, M., Dhandapani, K., 2010. LS-DYNA Implemented Multi-layer Fabric Material Model Development for Engine Fragment Mitigation. Phase III, Part 2: Fabric Material Tests. FAA Report DOT/FAA/AR-10/23.

Roache, P.J., 1998. Verification and Validation in Computational Science and Engineering. Hermosa Publishers, Albuquerque, New Mexico.

Stahlecker, Z., Mobasher, B., Rajan, S.D., Pereira, J.M., 2009. Development of reliable modeling methodologies for fan blade-out containment analysis. Part II: finite element analysis. Journal of Impact Engineering 36 (3), 447−459.

Vokshi, E., 2011. Experimental Investigation and Development of Finite Element Model for Knife and Spike Impact (MS thesis). Department of Civil and Environmental Engineering, Arizona State University, Tempe, AZ.

Zhu, D., Mobasher, B., Rajan, S.D., Peled, A., Mignolet, M., 2011a. Modal analysis of a servo-hydraulic high speed testing machine and its application to dynamic tensile testing at an intermediate strain rate. Experimental Mechanics 51 (8). http://dx.doi.org/10.1007/s11340-010-9443-2.

Zhu, D., Mobasher, B., Rajan, S.D., 2011b. Dynamic testing of Kevlar 49 fabrics. ASCE Journal of Materials in Civil Engineering 23 (3), 230−239.

Zhu, D., Mobasher, B., Rajan, S.D., 2013a. Mechanical behaviors of Kevlar 49 fabric subjected to uniaxial, biaxial tension and in-plane large shear deformation. Composites Science and Technology 74, 121−130.

Zhu, D., Vaidya, A., Mobasher, B., Rajan, S.D., 2013b. Finite element modeling of ballistic impact on multi-layer Kevlar 49 fabrics. Composites Part B 56, 254−262.

Index

Note: 'Page numbers followed by "f" indicate figures and "t" indicate tables.'

C

G

CPI Antony Rowe
Chippenham, UK
2016-02-23 09:49